Cold and Hot Forging

Fundamentals and Applications

Edited by
Taylan Altan, ERC/NSM, Ohio State University
Gracious Ngaile, North Carolina State University
Gangshu Shen, Ladish Company, Inc.

Materials Park, Ohio 44073-0002
www.asminternational.org

First printing, February 2005
Second printing, August 2005
Third printing, August 2007

Comments, criticisms, and suggestions are invited, and should be forwarded to ASM International.
Prepared under the direction of the ASM International Technical Books Committee (2004–2005), Yip-Wah Chung, FASM, Chair.

ASM International staff who worked on this project include Scott Henry, Senior Manager of Product and Service Development; Bonnie Sanders, Manager of Production; Carol Polakowski, Production Supervisor; and Pattie Pace, Production Coordinator.

Library of Congress Cataloging-in-Publication Data

Cold and hot forging : fundamentals and applications / edited by Taylan Altan, Gracious Ngaile, Gangshu Shen.
p. cm.
Includes bibliographical references and index.
ISBN-13: 978-0-87170-805-2
ISBN-10: 0-87170-805-1
1. Forging. I. Altan, Taylan. II. Ngaile, Gracious. III. Shen, Gangshu.
TS225.C63 2004
671.3 32—dc22 2004055439

SAN: 204-7586

ASM International®
Materials Park, OH 44073-0002
www.asminternational.org

Printed in the United States of America

Contents

Preface

Among all manufacturing processes, forging technology has a special place because it helps to produce parts of superior mechanical properties with minimum waste of material. In forging, the starting material has a relatively simple geometry; this material is plastically deformed in one or more operations into a product of relatively complex configuration. Forging to net or to net shape dimensions drastically reduces metal removal requirements, resulting in significant material and energy savings. Forging usually requires relatively expensive tooling. Thus, the process is economically attractive when a large number of parts must be produced and/or when the mechanical properties required in the finished product can be obtained only by a forging process.

The ever-increasing costs of material, energy, and, especially, manpower require that forging processes and tooling be designed and developed with minimum amount of trial and error with shortest possible lead times. Therefore, to remain competitive, the cost-effective application of computer-aided techniques, i.e., CAD, CAM, CAE, and, especially, finite element analysis (FEA)-based computer simulation is an absolute necessity. The practical use of these techniques requires a thorough knowledge of the principal variables of the forging process and their interactions. These variables include: a) the flow behavior of the forged material under processing conditions, b) die geometry and materials, c) friction and lubrication, d) the mechanics of deformation, i.e., strains and stresses, e) the characteristics of the forging equipment, f) the geometry, tolerances, surface finish and mechanical properties of the forging, and g) the effects of the process on the environment.

There are many excellent handbooks and technical papers on the technology of the forging. These principles are reviewed briefly in this book, but major emphasis is on the latest developments in the design of forging operations and dies. Thus, process modeling using FEA has been discussed in all appropriate chapters. The subject is introduced in Chapter 1 with a discussion of the position of metal forming processes in manufacturing. Chapter 2 considers forging process as a system consisting of several variables that interact with one another. This chapter also includes an overall review of the forging operations. The fundamentals of plastic deformation, i.e., metal flow, flow stress of materials, testing methods to determine materials properties, and flow rules are discussed in Chapters 3, 4, and 5. Chapters 6 and 8 cover the significant variables of the forging process such as friction, lubrication, and temperatures. Chapter 9 is devoted to approximate methods for analyzing simple forging operations. Chapters 10 through 13 discuss forging machines, including machines for shearing and pre-forming or materials distribution. Process and die design, methods for estimating forging loads, and the application of FEA-based process modeling in hot forging are discussed in Chapters 14, 15, and 16.

Chapters 17 and 18 cover cold and warm forging, including the application of FEA simulation in these processes. Microstructure modeling, using forging of high temperature alloys as example, is covered in Chapter 19, while Chapter 20 is devoted to isothermal and hot die forging of aerospace alloys. Die materials, die manufacturing, and die wear in hot and cold forging are discussed in Chapters 21 and 22.

Finally, Chapter 23 reviews the near-net shape forging technology, including enclosed die forging, multiple-action tooling, and the most recent developments in forging presses. This chapter also discusses briefly the future of forging technology in the global economy, the importance of information technology in the forge shop, and, finally, the need to continuously acquire knowledge on new methods and techniques to remain competitive.

Several chapters of the book (Chapters 4, 6, 7, 14, 15 and 17) contain appendixes that consist of presentation slides and computer animations. The animations represent the results of FEA simulations for various forging operations. They are given in a CD that is included with this book. The reader is encouraged to use the CD and these appendixes in order to understand better and easier some of the fundamental issues discussed in corresponding chapters.

The preparation of this book has been supported partially by the Jacob Wallenberg Foundation Prize, awarded to Dr. Taylan Altan by the Royal Swedish Academy of Engineering Sciences. The staff and the students of the Engineering Research Center for Net Shape Manufacturing (ERC/NSM) of The Ohio State University contributed significantly to the preparation of the book. Specifically, Mr. Pinak Barve, Graduate Research Associate, provided valuable assistance in preparing the text and the figures. Considerable information has been supplied by a large number of companies that support the forging research and development at the ERC/NSM. On behalf of the authors and the editors, I would like to thank all who made our work so much easier. Finally, I would like to thank my wife, Susan Altan, who has offered me enormous support and encouragement throughout the preparation of this book.

Taylan Altan
December 2004

CHAPTER 1

Metal Forming Processes in Manufacturing

Manas Shirgaokar

1.1 Classification of Manufacturing Processes

The term *metal forming* refers to a group of manufacturing methods by which the given material, usually shapeless or of a simple geometry, is transformed into a useful part without change in the mass or composition of the material. This part usually has a complex geometry with well-defined (a) shape, (b) size, (c) accuracy and tolerances, (d) appearance, and (e) properties.

The manufacture of metal parts and assemblies can be classified, in a simplified manner, into five general areas:

- *Primary shaping* processes, such as casting, melt extrusion, die casting, and pressing of metal powder. In all these processes, the material initially has no shape but obtains a well-defined geometry through the process.
- *Metal forming* processes such as rolling, extrusion, cold and hot forging, bending, and deep drawing, where metal is formed by plastic deformation.
- *Metal cutting* processes, such as sawing, turning, milling and broaching where removing metal generates a new shape.
- *Metal treatment* processes, such as heat treating, anodizing and surface hardening, where the part remains essentially unchanged in shape but undergoes change in properties or appearance.
- *Joining* processes, including (a) metallurgical joining, such as welding and diffusion bonding, and (b) mechanical joining, such as riveting, shrink fitting, and mechanical assembly. Metallurgical joining processes, such as welding, brazing, and soldering, form a permanent and robust joint between components. Mechanical joining processes, such as riveting and mechanical assembly, bring two or more parts together to build a subassembly that can be disassembled conveniently.

Among all manufacturing processes, metal forming technology has a special place because it helps to produce parts of superior mechanical properties with minimum waste of material. In metal forming, the starting material has a relatively simple geometry. The material is plastically deformed in one or more operations into a product of relatively complex configuration. Forming to near-net- or to net-shape dimensions drastically reduces metal removal requirements, resulting in significant material and energy savings. Metal forming usually requires relatively expensive tooling. Thus, the process is economically attractive only when a large number of parts must be produced and/or when the mechanical properties required in the finished product can be obtained only by a forming process.

Metal forming includes a large number of manufacturing processes producing industrial products as well as military components and consumer goods. These processes include (a) massive forming operations such as forging, rolling, and drawing, and (b) sheet forming processes, such as brake forming, deep drawing, and stretch forming. Unlike machining, metal forming processes do not involve extensive metal removal to achieve the desired shape of the workpiece. Forming processes are frequently used together with other manufacturing processes, such as machining, grinding, and heat treating, in order to complete the transformation from the raw material to the finished and assembly-ready part. Desirable material properties for forming include low yield strength and high ductility. These properties are affected by temperature and rate of deformation (strain rate). When the work temperature is raised, ductility is increased and yield strength is decreased. The effect of temperature gives rise to distinctions among cold forming (workpiece initially at room temperature), warm forming (workpiece heated above room temperature, but below the recrystallization temperature of the workpiece material), and hot forming (workpiece heated above the recrystallization temperature). For example, the yield stress of a metal increases with increasing strain (deformation) during cold forming. In hot forming, however, the yield stress, in general, increases with strain (deformation) rate.

Forming processes are especially attractive in cases where:

- The part geometry is of moderate complexity and the production volumes are large, so that tooling costs per unit product can be kept low (e.g., automotive applications).
- The part properties and metallurgical integrity are extremely important (e.g., load-carrying aircraft, jet engine, and turbine components).

The design, analysis, and optimization of forming processes require:

- Engineering knowledge regarding metal flow, stresses, and heat transfer
- Technological information related to lubrication, heating and cooling techniques, material handling, die design, and forming equipment [Altan et al., 1983]

The development in forming technology has increased the range of shapes, sizes, and properties of the formed products enabling them to have various design and performance requirements. Formed parts are required specifically when strength, reliability, economy, and resistance to shock and fatigue are essential. The products can be determined from materials with the required temperature performance, ductility, hardness, and machinability [ASM Handbook].

1.2 Characteristics of Manufacturing Processes

There are four main characteristics of any manufacturing process—namely, geometry, tolerances, production rates, and human and environmental factors.

1.2.1 Geometry

Each manufacturing process is capable of producing a family of geometries. Within this family there are geometries, which can be produced only with extraordinary cost and effort. For example, the forging process allows production of parts, which can be easily removed from a die set, that is, upper and lower die. By use of a "split die" design, it is possible to manufacture forgings with undercuts and with more complex shapes.

1.2.2 Tolerances

No variable, especially no dimensional variable, can be produced exactly as specified by the designer. Therefore, each dimension is associated with a tolerance. Each manufacturing process allows certain dimensional tolerances and surface finishes to be obtained. The quality of these variables can always be improved by use of more sophisticated variations of the process and by means of new developments. For example, through use of the lost-wax vacuum casting process, it is possible to obtain much more complex parts with tighter tolerances than are possible with ordinary sand casting methods. Dimensional tolerances serve a dual purpose. First, they allow proper functioning of the manufactured part: for example, an automotive brake drum must be round, within limits, to avoid vibrations and to ensure proper functioning of the brakes. The second role of dimensional tolerances is to provide interchangeability. Without interchangeability—the ability to replace a defective part or component (a bearing, for example) with a new one, manufactured by a different supplier—modern mass production would be unthinkable. Figure 1.1 shows the dimensional accuracy that is achievable by different processes. The values given in the figure must be considered as guidance values only.

Forming tolerances represent a compromise between the accuracy desired and the accuracy that can be economically obtained. The accuracy obtained is determined by several factors such as the initial accuracy of the forming dies and tooling, the complexity of the part, the type of material being formed, and the type of forming equipment that is used. Another factor determining the forming accuracy is the type of part being produced.

Manufacturing costs are directly proportional to tolerances and surface finish specifications. Under typical conditions, each manufacturing process is capable of producing a part to a certain surface finish and tolerance range without extra expenditure. Some general guidance on surface finish and tolerance range is given in Fig. 1.2. The tolerances given apply to a 25 mm (1 in.) dimension. For larger or smaller dimensions, they do not necessarily increase or decrease linearly. In a production situation it is best to take the recommendations published by various industry associations or individual companies. Surface roughness in Fig. 1.2 is given in terms of R_a (arithmetic average). In many applications the texture (lay) of the surface is also important, and for a given R_a value, different processes may result in quite different finishes.

It used to be believed that cost tends to rise exponentially with tighter tolerances and surface finish. This is true only if a process sequence involving processes and machine tools of limited capability is used to achieve these tolerances. There are, however, processes and machine tools of inherently greater accuracy and better surface finish. Thus, higher-quality products can be obtained with little extra cost and, if the application justifies it, certainly with greater competitiveness. Still, a fundamental rule of the cost-conscious designer is to specify the loosest possible tolerances and coarsest surfaces that still accomplish the intended function. The specified tolerances should, if possible, be within the range obtainable by the intended manufacturing process (Fig. 1.2) so as to avoid additional finishing operations [Schey et al., 2000].

1.2.3 Production Rate

The rate of production that can be attained with a given manufacturing operation is probably the most significant feature of that operation, because it indicates the economics of and the achievable productivity with that manufacturing operation. In industrialized countries, manufacturing industries represent 25 to 30% of gross national product. Consequently, manufacturing productivity, i.e., production of discrete parts, assemblies, and products per unit time, is the single most important factor that influences the standard of living in a country as well as that country's competitive position in international trade in manufactured goods.

The rate of production or manufacturing productivity can be increased by improving existing

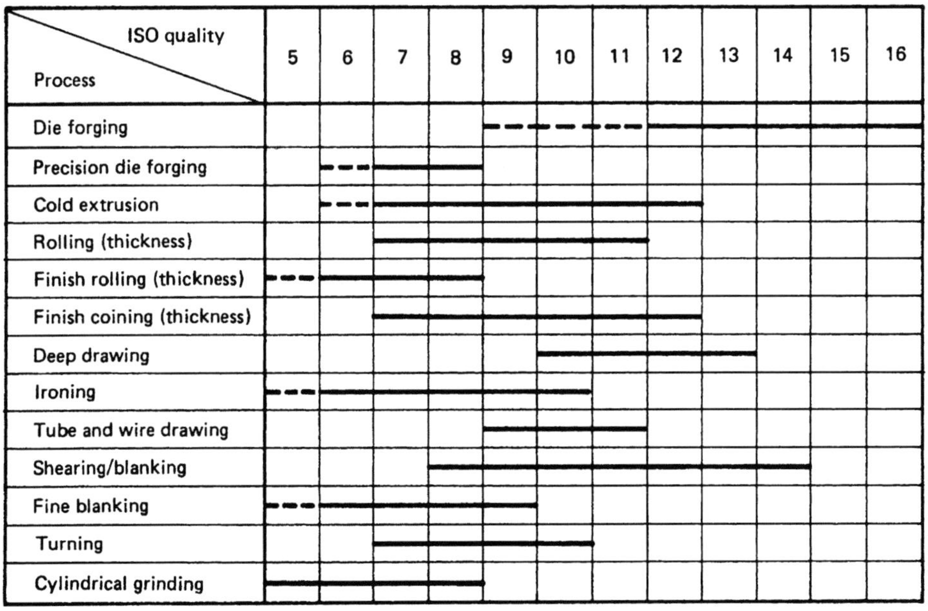

Process \ ISO quality	5	6	7	8	9	10	11	12	13	14	15	16
Die forging					- - -	- - -	- - -	——	——	——	——	——
Precision die forging		- - -	——	——								
Cold extrusion		- - -	——	——	——	——	——	——				
Rolling (thickness)			——	——	——	——	——					
Finish rolling (thickness)	- - -	——	——	——								
Finish coining (thickness)			——	——	——	——	——	——				
Deep drawing						——	——	——	——			
Ironing	- - -	——	——	——	——	——						
Tube and wire drawing					——	——	——					
Shearing/blanking				——	——	——	——	——	——	——		
Fine blanking	- - -	——	——	——	——							
Turning			——	——	——	——						
Cylindrical grinding	——	——	——	——								

Fig. 1.1 Approximate values of dimensional accuracies achievable in various processes. [Lange et al., 1985]

manufacturing processes and by introducing new machines and new processes, all of which require new investments. However, the most important ingredient for improving productivity lies in human and managerial resources, because good decisions regarding investments (when, how much, and in what) are made by people who are well trained and well motivated. As a result, the present and future manufacturing productivity in a plant, an industry, or a nation depends not only on the level of investment in new plants and machinery, but also on the level of training and availability of manufacturing engineers and specialists in that plant, industry, or nation.

1.2.4 *Environmental Factors*

Every manufacturing process must be examined in view of (a) its effects on the environment, i.e., in terms of air, water, and noise pollution, (b) its interfacing with human resources, i.e., in terms of human safety, physiological effects, and psychological effects, and (c) its use of energy and material resources, particularly in view of the changing world conditions concerning scarcity of energy and materials. Consequently, the introduction and use of a manufacturing process must also be preceded by a consideration of these environmental factors.

1.3 Metal Forming Processes in Manufacturing

Metal forming includes (a) massive forming processes such as forging, extrusion, rolling, and drawing and (b) sheet forming processes such as brake forming, deep drawing, and stretch forming. Among the group of manufacturing processes discussed earlier, metal forming represents a highly significant group of processes for producing industrial and military components and consumer goods.

The following list outlines some of the important areas of application of workpieces produced by metal forming, underlining their technical significance [Lange et al., 1985]:

- Components for automobiles and machine tools as well as for industrial plants and equipment

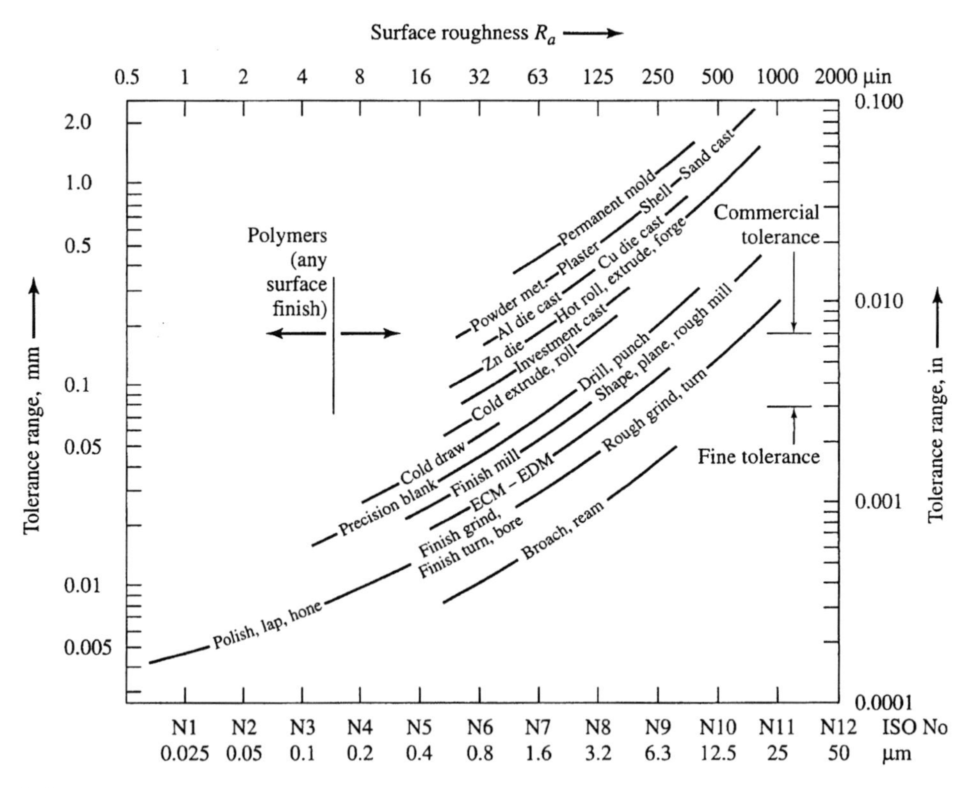

Fig. 1.2 Surface finish and tolerance range for various manufacturing processes. [Schey et al., 2000]

- Hand tools, such as hammers, pliers, screwdrivers, and surgical instruments
- Fasteners, such as screws, nuts, bolts, and rivets
- Containers, such as metal boxes, cans, and canisters
- Construction elements used in tunneling, mining, and quarrying (roofing and walling elements, pit props, etc.)
- Fittings used in the building industry, such as for doors and windows

A common way of classifying metal forming processes is to consider cold (room temperature) and hot (above recrystallization temperature) forming. Most materials behave differently under different temperature conditions. Usually, the yield stress of a metal increases with increasing strain (or deformation) during cold forming and with increasing strain rate (or deformation rate) during hot forming. However, the general principles governing the forming of metals at various temperatures are basically the same; therefore, classification of forming processes based on initial material temperature does not contribute a great deal to the understanding and improvement of these processes. In fact, tool design, machinery, automation, part handling, and lubrication concepts can be best considered by means of a classification based not on temperature, but rather on specific input and output geometries and material and production rate conditions.

Complex geometries, in both massive and sheet forming processes, can be obtained equally well by hot or cold forming. Of course, due to the lower yield strength of the deforming material at elevated temperatures, tool stresses and machine loads are, in a relative sense, lower in hot forming than in cold forming.

Forming is especially attractive in cases where (a) the part geometry is of moderate complexity and the production volumes are large, so that tooling costs per unit product can be kept low—for example, in automotive applications; and (b) the part properties and metallurgical integrity are extremely important, in examples such as load-carrying aircraft and jet engine and turbine components.

The design, analysis, and optimization of forming processes require (a) analytical knowledge regarding metal flow, stresses, and heat transfer as well as (b) technological information related to lubrication, heating, and cooling techniques; material handling; die design and manufacture; and forming equipment. A considerable amount of information on the general aspects of metal forming is available in the literature.

REFERENCES

[Altan et al., 1983]: Altan, T., Oh, S.-I., Gegel, H.L., *Metal Forming Fundamentals and Applications,* ASM International, 1983.

[ASM Handbook]: *Forming and Forging,* Vol 14, *ASM Handbook,* ASM International, 1988, p 6.

[Lange et al., 1985]: Lange, K., et al., *Handbook of Metal Forming,* McGraw-Hill, 1985, p 2.3, 9.19.

[Schey et al., 2000]: Schey, J.A., et al., *Introduction to Manufacturing Processes,* McGraw-Hill, 2000, p 67–69.

SELECTED REFERENCES

[Altan, 2002]: Altan, T., "Short Course on Near Net Shape Cold, Warm and Hot Forging Without Flash," Engineering Research Center for Net Shape Manufacturing, The Ohio State University, 2002.

[Kalpakjian et al., 2001]: Kalpakjian, S., Schmid, S., *Manufacturing Engineering and Technology,* Prentice Hall, 2001.

[SME Handbook, 1989]: *Tool and Manufacturers Engineering Handbook, Desk Edition (1989),* 4th ed., Society of Manufacturing Engineers, 1989, p 15-8.

CHAPTER 2

Forging Processes: Variables and Descriptions

Manas Shirgaokar

2.1 Introduction

In forging, an initially simple part—a billet, for example—is plastically deformed between two tools (or dies) to obtain the desired final configuration. Thus, a simple part geometry is transformed into a complex one, whereby the tools "store" the desired geometry and impart pressure on the deforming material through the tool/material interface. Forging processes usually produce little or no scrap and generate the final part geometry in a very short time, usually in one or a few strokes of a press or hammer. As a result, forging offers potential savings in energy and material, especially in medium and large production quantities, where tool costs can be easily amortized. In addition, for a given weight, parts produced by forging exhibit better mechanical and metallurgical properties and reliability than do those manufactured by casting or machining.

Forging is an experience-oriented technology. Throughout the years, a great deal of know-how and experience has been accumulated in this field, largely by trial-and-error methods. Nevertheless, the forging industry has been capable of supplying products that are sophisticated and manufactured to very rigid standards from newly developed, difficult-to-form alloys.

The physical phenomena describing a forging operation are difficult to express with quantitative relationships. The metal flow, the friction at the tool/material interface, the heat generation and transfer during plastic flow, and the relationships between microstructure/properties and process conditions are difficult to predict and analyze. Often in producing discrete parts, several forging operations (preforming) are required to transform the initial "simple" geometry into a "complex" geometry, without causing material failure or degrading material properties. Consequently, the most significant objective of any method of analysis is to assist the forging engineer in the design of forging and/or preforming sequences. For a given operation (preforming or finish forging), such design essentially consists of (a) establishing the kinematic relationships (shape, velocities, strain rates, strains) between the deformed and undeformed part, i.e., predicting metal flow, (b) establishing the limits of formability or producibility, i.e., determining whether it is possible to form the part without surface or internal failure, and (c) predicting the forces and stresses necessary to execute the forging operation so that tooling and equipment can be designed or selected.

For the understanding and quantitative design and optimization of forging operations it is useful to (a) consider forging processes as a system and (b) classify these processes in a systematic way [Altan et al., 1983].

2.2 Forging Operation as a System

A forging system comprises all the input variables such as the billet or blank (geometry and material), the tooling (geometry and material), the conditions at the tool/material interface, the mechanics of plastic deformation, the equipment used, the characteristics of the final product, and finally the plant environment where the process is being conducted.

The "systems approach" in forging allows study of the input/output relationships and the effect of the process variables on product quality and process economics. Figure 2.1 shows the different components of the forging system. The key to a successful forging operation, i.e., to obtaining the desired shape and properties, is the understanding and control of the metal flow. The direction of metal flow, the magnitude of deformation, and the temperatures involved greatly influence the properties of the formed components. Metal flow determines both the mechanical properties related to local deformation and the formation of defects such as cracks and folds at or below the surface. The local metal flow is in turn influenced by the process variables summarized below:

Billet

- Flow stress as a function of chemical composition, metallurgical structure, grain size, segregation, prior strain history, temperature of deformation, degree of deformation or strain, rate of deformation or strain, and microstructure
- Forgeability as a function of strain rate, temperature, deformation rate
- Surface texture
- Thermal/physical properties (density, melting point, specific heat, thermal conductivity and expansion, resistance to corrosion and oxidation)
- Initial conditions (composition, temperature, history/prestrain)
- Plastic anisotropy
- Billet size and thickness

Tooling/Dies

- Tool geometry
- Surface conditions, lubrication
- Material/heat treatment/hardness
- Temperature

Conditions at the Die/Billet Interface

- Lubricant type and temperature
- Insulation and cooling characteristics of the interface layer
- Lubricity and frictional shear stress
- Characteristics related to lubricant application and removal

Deformation Zone

- The mechanics of deformation, model used for analysis
- Metal flow, velocities, strain, strain rate (kinematics)
- Stresses (variation during deformation)
- Temperatures (heat generation and transfer)

Equipment

- Speed/production rate
- Binder design and capabilities
- Force/energy capabilities
- Rigidity and accuracy

Product

- Geometry
- Dimensional accuracy/tolerances
- Surface finish
- Microstructure, metallurgical and mechanical properties

Environment

- Available manpower
- Air, noise, and wastewater pollution
- Plant and production facilities and control

2.2.1 *Material Characterization*

For a given material composition and deformation/heat treatment history (microstructure), the flow stress and the workability (or forgeability) in various directions (anisotropy) are the most important material variables in the analysis of a metal forging process.

For a given microstructure, the flow stress, $\bar{\sigma}$, is expressed as a function of strain, $\bar{\varepsilon}$, strain rate, $\dot{\bar{\varepsilon}}$, and temperature, T:

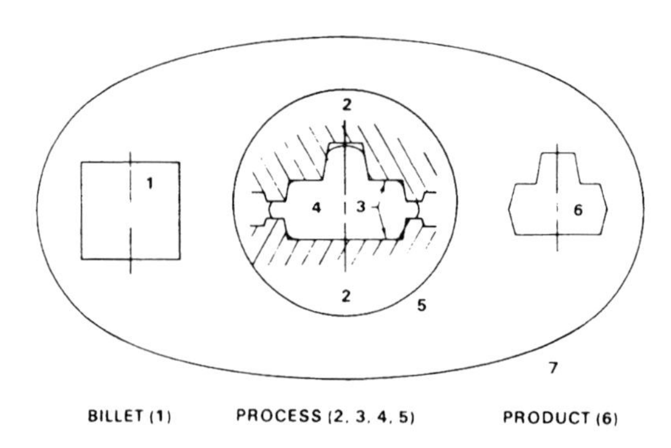

Fig. 2.1 One-blow impression-die forging considered as a system: (1) billet, (2) tooling, (3) tool/material interface, (4) deformation zone, (5) forging equipment, (6) product, (7) plant environment

$$\bar{\sigma} = f(\bar{\varepsilon}, \dot{\bar{\varepsilon}}, T) \quad \text{(Eq 2.1)}$$

To formulate the constitutive equation (Eq 2.1), it is necessary to conduct torsion, plane-strain compression, and uniform axisymmetric compression tests. During any of these tests, plastic work creates a certain increase in temperature, which must be considered in evaluating and using the test results.

Workability, forgeability, or formability is the capability of the material to deform without failure; it depends on (a) conditions existing during deformation processing (such as temperature, rate of deformation, stresses, and strain history) and (b) material variables (such as composition, voids, inclusions, and initial microstructure). In hot forging processes, temperature gradients in the deforming material (for example, due to local die chilling) also influence metal flow and failure phenomena.

2.2.2 *Tooling and Equipment*

The selection of a machine for a given process is influenced by the time, accuracy, and load/energy characteristics of that machine. Optimal equipment selection requires consideration of the entire forging system, including lot size, conditions at the plant, environmental effects, and maintenance requirements, as well as the requirements of the specific part and process under consideration.

The tooling variables include (a) design and geometry, (b) surface finish, (c) stiffness, and (d) mechanical and thermal properties under conditions of use.

2.2.3 *Friction and Lubrication at the Die/Workpiece Interface*

The mechanics of interface friction are very complex. One way of expressing friction quantitatively is through a friction coefficient, μ, or a friction shear factor, m. Thus, the frictional shear stress, τ, is:

$$\tau = \mu\sigma_n \quad \text{(Eq 2.2)}$$

or

$$\tau = f\bar{\sigma} = \frac{m}{\sqrt{3}}\bar{\sigma} \quad \text{(Eq 2.3)}$$

where σ_n is the normal stress at the interface, $\bar{\sigma}$ is the flow stress of the deforming material and f is the friction factor ($f = m/\sqrt{3}$). There are various methods of evaluating friction, i.e., estimating the value of μ or m. In forging, the most commonly used tests are the ring compression test, spike test, and cold extrusion test.

2.2.4 *Deformation Zone/Mechanics of Deformation*

In forging, material is deformed plastically to generate the shape of the desired product. Metal flow is influenced mainly by (a) tool geometry, (b) friction conditions, (c) characteristics of the stock material, and (d) thermal conditions existing in the deformation zone. The details of metal flow influence the quality and the properties of the formed product and the force and energy requirements of the process. The mechanics of deformation, i.e., the metal flow, strains, strain rates, and stresses, can be investigated by using one of the approximate methods of analysis (e.g., finite-element analysis, finite difference, slab, upper bound, etc.).

2.2.5 *Product Geometry and Properties*

The macro- and microgeometry of the product, i.e., its dimensions and surface finish, are influenced by the process variables. The processing conditions (temperature, strain, strain rate) determine the microstructural variations taking place during deformation and often influence the final product properties. Consequently, a realistic systems approach must include consideration of (a) the relationships between properties and microstructure of the formed material and (b) the quantitative influences of process conditions and heat treatment schedules on microstructural variations.

2.3 Types of Forging Processes

There are a large number of forging processes that can be summarized as follows:

- Closed/impression die forging with flash
- Closed/impression die forging without flash
- Electro-upsetting
- Forward extrusion
- Backward extrusion
- Radial forging
- Hobbing
- Isothermal forging
- Open-die forging

- Orbital forging
- Powder metal (P/M) forging
- Upsetting
- Nosing
- Coining

2.3.1 Closed-Die Forging with Flash (Fig. 2.2a and 2.2b)

Definition. In this process, a billet is formed (hot) in dies (usually with two halves) such that the flow of metal from the die cavity is restricted. The excess material is extruded through a restrictive narrow gap and appears as flash around the forging at the die parting line.

Equipment. Anvil and counterblow hammers, hydraulic, mechanical, and screw presses.

Materials. Carbon and alloy steels, aluminum alloys, copper alloys, magnesium alloys, beryllium, stainless steels, nickel alloys, titanium and titanium alloys, iron and nickel and cobalt superalloys, niobium and niobium alloys, tantalum and tantalum alloys, molybdenum and molybdenum alloys, tungsten alloys.

Process Variations. Closed-die forging with lateral flash, closed-die forging with longitudinal flash, closed-die forging without flash.

Application. Production of forgings for automobiles, trucks, tractors, off-highway equipment, aircraft, railroad and mining equipment, general mechanical industry, and energy-related engineering production.

2.3.2 Closed-Die Forging without Flash (Fig. 2.3)

Definition. In this process, a billet with carefully controlled volume is deformed (hot or cold) by a punch in order to fill a die cavity without any loss of material. The punch and the die may be made of one or several pieces.

Equipment. Hydraulic presses, multiram mechanical presses.

Materials. Carbon and alloy steels, aluminum alloys, copper alloys.

Process Variations. Core forging, precision forging, cold and warm forging, P/M forging.

Application. Precision forgings, hollow forgings, fittings, elbows, tees, etc.

2.3.3 Electro-Upsetting (Fig. 2.4)

Definition. Electro-upsetting is the hot forging process of gathering a large amount of material at one end of a round bar by heating the bar end electrically and pushing it against a flat anvil or shaped die cavity.

Equipment. Electric upsetters.

Materials. Carbon and alloy steels, titanium.

Application. Preforms for finished forgings.

2.3.4 Forward Extrusion (Fig. 2.5)

Definition. In this process, a punch compresses a billet (hot or cold) confined in a container so that the billet material flows through a die in the same direction as the punch.

Equipment. Hydraulic and mechanical presses.

Materials. Carbon and alloy steels, aluminum alloys, copper alloys, magnesium alloys, titanium alloys.

Process Variations. Closed-die forging without flash, P/M forging.

Application. Stepped or tapered-diameter solid shafts, tubular parts with multiple diameter

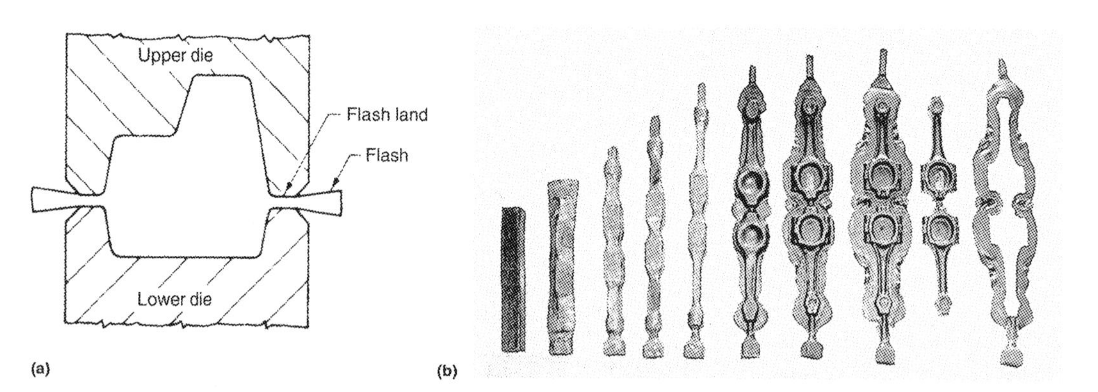

Fig. 2.2 Closed-die forging with flash. (a) Schematic diagram with flash terminology. (b) Forging sequence in closed-die forging of connecting rods

holes that are cylindrical, conical, or other non-round shapes.

2.3.5 *Backward Extrusion (Fig. 2.5)*

Definition. In this process, a moving punch applies a steady pressure to a slug (hot or cold) confined in a die and forces the metal to flow around the punch in a direction opposite the direction of punch travel (Fig. 2.5).

Equipment. Hydraulic and mechanical presses.

Materials. Carbon and alloy steels, aluminum alloys, copper alloys, magnesium alloys, titanium alloys.

Process Variations. Closed-die forging without flash, P/M forging.

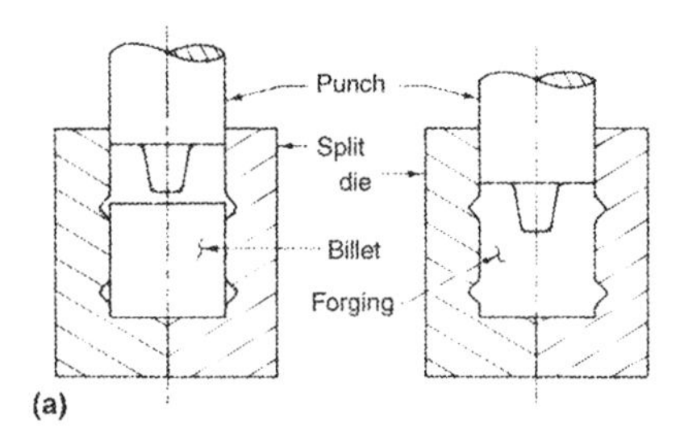

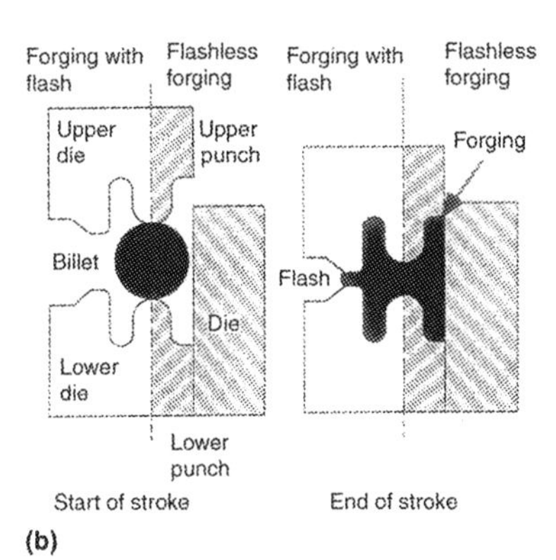

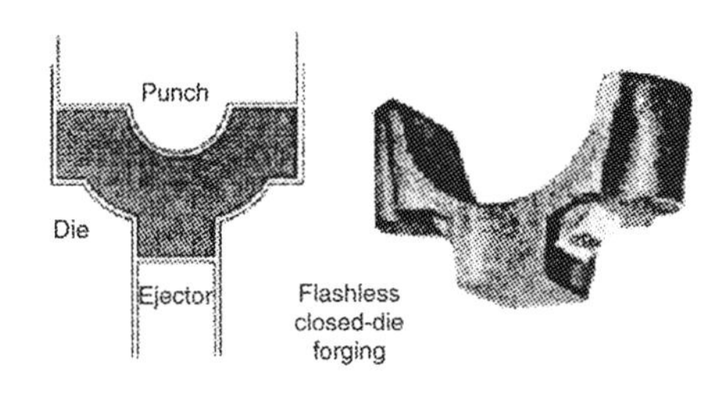

Fig. 2.3 Closed-die forging without flash

Application. Hollow parts having a closed end, cupped parts with holes that are cylindrical, conical, or of other shapes.

2.3.6 *Radial Forging (Fig. 2.6)*

Definition. This hot or cold forging process utilizes two or more radially moving anvils or dies for producing solid or tubular components with constant or varying cross sections along their length.

Equipment. Radial forging machines.

Materials. Carbon and alloy steels, titanium alloys, tungsten, beryllium, and high-temperature superalloys.

Process Variations. Rotary swaging.

Application. This is a technique that is used to manufacture axisymmetrical parts. Reducing the diameters of ingots and bars, forging of stepped shafts and axles, forging of gun and rifle barrels, production of tubular components with and without internal profiles.

2.3.7 *Hobbing (Fig. 2.7)*

Definition. Hobbing is the process of indenting or coining an impression into a cold or hot die block by pressing with a punch.

Equipment. Hydraulic presses, hammers.

Materials. Carbon and alloy steels.

Process Variations. Die hobbing, die typing.

Application. Manufacture of dies and molds with relatively shallow impressions.

2.3.8 *Isothermal Forging (Fig. 2.8)*

Definition. Isothermal forging is a forging process where the dies and the forging stock are at approximately the same high temperature.

Equipment. Hydraulic presses.

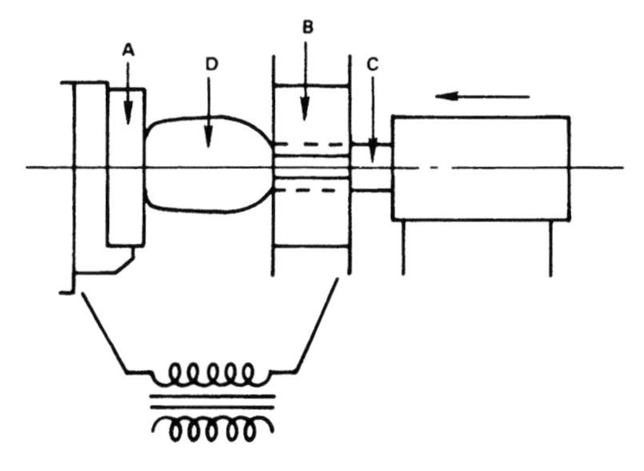

Fig. 2.4 Electro-upsetting. A, anvil electrode; B, gripping electrode; C, workpiece; D, upset end of workpiece

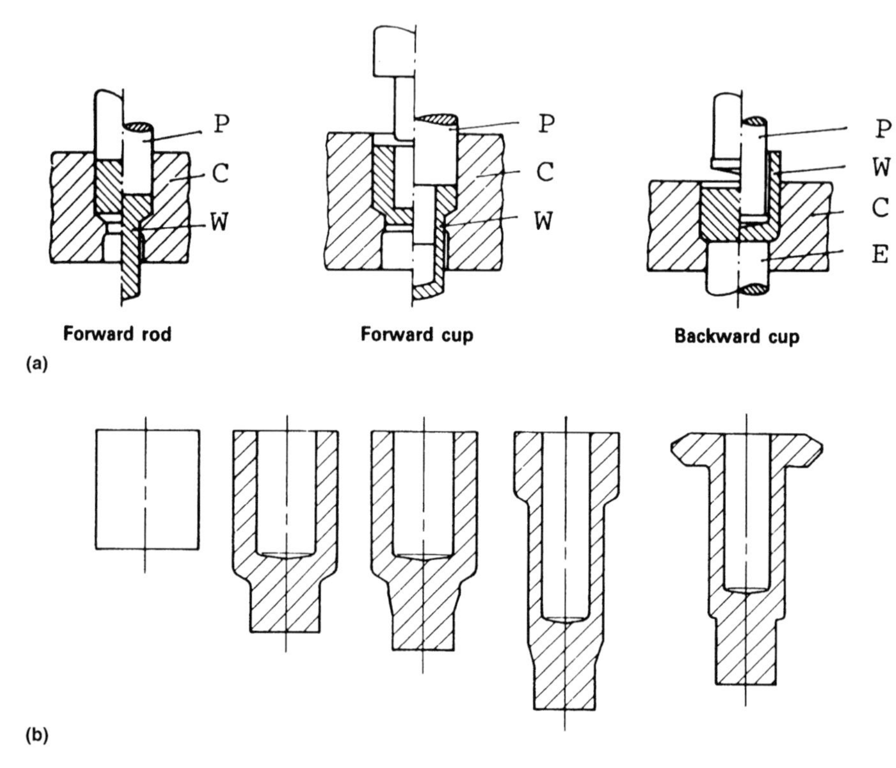

Fig. 2.5 Forward and backward extrusion processes. (a) Common cold extrusion processes (P, punch; W, workpiece; C, container; E, ejector). [Feldman, 1977]. (b) Example of a component produced using forward rod and backward extrusion. Left to right: sheared blank, simultaneous forward rod and backward cup extrusion, forward extrusion, backward cup extrusion, simultaneous upsetting of flange and coining of shoulder. [Sagemuller, 1968]

Materials. Titanium alloys, aluminum alloys.

Process Variations. Closed-die forging with or without flash, P/M forging.

Application. Net- and near-net shape forgings for the aircraft industry.

2.3.9 Open-Die Forging (Fig. 2.9)

Definition. Open-die forging is a hot forging process in which metal is shaped by hammering or pressing between flat or simple contoured dies.

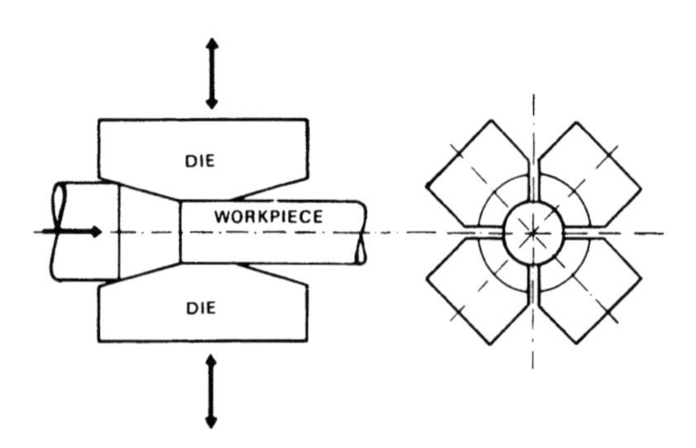

Fig. 2.6 Radial forging of a shaft

Equipment. Hydraulic presses, hammers.

Materials. Carbon and alloy steels, aluminum alloys, copper alloys, titanium alloys, all forgeable materials.

Process Variations. Slab forging, shaft forging, mandrel forging, ring forging, upsetting between flat or curved dies, drawing out.

Application. Forging ingots, large and bulky forgings, preforms for finished forgings.

2.3.10 Orbital Forging (Fig. 2.10)

Definition. Orbital forging is the process of forging shaped parts by incrementally forging

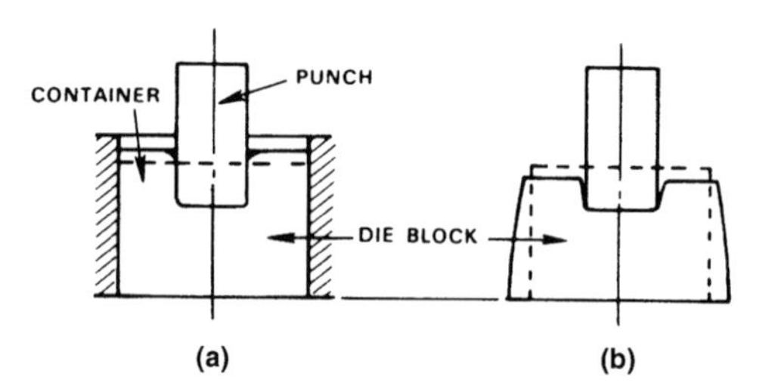

Fig. 2.7 Hobbing. (a) In container. (b) Without restriction

(hot or cold) a slug between an orbiting upper die and a nonrotating lower die. The lower die is raised axially toward the upper die, which is fixed axially but whose axis makes orbital, spiral, planetary, or straight-line motions.

Equipment. Orbital forging presses.

Materials. Carbon and low-alloy steels, aluminum alloys and brasses, stainless steels, all forgeable materials.

Process Variations. This process is also called rotary forging, swing forging, or rocking die forging. In some cases, the lower die may also rotate.

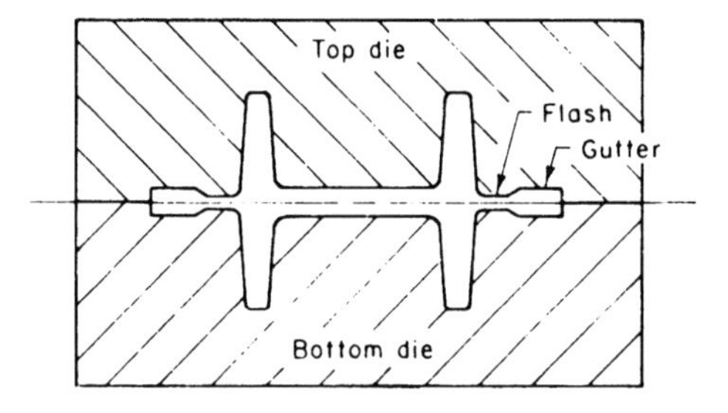

Fig. 2.8 Isothermal forging with dies and workpiece at approximately the same temperature

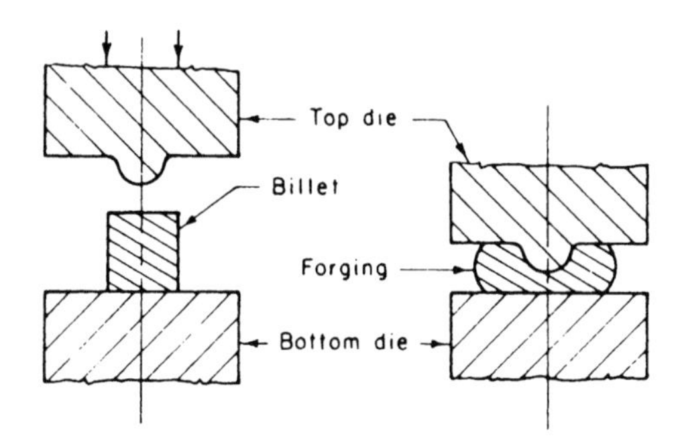

Fig. 2.9 Open-die forging

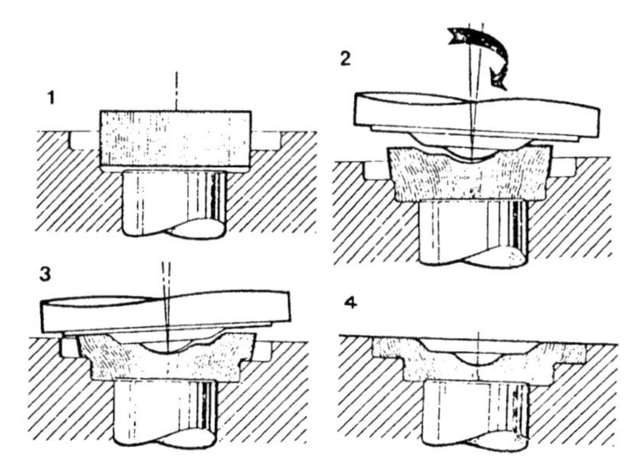

Fig. 2.10 Stages in orbital forging

Application. Bevel gears, claw clutch parts, wheel disks with hubs, bearing rings, rings of various contours, bearing-end covers.

2.3.11 Powder Metal (P/M) Forging (Fig.2.11)

Definition. P/M forging is the process of closed-die forging (hot or cold) of sintered powder metal preforms.

Equipment. Hydraulic and mechanical presses.

Materials. Carbon and alloy steels, stainless steels, cobalt-base alloys, aluminum alloys, titanium alloys, nickel-base alloys.

Process Variations. Closed-die forging without flash, closed-die forging with flash.

Application. Forgings and finished parts for automobiles, trucks, and off-highway equipment.

2.3.12 Upsetting or Heading (Fig. 2.12)

Definition. Upsetting is the process of forging metal (hot or cold) so that the cross-sectional area of a portion, or all, of the stock is increased.

Equipment. Hydraulic, mechanical presses, screw presses; hammers, upsetting machines.

Materials. Carbon and alloy steels, stainless steels, all forgeable materials.

Process Variations. Electro-upsetting, upset forging, open-die forging.

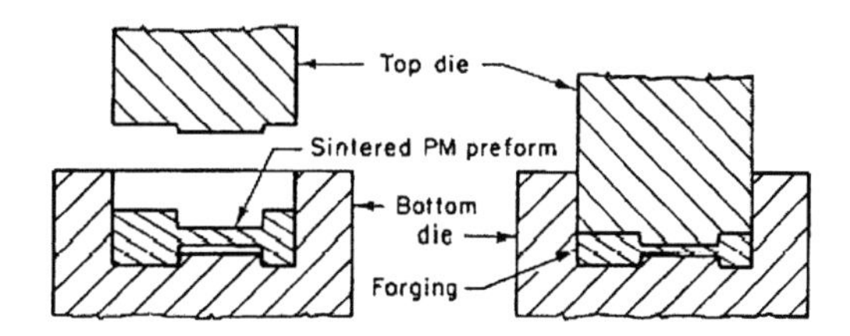

Fig. 2.11 Powder metal (P/M) forging

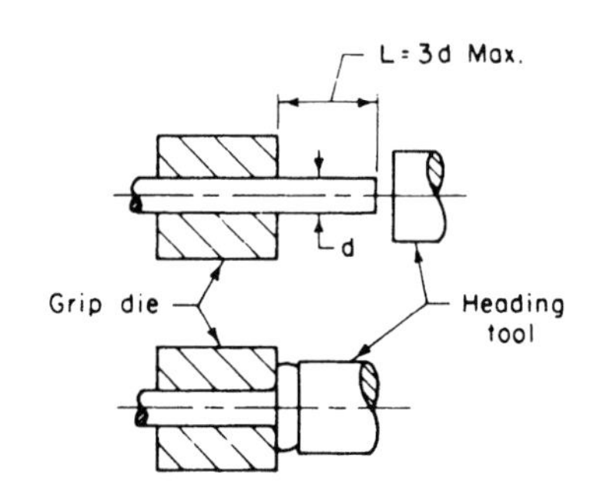

Fig. 2.12 Upset forging

Application. Finished forgings, including nuts, bolts; flanged shafts, preforms for finished forgings.

2.3.13 Nosing (Fig. 2.13)

Definition. Nosing is a hot or cold forging process in which the open end of a shell or tubular component is closed by axial pressing with a shaped die.

Equipment. Mechanical and hydraulic presses, hammers.

Materials. Carbon and alloy steels, aluminum alloys, titanium alloys.

Process Variations. Tube sinking, tube expanding.

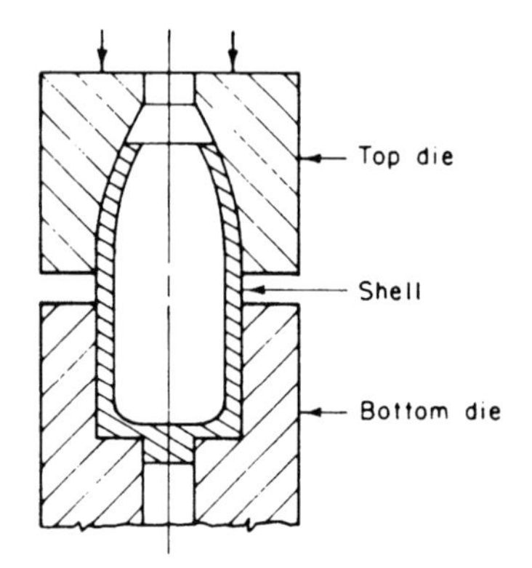

Fig. 2.13 Nosing of a shell

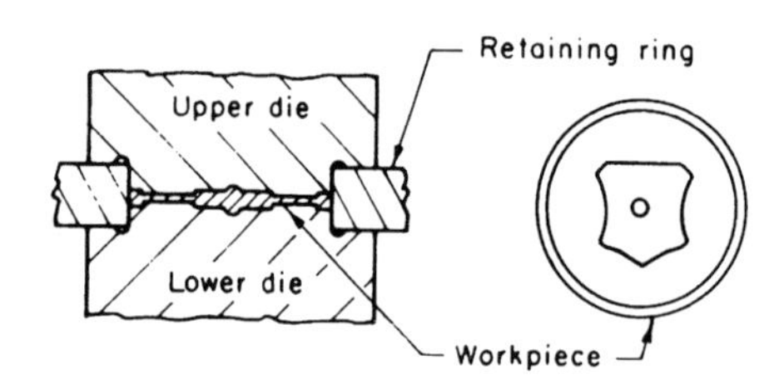

Fig. 2.14 Coining operation

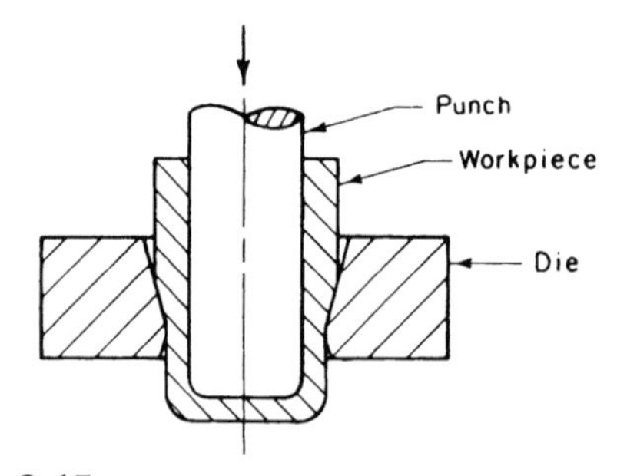

Fig. 2.15 Ironing operation

Applications. Forging of open ends of ammunition shells; forging of gas pressure containers.

2.3.14 Coining (Fig. 2.14)

Definition. In sheet metal working, coining is used to form indentations and raised sections in the part. During the process, metal is intentionally thinned or thickened to achieve the required indentations or raised sections. It is widely used for lettering on sheet metal or components such as coins. Bottoming is a type of coining process where bottoming pressure causes reduction in thickness at the bending area.

Equipment. Presses and hammers.

Materials. Carbon and alloy steels, stainless steels, heat-resistant alloys, aluminum alloys, copper alloys, silver and gold alloys.

Process Variations. Coining without flash, coining with flash, coining in closed die, sizing.

Applications. Metallic coins; decorative items, such as patterned tableware, medallions and metal buttons; sizing of automobile and aircraft engine components.

2.3.15 Ironing (Fig. 2.15)

Definition. Ironing is the process of smoothing and thinning the wall of a shell or cup (cold or hot) by forcing the shell through a die with a punch.

Equipment. Mechanical presses and hydraulic presses.

Materials. Carbon and alloy steels, aluminum and aluminum alloys, titanium alloys.

Applications. Shells and cups for various uses.

REFERENCES

[Altan et al., 1983]: Altan, T., Oh, S.-I., Gegel, H.L., *Metal Forming Fundamentals and Applications,* ASM International, 1983.

[Feldman, 1977]: Feldman, H.D., *Cold Extrusion of Steel,* Merkblatt 201, Düsseldorf, 1977 (in German).

[Sagemuller, 1968]: Sagemuller, Fr., "Cold Impact Extrusion of Large Formed Parts," *Wire,* No. 95, June 1968, p 2.

SELECTED REFERENCES

[Altan, 2002]: Altan, T., "The Greenfield Coalition Modules," Engineering Research Center for Net Shape Manufacturing, The Ohio State University, 2002.

[ASM, 1989]: *Production to Near Net Shape Source Book,* American Society for Metals 1989, p 33–80.

[ASM Handbook]: *Forming and Forging,* Vol 14, *ASM Handbook,* ASM International, 1988, p 6.

[Kalpakjian, 1984]: Kalpakjian, S., *Manufacturing Processes for Engineering Materials,* Addison-Wesley, 1984, p 381–409.

[Lange et al., 1985]: Lange, K., et al., *Handbook of Metal Forming,* McGraw-Hill, 1985, p 2.3, 9.19.

[Lindberg, 1990]: Lindberg, *Processes and Materials of Manufacture,* 4th ed., Allyn and Bacon, 1990, p 589–601.

[Niebel et al., 1989]: Niebel, B.W., Draper, A.B., Wysk, R.A., *Modern Manufacturing Process Engineering,* 1989, p 403–425.

[Schuler Handbook, 1998]: Schuler, *Metal Forging Handbook,* Springer, Goppingen, Germany, 1998.

[SME Handbook, 1989]: *Tool and Manufacturers Engineering Handbook, Desk Edition (1989),* 4th ed., Society of Manufacturing Engineers, 1989, p 15-8.

CHAPTER 3

Plastic Deformation: Strain and Strain Rate

Manas Shirgaokar
Gracious Ngaile

3.1 Introduction

The purpose of applying the plasticity theory in metal forming is to investigate the mechanics of plastic deformation in metal forming processes. Such investigation allows the analysis and prediction of (a) metal flow (velocities, strain rates, and strains), (b) temperatures and heat transfer, (c) local variation in material strength or flow stress, and (d) stresses, forming load, pressure, and energy. Thus, the mechanics of deformation provide the means for determining how the metal flows, how the desired geometry can be obtained by plastic forming, and what are the expected mechanical properties of the part produced by forming.

In order to arrive at a manageable mathematical description of the metal deformation, several simplifying (but reasonable) assumptions are made:

- Elastic deformations are neglected. However, when necessary, elastic recovery (for example, in the case of springback in bending) and elastic deflection of the tooling (in the case of precision forming to very close tolerances) must be considered.
- The deforming material is considered to be in continuum (metallurgical aspects such as grains, grain boundaries, and dislocations are not considered).
- Uniaxial tensile or compression test data are correlated with flow stress in multiaxial deformation conditions.
- Anisotropy and Bauschinger effects are neglected.
- Volume remains constant.
- Friction is expressed by a simplified expression such as Coulomb's law or by a constant shear stress. This is discussed later.

3.2 Stress Tensor

Consider a general case, where each face of a cube is subjected to the three forces F_1, F_2, and F_3 (Fig. 3.1). Each of these forces can be resolved into the three components along the three coordinate axes. In order to determine the stresses along these axes, the force components are divided by the area of the face upon which they act, thus giving a total of nine stress components, which define the total state of stress on this cuboidal element.

This collection of stresses is referred to as the stress tensor (Fig. 3.1) designated as σ_{ji} and is expressed as:

$$\sigma_{ij} = \begin{vmatrix} \sigma_{xx} & \sigma_{yx} & \sigma_{zx} \\ \sigma_{xy} & \sigma_{yy} & \sigma_{zy} \\ \sigma_{xz} & \sigma_{yz} & \sigma_{zz} \end{vmatrix}$$

A normal stress is indicated by two identical subscripts, e.g., σ_{xx}, while a differing pair indicates a shear stress. This notation can be simplified by denoting the normal stresses by a single subscript and shear stresses by the symbol τ. Thus one will have $\sigma_{xx} \equiv \sigma_x$ and $\sigma_{xy} \equiv \tau_{xy}$.

In case of equilibrium, $\sigma_{xy} = \sigma_{yx}$, thus implying the absence of rotational effects around any axis. The nine stress components then reduce to six independent components. A sign convention is required to maintain consistency throughout the use of these symbols and principles. The stresses shown in Fig. 3.1 are considered to be positive, thus implying that positive normal stresses are tensile and negative ones are compressive. The shear stresses acting along the directions shown in Fig. 3.1 are considered to be positive. The double suffix has the following physical meaning [Hosford & Caddell, 1983]:

- Suffix i denotes the normal to the plane on which a component acts, whereas the suffix j denotes the direction along which the component force acts. Thus σ_{yy} arises from a force acting in the positive y direction on a plane whose normal is in the positive y direction. If it acted in the negative y direction then this force would be compressive instead of tensile.
- A positive component is defined by a combination of suffixes where either both i and j are positive or both are negative.
- A negative component is defined by a combination of suffixes in which either one of i or j is negative.

3.3 Properties of the Stress Tensor

For a general stress state, there is a set of coordinate axes (1, 2, and 3) along which the shear stresses vanish. The normal stresses along these axes, viz. σ_1, σ_2, and σ_3, are called the principal stresses. Consider a small uniformly stressed block on which the full stress tensor is acting in equilibrium and assume that a small corner is cut away (Fig. 3.2). Let the stress acting normal to the triangular plane of section be a principal stress, i.e., let the plane of section be a principal plane.

The magnitudes of the principal stresses are determined from the following cubic equation developed from a series of force balances:

$$\sigma_i^3 - I_1\sigma_i^2 - I_2\sigma_i - I_3 = 0 \qquad \text{(Eq 3.1)}$$

where

$$I_1 = \sigma_{xx} + \sigma_{yy} + \sigma_{zz}$$

$$I_2 = -\sigma_{xx}\sigma_{yy} - \sigma_{yy}\sigma_{zz} - \sigma_{zz}\sigma_{xx} + \sigma_{xy}^2 + \sigma_{yz}^2 + \sigma_{zx}^2$$

$$I_3 = \sigma_{xx}\sigma_{yy}\sigma_{zz} + 2\sigma_{xy}\sigma_{yz}\sigma_{zx} - \sigma_{xx}\sigma_{yz}^2 - \sigma_{yy}\sigma_{zx}^2 - \sigma_{zx}\sigma_{xy}^2$$

In terms of principal stresses the above equations become:

$$I_1 = \sigma_1 + \sigma_2 + \sigma_3 \qquad \text{(Eq 3.1a)}$$

$$I_2 = -(\sigma_1\sigma_2 + \sigma_2\sigma_3 + \sigma_3\sigma_1) \qquad \text{(Eq 3.1b)}$$

$$I_3 = \sigma_1\sigma_2\sigma_3 \qquad \text{(Eq 3.1c)}$$

The coefficients I_1, I_2, and I_3 are independent of the coordinate system chosen and are hence

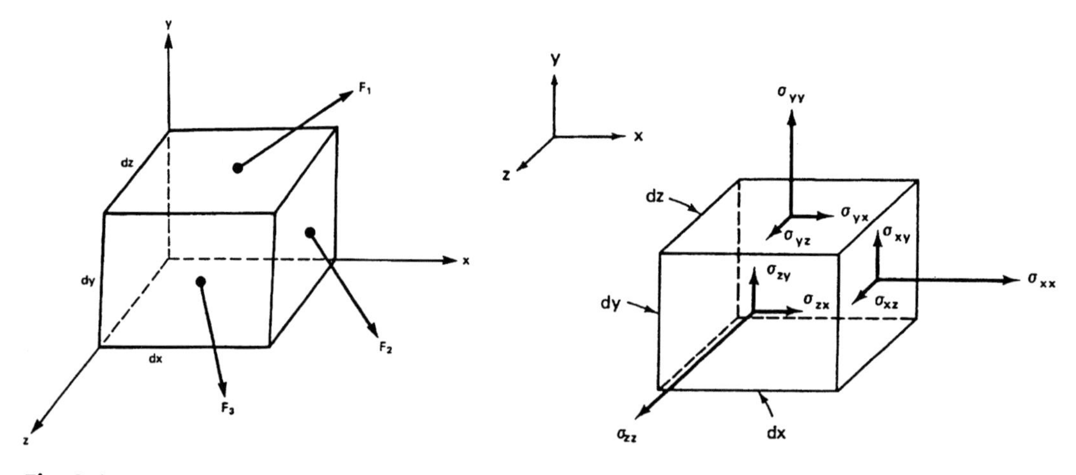

Fig. 3.1 Forces and the stress components as a result of the forces. [Hosford & Caddell, 1983]

called invariants. Consequently, the principal stresses for a given stress state are unique. The three principal stresses can only be determined by finding the three roots of the cubic equation. The invariants are necessary in determining the onset of yielding.

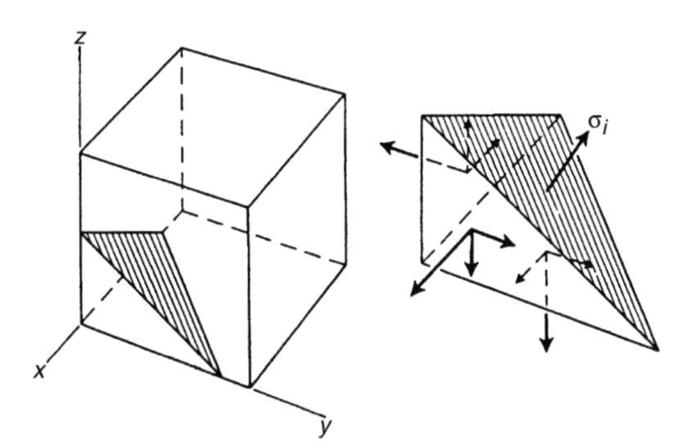

Fig. 3.2 Equilibrium in a three-dimensional stress state. [Backofen, 1972]

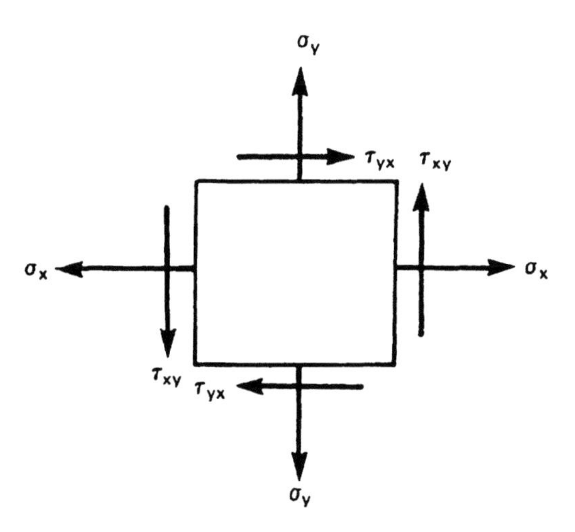

Fig. 3.3 Stresses in the x-y plane. [Hosford & Caddell, 1983]

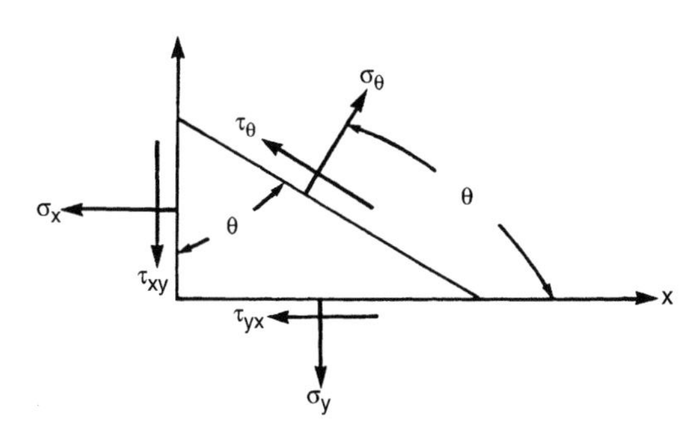

Fig. 3.4 Cut at an arbitrary angle θ in the x-y plane. [Hosford & Caddell, 1983]

3.4 Plane Stress or Biaxial Stress Condition

Consider Fig. 3.1 with the nine stress components and assume that any one of the three reference planes (x, y, z) vanishes (Fig. 3.3). Assuming that the z plane vanishes, one has $\sigma_z = \tau_{zy} = \tau_{zx}$ and a biaxial state of stress exists. To study the variation of the normal and shear stress components in the x-y plane, a cut is made at some arbitrary angle θ as shown in Fig. 3.4, and the stresses on this plane are denoted by σ_θ and τ_θ.

$$\sigma_\theta = \frac{\sigma_x + \sigma_y}{2} + \frac{\sigma_x - \sigma_y}{2} \cos 2\theta + \tau_{xy} \sin 2\theta \qquad \text{(Eq 3.2)}$$

$$\tau_\theta = -\left(\frac{\sigma_x - \sigma_y}{2}\right) \sin 2\theta + \tau_{xy} \cos 2\theta \qquad \text{(Eq 3.3)}$$

The two principal stresses in the x-y plane are the values of σ_θ on planes where the shear stress $\tau_\theta = 0$. Thus, under this condition,

$$\tan 2\theta = \frac{2\tau_{xy}}{\sigma_x - \sigma_y} \qquad \text{(Eq 3.4)}$$

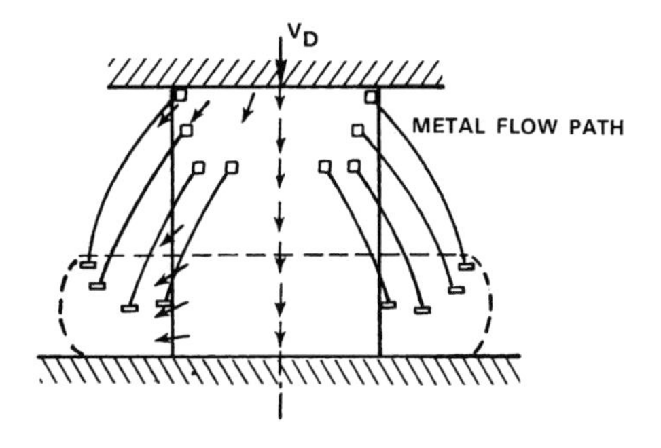

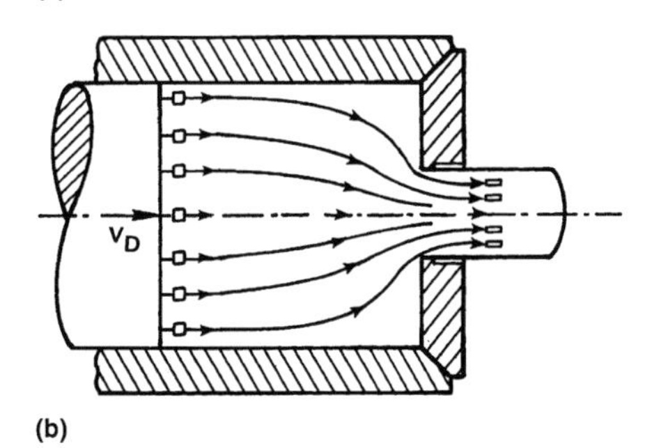

Fig. 3.5 Metal flow in certain forming processes. (a) Non-steady-state upset forging. (b) Steady-state extrusion. [Lange, 1972]

Thus, with the values of sin 2θ and cos 2θ the equation becomes

$$\sigma_1, \sigma_2 = \frac{1}{2}(\sigma_x + \sigma_y)$$

$$\pm \frac{1}{2}[(\sigma_x - \sigma_y)^2 + 4\tau_{xy}^2]^{1/2} \quad \text{(Eq 3.5)}$$

To find the planes where the shear stress τ_θ is maximum, differentiate Eq 3.3 with respect to θ and equate to zero. Thus,

$$\tau_{max} = \frac{1}{2}[(\sigma_x - \sigma_y)^2 + 4\tau_{xy}^2]^{1/2} \quad \text{(Eq 3.6)}$$

3.5 Local Deformations and the Velocity Field

The local displacement of the volume elements is described by the velocity field, e.g., velocities, strain rates, and strains (Fig. 3.5). To simplify analysis, it is often assumed that the velocity field is independent of the material properties. Obviously, this is not correct.

3.6 Strains

In order to investigate metal flow quantitatively, it is necessary to define the strains (or deformations), strain rates (deformation rates), and velocities (displacements per unit time). Figure 3.6 illustrates the deformation of an infinitesimal rectangular block, abcd, into a parallelogram, a′b′c′d′, after a small amount of plastic deformation. Although this illustration is in two dimensions, the principles apply also to three-dimensional cases.

The coordinates of a point are initially x and y (and z in three dimensions). After a small deformation, the same point has the coordinates x′ and y′ (and z′ in 3-D). By neglecting the higher-order components, one can determine the magnitude of the displacement of point b, u_{bx}, as a function of the displacement of point a. This value, u_{bx}, is different from the displacement of point a, u_x, about the variation of the function u_x over the length dx, i.e.,

$$u_{bx} = u_x + \frac{\partial u_x}{\partial x} dx \quad \text{(Eq 3.7)}$$

Note that u_x also depends on y and z.

The relative elongation of length ab (which is originally equal to dx), or the strain in the x direction, ε_x, is now:

$$\varepsilon_x = \frac{(u_{bx} - u_x)}{dx}$$

or

$$\varepsilon_x = \left(u_x + \frac{\partial u_x}{\partial u} dx - u_x\right) \Big/ dx = \frac{\partial u_x}{\partial x} \quad \text{(Eq 3.8a)}$$

Similarly, in the y and z directions,

$$\varepsilon_y = \frac{\partial u_y}{\partial y}; \ \varepsilon_z = \frac{\partial u_z}{\partial z} \quad \text{(Eq 3.8b)}$$

The angular variations due to the small deformation considered in Fig. 3.6 are infinitesimally small. Therefore, tan $\alpha_{xy} = \alpha_{xy}$ and tan $\alpha_{yx} = \alpha_{yx}$. Thus:

$$\alpha_{xy} = (u_{by} - u_y)/(u_{bx} + dx - u_x) \quad \text{(Eq 3.9)}$$

The expression for u_{bx} is given in Eq 3.7, and that for u_{by} can be obtained similarly as:

$$u_{by} = u_y + \frac{\partial u_y}{\partial y} dy \quad \text{(Eq 3.10)}$$

Using Eq 3.7 and 3.10, and considering that $\varepsilon_x = \partial u_x/\partial_x$ is considerably smaller than 1, Eq 3.9 leads to:

$$\alpha_{xy} = \frac{\partial u_y}{\partial x} \quad \text{(Eq 3.11a)}$$

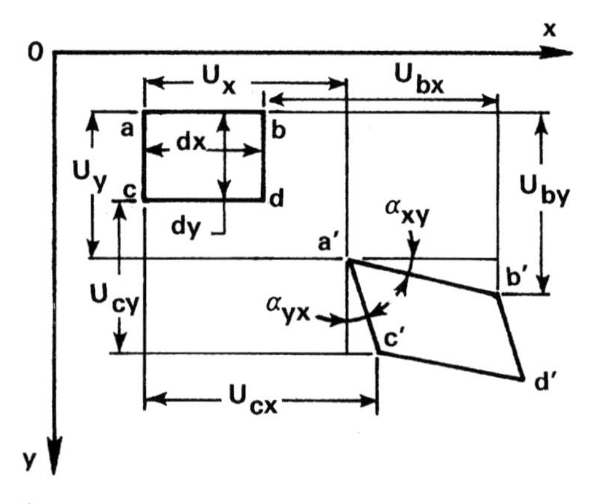

Fig. 3.6 Displacement in the x-y plane. [Altan et al., 1983]

and similarly,

$$\alpha_{yx} = \frac{\partial u_x}{\partial y} \qquad \text{(Eq 3.11b)}$$

Thus, the total angular deformation in the xy plane, or the shear strain, γ_{xy}, is:

$$\gamma_{xy} = \alpha_{xy} + \alpha_{yx} = \frac{\partial u_y}{\partial x} + \frac{\partial u_x}{\partial y} \qquad \text{(Eq 3.12a)}$$

Similarly:

$$\gamma_{yz} = \frac{\partial u_y}{\partial z} + \frac{\partial u_z}{\partial y} \qquad \text{(Eq 3.12b)}$$

and

$$\gamma_{xz} = \frac{\partial u_z}{\partial x} + \frac{\partial u_x}{\partial z} \qquad \text{(Eq 3.12c)}$$

3.7 Velocities and Strain Rates

The distribution of velocity components (v_x, v_y, v_z) within a deforming material describes the metal flow in that material. The velocity is the variation of the displacement in time or in the x, y, and z directions [Backofen, 1972, and Rowe, 1977].

$$v_x = \frac{\partial u_x}{\partial t};\ v_y = \frac{\partial u_y}{\partial t};\ v_z = \frac{\partial u_z}{\partial t} \qquad \text{(Eq 3.13)}$$

The strain rates, i.e., the variations in strain with time, are:

$$\dot{\varepsilon}_x = \frac{\partial \varepsilon_x}{\partial t} = \frac{\partial}{\partial t}\frac{\partial (u_x)}{\partial x} = \frac{\partial}{\partial x}\left(\frac{\partial u_x}{\partial t}\right) = \frac{\partial v_x}{\partial x}$$

Similarly,

$$\dot{\varepsilon}_x = \frac{\partial v_x}{\partial x};\ \dot{\varepsilon}_y = \frac{\partial v_y}{\partial y};\ \dot{\varepsilon}_z = \frac{\partial v_z}{\partial z} \qquad \text{(Eq 3.14a)}$$

$$\dot{\gamma}_{xy} = \frac{\partial v_x}{\partial y} + \frac{\partial v_y}{\partial x} \qquad \text{(Eq 3.14b)}$$

$$\dot{\gamma}_{yz} = \frac{\partial v_y}{\partial z} + \frac{\partial v_z}{\partial y} \qquad \text{(Eq 3.14c)}$$

$$\dot{\gamma}_{xz} = \frac{\partial v_x}{\partial z} + \frac{\partial v_z}{\partial x} \qquad \text{(Eq 3.14d)}$$

The state of deformation in a plastically deforming metal is fully described by the displacements, u, velocities, v, strains, ε, and strain rates, $\dot{\varepsilon}$ (in an x, y, z coordinate system). It is possible to express the same values in an x′, y′, z′ system, provided that the angle of rotation from x, y, z to x′, y′, z′ is known. Thus, in every small element within the plastically deforming body, it is possible to orient the coordinate system such that the element is not subjected to shear but only to compression or tension. In this case, the strains γ_{xy}, γ_{yz}, γ_{xz} all equal zero, and the element deforms along the principal axes of deformation.

In uniaxial tension and compression tests (no necking, no bulging), deformation is also in the directions of the principal axes.

The assumption of volume constancy made earlier neglects the elastic strains. This assumption is reasonable in most forming processes where the amount of plastic strain is much larger than the amount of elastic strain. This assumption can also be expressed, for the deformation along the principal axes, as follows:

$$\varepsilon_x + \varepsilon_y + \varepsilon_z = 0 \qquad \text{(Eq 3.15)}$$

and

$$\dot{\varepsilon}_x + \dot{\varepsilon}_y + \dot{\varepsilon}_z = 0 \qquad \text{(Eq 3.16)}$$

3.8 Homogeneous Deformation

Figure 3.7 considers "frictionless" upset forging of a rectangular block. The upper die is moving downward at velocity V_D. The coordinate axes x, y, and z have their origins on the lower platen, at the center of the lower rectangular surface.

The initial and final dimensions of the block are designated by the subscripts 0 and 1, respectively. The instantaneous height of the block during deformation is h. The velocity components v_x, v_y, and v_z, describing the motion of each particle within the deforming block, can be expressed as the linear function of the coordinates x, y, and z as follows:

$$v_x = \frac{V_D x}{2h};\ v_y = \frac{V_D y}{2h};\ v_z = -\frac{V_D z}{h} \qquad \text{(Eq 3.17)}$$

In order to demonstrate that the velocity field described by Eq 3.17 is acceptable, it is necessary to prove that these velocities satisfy (a) the volume constancy and (b) the boundary conditions [Johnson et al., 1975].

Satisfaction of the boundary conditions can be shown by considering the initial shape on the block before deformation (Fig. 3.7). At the origin of the coordinates, all the velocities must be equal to zero. This condition is satisfied because, at the origin, for $x = y = z = 0$, one has, from Eq 3.17, $v_x = v_y = v_z = 0$. At the boundaries:

At $x = l_o/2$; velocity in the x direction:

$$v_{xo} = V_D l_o / 4h_o \quad \text{(Eq 3.18a)}$$

At $y = w_o/2$; velocity in the y direction:

$$v_{yo} = V_D w_o / 4h_o \quad \text{(Eq 3.18b)}$$

At $z = h_o$; velocity in the z direction:

$$v_{zo} = -V_D \quad \text{(Eq 3.18c)}$$

It can be shown easily that the volume constancy is also satisfied. At the start of deformation, the upper volume rate or the volume per unit time displaced by the motion of the upper die is:

$$\text{Volume rate} = V_D w_o h_o \quad \text{(Eq 3.19)}$$

The volumes per unit time moved toward the sides of the rectangular block are:

$$2v_{xo}h_o w_o + 2v_{yo}l_o h_o \quad \text{(Eq 3.20)}$$

Using the values of v_{xo} and v_{yo} given by the Eq 3.18(a) and 3.18(b), Eq 3.20 gives:

$$\text{Volume rate} = 2h_o(w_o V_D l_o + l_o V_D w_o)/4h_o \quad \text{(Eq 3.21a)}$$

or

$$\text{Volume rate} = V_D w_o h_o \quad \text{(Eq 3.21b)}$$

The quantities given by Eq 3.19 and 3.21 are equal; i.e., the volume constancy condition is satisfied. The strain rates can now be obtained from the velocity components given by Eq 3.17.

$$\dot{\varepsilon}_x = \frac{\partial v_x}{\partial x} = \frac{V_D}{2h} \quad \text{(Eq 3.22a)}$$

Similarly:

$$\dot{\varepsilon}_y = \frac{V_D}{2h}; \ \dot{\varepsilon}_z = -\frac{V_D}{h} \quad \text{(Eq 3.22b)}$$

It can be easily seen that:

$$\dot{\gamma}_{xy} = \dot{\gamma}_{xz} = \dot{\gamma}_{yz} = 0$$

In homogeneous deformation, the shear strain rates are equal to zero. The strains can be obtained by integration with respect to time, t.

In the height direction:

$$\varepsilon_z = \int_{t_o}^{t_1} \dot{\varepsilon}_z dt = \int_{t_o}^{t_1} -\frac{V_D}{h} dt \quad \text{(Eq 3.23)}$$

For small displacements, $dh = -V_D dt$. Thus, Eq 3.23 gives:

$$\varepsilon_h = \varepsilon_z = \int_{h_o}^{h_1} \frac{dh}{h} = \ln \frac{h_1}{h_o} \quad \text{(Eq 3.24a)}$$

Other strains can be obtained similarly:

$$\varepsilon_l = \varepsilon_x = \ln \frac{l_1}{l_o}; \ \varepsilon_b = \varepsilon_y = \ln \frac{w_1}{w_o} \quad \text{(Eq 3.24b)}$$

Volume constancy in terms of strains can be verified from:

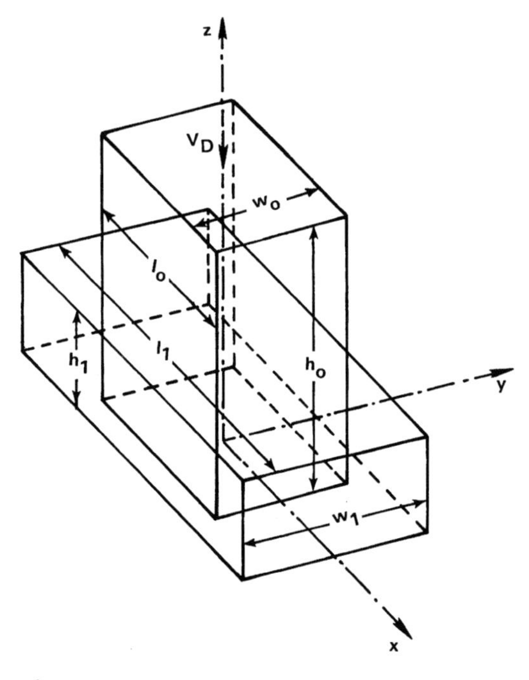

Fig. 3.7 Homogeneous deformation in frictionless upset forging

$$V = h_o w_o l_o = h_1 w_1 l_1 \rightarrow \frac{h_o w_o l_o}{h_1 w_1 l_1} = 1$$

or, taking the natural logarithm

$$\ln \frac{h_o}{h_1} + \ln \frac{b_o}{b_1} + \ln \frac{l_o}{l_1} = \varepsilon_h + \varepsilon_b + \varepsilon_l = 0 \qquad \text{(Eq 3.25)}$$

3.9 Plastic (True) Strain and Engineering Strain (Fig. 3.8)

The results of Eq 3.24 can also be obtained through a different approach. In the theory of strength of materials—during uniform elongation in tension, for example—the infinitesimal engineering strain, de, is considered with respect to the original length, l_0, or:

$$de = \frac{dl}{l_o} \rightarrow e = \int_{l_o}^{l_1} \frac{dl}{l_o} = \frac{l_1 - l_o}{l_o} \qquad \text{(Eq 3.26)}$$

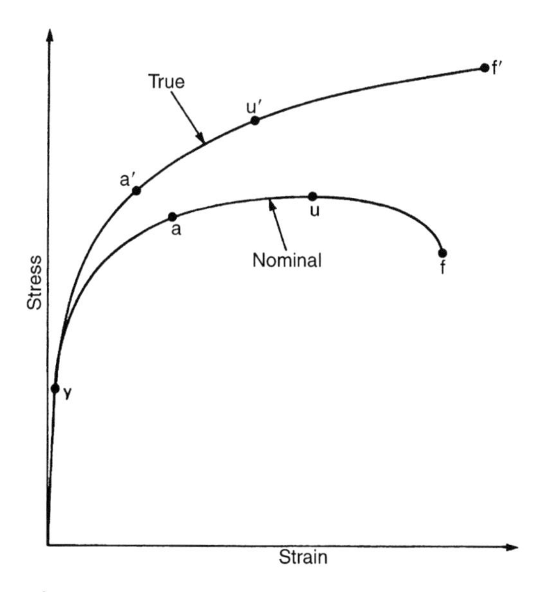

Fig. 3.8 Comparison of engineering and true stress-strain curve. [Hosford & Caddell, 1983]

In the theory of metal forming plasticity, the initial condition cannot be used as a frame of reference; therefore, the change in the length must be related to instantaneous length, or:

$$d\varepsilon = \frac{dl}{l} \rightarrow \varepsilon = \int_{l_o}^{l_1} \frac{dl}{l} = \ln \frac{l_1}{l_o} \qquad \text{(Eq 3.27)}$$

Equations 3.26 and 3.27 give:

$$\varepsilon = \ln \frac{l_1}{l_o} = \ln (e + 1) \qquad \text{(Eq 3.28)}$$

The relations between ε and e can be illustrated by considering the following example uniform deformations, where a bar is uniformly (or homogeneously) compressed to half its original length or is elongated to twice its original length:

	Compression for $l_1 = l_o/2$	Tension for $l_1 = 2l_o$
$\varepsilon = \ln \frac{l_1}{l_o}$	−0.693	+0.693
$\varepsilon = \frac{l_1 - l_o}{l_o}$	−0.5	+1

REFERENCES

[Altan et al., 1983]: Altan, T., Oh, S.-I., Gegel, H.L., *Metal Forming Fundamentals and Applications,* ASM International, 1983.

[Backofen, 1972]: Backofen, W., *Deformation Processing,* Addison-Wesley, 1972.

[Hosford & Caddell, 1983]: Hosford, W.F., Caddell, R.M., *Metal Forming: Mechanics and Metallurgy,* Prentice-Hall, 1983.

[Johnson et al., 1975]: Johnson, W., Mellor, P.B., *Engineering Plasticity,* Van Nostrand Reinhold Co., London, 1975.

[Lange, 1972]: Lange, K., Ed., *Study Book of Forming Technology,* (in German), Vol 1, *Fundamentals,* Springer-Verlag, 1972.

[Rowe, 1977]: Rowe, G.W., *Principles of Industrial Metalworking Processes,* Edward Arnold Publishers, London, 1975.

CHAPTER 4

Flow Stress and Forgeability

Manas Shirgaokar

4.1 Introduction

In order to understand the forces and stresses involved in metal forming processes it is necessary to (a) become familiar with the concept of flow stress and (b) start with the study of plastic deformation under conditions where a simple state of stress exists.

For studying the plastic deformation behavior of a metal it is appropriate to consider homogeneous or uniform deformation conditions. The yield stress of a metal under uniaxial conditions, as a function of strain, strain rate, and temperature, can also be considered as the "flow stress." The metal starts flowing or deforming plastically when the applied stress (in uniaxial tension without necking and in uniaxial compression without bulging) reaches the value of the yield stress or flow stress. The flow stress is very important because in metal forming processes the loads and stresses are dependent on (a) the part geometry, (b) friction, and (c) the flow stress of the deforming material. The flow stress of a metal is influenced by:

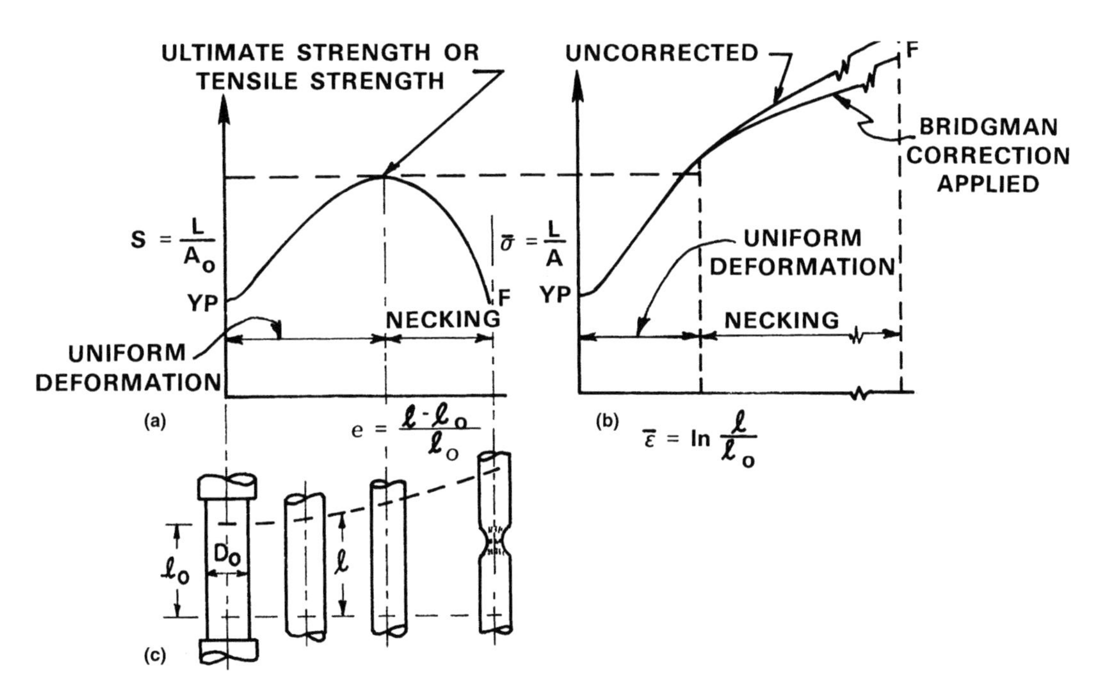

Fig. 4.1 Representation of data in tensile test. (a) Engineering stress-strain curve. (b) True stress-strain curve. (c) Schematic of dimensional change of the specimen during the test. [Thomsen et al., 1965]

- Factors unrelated to the deformation process, such as chemical composition, metallurgical structure, phases, grain size, segregation, and prior strain history
- Factors explicitly related to the deformation process, such as temperature of deformation, degree of deformation or strain, and rate of deformation or strain rate

Thus, the flow stress, $\bar{\sigma}$, can be expressed as a function of the temperature, θ, strain, ε, the strain rate, $\dot{\bar{\varepsilon}}$, and the microstructure, S. For a given microstructure, i.e., heat treatment and prior deformation history:

$$\bar{\sigma} = f(\theta, \bar{\varepsilon}, \dot{\bar{\varepsilon}}) \quad \text{(Eq 4.1)}$$

In hot forming of metals at temperatures above the recrystallization temperature the effect of strain on flow stress is insignificant and the influence of strain rate (i.e., rate of deformation) becomes increasingly important. Conversely, at room temperature (i.e., in cold forming) the effect of strain rate on flow stress is negligible. The degree of dependency of flow stress on temperature varies considerably among different materials. Therefore, temperature variations during the forming process can have different effects on load requirements and metal flow for different materials. The increase in the flow stress for titanium alloy Ti-8Al-1Mo-1V that

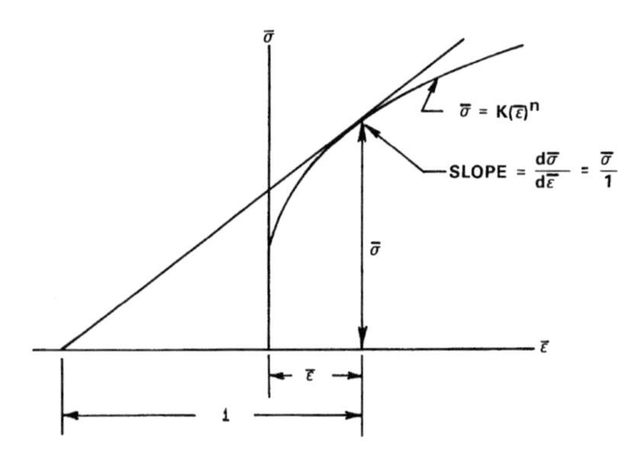

Fig. 4.2 Schematic representation of condition of necking in simple tension. [Thomsen et al., 1965]

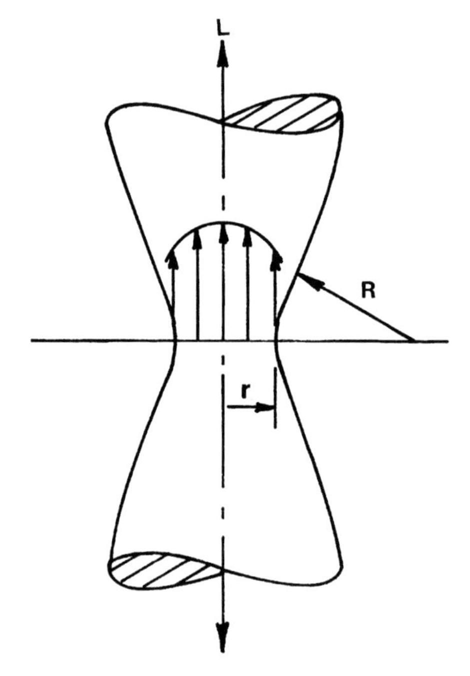

Fig. 4.3 Axial stress distribution in the necked portion of a tensile specimen. [Thomsen et al., 1965]

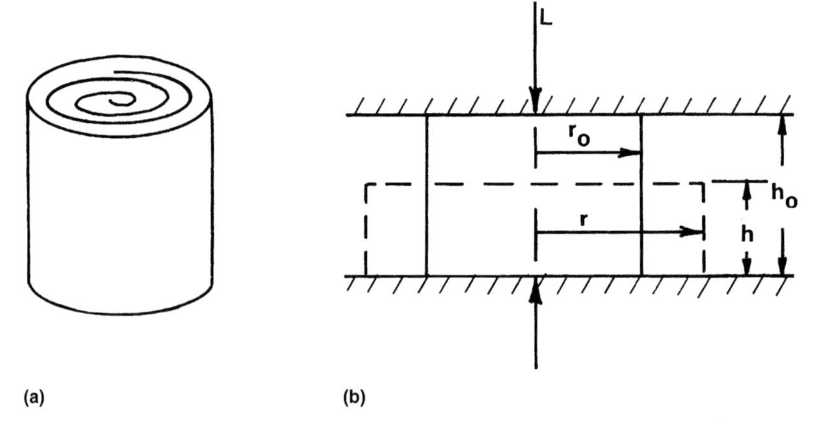

Fig. 4.4 Compression test specimen. (a) View of specimen, showing lubricated shallow grooves on the ends. (b) Shape of the specimen before and after the test

would result from a drop of 100 °F (55 °C) in the hot forging temperature (from 1700 to 1600 °F, or 925 to 870 °C) is about 40%. The same temperature drop in the hot working range of AISI type 4340 steel would result in a 15% increase in the flow stress [Altan et al., 1973].

To be useful in metal forming analyses, the flow stresses of metals should be determined experimentally for the strain, strain rate, and temperature conditions that exist during the forming processes. The most commonly used methods for determining flow stress are the tensile, uniform compression and torsion tests.

4.2 Tensile Test

The tensile test is commonly used for determining the mechanical properties of metals. However, the properties determined from this test are useful for designing components and not for producing parts by metal forming processes. The reason for this is that the tensile test data is valid for relatively small plastic strains. Flow stress data should be valid for a large range of plastic strains encountered in metal forming processes so that this data is useful in metal forming analysis.

Two methods of representing flow stress data are illustrated in Fig. 4.1 [Thomsen et al., 1965]. In the classical engineering stress-strain diagram (Fig. 4.1a), the stress is obtained by dividing the instantaneous tensile load, L, by the original cross-sectional area of the specimen, A_o. The stress is then plotted against the engineering strain, $e = (l - l_o)/l_o$. During deformation, the specimen elongates initially in a uniform fashion. When the load reaches its maximum value, necking starts and the uniform uniaxial stress condition ceases to exist. Deformation is then concentrated only in the neck region while the rest of the specimen remains rigid.

Figure 4.1(b) illustrates the true stress-strain representation of the same tensile test data. In this case, before necking occurs, the following relationships are valid:

$$\bar{\sigma} = \text{true stress (flow stress)} = \text{instantaneous load/instantaneous area} = L/A \quad \text{(Eq 4.2)}$$

and

$$\bar{\varepsilon} = \text{true strain} = \ln\left(\frac{l}{l_o}\right) = \ln\left(\frac{A_o}{A}\right) \quad \text{(Eq 4.3)}$$

The instantaneous load in tension is given by $L = A\bar{\sigma}$. The criterion for necking can be formulated as the condition that L be maximum or that:

$$\frac{dL}{d\bar{\varepsilon}} = 0 \quad \text{(Eq 4.4)}$$

Near but slightly before the attainment of maximum load, the uniform deformation conditions, i.e., Eq 4.2 and 4.3 are valid [Thomsen et al., 1965]. From Eq 4.3:

$$A = A_o(e)^{-\bar{\varepsilon}}$$

or

$$L = A\bar{\sigma} = A_o\bar{\sigma}(e)^{-\bar{\varepsilon}} \quad \text{(Eq 4.5)}$$

Combining Eq 4.4 and 4.5 results in:

$$\frac{dL}{d\bar{\varepsilon}} = 0 = A_o\left(\frac{d\bar{\sigma}}{d\bar{\varepsilon}}(e)^{-\bar{\varepsilon}} - \bar{\sigma}(e)^{-\bar{\varepsilon}}\right) \quad \text{(Eq 4.6)}$$

or

$$\frac{d\bar{\sigma}}{d\bar{\varepsilon}} = \bar{\sigma} \quad \text{(Eq 4.7)}$$

Fig. 4.5 Compression test tooling. [Dixit et al., 2002]

As is discussed later, very often the flow stress curve obtained at room temperature can be expressed in the form of an exponential equation or power law:

$$\bar{\sigma} = K(\bar{\varepsilon})^n \quad \text{(Eq 4.8)}$$

where K and n are constants.

Combining Eq 4.7 and 4.8 results in:

$$\frac{d\bar{\sigma}}{d\bar{\varepsilon}} = Kn(\bar{\varepsilon})^{n-1} = \bar{\sigma} = K(\bar{\varepsilon})^n \quad \text{(Eq 4.9)}$$

or

$$\bar{\varepsilon} = n \quad \text{(Eq 4.10)}$$

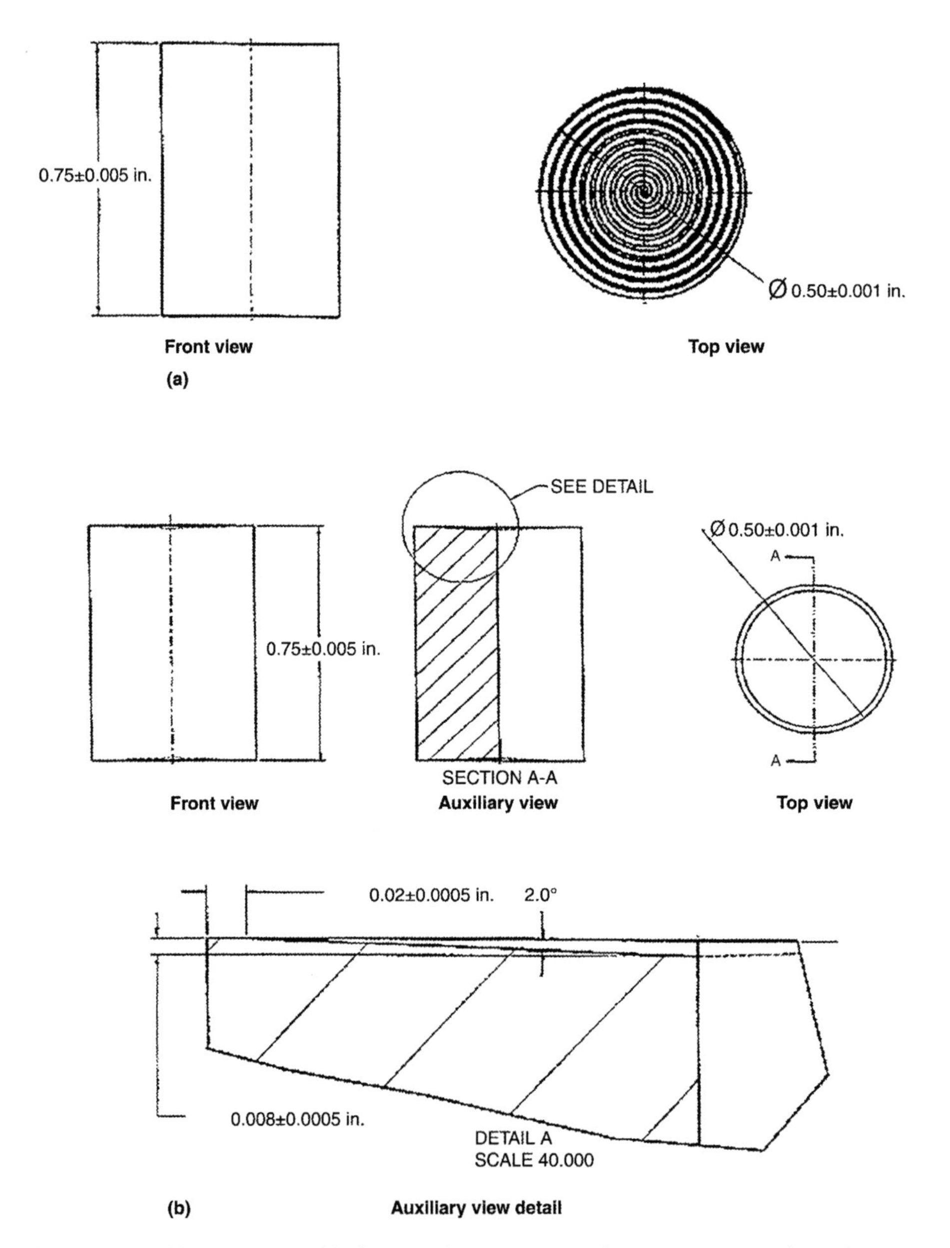

Fig. 4.6 Dimensions of the specimens used for flow stress determination using the compression test at the ERC/NSM. (a) Specimen with spiral groove. (b) Rastegaev specimen. [Dahl et al., 1999]

This condition is shown schematically in Fig. 4.2. From this figure and from Eq 4.10, it is evident that at low forming temperatures, where Eq 4.8 is valid, a material with a large n or strain hardening exponent, has greater formability; i.e., it sustains a large amount of uniform deformation in tension than a material with a smaller n. It should be noted, however, that this statement is not correct for materials and conditions where the flow stress cannot be expressed by Eq 4.8.

The calculation of true stress after the necking strain (Fig. 4.1b) requires a correction because a triaxial state of stress is induced. Such a correction, derived by Bridgman, is given by:

$$\sigma_s = \bar{\sigma} = \frac{L}{\pi r^2}\left[\left(1 + \frac{2R}{r}\right)\ln\left(1 + \frac{r}{2R}\right)\right]^{-1} \quad \text{(Eq 4.11)}$$

The quantities r and R are defined in Fig. 4.3. It can be clearly seen that, for evaluation of Eq 4.11, the values of r and R must be measured continuously during the test. This is quite cumbersome and prone to error. Therefore, other tests, which provide the true stress-strain data at larger strains relative to the tensile test, are used for metal forming applications.

4.3 Compression Test

The compression test is used to determine the flow stress data (true-stress/true-strain relationships) for metals at various temperatures and strain rates. In this test, the flat platens and the cylindrical sample are maintained at the same temperature so that die chilling, with its influence on metal flow, is prevented. To be applicable without corrections or errors, the cylindrical sample must be upset without any barreling; i.e., the state of uniform stress in the sample must be maintained as shown in Fig. 4.4. Barreling is prevented by using adequate lubrication, e.g., Teflon or machine oil at room temperature and at hot working temperatures, graphite in oil for aluminum alloys, and glass for steel, titanium, and high-temperature alloys. The load and displacement, or sample height, are measured during the test. From this information the flow stress is calculated at each stage of deformation, or for increasing strain. Figure 4.5 shows the tooling used for compression tests conducted at the Engineering Research Center for Net Shape Manufacturing (ERC/NSM) of the Ohio State University [Dixit et al., 2002].

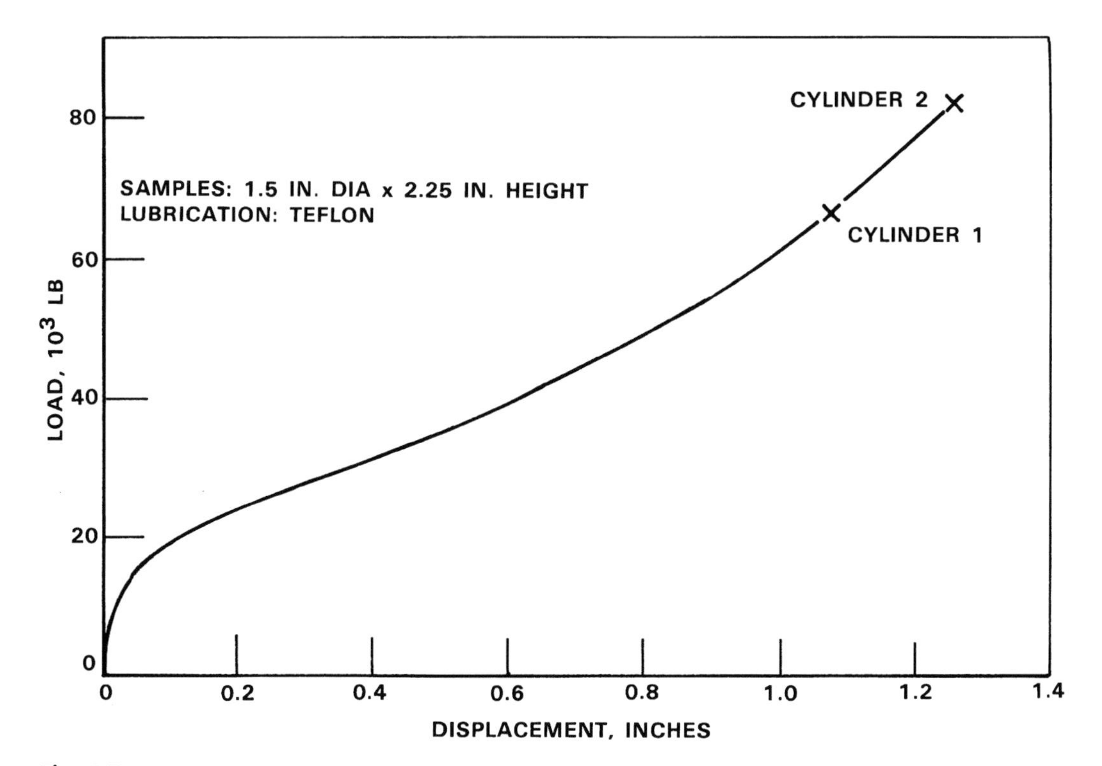

Fig. 4.7 Load-displacement curve obtained in uniform upsetting of annealed 1100 aluminum cylinders. [Lee et al., 1972]

Similar to the uniform elongation portion of the tensile test, the following relationships are valid for the uniform compression test:

$$\bar{\varepsilon} = \ln \frac{h_o}{h} = \ln \frac{A}{A_o} \quad \text{(Eq 4.12)}$$

$$\bar{\sigma} = \frac{L}{A} \quad \text{(Eq 4.13)}$$

$$A = A_o(e)^{\bar{\varepsilon}} \quad \text{(Eq 4.14)}$$

$$\dot{\bar{\varepsilon}} = \frac{d\bar{\varepsilon}}{dt} = \frac{dh}{hdt} = \frac{V}{h} \quad \text{(Eq 4.15)}$$

where V is instantaneous deformation velocity; h_o and h are initial and instantaneous heights, respectively, and A_o and A are initial and instantaneous surface areas, respectively.

As discussed earlier the flow stress values determined at high strains in the tensile test require a correction because of necking. Therefore, the compression test, which can be conducted without barreling up to about 50% reduction in height ($\bar{\varepsilon}$ = 0.693 or more), is widely used to obtain flow stress data for metal forming applications.

At room temperature, the flow stresses of most metals (except that of lead) are only slightly strain-rate dependent. Therefore, any testing machine or press can be used for the compression test, regardless of its ram speed. Adequate lubrication of the platens is usually accomplished by (a) using lubricants such as Teflon, molybdenum disulfide, or high-viscosity oil and (b) by using Rastegaev specimens (Fig. 4.6) or specimens with spiral grooves machined on both the flat surfaces of the specimen to hold the lubricant (Fig. 4.6). A typical load-displacement curve obtained in the uniform compression test of aluminum alloy (Al 1100, annealed) at room temperature in a testing machine is shown in Fig. 4.7. The $\bar{\sigma}$-$\bar{\varepsilon}$ data obtained from this curve are shown in Fig. 4.8.

At hot working temperatures, i.e., above the recrystallization temperature, the flow stresses of nearly all metals are very much strain-rate dependent. Therefore, whenever possible, hot compression tests are conducted on a machine that provides a velocity-displacement profile such that the condition $\dot{\bar{\varepsilon}}$ = velocity/sample

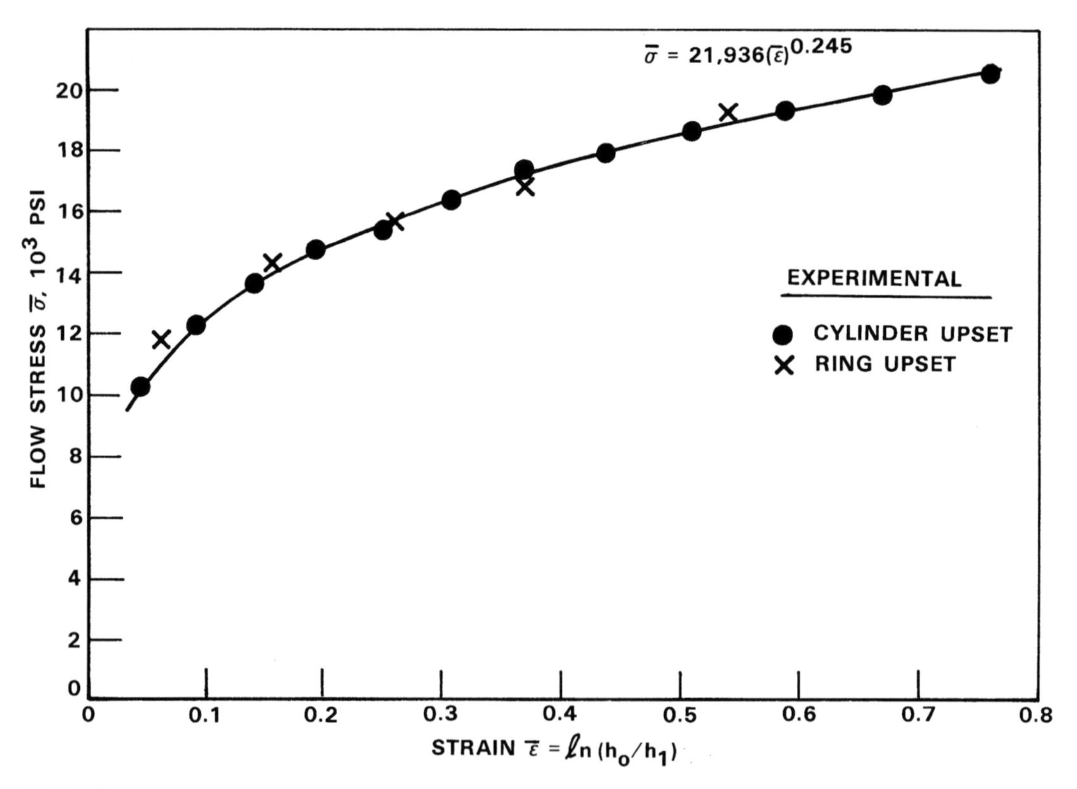

Fig. 4.8 Flow stress-strain curve for annealed 1100 aluminum obtained from uniform cylinder and ring upset tests. [Lee et al., 1972]

height can be maintained throughout the test. Mechanical cam-activated presses called plastometer or hydraulic programmable testing machines (MTS, for example) are used for this purpose. In order to maintain nearly isothermal and uniform compression conditions, the test is conducted in a furnace or a fixture such as that shown in Fig. 4.9. The specimens are lubricated with appropriate lubricants—for example, oil graphite for temperatures up to 800 °F (425 °C) and glass for temperatures up to 2300 °F (1260 °C). The fixture and the specimens are heated to the test temperature and then the test is initiated. Examples of hot-formed compression samples are shown in Fig. 4.10. Examples of high-temperature $\bar{\sigma}$-$\bar{\varepsilon}$ data are given in Fig. 4.11 and 4.12.

4.3.1 *Specimen Preparation*

There are two machining techniques that can be used for preparing the specimens for the compression test, viz. the spiral specimen (Fig. 4.6a) and the Rastegaev specimen (Fig. 4.6b). The specimens shown are of standard dimensions used for the compression test. The spiral grooves and the recesses of the Rastegaev specimen serve the purpose of retaining the lubricant at the tool/workpiece interface during compression thus preventing barreling. It has been determined through tests conducted at the ERC/NSM that Rastegaev specimens provide better lubrication and hold their form better during testing compared to the spiral grooved specimens. The specifications for the specimens and the test conditions are [Dahl et al., 1999]:

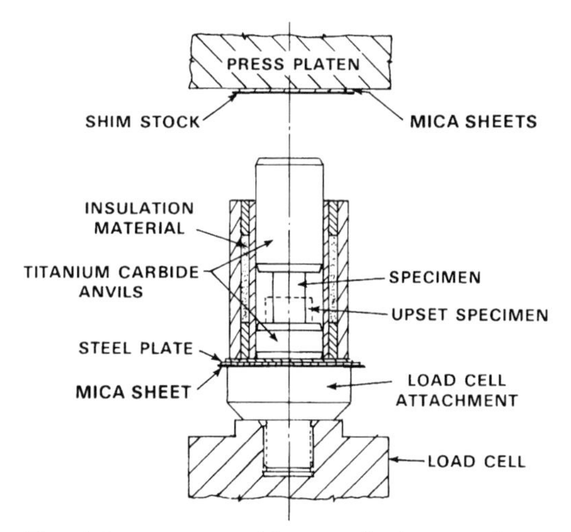

Fig. 4.9 Press setup and fixture used in heating and compression of cylinders and rings

Specimen with spiral grooves (Fig. 4.6a):

- Solid cylinder (diameter = $0.5^{\pm 0.001}$ in., length = $0.75^{\pm 0.005}$ in.).
- Ends should be flat and parallel within 0.0005 in./in.
- Surface should be free of grooves, nicks and burrs.
- Spiral grooves machined at the flat ends of the specimen with approximately 0.01 in. depth.

Rastegaev specimen (Fig. 4.6b):

- Flat recesses at the ends should be filled with lubricant.
- Dimensions t_0 = $0.008^{\pm 0.0005}$ in. and u_o = $0.02^{\pm 0.0005}$ in. at the end faces have a significant effect on the lubrication conditions.
- Rastegaev specimen ensures good lubrication up to high strains of about 0.8 to 1, so that the specimen remains cylindrical (due to radial pressure that the lubricant exerts on the ring).
- t_o/u_o = 0.4 (Fig. 4.6b) for steels (optimum value at which the specimen retains cylin-

Fig. 4.10 Uniform compression samples before and after deformation (left to right: AISI 1018 steel, INCO 718, Ti-6Al-4V)

drical shape up to maximum strain before bulging occurs).

4.3.2 *Parallelism of the Press (or Testing Machine) Slides*

In a compression test, load is applied on the billet using flat dies. In order to ensure that a uniaxial state of stress exists during the experiment, the load applied should be perpendicular to the axis of the cylindrical specimen. This calls for measurement of the parallelism of the platens of the press. A commonly used technique for parallelism measurement involves compressing lead billets of the same height. The difference in the heights of the lead billets is an indication of the parallelism of the platens. Lead is used since it is soft and deforms easily at room temperature. The procedure followed for determining the parallelism for recent tests conducted at the ERC/NSM is described below [Dixit et al., 2002]:

1. Lead bar of 1 in. diameter was cut into approximately 1 in. length. The height of each specimen was noted and an average value was calculated (Table 4.1).
2. The specimen were numbered and positioned on the compression test die (Fig. 4.13 and 4.14). The distance between them was measured.
3. The samples were compressed in the tooling (Fig. 4.14). The final heights of the lead blocks were determined using a digital caliper. They are tabulated in Table 4.1.
4. From the difference in the height of two specimens and the distance between their locations, the parallelism was determined as shown in Table 4.2. For example, for specimens 1 and 2, the difference in final height was 0.386 mm. This value divided by the distance between their locations (60.2 mm) gave the ratio 0.0064 mm/mm (Table 4.2). From the data summarized in Table 4.2 and the ex-

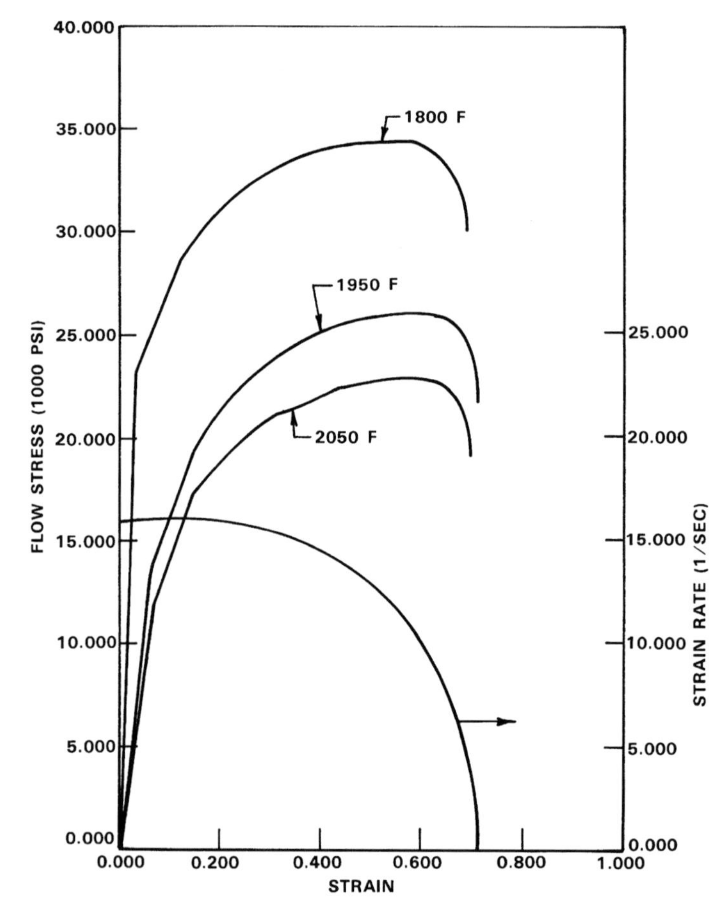

Fig. 4.11 Flow stress versus strain and strain rate versus strain, for type 403 stainless steel at 1800, 1950, and 2050 °F (980, 1065, and 1120 °C) (tests were conducted in a mechanical press where strain rate was not constant). [Douglas et al., 1975]

periments, it was concluded that a parallelism less than 0.01 was acceptable for conducting reliable compression tests.

4.3.3 *Errors in the Compression Test*

Errors in the determination of flow stress by the compression test can be classified in three categories [Dahl et al., 1999]:

- Errors in the displacement readings, which result in errors in the calculated strain
- Errors in the load readings, which result in errors in the calculated stress
- Errors in the processing of the data due to barreling of the test specimens

The first and second type errors may be reduced or eliminated by careful calibration of the transducers and data acquisition equipment. However, barreling of the test specimens during compression cannot be entirely eliminated because there is always friction between the specimen and the tools.

4.3.4 *Determination of Error in Flow Stress Due to Barreling*

The maximum error in determining flow stress may be the result of friction. In order to correct the flow stress curve and to determine the percentage error in flow stress, finite element (FE) analysis is used. The amount of barreling (Fig. 4.15 and 4.16) of different specimens expressed by (H_2 − H_1) for the given height reductions during a particular compression test is given in Table 4.3. Figure 4.16(a) shows the effect of friction on the end face of the billet.

Figure 4.17 shows the load stroke curves obtained from FE simulations for different values of shear friction factors (m) and from experiment for one specimen. When the load stroke curves are compared it can be seen that simu-

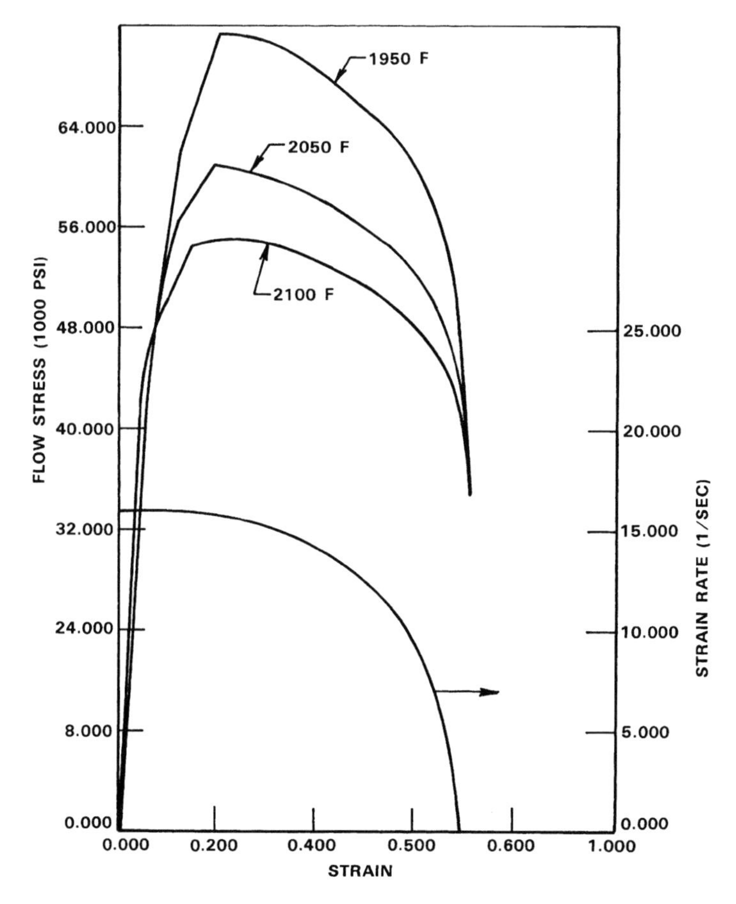

Fig. 4.12 Flow stress versus strain and strain rate versus strain, for Waspaloy at 1950, 2050, and 2100 °F (1065, 1120, and 1150 °C) (tests were conducted in a mechanical press where strain rate was not constant). [Douglas et al., 1975]

lations slightly overpredict the load. It should be noted that the difference in the load remains the same throughout the stroke.

The stress obtained from finite element simulations with shear friction factor, m, greater than zero is called apparent flow stress. Apparent flow stress curves can be used to determine the error in flow stress obtained in experiments due to barreling at higher strains (e.g., strain = 1.0). An "apparent flow stress" curve can be calculated for a given value of shear friction factor m, as follows [Dixit et al., 2002]. At several reductions in height:

- The value of load and the associated diameters (H_1 and H_2 in Fig. 4.15) are noted. A mean diameter is calculated as: $(H_1 + H_2)/2$.
- The cross-sectional area is calculated using the mean diameter.

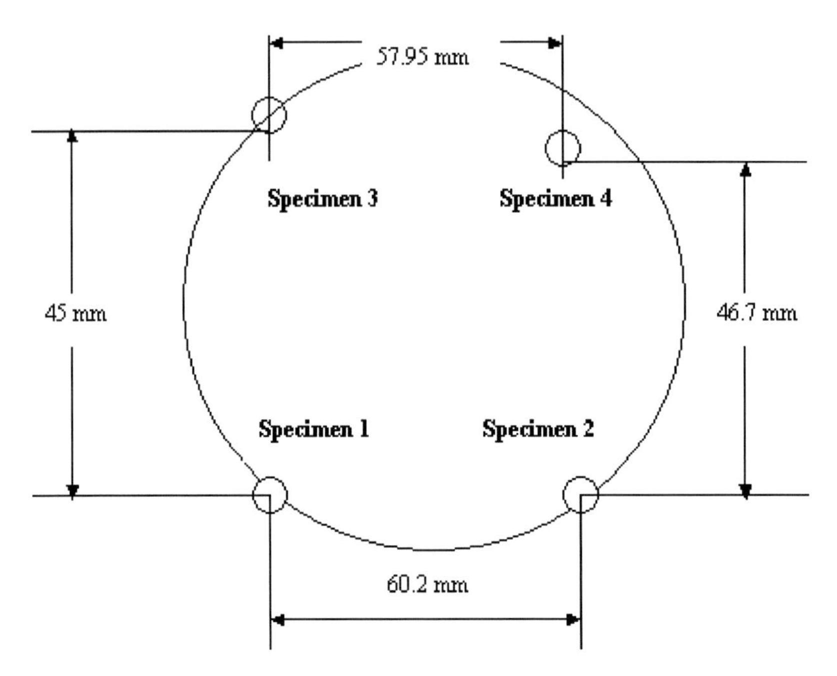

Fig. 4.13 Distance between the billets that were placed inside the press. [Dixit et al., 2002]

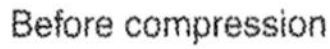

Before compression

After compression

Fig. 4.14 Lead samples on the compression test die. [Dixit et al., 2002]

Table 4.1 Height of the lead specimens used in tests conducted at the ERC/NSM

	Specimen No. 1	Specimen No. 2	Specimen No. 3	Specimen No. 4
Initial height, in. (mm)				
	0.9933 (25.23)	1.0071 (25.58)	1.0181 (25.86)	1.0185 (25.87)
	0.9894 (25.13)	1.0075 (25.59)	1.0157 (25.8)	1.0197 (25.9)
	0.9929 (25.22)	1.0071 (25.58)	1.0169 (25.83)	1.0236 (26)
	0.9913 (25.18)	1.0075 (25.59)	1.0177 (25.85)	1.0232 (25.99)
	0.9937 (25.24)	1.0067 (25.57)	1.0169 (25.83)	1.0217 (25.95)
Average	**0.992 (25.2)**	**1.0072 (25.582)**	**1.0171 (25.834)**	**1.0213 (25.942)**
Final height, in. (mm)				
	0.5594 (14.21)	0.5594 (14.21)	0.5591 (14.2)	0.5591 (14.2)
	0.5610 (14.22)	0.5610 (14.22)	0.5587 (14.19)	0.5587 (14.19)
	0.5610 (14.22)	0.5594 (14.21)	0.5594 (14.21)	0.5594 (14.21)
Average	**0.5597 (14.217)**	**0.5596 (14.213)**	**0.5591 (14.2)**	**0.5591 (14.2)**
Difference in height	**0.4324 (10.983)**	**0.4476 (11.369)**	**0.4580 (11.634)**	**0.4623 (11.742)**

Source: [Dixit et al., 2002]

- The "apparent stress" is calculated using the load at that height reduction and cross-sectional area (= Load/Area).
- The value of strain is calculated as $\log_e$ (original height/instantaneous height).
- A particular value of shear friction factor, m, results in an "apparent flow stress" that is higher in magnitude than the value obtained with zero friction. A graph of stress versus strain plotted for different values of shear friction factor m can be drawn as shown in Fig. 4.18. As the value of m increases, the "apparent flow stress" increases.
- The barreling of the specimen at a strain of 1.0 is noted down.

The "apparent flow stress" curves obtained above can be used to calculate the error in flow stress obtained due to nonhomogenous deformation (barreling) during the tests at a strain of 1.0 as follows:

1. Conduct cylinder compression tests until a strain of 1.0.
2. Determine the amount of barreling in the specimen at a strain of 1.0.
3. By comparing the barreling of the actual specimen with the "apparent flow stress" curves given in Fig. 4.18, the value of stress in the experiments can be noted.
4. Thus, the error in flow stress obtained from an experiment can be calculated with respect to the stress in the curve with m = 0 (Fig. 4.18) as a percentage value at a strain of 1.0.

If needed, the "apparent flow stress" curves can be generated for higher strains and the procedure can be repeated for estimating stress at that particular higher strain.

4.4 Ring Test

The ring test consists of compressing a flat ring shaped specimen to a known reduction (Fig. 4.19). The changes in the external and internal diameters of the ring are very much dependent on the friction at the tool/specimen interface [Lee et al., 1972]. If the friction were equal to zero, the ring would deform in the same way as a solid disk, with each element flowing radially outward at a rate proportional to its distance from the center. With increasing deformation, the internal diameter of the ring is reduced if friction is large and is increased if friction is low. Thus, the change in the internal diameter represents a simple method for evaluating interface friction. This method of estimating the friction factor using the ring test is discussed in detail in a later chapter.

The ring test can also be used for determining $\bar{\sigma}$-$\bar{\varepsilon}$ data for practical applications. For this purpose it is necessary to perform an analysis or a mathematical simulation of the ring test. This simulation allows the prediction of a load-stroke curve, if the $\bar{\sigma}$-$\bar{\varepsilon}$ and the friction are unknown. Inversely, by using this mathematical model of the ring test, it is possible to calculate the $\bar{\sigma}$-$\bar{\varepsilon}$ curve if the load-stroke curve and the friction are known.

Table 4.2 Parallelism between different points that are shown in Fig. 4.13

Parallelism between points 1 and 2 (mm/mm)	0.0064
Parallelism between points 2 and 4 (mm/mm)	0.0080
Parallelism between points 3 and 4 (mm/mm)	0.0019
Parallelism between points 3 and 1 (mm/mm)	0.0145
Parallelism between points 1 and 4 (mm/mm)	0.0096
Parallelism between points 2 and 3 (mm/mm)	0.0035

Source: [Dixit et al., 2002]

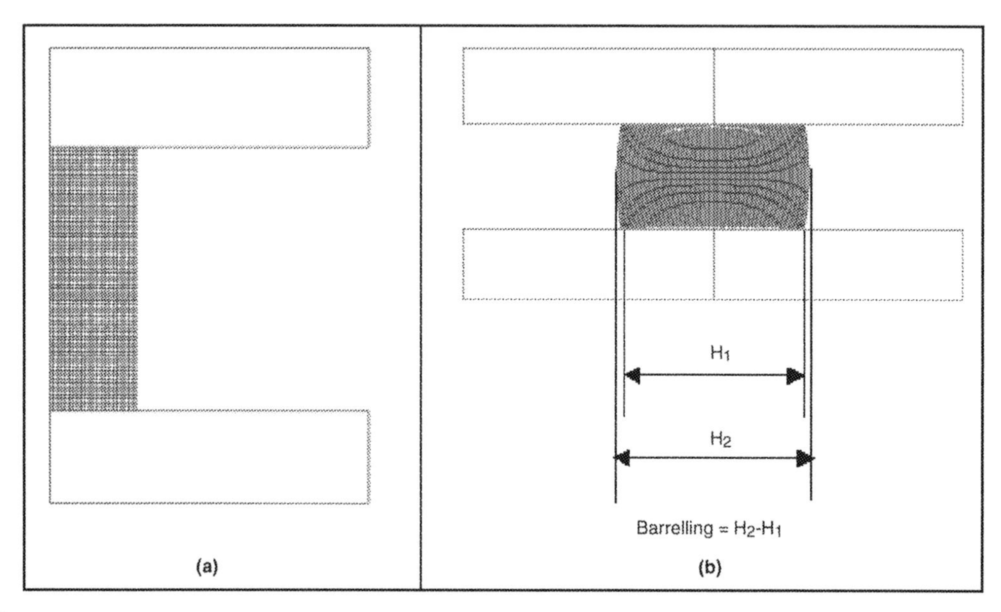

Fig. 4.15 Finite element model. (a) Before compression. (b) Barreling at 50% reduction. [Dixit et al., 2002]

4.5 Torsion Test

The torsion test can be used to obtain the $\bar{\sigma}$ data at higher strains up to $\bar{\varepsilon} = 2$ to 4. Therefore, it is used when $\bar{\sigma}$ must be known for forming operations such as extrusion, radial forging, or pilger rolling, where large strains are present. In the torsion test, a notched tube (internal radius = r, wall thickness at the notched portion = t, and gage length = l) is twisted at a given rotational speed; the torque T and the number of rotations, θ (in radians), are measured. The average shear stress, τ, in the gage section is given by:

$$\tau = \frac{T}{2\pi r^2 t} \quad \text{(Eq 4.16)}$$

The shear strain, γ, is:

$$\gamma = \frac{r\theta}{l} \quad \text{(Eq 4.17)}$$

Torsion test results can be correlated with the uniform tensile or compression results as follows:

$$\tau = \frac{\bar{\sigma}}{\sqrt{3}} \quad \text{(Eq 4.18)}$$

and

$$\gamma = \int dy = \sqrt{3}\int d\bar{\varepsilon} = \sqrt{3}\varepsilon \quad \text{(Eq 4.19)}$$

Equation 4.18 and 4.19 are obtained from the von Mises flow rule, which is discussed later.

4.6 Representation of Flow Stress Data

A typical $\bar{\sigma}$-$\bar{\varepsilon}$ curve, obtained at temperatures below the recrystallization temperature, i.e., in the cold forming range, is similar to that shown in Fig. 4.7. Here the strain hardening is pronounced and the $\bar{\sigma}$ for most materials is not ap-

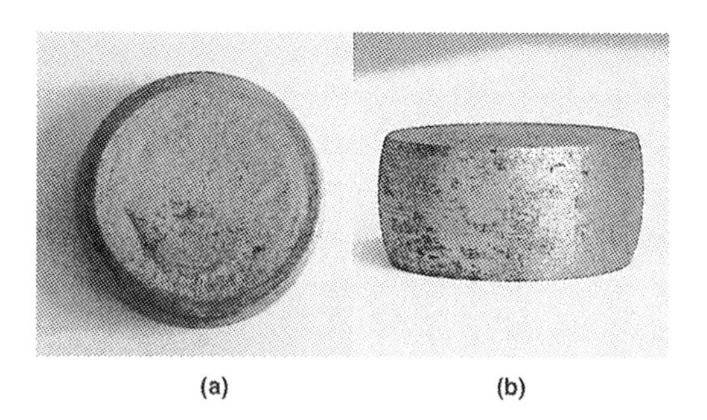

Fig. 4.16 Compression test specimen showing the effects of barreling. (a) Top view. (b) Front view. [Dixit et al., 2002]

preciably affected by $\dot{\bar{\varepsilon}}$. At hot working temperatures, most $\bar{\sigma}$-$\bar{\varepsilon}$ curves are similar to those given in Fig. 4.11 and 4.12. Obviously, values of $\bar{\sigma}$ are higher for stronger materials. At cold forming temperatures, $\bar{\sigma}$ increases with increasing $\bar{\varepsilon}$ and reaches a saturation stress at values of $\bar{\varepsilon}$ larger than 0.8 or 1.0. At hot working temperatures, $\bar{\sigma}$ increases with increasing $\dot{\bar{\varepsilon}}$ and with decreasing temperature, θ. At constant $\dot{\bar{\varepsilon}}$, $\bar{\sigma}$ versus $\bar{\varepsilon}$ increases first, then decreases because of internal heat generation and thermal softening. In all tests, the test temperature is not constant in a strict sense. Because of plastic deformation, a temperature increase, $\Delta\theta$, takes place. This can be estimated as:

$$\Delta\theta = \frac{A\bar{\varepsilon}\bar{\sigma}}{c\rho} \qquad \text{(Eq 4.20)}$$

where A is a conversion factor, c is the heat capacity and ρ is density.

Most materials, when tested at room temperature in the work hardening range, are not affected by moderate strain rates; hence, the speed of loading need not be controlled too closely. Approximate stress-strain relationships for a limited region of strain can often be given by exponential equation of the form:

$$\bar{\sigma} = K(\bar{\varepsilon})^n \qquad \text{(Eq 4.21)}$$

where K and n are constants.

Equation 4.21 is illustrated graphically in Fig. 4.20 [Thomsen et al., 1965]. The slope of the curve on the log-log coordinates is n, and $K = \bar{\sigma}$ when $\bar{\varepsilon} = 1$. It may be noted from the schematic diagram of Fig. 4.20 that at small strains, an experimentally determined curve may depart from the curve given by Eq 4.14. In that case, other values on n and K may be specified for different ranges of effective strain. Typical values of n and K are given in Tables 4.4 to 4.6 for various metals. It should be noted that other forms of stress-strain curves for room-temperature forming, i.e., cold forming have been suggested. Some of these are:

$$\text{Ludwik: } \bar{\sigma} = a + b(\bar{\varepsilon})^c \qquad \text{(Eq 4.22)}$$

Table 4.3 Barreling of the compression test specimen

	Diameter at the top surface H_1		Diameter at center H_2		Barreling $H_2 - H_1$		
Specimen No.	in.	mm	in.	mm	in.	mm	Strain
1	0.3508	8.91	0.3618	9.19	0.0110	0.28	0.550
2	0.3890	9.88	0.3929	9.98	0.0039	0.1	0.710
3	0.3283	8.34	0.3311	8.41	0.0276	0.07	0.370
4	0.4799	12.19	0.4846	12.31	0.0047	0.12	1.120
5	0.3980	10.11	0.4031	10.24	0.0051	0.13	0.736
6	0.4008	10.18	0.4114	10.45	0.0106	0.27	0.785
7	0.4035	10.25	0.4083	10.37	0.0047	0.12	0.790
8	0.3984	10.12	0.4067	10.33	0.0083	0.21	0.785

Source: [Dixit et al., 2002]

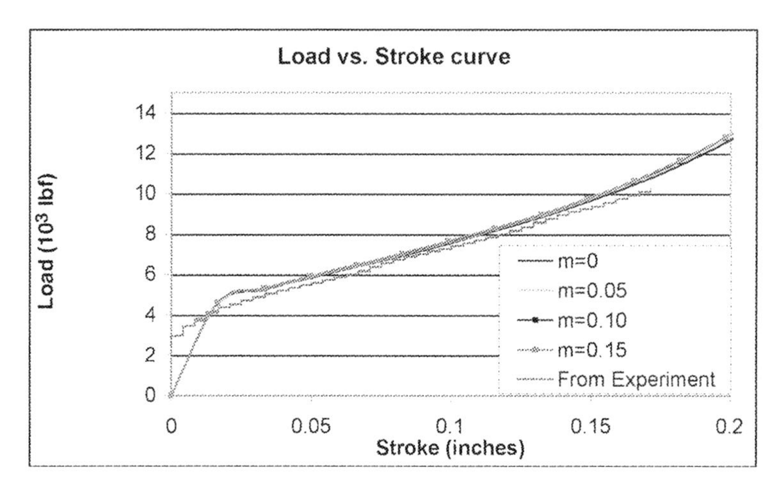

Fig. 4.17 Load stroke curves obtained from experiment and finite element simulations. [Dixit et al., 2002]

where a, b, and c are arbitrary constants. This form approximates the stress-strain curves for annealed materials, but tends to underestimate the stress where strains are small (<0.2) and to overestimate the actual stress for larger strains. For heavily prestrained materials, $c \cong 1$.

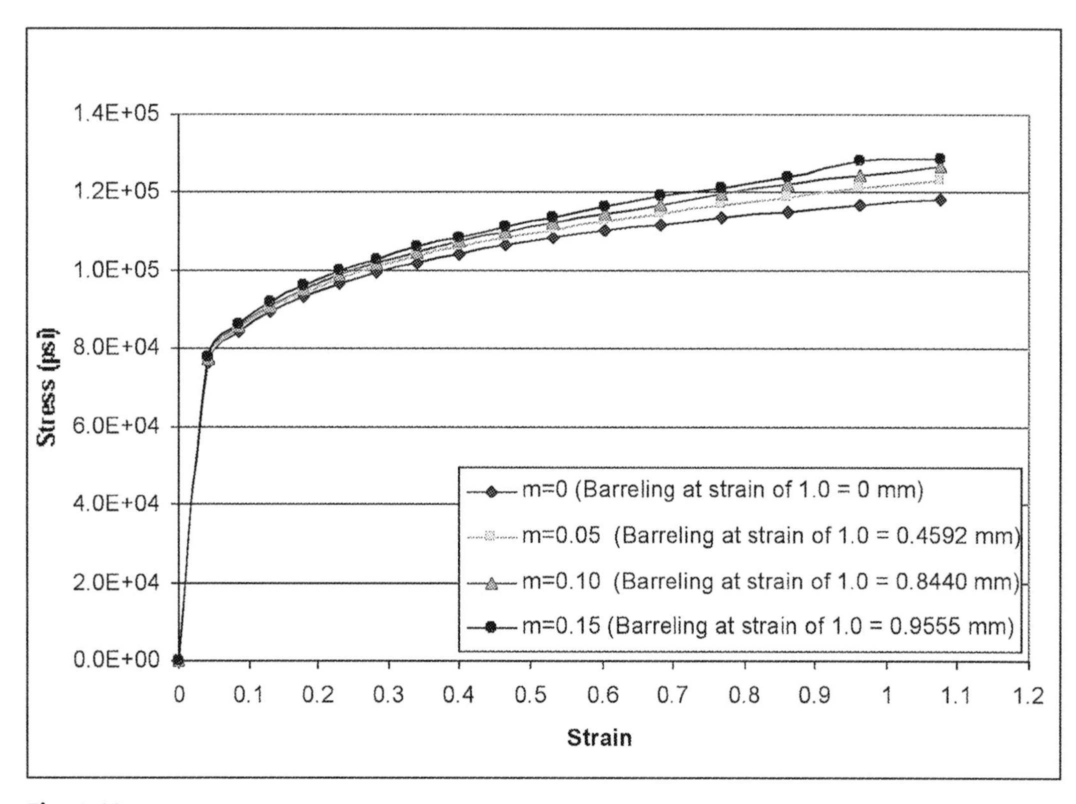

Fig. 4.18 Apparent flow stress for curves for tested specimens. [Dixit et al., 2002]

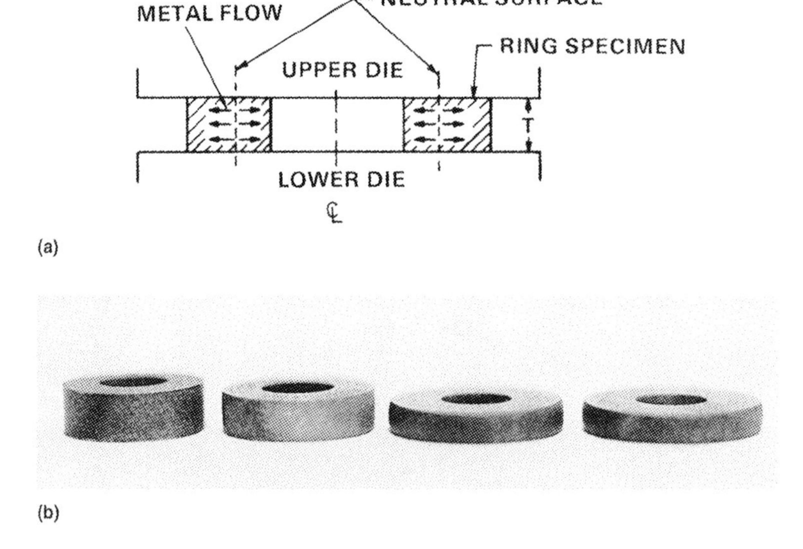

Fig. 4.19 The ring test. (a) Schematic of metal flow. (b) Example rings upset to various reductions in height

Voce: $\bar{\sigma} = a + [b - a]*[1 - \exp(-c\bar{\varepsilon})]$ (Eq 4.23)

This gives a good fit, but is not suitable for use in analysis because of its complexity.

Swift: $\bar{\sigma} = c(a + \bar{\varepsilon})^n$ (Eq 4.24)

This is a more realistic equation than Eq 4.21. However, algebraic manipulations resulting from such an expression may be difficult.

For strain-rate-sensitive materials, the most commonly used expression is:

$$\bar{\sigma} = C(\dot{\bar{\varepsilon}})^m \qquad \text{(Eq 4.25)}$$

The coefficients C and m of this curve would be obtained at various temperatures and strains, so that C and m would have different values at a given temperature for various strains. As examples, C and m values for some metals are given in Tables 4.7 through 4.11 [Altan et al., 1973]. For predicting forces and stresses in practical forming operations, very often it is sufficient to specify an average or maximum value of $\bar{\sigma}$ to be used in equations for predicting the maximum forming load. In such practical cases, use of a constant average value for $\bar{\sigma}$ is justified. If $\bar{\varepsilon}$ and $\dot{\bar{\varepsilon}}$ are not accurately known, then the

Table 4.4 Summary of K and n values describing the flow stress-strain relation, $\bar{\sigma} = K(\bar{\varepsilon})^n$, for various steels

Steel	Composition(a), %												Material history(b)	Temperature		Strain rate, 1/s	Strain range	K, 10^3 psi	n
	C	Mn	P	S	Si	N	Al	V	Ni	Cr	Mo	W		F	C				
Armco iron	0.02	0.03	0.021	0.010	Tr								A	68	20	(c)	0.1–0.7	88.2	0.25
1006	0.06	0.29	0.02	0.042	Tr	0.004							A	68	20	(c)	0.1–0.7	89.6	0.31
1008	0.08	0.36	0.023	0.031	0.06	0.007							A	68	20	(c)	0.1–0.7	95.3	0.24
	0.07	0.28			0.27								A	68	20	(c)	0.1–0.7	95.3	0.17
1010	0.13	0.31	0.010	0.022	0.23	0.004							A	68	20	(c)	0.1–0.7	103.8	0.22
1015	0.15	0.40	0.01	0.016	Tr								F,A	32	0	30	0.2–0.7	91.4	0.116
1015	0.15	0.40	0.01	0.016	Tr								F,A	390	200	30	0.2–0.6	73.7	0.140
1015(d)	0.15	0.40	0.045	0.045	0.25								A	68	20	1.6		113.8	0.10
1015(d)	0.15	0.40	0.045	0.045	0.25								A	572	300	1.6		115.2	0.11
1020	0.22	0.44	0.017	0.043	Tr	0.005							A	68	20	(c)	0.1–0.7	108.1	0.20
1035	0.36	0.69	0.025	0.032	0.27	0.004							A	68	20	(c)	0.1–0.7	130.8	0.17
													A	68	20	1.6		139.4	0.11
													A	572	300	1.6		122.3	0.16
1045(d)	0.45	0.65	0.045	0.045	0.25								A	68	20	1.6		147.9	0.11
													A	68	20	1.5		137.9	0.14
													A	572	300	1.6		126.6	0.15
1050(e)	0.51	0.55	0.016	0.041	0.28	0.0062	0.03						A	68	20	(c)	0.1–0.7	140.8	0.16
1060													A	68	20	1.6		163.5	0.09
													A	68	20	1.5		157.8	0.12
2317(e)	0.19	0.55	0.057	0.023	0.26	0.016							A	68	20	(c)	0.2–1.0	111.2	0.170
5115	0.14	0.53	0.028	0.027	0.37					0.71			A	68	20	(c)	0.1–0.7	115.2	0.18
													A	68	20	1.6		123.7	0.09
													A	572	300	1.6		102.4	0.15
5120(e)	0.18	1.13	0.019	0.023	0.27					0.86			A	68	20	(c)	0.1–0.7	126.6	0.18
													A	68	20	1.6		116.6	0.09
													A	572	300	1.6		98.1	0.16
5140	0.41	0.67	0.04	0.019	0.35					1.07			A	68	20	(c)	0.1–0.7	125.1	0.15
													A	68	20	1.6		133.7	0.09
													A	572	300	1.6		112.3	0.12
D2 tool steel(e)	1.60	0.45			0.24			0.46		11.70	0.75	0.59	A	68	20	(c)	0.2–1.0	191.0	0.157
L6 tool steel	0.56							0.14	1.60	1.21	0.47		A	68	20	(c)	0.2–1.0	170.2	0.128
W1-1.0C special	1.05	0.21			0.16								A	68	20	(c)	0.2–1.0	135.6	0.179
302 SS	0.08	1.06	0.037	0.005	0.49				9.16	18.37			HR,A	32	0	10	0.25–0.7	186.7	0.295
													HR,A	390	200	30	0.25–0.7	120.8	0.278
													HR,A	750	400	30	0.25–0.7	92.7	0.279
302 SS	0.053	1.08	0.027	0.015	0.27				10.2	17.8			A	68	20	(c)	0.1–0.7	210.5	0.6
304 SS(e)	0.030	1.05	0.023	0.014	0.47				10.6	18.7			A	68	20	(c)	0.1–0.7	210.5	0.6
316 SS	0.055	0.92	0.030	0.008	0.49				12.9	18.1	2.05		A	68	20	(c)	0.1–0.7	182.0	0.59
410 SS	0.093	0.31	0.026	0.012	0.33					13.8			A	68	20	(c)	0.1–0.7	119.4	0.2
													A	68	20	1.6		137.9	0.09
431 SS	0.23	0.38	0.020	0.006	0.42				1.72	16.32			A	68	20	(c)	0.1–0.7	189.1	0.11

(a) Tr = trace. (b) A = annealed, F = forged, HR = hot rolled. (c) Low-speed testing machine; no specific rate given. (d) Composition given is nominal (analysis not given in original reference). (e) Approximate composition.

value of C in Eq 4.25 may be used as an approximation for $\bar{\sigma}$. Such values for hot working temperatures are given for a few materials in Tables 4.12 and 4.13 [Douglas et al., 1975]. The data given in Table 4.10 were obtained from uniform isothermal compression tests, whereas that in Table 4.13 were obtained from nonisothermal ring tests. In these tests the ring dimensions are also important, because the average ring specimen temperature varies during the test and is influenced by the heat transfer and thickness of the ring.

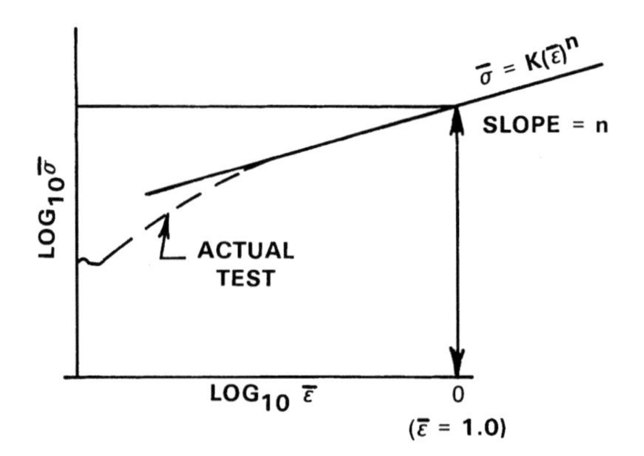

Fig. 4.20 Effective stress versus effective strain curve in log-log scale

Table 4.5 Summary of K and n values describing the flow stress-strain relation, $\bar{\sigma} = K(\bar{\varepsilon})^n$, for various aluminum alloys

	Composition, %											Temperature		Strain			
Alloy	Al	Cu	Si	Fe	Mn	Mg	Zn	Ti	Cr	Pb	Material history(a)	F	C	rate, 1/s	Strain range	K, 10^3 psi	n
1100	99.0	0.10	0.15	0.50	0.01	0.01					CD,A	32	0	10	0.25–0.7	25.2	0.304
1100	Rem	0.01	0.10	0.16	0.01	0.01	0.03				A	68	20	(b)	0.2–1.0	17.3	0.297
EC	99.5	0.01	0.092	0.23	0.026	0.033	0.01				A(c)	68	20	4	0.2–0.8	22.4	0.204
2017	Rem	4.04	0.70	0.45	0.55	0.76	0.22			0.06	A	68	20	(b)	0.2–1.0	45.2	0.180
2024(d)	Rem	4.48	0.60	0.46	0.87	1.12	0.20			0.056	A	68	20	(b)	0.2–1.0	56.1	0.154
5052	Rem	0.068	0.10	0.19	0.04	2.74	0.01	0.003			A(e)	68	20	4	0.2–0.8	29.4	0.134
5052(d)	Rem	0.09	0.13	0.16	0.23	2.50	0.05				A	68	20	(b)	0.2–1.0	55.6	0.189
5056	Rem	0.036	0.15	0.22	0.04	4.83	0.01		0.14		A(e)	68	20	4	0.2–0.7	57.0	0.130
5083	Rem	0.01	0.10	0.16	0.77	4.41	0.01	0.002	0.13		A	68	20	4	0.2–0.8	65.2	0.131
5454	Rem	0.065	0.12	0.18	0.81	2.45	<0.01	0.002			A(e)	68	20	4	0.2–0.8	49.9	0.137
6062	Rem	0.03	0.63	0.20	0.63	0.68	0.065	0.08				68	20	(b)	0.2–1.0	29.7	0.122

(a) CD = cold drawn, A = annealed. (b) Low-speed testing machine; no specific rate given. (c) Annealed for 4 h at 752 F (400 C). (d) Approximate composition. (e) Annealed for 4 h at 788 F (420 C)

Table 4.6 Summary of K and n values describing the flow stress-strain relation, $\bar{\sigma} = K(\bar{\varepsilon})^n$, for various copper alloys

	Composition(b), %									Temperature			Strain			
Alloy(a)	Cu	Si	Fe	Sb	Sn	Zn	S	Pb	Ni	F	C	Material history(c)	rate, 1/s	Strain range	K, 10^3 psi	n
CDA110	99.94		0.0025	0.0003			0.0012	0.0012	0.001	64	18	HR,A	2.5	0.25–0.7	65.5	0.328
CDA110										68	20	F	(d)	0.2–1.0	54.0	0.275
CDA230	84.3					15.7				68	20	A	(d)	0.2–1.0	76.7	0.373
CDA260	70.8					29.2				68	20	A	(d)	0.2–1.0	98.1	0.412
CDA260	70.05		Tr		Tr	Rem				390	200	HR,A		0.25–0.7	71.7	0.414
CDA272	63.3					36.7				68	20	A	(d)	0.2–1.0	103.9	0.394
CDA377	58.6		Tr			39.6		1.7		68	20	A	(d)	0.2–1.0	115.3	0.334
CDA521(e)	91.0				9.0					68	20	F	(d)	0.2–1.0	130.8	0.486
CDA647	97.0	0.5							2.0	68	20	F	(d)	0.2–1.0	67.2	0.282
CDA757	65.1					22.4		<0.05	12.4	68	20	A	(d)	0.2–1.0	101.8	0.401
CDA794	61.7		Tr			20.6		Tr	17.5	68	20	A	(d)	0.2–1.0	107.0	0.336

(a) CDA = Copper Development Association. (b) Tr = trace. (c) HR = hot rolled, A = annealed, F = forged. (d) Low-speed testing machine; no specific rate given. (e) Approximate composition.

Table 4.7 Summary of C (ksi) and m values describing the flow stress-strain rate relation, $\bar{\sigma} = C(\dot{\bar{\varepsilon}})^m$, for steels at various temperatures (C is in 10^3 psi)

Steel	Material history	Strain rate range, 1/s	Strain	C	m	C	m	C	m	C	m	C	m
			Test temperature, F (C):	1110 (600)		1470 (800)		1830 (1000)		2190 (1200)			
1015	Forged,	0.2–30	0.2	36.8	0.112								
0.15 C, trace Si, 0.40	annealed		0.25			19.9	0.105	17.0	0.045	7.2	0.137		
Mn, 0.01 P, 0.016 S			0.4	40.6	0.131								
			0.5			21.5	0.104	18.8	0.058	6.8	0.169		
			0.6	40.0	0.121								
			0.7	39.5	0.114	21.1	0.109	18.3	0.068	5.7	0.181		
			Test temperature, F (C):	1650 (900)		1830 (1000)		2010 (1100)		2190 (1200)			
1016	Hot rolled,	1.5–100	0.10	16.6	0.092	13.4	0.100	9.9	0.124	7.5	0.143		
0.15 C, 0.12 Si, 0.68	annealed		0.30	22.7	0.082	18.2	0.085	13.3	0.115	9.4	0.153		
Mn, 0.034 S, 0.025 P			0.50	23.7	0.087	18.2	0.105	12.7	0.146	8.5	0.191		
			0.70	23.1	0.099	16.1	0.147	11.9	0.166	7.5	0.218		
1016	Hot rolled,		0.05	11.8	0.133	10.7	0.124	9.0	0.117	6.4	0.150		
0.15 C, 0.12 Si, 0.68	annealed		0.1	16.5	0.099	13.7	0.099	9.7	0.130	7.1	0.157		
Mn, 0.034 S, 0.025 P			0.2	20.8	0.082	16.5	0.090	12.1	0.119	9.1	0.140		
			0.3	22.8	0.085	18.2	0.088	13.4	0.109	9.5	0.148		
			0.4	23.0	0.084	18.2	0.098	12.9	0.126	9.1	0.164		
			0.5	23.9	0.088	18.1	0.109	12.5	0.141	8.2	0.189		
			0.6	23.3	0.097	16.9	0.127	12.1	0.156	7.8	0.205		
			0.7	22.8	0.104	17.1	0.127	12.4	0.151	8.1	0.196		
			Test temperature, F (C):	1600 (870)		1800 (980)		2000 (1090)		2200 (1205)		2150 (1180)	
1018				25.2	0.07	15.8	0.152	11.0	0.192	9.2	0.20		
1025	Forged,	3.5–30	0.25			33.7	0.004	16.2	0.075	9.3	0.077		
0.25 C, 0.08 Si, 0.45	annealed		0.50			41.4	−0.032	17.2	0.080	9.6	0.094		
Mn, 0.012 P, 0.025 S			0.70			41.6	−0.032	17.5	0.082	8.8	0.105		
1043	Hot rolled, as received	0.1–100	0.3/0.5/0.7									10.8	0.21
			Test temperature, F (C):	1650 (900)		1830 (1000)		2010 (1100)		2190 (1200)			
1045(a)			0.05	25.4	0.080	15.1	0.089	11.2	0.100	8.0	0.175		
0.46 C, 0.29 Si, 0.73			0.10	28.9	0.082	18.8	0.103	13.5	0.125	9.4	0.168		
Mn, 0.018 P, 0.021 S,			0.20	33.3	0.086	22.8	0.108	15.4	0.128	10.5	0.167		
0.08 Cr, 0.01 Mo,			0.30	35.4	0.083	24.6	0.110	15.8	0.162	10.8	0.180		
0.04 Ni			0.40	35.4	0.105	24.7	0.134	15.5	0.173	10.8	0.188		
			Test temperature, F (C):	1110 (600)		1470 (800)		1830 (1000)		2190 (1200)			
1055	Forged,	3.5–30				29.4	0.087	14.9	0.126	7.4	0.145		
0.55 C, 0.24 Si, 0.73	annealed					32.5	0.076	13.3	0.191	7.4	0.178		
Mn, 0.014 P, 0.016 S						32.7	0.066	11.5	0.237	6.4	0.229		
			Test temperature, F (C):	1650 (900)		1830 (1000)		2010 (1100)		2190 (1200)			
1060(a)	Hot rolled,	1.5–100	0.10	18.5	0.127	13.3	0.143	10.1	0.147	7.4	0.172		
0.56 C, 0.26 Si, 0.28	annealed		0.30	23.3	0.114	16.9	0.123	12.6	0.135	8.9	0.158		
Mn, 0.014 S, 0.013 P,			0.50	23.3	0.118	16.4	0.139	12.0	0.158	8.6	0.180		
0.12 Cr, 0.09 Ni			0.70	21.3	0.132	14.9	0.161	10.4	0.193	7.8	0.207		
1060(a)			0.05	16.2	0.128	10.8	0.168	8.7	0.161	6.5	0.190		
0.56 C, 0.26 Si, 0.28			0.10	18.3	0.127	13.2	0.145	10.1	0.149	7.5	0.165		
Mn, 0.014 S, 0.013			0.20	21.8	0.119	16.1	0.125	12.1	0.126	8.5	0.157		
P, 0.12 Cr, 0.09 Ni			0.30	23.3	0.114	17.1	0.125	12.8	0.132	8.8	0.164		
			0.40	23.7	0.112	16.8	0.128	12.5	0.146	8.8	0.171		
			0.50	23.6	0.110	16.6	0.133	12.7	0.143	8.7	0.176		
			0.60	22.8	0.129	17.1	0.127	11.7	0.169	8.4	0.189		
			0.70	21.3	0.129	16.2	0.138	10.7	0.181	7.8	0.204		
1095(a)	Hot rolled,	1.5–100	0.10	18.3	0.146	13.9	0.143	9.8	0.159	7.1	0.184		
100 C, 0.19 Si, 0.17	annealed		0.30	21.9	0.133	16.6	0.132	11.7	0.147	8.0	0.183		
Mn, 0.027 S, 0.023			0.50	21.8	0.130	15.7	0.151	10.6	0.176	7.3	0.209		
P, 0.10 Cr, 0.09 Ni			0.70	21.0	0.128	13.6	0.179	9.7	0.191	6.5	0.232		
			Test temperature, F (C):	1705 (930)		1830 (1000)		1940 (1060)		2075 (1135)		2190 (1200)	
1115	Hot rolled,	4.4–23.1	0.105	16.3	0.088	13.0	0.108	10.9	0.112	9.1	0.123	7.6	0.116
0.17 C, 0.153 Si, 0.62	as		0.223	19.4	0.084	15.6	0.100	12.9	0.107	10.5	0.129	8.6	0.122
Mn, 0.054 S, 0.032 P	received		0.338	20.4	0.094	17.3	0.090	14.0	0.117	11.2	0.138	8.8	0.141
			0.512	20.9	0.099	18.0	0.093	14.4	0.127	11.0	0.159	8.3	0.173
			0.695	20.9	0.105	16.9	0.122	13.6	0.150	9.9	0.198	7.6	0.196

(continued)

(a) Approximate composition.

Table 4.7 (continued)

Steel	Material history	Strain rate range, 1/s	Strain	C	m	C	m	C	m	C	m	C	m
			Test temperature, F (C):	1650 (900)		1830 (1000)		2010 (1100)		2190 (1200)			
Alloy steel			0.05	16.6	0.102	12.2	0.125	9.4	0.150	7.4	0.161		
0.35 C, 0.27 Si, 1.49			0.10	19.9	0.091	14.8	0.111	11.5	0.121	8.1	0.149		
Mn, 0.041 S, 0.037			0.20	23.0	0.094	17.6	0.094	13.5	0.100	9.4	0.139		
P, 0.03 Cr, 0.11 Ni,			0.30	24.9	0.092	19.1	0.093	14.4	0.105	10.2	0.130		
0.28 Mo			0.40	26.0	0.088	19.6	0.095	14.5	0.112	10.4	0.139		
			0.50	25.9	0.091	19.6	0.100	14.4	0.112	10.1	0.147		
			0.60	25.9	0.094	19.5	0.105	14.2	0.122	9.7	0.159		
			0.70	25.5	0.099	19.2	0.107	13.9	0.126	9.2	0.165		
4337(a)	Hot rolled,	1.5–100	0.10	22.1	0.080	16.6	0.109	12.1	0.115	8.2	0.165		
0.35 C, 0.27 Si, 0.66	annealed		0.30	28.1	0.077	20.8	0.098	15.0	0.111	10.7	0.138		
Mn, 0.023 S, 0.029			0.50	29.2	0.075	21.8	0.096	15.7	0.112	11.3	0.133		
P, 0.59 Cr, 2.45 Ni,			0.70	28.1	0.080	21.3	0.102	15.5	0.122	11.3	0.135		
0.59 Mo													
926(a)	Hot rolled,	1.5–100	0.10	22.9	0.109	17.1	0.106	11.8	0.152	8.6	0.168		
0.61 C, 1.58 Si, 0.94	annealed		0.30	28.2	0.101	20.4	0.106	14.3	0.140	10.1	0.162		
Mn, 0.038 S, 0.035			0.50	27.8	0.104	20.0	0.120	13.8	0.154	9.1	0.193		
P, 0.12 Cr, 0.27 Ni,			0.70	25.8	0.112	18.2	0.146	11.8	0.179	7.5	0.235		
0.06 Mo													
50100(a)			0.05	16.1	0.155	12.4	0.155	8.2	0.175	6.3	0.199		
1.00 C, 0.19 Si, 0.17			0.10	18.6	0.145	14.1	0.142	9.5	0.164	6.8	0.191		
Mn, 0.027 S, 0.023			0.20	20.9	0.135	15.9	0.131	11.4	0.141	8.1	0.167		
P, 0.10 Cr, 0.09 Ni			0.30	21.8	0.135	16.6	0.134	11.7	0.142	8.0	0.174		
			0.40	22.0	0.134	16.8	0.134	11.2	0.155	8.4	0.164		
			0.50	21.5	0.131	15.6	0.150	11.1	0.158	7.4	0.199		
			0.60	21.3	0.132	14.6	0.163	10.0	0.184	7.0	0.212		
			0.70	20.9	0.131	13.5	0.176	9.7	0.183	6.7	0.220		
52100	Hot rolled,	1.5–100	0.10	20.9	0.123	14.3	0.146	9.5	0.169	6.7	0.203		
1.06 C, 0.22 Si, 0.46	annealed		0.30	25.5	0.107	17.7	0.127	12.0	0.143	8.3	0.171		
Mn, 0.019 S, 0.031			0.50	25.9	0.107	17.7	0.129	12.3	0.143	8.3	0.178		
P, 1.41 Cr, 0.17 Ni			0.70	23.3	0.131	16.8	0.134	12.0	0.148	7.7	0.192		
			Test temperature, F(C):	1650 (900)		1830 (1000)		2010 (1100)		2190 (1200)			
Mn-Si steel			0.05	19.2	0.117	14.8	0.119	9.7	0.172	7.5	0.181		
0.61 C, 1.58 Si, 0.94			0.10	22.6	0.112	17.1	0.108	11.8	0.151	8.7	0.166		
Mn, 0.038 S, 0.035			0.20	25.7	0.108	19.5	0.101	13.5	0.139	9.7	0.160		
P, 0.12 Cr, 0.27 Ni,			0.30	27.6	0.108	20.5	0.109	14.8	0.126	10.0	0.161		
0.06 Mo			0.40	27.6	0.114	20.2	0.114	14.4	0.141	9.5	0.179		
			0.50	27.2	0.113	19.8	0.125	14.1	0.144	9.1	0.188		
			0.60	26.0	0.121	18.8	0.137	12.8	0.162	8.2	0.209		
			0.70	24.7	0.130	17.8	0.152	11.9	0.178	7.5	0.228		
Cr-Si steel			0.05	19.9	0.118	23.9	0.104	15.1	0.167	10.0	0.206		
0.47 C, 3.74 Si, 0.58			0.10	19.9	0.136	25.6	0.120	16.8	0.162	11.1	0.189		
Mn, 8.20 Cr 20 Ni			0.20	19.9	0.143	27.6	0.121	18.5	0.153	11.9	0.184		
			0.30	19.9	0.144	28.4	0.119	19.1	0.148	12.1	0.182		
			0.40	19.3	0.150	28.2	0.125	18.9	0.150	12.1	0.178		
			0.50	18.5	0.155	26.6	0.132	18.5	0.155	11.8	0.182		
			0.60	17.5	0.160	25.2	0.142	17.5	0.160	11.5	0.182		
			0.70	16.1	0.163	23.3	0.158	16.1	0.162	10.7	0.199		
D3(a)	Hot rolled,	1.5–100	0.10	39.2	0.087	29.0	0.108	21.0	0.123	14.6	0.121		
2.23 C, 0.43 Si, 0.37	annealed		0.30	43.7	0.087	30.4	0.114	21.0	0.139	13.9	0.130		
Mn, 13.10 Cr, 0.33			0.50	39.7	0.101	27.1	0.125	18.4	0.155	12.2	0.124		
Ni			0.70	33.3	0.131	22.5	0.145	15.3	0.168	10.7	0.108		
			Test temperature, F (C):	1290 (700)		1510 (820)		1650 (900)		1830 (1000)			
H-13		290–906	0.1	19.1	0.232	10.2	0.305	6.0	0.373	4.8	0.374		
0.39 C, 1.02 Si, 0.60			0.2	30.1	0.179	13.7	0.275	8.2	0.341	9.0	0.295		
Mn, 0.016 P, 0.020			0.3	31.0	0.179	15.1	0.265	10.8	0.305	11.6	0.267		
S, 5.29 Cr, 0.04 Ni,			0.4	25.9	0.204	12.3	0.295	12.5	0.287	11.8	0.269		
1.35 Mo, 0.027 N,													
0.83 V													
			Test temperature, F (C):	1650 (900)		1830 (1000)		2010 (1100)		2190 (1200)			
H-26(a)	Hot rolled,	1.5–100	0.10	46.7	0.058	37.4	0.072	26.2	0.106	18.7	0.125		
0.80 C, 0.28 Si, 0.32	annealed		0.30	49.6	0.075	38.1	0.087	26.0	0.121	18.3	0.140		
Mn, 4.30 Cr, 0.18			0.50	44.6	0.096	33.7	0.102	23.6	0.131	16.2	0.151		
Ni, 0.55 Mo, 18.40			0.70	39.1	0.115	27.9	0.124	20.1	0.149	13.8	0.162		
W, 1.54 V													

(continued)

(a) Approximate composition.

Table 4.7 (continued)

Steel	Material history	Strain rate range, 1/s	Strain	C	m	C	m	C	m	C	m	C	m
		Test temperature, F (C):		1110 (600)		1470 (800)		1830 (1000)		2190 (1200)			
301 SS(a)	Hot rolled,	0.8–100	0.25			40.5	0.051	16.3	0.117	7.6	0.161		
0.08 C, 0.93 Si, 1.10	annealed		0.50			39.3	0.062	17.8	0.108	7.6	0.177		
Mn, 0.009 P, 0.014			0.70			37.8	0.069	17.4	0.102	6.6	0.192		
S, 16.99 Cr, 6.96 Ni,													
0.31 Mo, 0.93 Al.													
0.02 N, 0.063 Se													
302 SS	Hot rolled,	310–460	0.25	26.5	0.147	25.1	0.129	11.0	0.206	4.6	0.281		
0.07 C, 0.71 Si, 1.07	annealed		0.40	31.3	0.153	30.0	0.121	13.5	0.188	4.7	0.284		
Mn, 0.03 P, 0.005			0.60	17.5	0.270	45.4	0.063	16.8	0.161	4.1	0.310		
S, 18.34 Cr, 9.56 Ni													
302 SS	Hot rolled,	0.2–30	0.25	52.2	0.031	36.6	0.042	23.1	0.040	12.8	0.082		
0.08 C, 0.49 Si, 1.06	annealed		0.40	58.9	0.022	40.4	0.032	24.7	0.050	13.6	0.083		
Mn, 0.037 P, 0.005			0.60	63.2	0.020	41.9	0.030	24.9	0.053	13.5	0.091		
S, 18.37 Cr, 9.16			0.70	64.0	0.023	42.0	0.031	24.7	0.052	13.4	0.096		
Ni													
		Test temperature, F (C):		1650 (900)		1830 (1000)		2010 (1100)		2190 (1200)			
302 SS		1.5–100	0.05	24.6	0.023	16.8	0.079	13.7	0.093	9.7	0.139		
0.07 C, 0.43 Si, 0.48			0.10	28.4	0.026	21.2	0.068	15.6	0.091	11.1	0.127		
Mn, 18.60 Cr, 7.70			0.20	33.6	0.031	25.2	0.067	18.1	0.089	12.5	0.120		
Ni			0.30	35.3	0.042	26.3	0.074	19.5	0.089	13.5	0.115		
			0.40	35.6	0.055	26.9	0.084	19.9	0.094	14.2	0.110		
			0.50	35.6	0.060	27.0	0.093	19.6	0.098	14.2	0.115		
			0.60	34.1	0.068	26.4	0.092	19.3	0.102	13.8	0.118		
			0.70	33.6	0.072	25.7	0.102	18.9	0.108	13.9	0.120		
		Test temperature, F (C):		1110 (600)		1470 (800)		1830 (1000)		2190 (1200)		1650 (900)	
309 SS	Hot drawn,	200–525	0.25			39.4	0.079			8.7	0.184		
0.13 C, 0.42 Si, 1.30	annealed		0.40			45.1	0.074			9.6	0.178		
Mn, 0.023 P, 0.008			0.60			48.1	0.076			9.5	0.185		
S, 22.30 Cr, 12.99 Ni													
310 SS	Hot drawn,	310–460	0.25	50.3	0.080	32.3	0.127	27.5	0.101	12.0	0.154		
0.12 C, 1.26 Si, 1.56	annealed		0.40	56.5	0.080	32.2	0.142	22.8	0.143	10.8	0.175		
Mn, 0.01 P, 0.009			0.60	61.8	0.067	21.9	0.212	9.7	0.284	4.5	0.326		
S, 25.49 Cr, 21.28 Ni													
316 SS	Hot drawn,	310–460	0.25	13.5	0.263	22.2	0.149	6.4	0.317	8.0	0.204		
0.06 C, 0.52 Si, 1.40	annealed		0.40	28.8	0.162	26.8	0.138	3.7	0.435	7.4	0.227		
Mn, 0.035 P, 0.005			0.60	39.3	0.128	30.1	0.133	6.1	0.365	6.5	0.254		
S, 17.25 Cr, 12.23 Ni,													
2.17 Mo													
403 SS	Hot rolled,	0.8–100	0.25			26.3	0.079	15.4	0.125	7.3	0.157		
0.16 C, 0.37 Si, 0.44	annealed		0.50			26.9	0.076	16.0	0.142	7.8	0.152		
Mn, 0.024 P, 0.007			0.70			24.6	0.090	15.3	0.158	7.5	0.155		
S, 12.62 Cr													
SS	Hot rolled,	0.8–100	0.25			28.7	0.082	17.2	0.082	11.9	0.079		
0.12 C, 0.12 Si, 0.29	annealed		0.50			29.1	0.093	20.7	0.073	11.6	0.117		
Mn, 0.014 P, 0.016			0.70			28.7	0.096	22.5	0.067	11.2	0.131		
S, 12.11 Cr, 0.50 Ni,													
0.45 Mo													
SS	Hot rolled,	3.5–30	0.25					19.5	0.099	8.9	0.128	28.3	0.114
0.08 C, 0.45 Si, 0.43	annealed		0.50					22.3	0.097	9.5	0.145	34.9	0.105
Mn, 0.031 P, 0.005			0.70					23.2	0.098	9.2	0.158	37.1	0.107
S, 17.38 Cr, 0.31 Ni													
		Test temperature, F (C):		1600 (870)		1700 (925)		1800 (980)		2000 (1095)		2100 (1150)	
Maraging 300				43.4	0.077	36.4	0.095	30.6	0.113	21.5	0.145	18.0	0.165
		Test temperature, F (C):		2200 (1205)									
Maraging 300				12.8	0.185								

(a) Approximate composition.

Table 4.8 Summary of C (ksi) and m values describing the flow stress-strain rate relation, $\bar{\sigma} = C(\dot{\bar{\varepsilon}})^m$, for aluminum alloys at various temperatures

Alloy	Material history	Strain rate range, 1/s	Strain	C	m	C	m	C	m	C	m	C	m
		Test temperature, F (C):		390 (200)		570 (300)		750 (400)		930 (500)		1110 (600)	
Super-pure 99.98 Al, 0.0017 Cu, 0.0026 Si, 0.0033 Fe, 0.006 Mn	Cold rolled, annealed 1/2 h at 1110 F	0.4–311	0.288	5.7	0.110	4.3	0.120	2.8	0.140	1.6	0.155	0.6	0.230
			2.88	8.7	0.050	4.9	0.095	2.8	0.125	1.6	0.175	0.6	0.215
		Test temperature, F (C):		465 (240)		645 (360)		825 (480)					
EC 0.01 Cu, 0.026 Mn, 0.033 Mg, 0.092 Si, 0.23 Fe, 0.01 Zn, 99.5 Al	Annealed 3h at 750 F	0.25–63	0.20	10.9	0.066	5.9	0.141	3.4	0.168				
			0.40	12.3	0.069	6.3	0.146	3.3	0.169				
			0.60	13.1	0.067	6.4	0.147	3.2	0.173				
			0.80	13.8	0.064	6.7	0.135	3.4	0.161				
		Test temperature, F (C):		390 (200)		750 (400)		930 (500)					
1100 99.0 Al (min), 0.10 Cu, 0.15 Si, 0.50 Fe, 0.01 Mn, 0.01 Mg	Cold drawn, annealed	0.25–40	0.25	9.9	0.066	4.2	0.115	2.1	0.211				
			0.50	11.6	0.071	4.4	0.132	2.1	0.227				
			0.70	12.2	0.075	4.5	0.141	2.1	0.224				
		Test temperature, F (C):		300 (150)		480 (250)		660 (350)		840 (450)		1020 (550)	
1100(a) 0.10 Cu, 0.20 Si, 0.02 Mn, 0.46 Fe, 0.01 Zn, Rem Al	Extruded, annealed 1 h at 750 F	4–40	0.105	11.4	0.022	9.1	0.026	6.3	0.055	3.9	0.100	2.2	0.130
			0.223	13.5	0.022	10.5	0.031	6.9	0.061	4.3	0.098	2.4	0.130
			0.338	15.0	0.021	11.4	0.035	7.2	0.073	4.5	0.100	2.5	0.141
			0.512	16.1	0.024	11.9	0.041	7.3	0.084	4.4	0.116	2.4	0.156
			0.695	17.0	0.026	12.3	0.041	7.4	0.088	4.3	0.130	2.4	0.155
		Test temperature, F (C):		390 (200)		750 (400)		930 (500)					
2017 94.95 Al, 3.50 Cu, 0.10 Si, 0.50 Fe, 0.50 Mn, 0.45 Mg	Cold drawn, annealed	0.2–30	0.250	34.5	0.014	14.8	0.110	5.8	0.126				
			0.500	32.2	−0.025	13.2	0.121	5.2	0.121				
			0.700	29.5	−0.038	12.5	0.128	5.1	0.119				
		Test temperature, F (C):		570 (300)		660 (350)		750 (400)		840 (450)		930 (500)	
2017(a) 0.89 Mg, 4.17 Cu, 0.61 Si, 0.41 Fe, 0.80 Mn, 0.052 Zn, 0.01 Pb, 92.9 Al	Solution treated 1 h at 950 F, water quenched annealed 4 h at 750 F	0.4–311	0.115	10.8	0.695	9.1	0.100	7.5	0.110	6.2	0.145	5.1	0.155
			2.660	10.0	0.100	9.2	0.100	7.7	0.080	6.8	0.090	4.6	0.155
		Test temperature, F (C):		465 (240)		645 (360)		825 (480)					
5052 0.068 Cu, 0.04 Mn, 2.74 Mg, 0.10 Si, 0.19 Fe, 0.01 Zn, 0.003 Ti, Rem Al	Annealed 3 h at 790 F	0.25–63	0.20	14.3	0.038	8.9	0.067	5.6	0.125				
			0.40	15.9	0.035	9.3	0.071	5.3	0.130				
			0.60	16.8	0.035	9.0	0.068	5.1	0.134				
			0.80	17.5	0.038	9.4	0.068	5.6	0.125				
5056 0.036 Cu, 0.04 Mn, 4.83 Mg, 0.15 Si, 0.22 Fe, 0.01 Zn, 0.14 Cr, Rem Al	Annealed 3 h at 790 F	0.25–63	0.20	42.6	−0.032	20.9	0.138	11.7	0.200				
			0.40	44.0	−0.032	20.8	0.138	10.5	0.205				
			0.60	44.9	−0.031	19.9	0.143	10.3	0.202				
			0.70	45.6	−0.034	20.3	0.144	10.3	0.203				
5083 0.01 Cu, 0.77 Mn, 4.41 Mg, 0.10 Si, 0.16 Fe, 0.01 Zn, 0.13 Cr, 0.002 Ti, Rem Al	Annealed 3 h at 790 F	0.25–63	0.20	43.6	−0.006	20.5	0.095	9.3	0.182				
			0.40	43.6	−0.001	19.7	0.108	8.3	0.208				
			0.60	41.9	0.003	18.8	0.111	8.5	0.201				
			0.80	40.2	0.002	19.1	0.105	9.7	0.161				
5454 0.065 Cu, 0.81 Mn, 2.45 Mg, 0.12 Si, 0.18 Fe, <0.01 Zn, 0.002 Ti, Rem Al	Annealed 3 h at 790 F	0.25–63	0.20	33.6	−0.005	16.8	0.093	10.8	0.182				
			0.40	36.0	−0.009	16.3	0.104	10.7	0.188				
			0.60	36.9	−0.009	16.0	0.102	10.0	0.191				
			0.80	37.0	−0.009	16.2	0.097	10.2	0.183				
		Test temperature, F (C):		750 (400)		840 (450)		930 (500)		1020 (550)			
7075(a) 89.6 Al, 1.31 Cu, 2.21 Mg, 0.21 Si, 0.30 Fe, 0.34 Mn, 5.75 Zn, 0.01 Pb	Solution treated 1 h at 870 F, water quenched, aged at 285 F for 16 h	0.4–311	0.115	10.0	0.090	6.0	0.135	3.9	0.150	2.9	0.170		
			2.66	9.7	0.115	6.2	0.120	4.8	0.115	2.7	0.115		

(a) Approximate composition.

Table 4.9 Summary of C (ksi) and m values describing the flow stress-strain rate relation, $\bar{\sigma} = C(\dot{\bar{\varepsilon}})^m$, for copper alloys at various temperatures

Alloy	Material history	Strain rate range, 1/s	Strain	C	m	C	m	C	m	C	m	C	m
		Test temperature, F (C):		570 (300)		840 (450)		1110 (600)		1380 (750)		1650 (900)	
Copper	Cold drawn,	4–40	0.105	20.2	0.016	17.0	0.010	12.7	0.050	7.6	0.096	4.7	0.134
0.018 P, 0.0010 Ni,	annealed 2 h		0.223	26.5	0.018	22.5	0.004	16.8	0.043	9.7	0.097	6.3	0.110
0.0003 Sn, 0.0002 Sb,	at 1110 F		0.338	30.2	0.017	25.1	0.008	18.9	0.041	10.0	0.128	6.1	0.154
0.0005 Pb 0.0010 Fe,			0.512	32.2	0.025	26.6	0.014	19.4	0.056	8.5	0.186	5.5	0.195
0.0020 Mn <0.0005			0.695	34.4	0.024	26.8	0.031	19.0	0.078	8.2	0.182	5.2	0.190
Mg, <0.0005 As,													
<0.0001 Bi, 0.0014													
S, less than 0.003 O_2,													
Se + Te not detected													
		Test temperature, F (C):		800 (427)									
OFHC Copper				26.7	0.0413								
		Test temperature, F (C):		750 (400)		930 (500)		1110 (600)					
CDA 110	Hot rolled,	0.25–40	0.25	23.0	0.046	12.9	0.136	6.6	0.160				
99.94 Cu, 0.0003 Sb,	annealed		0.50	27.4	0.049	13.7	0.150	6.9	0.168				
0.0012 Pb 0.0012 S,			0.70	28.8	0.057	13.3	0.165	6.8	0.176				
0.0025 Fe, 0.001 Ni													
		Test temperature, F (C):		390 (200)		750 (400)		1110 (600)		1470 (800)			
CDA 220	Extruded, cold	0.1–10	0.25	41.0	0.017	34.1	0.018	22.6	0.061	11.2	0.134		
90.06 Cu, 0.033 Fe,	drawn 30%;		0.50	46.7	0.029	39.9	0.032	24.4	0.084	11.0	0.156		
0.004 Pb, 0.003 Sn,	annealed 650		0.70	48.1	0.034	40.7	0.024	24.6	0.086	11.4	0.140		
Rem Zn	C, 90 min												
CDA 260	Hot rolled,	3.5–30	0.25			34.9	0.036	16.0	0.194	7.1	0.144		
70.05 Cu, trace Fe +	annealed		0.50			42.3	0.031	14.8	0.237	7.0	0.148		
Sn, Rem Zn			0.70			42.4	0.045	14.3	0.228	6.3	0.151		
CDA 280	Hot rolled,	3.5–30	0.25	49.0	0.028	26.9	0.083	7.6	0.189	3.1	0.228		
60.44 Cu, 0.01 Pb,	annealed		0.50	58.6	0.027	28.6	0.075	5.4	0.281	2.8	0.239		
0.02 Fe, trace Sn,			0.70	60.3	0.027	26.7	0.081	4.7	0.291	2.7	0.220		
Rem Zn													
CDA 365	Hot rolled,	3.5–30	0.25	45.8	0.038	28.6	0.065	9.8	0.106	2.4	0.166		
59.78 Cu, 0.90 Pb,	annealed		0.50	57.2	0.032	28.9	0.085	8.5	0.137	2.1	0.197		
0.02 Fe, trace Sn,			0.70	59.1	0.035	26.6	0.078	8.4	0.113	1.8	0.222		
Rem Zn													

Table 4.10 Summary of C (ksi) and m values describing the flow stress-strain rate relation, $\bar{\sigma} = C(\dot{\bar{\varepsilon}})^m$, for titanium alloys at various temperatures

Alloy	Material history	Strain rate range, 1/s	Strain	C	m	C	m	C	m	C	m	C	m	C	m	C	m
		Test temperature, F (C):		68 (20)		392 (200)		752 (400)		1112 (600)		1472 (800)		1652 (900)		1832 (1000)	
Type 1	Annealed	0.25–16.0	0.2	92.8	0.029	60.9	0.046	39.8	0.074	25.3	0.097	12.8	0.167	5.4	0.230	3.0	0.387
0.04 Fe, 0.02	15 min at		0.4	113.7	0.029	73.3	0.056	48.8	0.061	29.6	0.115	14.6	0.181	5.5	0.248	3.6	0.289
C, 0.005 H_2,	1200 F in		0.6	129.6	0.028	82.2	0.056	53.9	0.049	32.1	0.105	14.9	0.195	5.5	0.248	3.5	0.289
0.01 N_2, 0.04	high		0.8	142.5	0.027	87.7	0.058	56.3	0.042	32.7	0.099	15.4	0.180	5.9	0.186	3.2	0.264
O_2, Rem Ti	vacuum		1.0	150.6	0.027	90.7	0.054	56.6	0.044	32.5	0.099	15.9	0.173	5.9	0.167	3.0	0.264
Type 2	Annealed	0.25–16.0	0.2	143.3	0.021	92.7	0.043	54.5	0.051	33.6	0.092	17.5	0.167	6.9	0.135	4.2	0.220
0.15 Fe, 0.02	15 min at		0.4	173.2	0.021	112.1	0.042	63.1	0.047	36.3	0.101	18.4	0.190	7.2	0.151	4.9	0.167
C, 0.005 H_2,	1200 F in		0.6	193.8	0.024	125.3	0.045	65.6	0.047	36.9	0.104	18.4	0.190	7.8	0.138	4.5	0.167
0.02 N_2, 0.12	high		0.8	208.0	0.023	131.9	0.051	66.0	0.045	37.0	0.089	18.4	0.190	7.6	0.106	3.9	0.195
O_2, Rem Ti	vacuum		1.0	216.8	0.023	134.8	0.056	65.3	0.045	36.9	0.092	18.6	0.190	6.8	0.097	3.7	0.167
		Test temperature, F (C):		1110 (600)		1290 (700)		1470 (800)		1650 (900)							
Unalloyed	Hot rolled,	0.1–10	0.25	23.4	0.062	14.3	0.115	8.2	0.236	1.8	0.324						
0.03 Fe,	annealed		0.50	27.9	0.066	17.8	0.111	10.0	0.242	2.1	0.326						
0.0084 N,	800 C,		0.70	30.1	0.065	20.0	0.098	12.2	0.185	2.5	0.316						
0.0025 H,	90 min																
Rem Ti																	
		Test temperature, F (C):		68 (20)		392 (200)		752 (400)		1112 (600)		1472 (800)		1652 (900)		1832 (1000)	
Ti-5Al-2.5 Sn	Annealed	0.25–16.0	0.1	173.6	0.046	125.6	0.028	97.6	0.028								
5.1 Al, 2.5 Sn,	30 min at		0.2	197.9	0.048	138.8	0.022	107.4	0.026	86.1	0.025	58.5	0.034	44.2	0.069	5.4	0.308
0.06 Fe, 0.03	1470 F in		0.3	215.6	0.046	147.4	0.021	112.5	0.027	92.8	0.020						
C, 0.01 H_2,	high		0.4	230.6	0.039	151.4	0.022	116.0	0.022	95.6	0.019	58.7	0.040	44.8	0.082	5.1	0.294
0.03 N_2, 0.1	vacuum		0.5							96.7	0.021						
O_2, Rem Ti			0.6							96.6	0.024	55.6	0.042	43.0	0.078	5.2	0.264
			0.8									50.2	0.033	39.1	0.073	5.2	0.264
			0.9									46.8	0.025				
			1.0											35.2	0.056	5.3	0.280
Ti-6Al-4V	Annealed	0.25–16.0	0.1	203.3	0.017	143.8	0.026	119.4	0.025								
6.4 Al, 4.0 V,	120 min		0.2	209.7	0.015	151.0	0.021	127.6	0.022	94.6	0.064	51.3	0.146	23.3	0.143	9.5	0.131
0.14 Fe, 0.05	at 1200 F		0.3	206.0	0.015	152.0	0.017	126.2	0.017	91.2	0.073						
C, 0.01 H_2,	in high		0.4					118.7	0.014	84.6	0.079	39.8	0.175	21.4	0.147	9.4	0.118
0.015 N_2, 0.1	vacuum		0.5							77.9	0.080						
O_2, Rem Ti			0.6									30.4	0.205	20.0	0.161	9.6	0.118
			0.8									26.6	0.199	19.5	0.172	9.3	0.154
			0.9									24.9	0.201				
			1.0											20.3	0.146	8.9	0.192
		Test temperature, F (C):		1550 (843)		1750 (954)		1800 (982)									
Ti-6Al-4V				38.0	0.064	12.3	0.24	9.4	0.29								
		Test temperature, F (C):		68 (20)		392 (200)		752 (400)		1112 (600)		1472 (800)		1652 (900)		1832 (1000)	
Ti-13V-11Cr-3Al	Annealed	0.25–16.0	0.1	173.1	0.041												
3.6 Al, 14.1 V,	30 min at		0.2	188.2	0.037	150.5	0.030	136.5	0.035	118.4	0.040	65.4	0.097	44.6	0.147	32.4	0.153
10.6 Cr, 0.27	1290 F in		0.3	202.3	0.034												
Fe, 0.02 C,	high		0.4	215.2	0.029	174.2	0.024	153.9	0.030	107.5	0.039	59.5	0.096	42.1	0.139	30.9	0.142
0.014 H_2, 0.03	vacuum		0.5	226.3	0.026	181.1	0.023										
N_2, 0.11 O_2,			0.6			183.5	0.026	147.9	0.046	92.8	0.045	56.7	0.088	40.9	0.127	29.2	0.155
Rem Ti			0.7			181.4	0.029										
			0.8					136.3	0.045	84.7	0.036	53.9	0.081	39.3	0.125	27.8	0.167
			0.9									52.9	0.080				
			1.0											38.8	0.127	28.0	0.159

Table 4.11 Summary of C (ksi) and m values describing the flow stress-strain rate relation, $\bar{\sigma} = C(\dot{\bar{\varepsilon}})^m$, for various materials

Alloy	Material history	Strain rate range, 1/s	Strain	C	m	C	m	C	m	C	m	C	m	C	m	C	m
		Test temperature, F (C):		**72 (22)**		**230 (110)**		**335 (170)**		**415 (215)**		**500 (260)**		**570 (300)**			
Lead			0.115	2.0	0.040	1.56	0.065	1.21	0.085	0.70	0.130	0.47	0.160	0.40	0.180		
99.98 Pb,			2.66	4.0	0.055	1.47	0.100	1.04	0.125	0.55	0.135	0.36	0.180	0.28	0.225		
0.003 Cu,																	
0.003 Fe,																	
0.002 Zn,																	
0.002 Ag																	
		Test temperature, F (C):		**390 (200)**		**570 (300)**		**750 (400)**		**930 (400)**							
Magnesium	Extruded,			(13)		(14)		(13)		(14)							
0.010 Al,	cold	0.1–10	0.25	19.1	0.069	9.8	0.215	4.1	0.263	1.7	0.337						
0.003 Zn,	drawn		0.50	17.2	0.093	8.4	0.211	4.0	0.234	1.7	0.302						
0.008 Mn,	15%,		0.70	15.5	0.094	8.3	0.152	4.3	0.215	2.1	0.210						
0.004 Si,	annealed																
0.003 Cu,	550 C 90																
0.0008 Ni,	min																
Rem Mg																	
		Test temperature, F (C):		**1975 (1080)**		**2030 (1166)**											
U-700				26.6	0.21	22.1	0.21										
		Test temperature, F (C):		**68 (20)**		**392 (200)**		**752 (400)**		**1112 (600)**		**1472 (800)**		**1652 (900)**		**1832 (1000)**	
Zirconium	Annealed	0.25–16.0	0.2	117.4	0.031	74.0	0.052	40.2	0.050	23.8	0.069	16.8	0.069	6.8	0.227	4.6	0.301
99.8 Zr, 0.009	15 min at		0.3	143.7	0.022	92.2	0.058										
Hf, 0.008 Al,	1380 F in		0.4	159.5	0.017	105.1	0.046	54.4	0.085	29.4	0.09	18.2	0.116	7.1	0.252	4.0	0.387
0.038 Fe,	high		0.5	169.3	0.017	112.8	0.041	58.2	0.093								
0.0006 H_2,	vacuum		0.6			118.5	0.042	60.2	0.095	31.3	0.089	18.8	0.118	7.2	0.264	4.0	0.387
0.0025 N_2,			0.7					61.9	0.095								
0.0825 O_2, 0.0			0.8							32.0	0.081	19.4	0.101	6.9	0.252	4.1	0.403
Ni			1.0							32.1	0.085	19.7	0.108	6.9	0.252	4.1	0.403
Zircaloy 2	Annealed	0.25–16.0	0.1	96.8	0.031	65.9	0.046										
98.35 Zr,	15 min at		0.2	136.9	0.025	105.8	0.035	58.3	0.065	30.4	0.049	16.6	0.147	7.5	0.325	3.9	0.362
0.015 Hf, 1.4	1380 F in		0.3	178.5	0.034	131.4	0.035	67.9	0.056								
Zn, 0.01 Al,	high		0.4	202.7	0.027	145.4	0.036	73.5	0.056	37.8	0.053	18.7	0.172	7.8	0.342	4.0	0.387
0.06 Fe, 0.045	vacuum		0.5			154.2	0.034	77.3	0.057								
Ni, 0.0006 H_2,			0.6					79.9	0.055	39.2	0.059	18.8	0.178	7.2	0.387	4.0	0.387
0.0023 N_2,			0.8							40.4	0.057	18.8	0.178	7.9	0.342	4.8	0.333
0.0765 O_2			1.0							40.7	0.053	18.8	0.178	8.5	0.310	4.8	0.333
		Test temperature, F (C):		**68 (20)**		**212 (100)**		**392 (200)**		**572 (300)**		**932 (500)**		**1292 (900)**		**1652 (900)**	
Uranium	Annealed 2	0.25–16.0	0.2	151.0	0.043	113.0	0.042	77.4	0.034	45.9	0.044	31.9	0.051	16.0	0.081	4.5	0.069
99.8 U,	hr at		0.4	173.9	0.033	132.7	0.049	91.0	0.031	53.3	0.047	33.1	0.059	16.1	0.089	4.5	0.069
0.0012 Mn,	1110 F in		0.6	184.9	0.023	143.1	0.047	98.1	0.032	56.0	0.056	33.4	0.054	16.1	0.089	4.5	0.069
0.0012 Ni,	high		0.8	189.8	0.018	149.5	0.048	102.0	0.036	58.3	0.057	33.3	0.049	16.2	0.097	4.5	0.069
0.00074 Cu,	vacuum		1.0							59.0	0.056	32.5	0.055	16.4	0.097	4.5	0.069
0.00072 Cr,																	
0.0001 Co,																	
0.0047 H_2,																	
0.0041 N_2,																	
0.1760 O_2																	
(free of																	
cadmium and																	
boron)																	

Table 4.12 Average flow stress values determined in the uniform compression test that might be used in practical load-predicting applications

Material	Flow stress, 10^3 psi	Temperature, F	Strain range (ln), h_0/h_1	Strain rate range, 1/s
403 stainless steel	33.0	1800	0.3–0.7	10.0–14.0
	25.0	1950	0.3–0.7	10.0–14.0
	21.0	2050	0.3–0.7	10.0–14.0
Waspaloy	62.0	1950	0.2–0.4	13.0–15.0
	56.0	1950	0.4–0.6	10.0–13.0
	52.0	2050	0.1–0.3	12.0–15.0
	48.0	2050	0.3–0.6	10.0–13.0
	46.0	2100	0.1–0.3	13.0–15.0
	42.0	2100	0.3–0.6	10.0–13.0
Ti-6Al-2Sn-4Zr-2Mo	56.0	1600	0.1–0.4	13.0–15.0
	52.0	1600	0.4–0.6	10.0–13.0
	52.0	1675	0.1–0.4	13.0–15.0
	46.0	1675	0.4–0.6	10.0–13.0
	38.0	1750	0.1–0.4	13.0–15.0
	34.0	1750	0.4–0.6	10.0–13.0
Inconel 718	54.0	2000	0.1–0.4	13.0–15.0
	48.0	2000	0.4–0.6	10.0–13.0
	46.0	2100	0.1–0.4	13.0–15.0
	40.0	2100	0.4–0.6	10.0–13.0
Ti-8Mo-8V-2Fe-3Al	40.0	1650	0.1–0.6	10.0–15.0
	28.0	1850	0.1–0.6	10.0–15.0
	24.0	2000	0.1–0.6	10.0–15.0
AISI 4340	25.0	1900	0.2–0.7	10.0–14.0
	21.0	2000	0.3–0.8	12.0–17.0

Table 4.13 Average flow stress values obtained from ring compression tests suggested for use in practical applications

Material	Flow stress(a), 10^3 psi	Temperature, F	Strain rate range, 1/s	Frictional shear factor, m	Contact time, s	Ring dimensions(b)
6061 Al	9	800	18–22	0.4	0.038	A
	9	800	15–17	0.31	0.047	B
	7	800	10–13	0.53	0.079	C
Ti-7Al-4Mo	48	1750	13	0.42	0.033	D
	30	1750	18–23	0.42	0.044	E
	30	1750	15–18	0.7	0.056	F
403 SS	37	1800	25–28	0.23	0.029	D
	33	1800	25–27	0.24	0.037	E
	33	1800	16–18	0.34	0.047	F
403 SS	32	1950	20	0.28	0.06	F
	28	1950	16	0.29	0.07	F
	25	2050	20	0.35	0.06	F
	19	2050	16	0.43	0.07	F
Waspaloy	55	2100	20	0.18	0.06	F
	50	2100	13–16	0.21–0.24	0.07–0.09	F
17-7PH SS	34	1950	13–20	0.22–0.28	0.06–0.09	F
	22	2100	16–20	0.35	0.06–0.07	F
	18	2100	13	0.31	0.09	F
Ti-6Al-4V	43	1700	20	0.30	0.06	F
	35	1700	13–16	0.29–0.34	0.07–0.09	F
	27	1750	16–20	0.32–0.46	0.06–0.07	F
	20	1750	13	0.38	0.09	F
Inconel 718	65	2000	16–20	0.17–0.18	0.06–0.07	F
	58	2000	13	0.18	0.09	F
	50	2100	20	0.33	0.06	F
	48	2100	13–16	0.29–0.30	0.07–0.09	F
Ti-8Al-1Mo-1V	50	1750	13–16	0.22–0.26	0.07–0.09	F
	47	1750	20	0.27	0.06	F
	40	1800	13–16	0.27–0.32	0.07–0.09	F
	27	1800	20	0.27	0.06	F
7075 Al	19	700	13–20	0.36–0.42	0.06–0.09	G
	16	800	13–20	0.31–0.49	0.06–0.09	G
Udimet	65	2050	14–17	0.4	(c)	F

(a) At 10 to 30% reduction. (b) Dimensions, OD:ID:thickness, in inches: A = 6:3:0.5, B = 6:3:1.0, C= 6:3:2.0, D = 3:1.5:0.25, E = 3:1.5:0.5, F = 3:1.5:1.0, G = 5:3:1. (c) Not measured.

REFERENCES

[Altan et al., 1973]: Altan, T., Boulger, F.W., "Flow Stress of Metals and Its Application in Metal Forming Analyses," *Trans. ASME, J. Eng. Ind.,* Nov 1973, p 1009.

[Dahl et al., 1999]: Dahl, C., Vazquez, V., Altan, T., "Determination of Flow Stress of 1524 Steel at Room Temperature using the Compression Test," Engineering Research Center for Net Shape Manufacturing, ERC/NSM-99-R-22, 1999.

[Dixit et al., 2002]: Dixit, R., Ngaile, G., Altan, T., "Measurement of Flow Stress for Cold Forging," Engineering Research Center for Net Shape Manufacturing, ERC/NSM-01-R-05, 2002.

[Douglas et al., 1975]: Douglas, J.R., Altan, T., "Flow Stress Determination of Metals at Forging Rates and Temperatures," *Trans. ASME, J. Eng. Ind.,* Feb 1975, p 66.

[Lee et al., 1972]: Lee, C.H., Altan, T., "Influence of Flow Stress and Friction upon Metal Flow in Upset Forging of Rings and Cylinders," *Trans. ASME, J. Eng. Ind.,* Aug 1972, p 775.

[Thomsen et al., 1965]: Thomsen, E.G., Yang, C.T., Kobayashi, S., *Mechanics of Plastic Deformation in Metal Processing,* The Macmillan Company, 1965.

SELECTED REFERENCES

[Altan et al., 1981]: Altan, T., Semiatin, S.L., Lahoti, G.D., "Determination of Flow Stress Data for Practical Metal Forming Analysis," *Ann. CIRP,* Vol 30 (No. 1), 1981, p 129.

[Altan et al., 1983]: Altan, T., Oh, S.-I., Gegel, H.L., *Metal Forming Fundamentals and Applications,* ASM International, 1983.

[Lahoti et al., 1975]: Lahoti, G.D., Altan, T., "Prediction of Temperature Distributions in Axisymmetric Compression and Torsion," *J. Eng. Mater. Technol.,* April 1975, p 113.

CHAPTER 5

Plastic Deformation: Complex State of Stress and Flow Rules

Gracious Ngaile

5.1 State of Stress

In a deforming object, different states of stress would exist depending on the loading conditions and boundary constraints. Figure 5.1 shows a cylinder upsetting process. The local state of stress in the deforming cylinder can be visualized by discretizing the object into small elements as shown in Fig. 5.1(a) and (b). The state of stress can also be presented in a matrix form, commonly known as stress tensor, as shown in Fig. 5.1(c) and 5.2.

A normal stress is indicated by two identical subscripts, e.g., σ_{xx}, while a shear stress is indicated by a differing pair, σ_{xy}. This notation can be simplified by denoting the normal stresses by a single subscript and shear stresses by the symbol τ. Thus one will have $\sigma_{xx} \equiv \sigma_x$ and $\sigma_{xy} \equiv \tau_{xy}$. In the case of rotational equilibrium, $\sigma_{xy} = \sigma_{yx}$, thus implying the absence of rotational effects around any axis. The nine stress components then reduce to six independent components.

For a general stress state, there is a set of coordinate axes (1, 2, and 3) along which the shear stresses vanish. The normal stresses along these axes, σ_1, σ_2, and σ_3, are called the principal stresses. The magnitudes of the principal stresses are determined from the following cubic equation derived from a series of force equilibrium equations [Backofen, 1972].

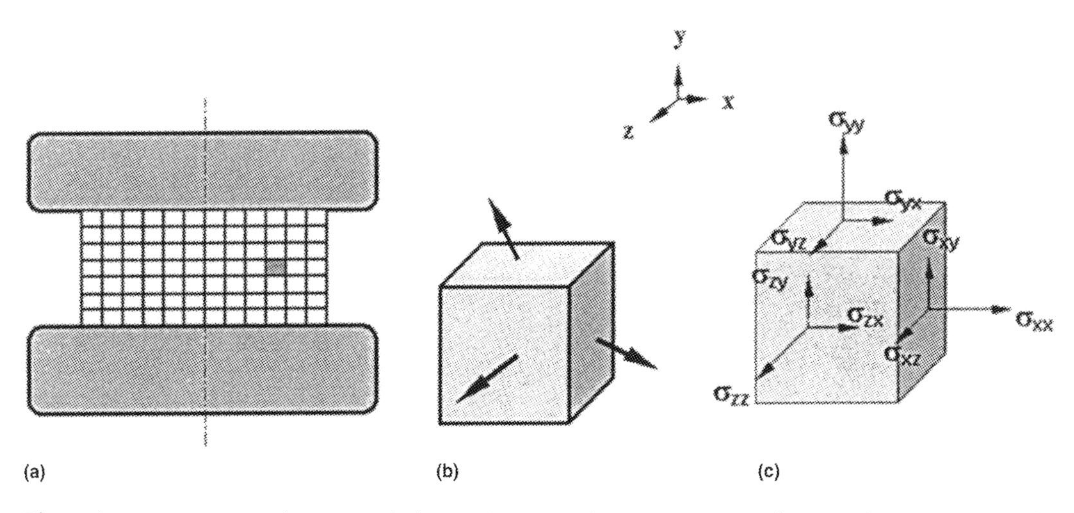

Fig. 5.1 Stress acting on an element. (a) Cylinder upsetting process. (b) Forces acting on an element. (c) Stress components acting on an element

$$\sigma_i^3 - I_1\sigma_i^2 - I_2\sigma_i - I_3 = 0 \quad \text{(Eq 5.1a)}$$

where

$$I_1 = \sigma_{xx} + \sigma_{yy} + \sigma_{zz}$$

$$I_2 = -\sigma_{xx}\,\sigma_{yy} - \sigma_{yy}\,\sigma_{zz} - \sigma_{zz}\,\sigma_{xx} + \sigma_{xy}^2 + \sigma_{yz}^2 + \sigma_{zx}^2$$

$$I_3 = \sigma_{xx}\,\sigma_{yy}\,\sigma_{zz} + 2\sigma_{xy}\,\sigma_{yz}\,\sigma_{zx} - \sigma_{xx}\,\sigma_{yz}^2 - \sigma_{yy}\sigma_{zx}^2 - \sigma_{zz}\sigma_{xy}^2$$

The coefficients I_1, I_2, and I_3 are independent of the coordinate system chosen and are hence called invariants. Consequently, the principal stresses for a given stress state are unique. The three principal stresses can only be determined by finding the three roots of the cubic equation. The first (linear) and second (quadratic) invariants have particular physical significance for the theory of plasticity [Kobayashi et al., 1989]. The invariants can also be expressed in terms of principal stresses.

$$I_1 = \sigma_1 + \sigma_2 + \sigma_3 \quad \text{(Eq 5.1b)}$$

$$I_2 = -(\sigma_1\sigma_2 + \sigma_2\sigma_3 + \sigma_3\sigma_1) \quad \text{(Eq 5.1c)}$$

$$I_3 = \sigma_1\,\sigma_2\,\sigma_3 \quad \text{(Eq 5.1d)}$$

5.2 Yield Criteria

A yield criterion is a law defining the limit of elasticity or the start of plastic deformation under any possible combination of stresses. It can be expressed by $f(\sigma_{ij}) = C$ (constant). For isotropic materials, plastic yielding can depend only on the magnitude of the principal stresses; i.e., the yield criteria is expressed in function of invariants I_1, I_2, and I_3 or $f(I_1, I_2, I_3) = C$.

In simple homogeneous (uniaxial) compression or tension, the metal flows plastically when the stress, σ, reaches the value of the flow stress, $\bar{\sigma}$, in other words the flow rule in uniaxial deformation is:

$$|\sigma| = \frac{F}{A} = \bar{\sigma} \quad \text{(Eq 5.2)}$$

where F and A are the instantaneous force and cross-sectional area on which the force acts. In a multiaxial state of stress, plastic flow (yielding) depends on a combination of all stresses [Thomsen et al., 1965]. Consider a metal plate subjected to tensile and compressive loading in the y-axis until the material yields. If the process is repeated using different specimens and by gradually varying the loading in the x and y directions (Fig. 5.3), and the stresses at the onset of yielding are plotted on the σ_1-σ_3 plane, then a yield locus of the schematic form shown in Fig. 5.4 will be obtained.

There are various yield criteria that have been proposed to date. However, this chapter discusses two major yield criteria that have been used extensively in the analysis of metal forming and forging.

- Tresca or shear stress criterion of yield or plastic flow
- von Mises or distortion energy criterion of yield or plastic flow

5.2.1 Tresca Yield Criterion

The Tresca yield criterion states that plastic flow starts when the maximum shear stress, τ_{max},

$$\sigma_{ij} = \begin{bmatrix} \sigma_{xx} & \sigma_{yx} & \sigma_{zx} \\ \sigma_{xy} & \sigma_{yy} & \sigma_{zy} \\ \sigma_{xz} & \sigma_{yz} & \sigma_{zz} \end{bmatrix} = \begin{bmatrix} \sigma_x & \tau_{yx} & \tau_{zx} \\ \tau_{xy} & \sigma_y & \tau_{zy} \\ \tau_{xz} & \tau_{yz} & \sigma_z \end{bmatrix}$$

Fig. 5.2 Stress tensor

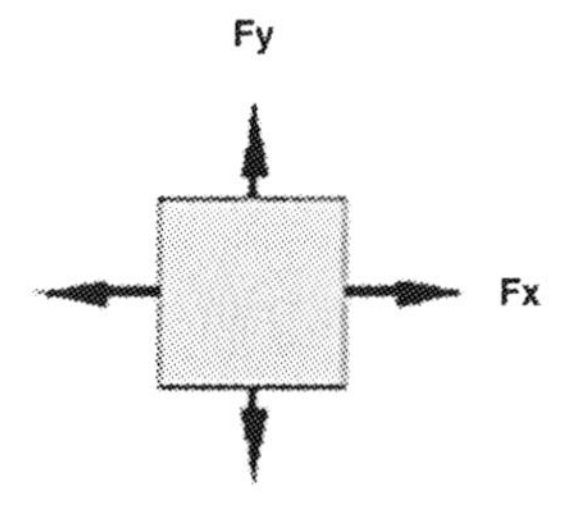

Fig. 5.3 Metal plate subjected to various loading conditions

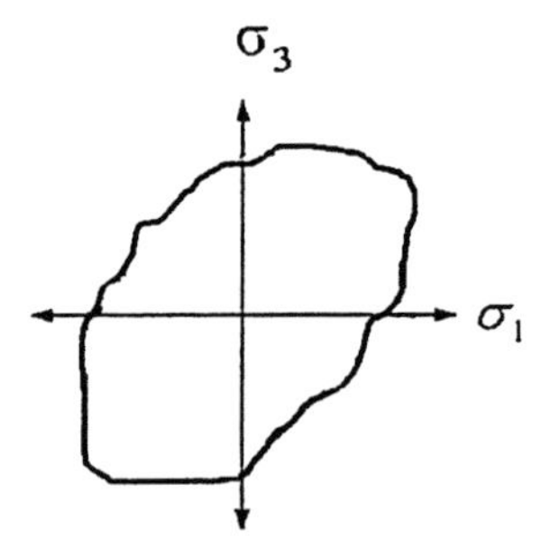

Fig. 5.4 Possible yield locus (schematic) showing enclosed elastic region

reaches a certain critical value, k, or when $|\tau_{max}| = k$, where k is the shear flow stress that is characteristic of a given material and its microstructure and depends on shear strain rate, strain, and deformation temperature. This yield criterion can easily be described by the aid of Mohr circles for stresses.

Figure 5.5 shows a Mohr circle that represents the stresses in a plane whose coordinate axes are chosen to be the shear stress τ (ordinate) and the normal stress σ (abscissa). In the physical x-y plane, the "principal" stresses are perpendicular to each other, and in the direction of "principal" stresses the shear stresses are zero, as can be seen in Fig. 5.1(b).

In the Mohr circle representation, i.e., in the τ-σ plane, the maximum principal stress, σ_1, and the minimum principal stress, σ_3, define the size of the Mohr circle (Fig. 5.5a). The subscripts 1 and 3 are arbitrary and indicate only that $\sigma_3 \leq \sigma_2 \leq \sigma_1$. As can be seen in Fig. 5.5, the largest shear stress, τ_{max}, acting on the τ-σ plane is given by one radius of the Mohr circle. Thus: $\tau_{max} = (\sigma_1 - \sigma_3)/2$. The states of stress, which cause plastic deformation, are illustrated with the Mohr circle of radius k.

Figures 5.6(a) and 5.6(b) show Mohr circles for uniaxial tension (no necking) and compression (no bulging). With $\sigma_2 = \sigma_3 = 0$ (uniaxial deformation), plastic flow starts when

$$\sigma_1 = \frac{F}{A} = \bar{\sigma} = 2k \text{ or } k = \frac{\bar{\sigma}}{2} \quad \text{(Eq 5.3)}$$

where F is the tensile or compressive force, A is the instantaneous cross-sectional area of the sample, and $\bar{\sigma}$ is the flow stress (or instantaneous yield stress).

If the principal stresses are arranged such that $\sigma_1 > \sigma_2 > \sigma_3$, then the Tresca yield criterion can be expressed as:

$$\sigma_1 - \sigma_3 = \bar{\sigma} \quad \text{(Eq 5.4)}$$

Equation 5.4 shows that, according to Tresca's rule, plastic flow or yielding starts if the difference of maximum (σ_1) and minimum (σ_3) principal stresses is equal to the flow stress, $\bar{\sigma}$. Figure 5.6 shows that the position of the Mohr circle, i.e., the hydrostatic stress, is not important for the start of plastic flow.

The mean principal stress is

$$\sigma_m = \frac{\sigma_1 + \sigma_2 + \sigma_3}{3} \quad \text{(Eq 5.5)}$$

and the hydrostatic pressure is

$$P = -\sigma_m \quad \text{(Eq 5.6)}$$

Figure 5.7 shows Mohr circles for biaxial and triaxial states of stress, respectively. Note that only principal stresses are shown in these figures. Details on Mohr circles and stress transformation can be found in solid mechanics textbooks.

For plane stress condition ($\sigma_2 = 0$), the Tresca's yield criterion can be graphically represented in two dimensions (Fig. 5.8). The hexagonal diagram encloses the elastic region.

5.2.2 *von Mises Yield Criterion*

The von Mises yield criterion considers all the stresses acting on the deforming body and can be expressed as follows. The start of plastic flow (yielding) must depend on a combination of normal and shear stresses, which does not change its value when transformed from one coordinate system into another [Lange, 1972] and [Johnson

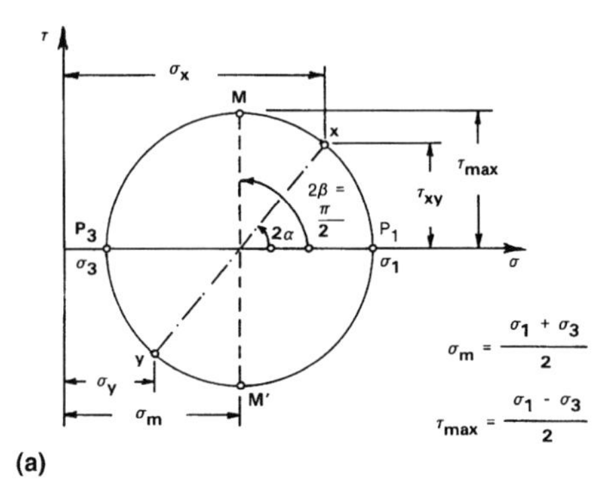

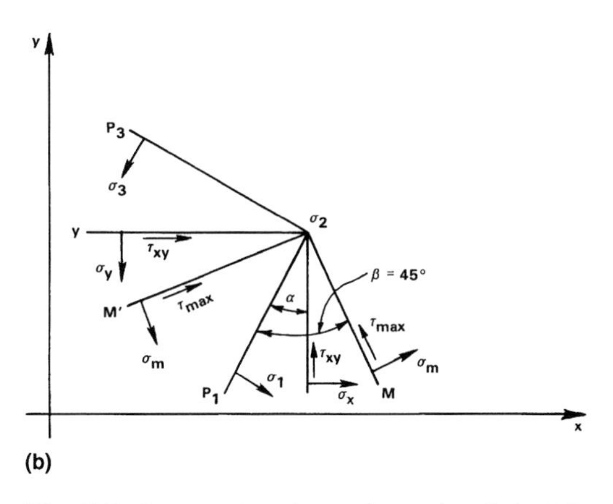

Fig. 5.5 Representation of state of stress through the Mohr circle

et al., 1975]. In terms of principal stresses, the von Mises rule is given by:

$$\left\{\frac{1}{2}\left[(\sigma_1 - \sigma_2)^2 + (\sigma_2 - \sigma_3)^2 + (\sigma_1 - \sigma_3)^2\right]\right\}^{1/2} = \bar{\sigma} \quad \text{(Eq 5.6a)}$$

In a general way:

$$\left\{\frac{1}{2}\left[(\sigma_x - \sigma_y)^2 + (\sigma_y - \sigma_z)^2 + (\sigma_z - \sigma_x)^2 + 3(\tau_{xy}^2 + \tau_{yz}^2 + \tau_{zx}^2)\right]\right\}^{1/2} = \bar{\sigma} \quad \text{(Eq 5.6b)}$$

or, it can be given by:

$$\left\{\frac{3}{2}\left[(\sigma_1 - \sigma_m)^2 + (\sigma_2 - \sigma_m)^2 + (\sigma_3 - \sigma_m)^2\right]\right\}^{1/2} = \bar{\sigma} \quad \text{(Eq 5.6c)}$$

where σ_m is the mean principal stress given by:

$$\sigma_m = \frac{\sigma_1 + \sigma_2 + \sigma_3}{3}$$

A physical interpretation of the von Mises yield criterion shows that the left side of Eq 5.6 is proportional to the energy that is stored in the elastically deformed material prior to yielding. This is the energy necessary for elastic volume change. The flow rule then states that plastic flow starts when this elastic energy reaches a critical value. That is why the von Mises rule is also called the "distortion energy criterion."

For the plane stress condition ($\sigma_2 = 0$), the von Mises yield locus takes the form of an elliptical curve as shown in Fig. 5.9.

5.2.3 Comparison of Tresca and von Mises Criteria

The comparison of Tresca and von Mises criteria can be expressed by superimposing the elliptical yield locus (von Mises) and hexagonal yield locus (Tresca) together as shown in Fig. 5.10. The shaded regions show the difference between the two yield criteria.

- Points A, B, C, D, and E shown in Fig. 5.10 are used to describe the similarities and differences of the two yield criteria.

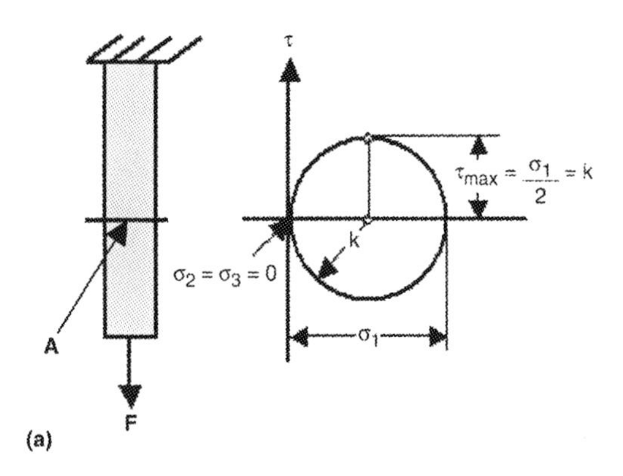

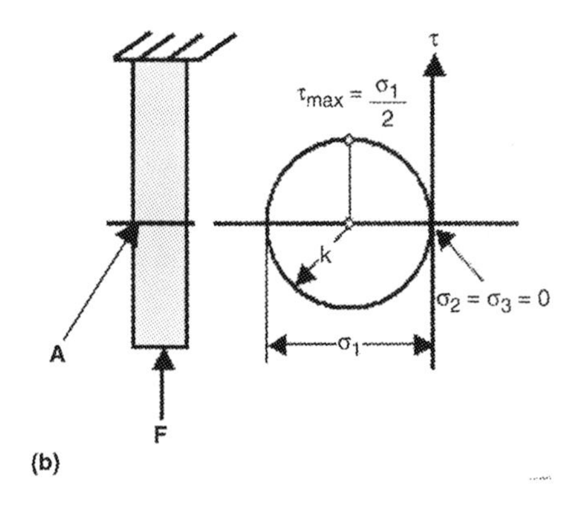

Fig. 5.6 Mohr circles. (a) Uniaxial tension. (b) Uniaxial compression

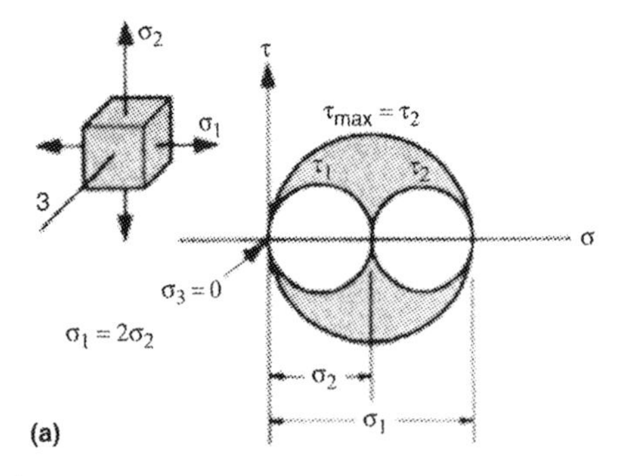

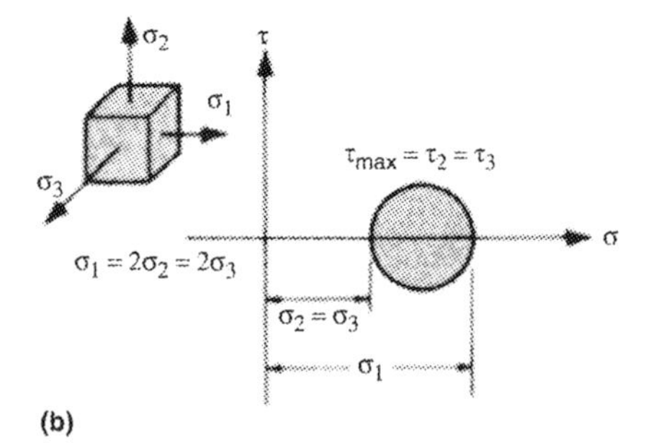

Fig. 5.7 Mohr circles. (a) Biaxial state of stress. (b) Triaxial state of stress. [Kalpakjian, 1997]

- In uniaxial tension or compression (points A and B), the von Mises and Tresca yield criteria exhibit the same values; i.e., when $\sigma_1 = F/A$ and $\sigma_2 = \sigma_3 = 0$, Eq 5.6(a) gives:

$$\sqrt{\left(\frac{1}{2}\, 2\sigma_1^2\right)} = \bar{\sigma}, \text{ or } \sigma_1 = \bar{\sigma} \qquad \text{(Eq 5.7)}$$

which is the same as that obtained from Eq 5.3.
- The state of stress for a balanced biaxial mode of deformation corresponds to point C in Fig. 5.10. For this condition, the Tresca and von Mises yield criteria exhibit the same results.
- In pure shear (point D in Fig. 5.10), the two yield criteria exhibit different yield stresses. Pure shear occurs when $\sigma_3 = -\sigma_1$, $\sigma_2 = 0$. Thus, from Eq 5.6(a) the von Mises yield criterion gives:

$$\bar{\sigma} = \sqrt{\left(\frac{1}{2}\,({\sigma_1}^2 + {\sigma_1}^2 + 4{\sigma_1}^2)\right)}$$
$$= \sqrt{3}\,\sigma_1 \cong 1.73\sigma_1 \qquad \text{(Eq 5.8)}$$

$$\tau_{max} = \sigma_1 = \frac{\bar{\sigma}}{\sqrt{3}} = 0.577\,\bar{\sigma} \qquad \text{(Eq 5.9)}$$

Using Eq 5.3, one finds that the Tresca criterion gives:

$$\bar{\sigma} = 2\sigma_1 \text{ and } \tau_{max} = \sigma_1 = \frac{\bar{\sigma}}{2} = 0.5\bar{\sigma} \qquad \text{(Eq 5.10)}$$

Thus, in pure shear, there is a 15% difference between values of τ_{max} obtained from the Tresca and von Mises yield criteria rules.

- For a plane strain condition (point E in Fig. 5.10), the stress required for deformation under Tresca's yield criterion in still σ_y. However, according to von Mises, the stress required is higher, 1.15 σ_y.

Experiments (with combined shear and tension) indicate that the von Mises rule is a better criterion (closer to reality) than Tresca's flow rule.

In three-dimensional space, the yield surfaces for von Mises and Tresca can be represented by inclined cylindrical and octagonal prisms, respectively (Fig. 5.11).

5.3 Flow Rules

When the stresses at a given point in the metal reach a certain level, as specified by a flow rule (Tresca or von Mises), then plastic flow, i.e., plastic deformation, starts. Similar to the Hooke's law, which gives the relationship between the stress and the corresponding deformation in the elastic range, analysis of plastic deformation requires a certain relation between the applied stresses and the velocity field (kinematics as described by velocity, ε and $\dot{\varepsilon}$ fields). Such a relation exists between the stresses (in principal axes) and strain rates:

$$\dot{\varepsilon}_1 = \lambda(\sigma_1 - \sigma_m) \qquad \text{(Eq 5.11a)}$$

$$\dot{\varepsilon}_2 = \lambda(\sigma_2 - \sigma_m) \qquad \text{(Eq 5.11b)}$$

$$\dot{\varepsilon}_3 = \lambda(\sigma_3 - \sigma_m) \qquad \text{(Eq 5.11c)}$$

Equations 5.11(a), (b), and (c) are called "plasticity equations" [Thomsen et al., 1965],

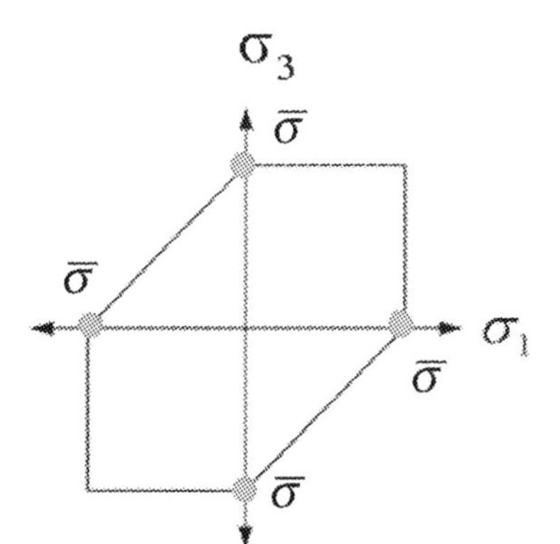

Fig. 5.8 Tresca's yield locus for plane stress condition

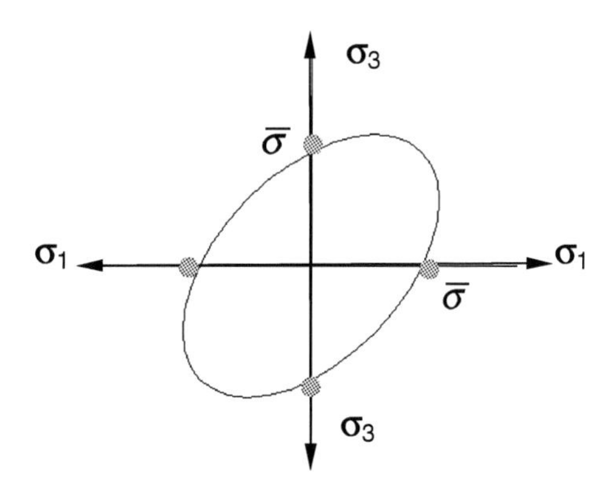

Fig. 5.9 von Mises yield locus for plane stress condition

[Lange, 1972], [Backofen, 1972]. The variable λ depends on direction of plastic flow, temperature, material, strain, and strain rate. The plasticity equations—for example, Eq 5.11(a)—can also be expressed in the form:

$$d\varepsilon_1 = \frac{3}{2}\frac{d\bar{\varepsilon}}{\bar{\sigma}}(\sigma_1 - \sigma_m) \quad \text{(Eq 5.12)}$$

where $\bar{\varepsilon}$ and $\bar{\sigma}$ denote effective stress and strain. Equation 5.11 can also be expressed as:

$$\dot{\varepsilon}_1 = \frac{3}{2}\frac{\dot{\bar{\varepsilon}}}{\bar{\sigma}}(\sigma_1 - \sigma_m) \quad \text{(Eq 5.13)}$$

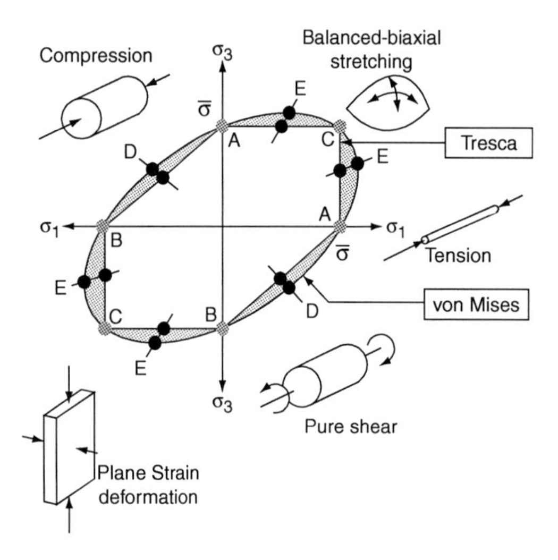

Fig. 5.10 Tresca and von Mises yield loci for the same value of $\bar{\sigma}$, showing several loading paths. Note: $\sigma^2 = 0$.

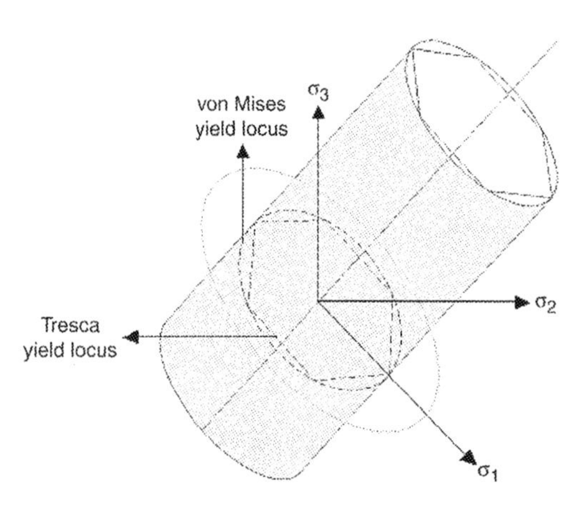

Fig. 5.11 Physical representation of von Mises and Tresca criterion in three dimensions

5.4 Power and Energy of Deformation

The plastic deformation processes are irreversible. The mechanical energy, consumed during deformation, is transformed largely into heat. It is useful to consider again the homogeneous deformation of a block (Fig. 5.12). The following relations, derived in Chapter 3, hold here also [Lange, 1972].

$$\varepsilon_h = \varepsilon_1 = \ln\frac{h}{h_o};\ \dot{\varepsilon}_1 = \frac{v_h}{h}$$

$$\varepsilon_2 = \ln\frac{w}{w_o};\ \dot{\varepsilon}_2 = \frac{v_w}{w}$$

$$\varepsilon_3 = \ln\frac{l}{l_o};\ \dot{\varepsilon}_3 = \frac{v_l}{l}$$

Following Fig. 5.12, the instantaneous power of deformation (force times velocity) is given by:

$$\begin{aligned} P &= \sigma_1 wlv_h + \sigma_2 hlv_w + \sigma_3 whv_l \\ &= \sigma_1 wlh\dot{\varepsilon}_1 + \sigma_2 wlh\dot{\varepsilon}_2 + \sigma_3 wlh\dot{\varepsilon}_3 \\ &= (\sigma_1\dot{\varepsilon}_1 + \sigma_2\dot{\varepsilon}_2 + \sigma_3\dot{\varepsilon}_3)\,V \end{aligned} \quad \text{(Eq 5.14)}$$

where V is the volume of the deforming block. It follows that the energy of deformation, E, is:

$$E = V\int_{t_0}^{t_1}(\sigma_1\dot{\varepsilon}_1 + \sigma_2\dot{\varepsilon}_2 + \sigma_3\dot{\varepsilon}_3)\,dt \quad (5.15)$$

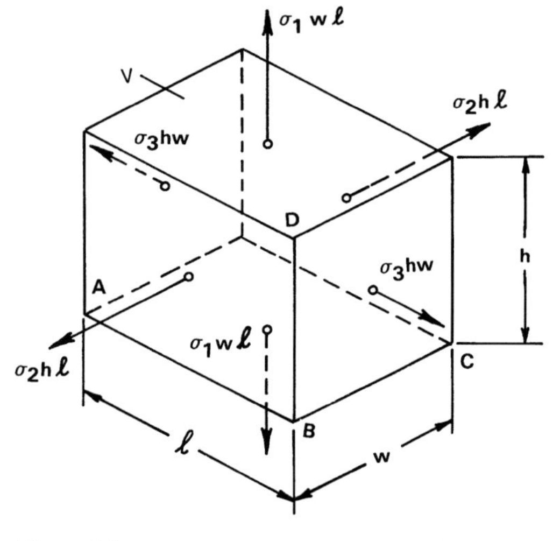

Fig. 5.12 Homogeneous deformation of a block

with $\dot{\varepsilon}\,dt = d\varepsilon$, Eq 5.15 can also be written as:

$$E = V\left(\int_0^{\varepsilon_1} \sigma_1 d\varepsilon_1 + \int_0^{\varepsilon_2} \sigma_2 d\varepsilon_2 + \int_0^{\varepsilon_3} \sigma_3 d\varepsilon_3\right) \quad \text{(Eq 5.16)}$$

5.5 Effective Strain and Effective Strain Rate

The flow stress, $\bar{\sigma}$, is determined from a uniaxial test (compression or homogeneous tension). Under multiaxial deformation conditions, it is necessary to relate uniaxial material behavior to multiaxial material behavior. Considering an element and the principal directions, the deformation energy, dW, expended during a time element Δt, is:

$$dW = (\sigma_1 d\varepsilon_1 + \sigma_2 d\varepsilon_2 + \sigma_3 d\varepsilon_3)V \quad \text{(Eq 5.17)}$$

or divided by dt, the deformation power, P, is:

$$P = \frac{dW}{dt} = (\sigma_1 \dot{\varepsilon}_1 + \sigma_2 \dot{\varepsilon}_2 + \sigma_3 \dot{\varepsilon}_3)V \quad \text{(Eq 5.18)}$$

The effective strain, $\bar{\varepsilon}$, and strain rate, $\dot{\bar{\varepsilon}}$ (both indicated with overbar), are defined as:

$$dW = \bar{\sigma}\, d\bar{\varepsilon} V \quad \text{(Eq 5.19)}$$

or

$$P = \bar{\sigma}\dot{\bar{\varepsilon}} V \quad \text{(Eq 5.20)}$$

Equations 5.18 and 5.20 give:

$$\bar{\sigma}\dot{\bar{\varepsilon}} = \sigma_1 \dot{\varepsilon}_1 + \sigma_2 \dot{\varepsilon}_2 + \sigma_3 \dot{\varepsilon}_3 \quad \text{(Eq 5.21)}$$

From volume constancy, it can be shown that:

$$\dot{\varepsilon}_1 + \dot{\varepsilon}_2 + \dot{\varepsilon}_3 = 0 \quad \text{(Eq 5.22a)}$$

or

$$\sigma_m(\dot{\varepsilon}_1 + \dot{\varepsilon}_2 + \dot{\varepsilon}_3) = 0 \quad \text{(Eq 5.22b)}$$

Equations 5.20 and 5.22 give:

$$\bar{\sigma}\dot{\bar{\varepsilon}} = \dot{\varepsilon}_1 (\sigma_1 - \sigma_m) + \dot{\varepsilon}_2 (\sigma_2 - \sigma_m) + \dot{\varepsilon}_3 (\sigma_3 - \sigma_m) \quad \text{(Eq 5.23)}$$

Using the one form of the von Mises rule, Eq 5.6(c), Eq 5.23 gives:

$$\dot{\bar{\varepsilon}} = \frac{\dot{\varepsilon}_1 (\sigma_1 - \sigma_m) + \dot{\varepsilon}_2 (\sigma_2 - \sigma_m) + \dot{\varepsilon}_3 (\sigma_3 - \sigma_m)}{\left\{\frac{3}{2}[(\sigma_1 - \sigma_m)^2 + (\sigma_2 - \sigma_m)^2 + (\sigma_3 - \sigma_m)^2]\right\}^{1/2}} \quad \text{(Eq 5.24)}$$

Using the plasticity equations, Eq 5.11(a), 5.11(b), and 5.11(c), which give $\sigma_1 - \sigma_m = \dot{\varepsilon}_1/\lambda$, etc., Eq 5.24 can be reduced to:

$$\dot{\bar{\varepsilon}} = \sqrt{\left(\frac{2}{3}(\dot{\varepsilon}_1^2 + \dot{\varepsilon}_2^2 + \dot{\varepsilon}_3^2)\right)} \quad \text{(Eq 5.25)}$$

or by integration, to:

$$\bar{\varepsilon} = \int_{t_0}^{t_1} \dot{\bar{\varepsilon}}\, dt \quad \text{(Eq 5.26)}$$

Equations 5.19 and 5.26 show how to calculate the effective strain rate and the effective strain in principal directions.

REFERENCES

[Backofen, 1972]: Backofen, W.A., *Deformation Processing,* Addison-Wesley, 1972.

[Johnson et al., 1975]: Johnson, W., Mellor, P.B., *Engineering Plasticity,* Van Nostrand Reinhold, London, 1975.

[Kalpakjian, 1997]: Kalpakjian, S., *Manufacturing Processes for Engineering Materials,* Addison-Wesley, 1997.

[Kobayashi et al., 1989]: Kobayashi, S., Oh, S., Altan, T., *Metal Forming and the Finite Element Method,* Oxford University Press, 1989.

[Lange, 1972]: Lange, K., Ed., *Study Book of Forming Technology,* (in German), Vol 1, *Fundamentals,* Springer-Verlag, 1972.

[Thomsen et al., 1965]: Thomsen, E.G., Yang, C.T., and Kobayashi, S., *Mechanics of Plastic Deformation in Metal Processing,* Macmillan, 1965.

CHAPTER 6

Temperature and Heat Transfer

Gangshu Shen

6.1 Introduction

In metal forming processes, both plastic deformation and friction contribute to heat generation. Approximately 90 to 95% of the mechanical energy involved in the process is transformed into heat [Farren et al., 1925]. In some continuous forming operations such as drawing and extrusion, performed at high speeds, temperature increases of several hundred degrees may be involved [Lahoti et al., 1978]. Heat generation is also significant in forgings produced in high-speed equipment such as mechanical press, screw press, and hammer. A part of generated heat remains in the deformed material, another part flows into the undeformed/less-deformed portion of the material where temperature is lower, while still an additional part may flow into the tooling. The temperatures developed during the forging operation influence lubrication conditions, tool life, as well as microstructure and properties of the forged part. With the finite element based process modeling, the heat generation during deformation and heat transfer before, during, and after deformation can all be calculated in a computer. To ensure accurate heat transfer calculation, correct workpiece and die interface heat transfer coefficient must be known. Using accurate process modeling, the influence of press speed, contact time, and heat transfer in metal forming can be evaluated.

6.2 Heat Generation and Heat Transfer in Metal Forming Processes

In metal forming, the magnitudes and distribution of temperatures depend mainly on:

- The initial workpiece and die temperatures
- Heat generation due to plastic deformation and friction at the workpiece/die interface
- Heat transfer between the workpiece and dies and between the workpiece and the environment (air or lubricant and coolant, etc.)

In processes such as forging and extrusion, the average instantaneous temperature in the deforming workpiece, θ_A, can be estimated by [Altan et al., 1970]:

$$\theta_A = \theta_W + \theta_D + \theta_F - \theta_T - \theta_R - \theta_C \qquad \text{(Eq 6.1)}$$

where θ_W is the initial workpiece temperature, θ_D is the temperature increase due to plastic deformation, θ_F is the temperature increase due to interface friction, θ_T is the temperature drop due to heat transfer into the dies, θ_R is the temperature drop due to radiation to the environment, and θ_C is the temperature drop due to convection to the environment.

The temperature increase due to the deformation, in a time interval Δt, is given by:

$$\theta_D = \frac{A\bar{\sigma}\dot{\bar{\varepsilon}}\Delta t}{c\rho}\beta = \frac{\bar{\sigma}\Delta\bar{\varepsilon}}{Jc\rho}\beta \qquad \text{(Eq 6.2)}$$

where $\bar{\sigma}$ is the flow stress of the workpiece, $\dot{\bar{\varepsilon}}$ is the effective strain rate, $\Delta\bar{\varepsilon}$ is the effective strain generated during Δt, A is a conversion factor between mechanical and thermal energies, c is the specific heat of the workpiece, ρ is the specific weight of the workpiece, and β is the fraction of deformation energy transformed into heat ($0 \leq \beta \leq 1$); usually, $\beta = 0.95$.

The temperature increase due to friction, θ_F, is given by:

$$\theta_F = \frac{Af\bar{\sigma}vF\Delta F}{c\rho\rho_a} \qquad \text{(Eq 6.3)}$$

where, in addition to the symbols already described, f is the friction factor at the workpiece/tool interface, such that frictional shear stress $\tau = f\bar{\sigma}$, v is the velocity at the workpiece/tool interface, and V_a is the volume of the workpiece which is subject to temperature increase.

6.3 Temperatures in Forging Operations

In forging, the metal flow is non steady state. Contact between the deforming metal and the dies is intermittent. The length of contact time and the nature of the heat transfer at the die/material interface influence temperatures very significantly. A simple example of an operation involving non-steady-state metal flow is the cold upsetting of a cylinder. In this process, a grid system is established for calculation of temperatures (Fig. 6.1). For various points, indicated in Fig. 6.1, temperatures were calculated for cold upsetting of a steel cylinder initially at room temperature [Lahoti et al., 1975].

The calculated results for the grid points $P_{1,1}$, $P_{1,5}$, and $P_{1,8}$ in Fig. 6.1 are compared with experimental data in Fig. 6.2. As expected, temperatures increase with increasing deformation.

In hot forging operations, the contact time under pressure between the deforming material and the dies is the most significant factor influencing temperature conditions. This is illustrated in Fig. 6.3, where the load-displacement curves are given for hot forging of a steel part using different types of forging equipment [Altan et al., 1973]. These curves illustrate that, due to strain rate and temperature effects, for the same forging process, different forging loads and energies are required by different machines. For the hammer, the forging load is initially higher due to strain-rate effects, but the maximum load is lower than for either hydraulic or screw presses. The reason for this is that in the presses the flash cools rapidly, whereas in the hammer the flash temperature remains nearly the same as the initial stock temperature.

Thus, in hot forging, not only the material and the formed shape but also the type of equipment used (rate of deformation and die chilling effects) determine the metal flow behavior and the forming load and energy required for the process. Surface tearing and cracking or development of shear bands in the formed material often can be explained by excessive chilling of the surface layers of the formed part near the die/material interface.

With the advancement of finite element modeling and the increase in computer speed, heat transfer in any forging and heat treatment condition can be simulated accurately in a very short time.

Figure 6.4 shows the temperature distribution at the end of a coupled deformation and heat transfer modeling of a Ti-64 cylinder upset test in a hydraulic press. The starting temperature of the Ti-64 workpiece was 1750 °F (955 °C), and the starting tool steel die temperature was 300 °F (150 °C). At the end of the cylinder upsetting, there was quite a temperature gradient inside both the upset cylinder (or pancake) and the dies. The temperature range of the pancake at the end of upsetting was 1044 to 1819 °F (560 to 990 °C). The temperature range of the dies at the end of upsetting was 298 to 723 °F (145 to 385 °C).

6.4 Measurement of Temperatures at the Die/Material Interface

Often it is desirable to measure the temperatures at the die/material interface in hot forging operations. A thermocouple for measuring in-

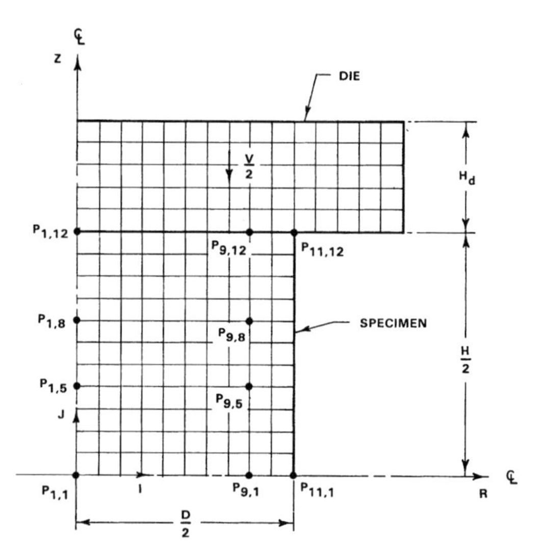

Fig. 6.1 Grid system for calculating velocity and temperature fields in cold upsetting of a cylinder. [Lahoti et al., 1975]

terface temperatures in hot forging must exhibit very fast response (a few milliseconds), accuracy, minimum interference with heat flow, and the ability to sustain high normal and shear stresses under high temperatures. Such thermocouples are available and were used for measuring die temperatures in forging of steel [Vigor et al., 1961]. The results are given in Fig. 6.5. These results clearly indicate that the temperature gradient is very large at the vicinity of the die/material interface.

In another study, fast response thermocouples were embedded in the bottom flat die of the upset tooling as shown in Fig. 6.6. The output of the thermocouples was recorded on a light beam oscillograph along with the load required to upset the specimens and the movements of the press ram as determined by the potentiometric displacement transducers [Altan et al., 1970]. Figure 6.6 shows the temperature-time traces for four thermocouples—two in the die and two in the 1020 steel billet forged at 2250 °F (1230 °C). These data show that the interface (or insulated) thermocouple measures the billet surface temperature, but only under load. Evidently, a high contact pressure is necessary to ensure good thermal contact between the billet surface and the thermocouple junction. After the load had been removed, the interface thermocouple indicated a decrease in temperature, while the thermocouple placed in the sample actually showed an increase in the temperature at the bottom of the forging. It is interesting that the rate of the temperature drop, as indicated by the interface thermocouple, decreases significantly once the upper ram and the pressure are removed. As can be expected, the rate of temperature drop further increases after the sample is removed from the die.

The results shown in Fig. 6.7 indicate that, during forging of steel at 2250 °F (1230 °C) with dies initially at 400 °F (205 °C), die surface temperatures can reach approximately 1200 °F (650 °C) in a fraction of a second while the billet

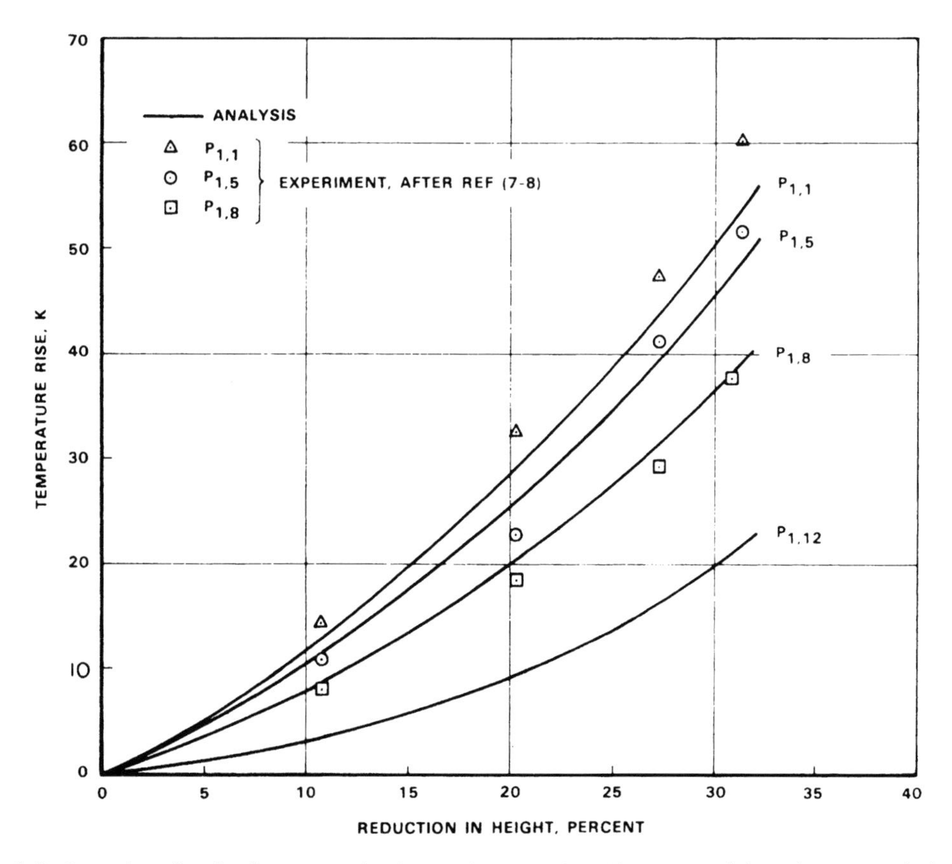

Fig. 6.2 Comparison of predicted temperature in axisymmetric compression with experimental data (refer to Fig. 6.1 for locations of grid points $P_{i,j}$) (material, AISI type 1015 steel; specimen dimensions, 20 mm diam by 30 mm high; initial temperatures, 293 K). [Lahoti et al., 1975]

temperature at the interface drops to 1450 °F (790 °C). These data agree with the measurements shown in Fig. 6.5. Similar measurement made during forging of aluminum alloy 6061 showed essentially the same trend of temperature variations with time. With the billet at 800 °F (430 °C) and the dies at 400 °F (205 °C), the interface temperature reached 700 °F (370 °C). These data demonstrate how rapidly temperatures change in hot forging, especially under pressure contact.

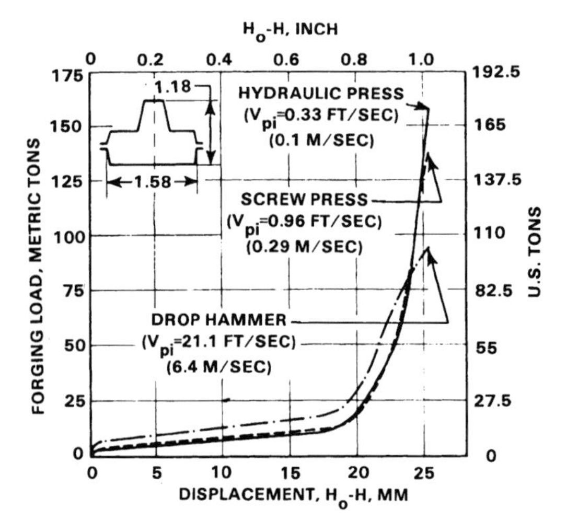

Fig. 6.3 Load-versus-displacement curves obtained in closed-die forging of an axisymmetric steel part (dimensions in inches) at 2012 °F (1100 °C) in three different machines with different initial velocities, $V_{p,i}$. [Altan et al., 1973]

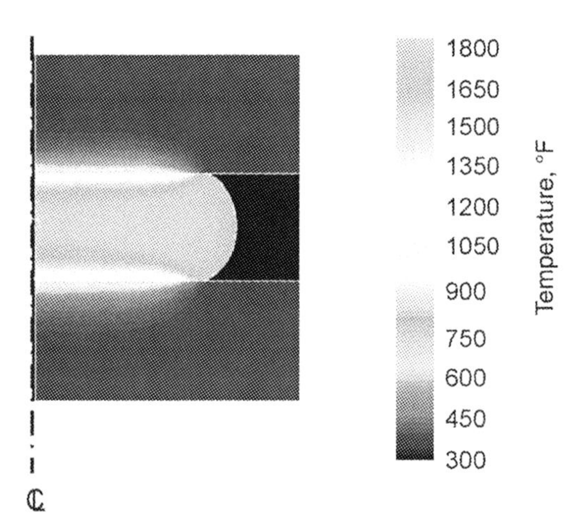

Fig. 6.4 The temperature distribution (in degree Fahrenheit) at the end of a Ti-64 cylinder upset test. The dimension of the cylinder, 1 in. (25 mm) diam by 1.5 in. (38 mm) height; starting temperature of cylinder, 1750 °F (955 °C); starting die temperature, 300 °F (150 °C); total reduction in height, 50%; temperature range of the pancake at the end of upsetting, 1044–1819 °F (560–990 °C); temperature range of the dies at the end of upsetting, 298–723 °F (145–385 °C).

6.5 Measurement of Interface Heat Transfer Coefficient

Interface heat transfer coefficient, h, determines the amount of heat transferred across an interface. Advances in the analysis of complicated forming processes such as nonisothermal forging have required that the interface heat transfer between objects be characterized. Thus, the numerical analysis, such as finite element method, can carry out quantitative calculations at the interface between the objects that are in contact. The interface heat transfer coefficient, h, for hot forging application, is measured experimentally in a couple of different ways: [Semiatin et al., 1987] [Burte et al., 1989]:

- **Two die tests,** in which two flat H13 tool steel dies were heated to different initial temperatures and brought together under varying pressure levels (h determined under nondeforming conditions)
- **Upset tests,** in which two dies were heated to the same temperature and used to upset a workpiece that had a same or higher temperature, which represented isothermal and nonisothermal forging conditions (h determined under deformation conditions)

Since interface friction also plays an important role in metal flow in the second test, the coupling determination of heat transfer and friction in one test is desired. Thus, ring tests were

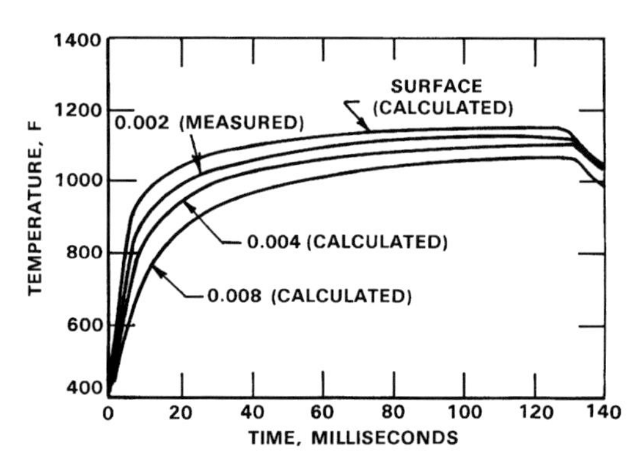

Fig. 6.5 Temperatures at the surface and at various depths in the forging die obtained in forging of 1040 steel without a lubricant (sample 1.125 in., or 29 mm, high; reduction in height, 75%; sample temperature, 2050 °F, or 1120 °C; die temperature, 450 °F, or 230 °C). [Vigor et al., 1961]

selected in Burte's experiments [Burte et al., 1989]. Using ring tests, the interface friction and interface heat transfer coefficients were determined at the same time. The effects of forging pressure, deformation rate, and lubrication on the heat transfer coefficient and the friction shear factor were evaluated simultaneously.

A schematic of ring compression tests used for the measurement of interface heat transfer coefficient is shown in Fig. 6.8. Two pairs of thermocouples were embedded in different depths of the bottom die. The detailed ring compression process was simulated using finite element model (FEM) package ALPID (a parent version of DEFORM™) [SFTC, 2002]. The elements were generated such that there were two nodes having the exact locations as the two pairs of the thermocouples in real die for tracking the temperature history. This kind of arrangement is no longer necessary at present time because the current DEFORM™ allows users to define any location of the workpiece or dies for tracking of the thermomechanical histories.

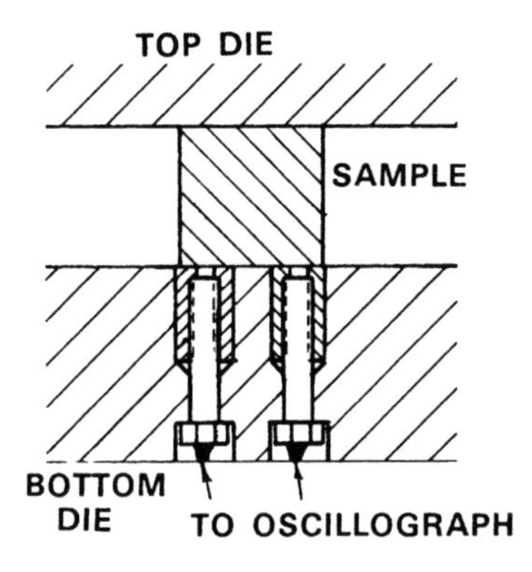

Fig. 6.6 Sketch of cross section through upset forging setup, showing location of fast response thermocouple in bottom forging die

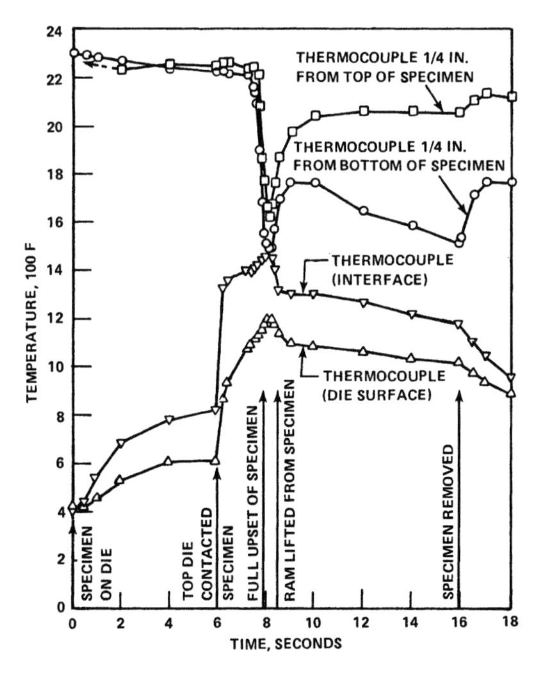

Fig. 6.7 Variations in temperatures at various locations in forging of 1020 steel billets (3 in. diam by 3 in. high, or 76 mm diam by 76 mm high) to 1 in. (25 mm) thickness at 2250 °F (1230 °C) between dies at 400 °F (205 °C).

Figure 6.9 illustrates the method for the measurement of interface heat transfer coefficient. The interface heat transfer coefficient was determined by calibration curves. The increase in the temperature of the bottom die (the instantaneous die temperature T_1 minus the initial die temperature T_{10}) versus time obtained from experiments was plotted as experimental data. On the same chart, the FEM generated calibration curves ($h = 0.0068$ btu/s/in.^{2}/°F, or 20 kW/m^2 · K, and $h = 0.0136$ btu/s/in.^{2}/°F, or 40 kW/m^2 · K) were also plotted. From the relative location between the calibration curve and experimental data the interface heat transfer coefficient was determined. The data shown in Fig. 6.8 were obtained from nonisothermal ring tests with 304 stainless steel. The process conditions used for the ring tests are shown in Table 6.1.

From the calibration curves and experimental data displayed in Fig. 6.9, an interface heat transfer coefficient of 0.0068 btu/s/in.^{2}/°F (20 kW/m^2 · K) was obtained.

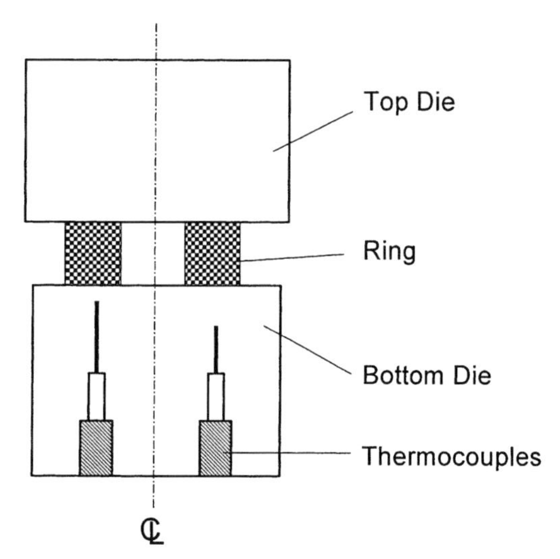

Fig. 6.8 Setup used in the ring test for the measurement of interface heat transfer coefficient. [Burte et al., 1989]

The following conclusions were drawn from these experiments conducted in references [Semiatin et al., 1987] and [Burte et al., 1989]:

- The interface heat transfer coefficient increases with forging pressures. When the workpiece is free resting on a die, the heat transfer coefficient is an order of magnitude smaller than during forging, when there is pressure at the die/workpiece interface.
- The value of the interface heat transfer coefficient is unchanged above a certain threshold pressure. This threshold value is approximately 2 ksi (14 MPa) in this test.
- The value of the interface heat transfer coefficient under deformation conditions is about 0.0068 btu/s/in.^{2}/°F (20 kW/m^2 · K) for all the combinations of workpiece/die material pairs used.
- The value of the interface heat transfer coefficient under free resting condition is about 0.00034 btu/s/in.^{2}/°F (1 kW/m^2 · K).

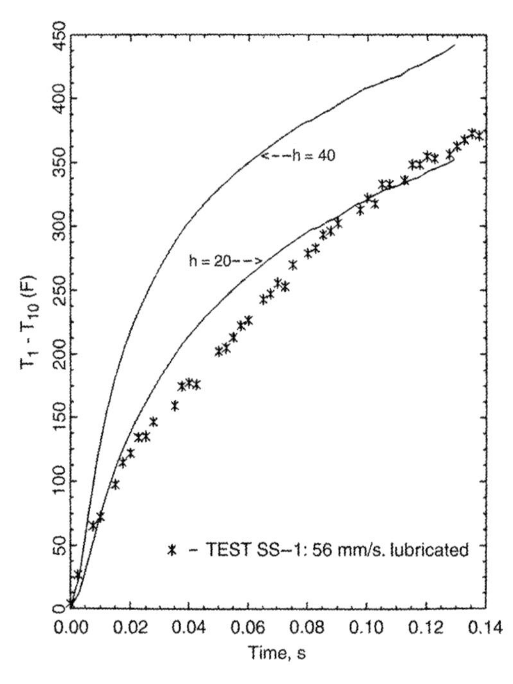

Fig. 6.9 The increase of the bottom die temperature (the instantaneous die temperature T_1 minus the initial die temperature T_{10}) versus time obtained from experiments and FEM generated calibration curves (h = 0.0068 btu/s/in.^{2}/°F, or 20 kW/m^2 · K, and h = 0.0136 btu/s/in.^{2}/°F, or 40 kW/m^2 · K) for the nonisothermal ring tests of 304 stainless steel. Lubricant, Deltaforge; ram speed, 2.2 in./s (56 mm/s); initial die temperature, T_{10} = 600 °F (316 °C); initial ring temperature, T_{20} = 2000 °F (1093 °C)

6.6 Influence of Press Speed and Contact Time on Heat Transfer

The heat transfer and hence the temperature history of the workpiece is also influenced by the forging equipment. Hydraulic press has lower speed and lower strain rate during deformation. Mechanical press and screw press have higher speed and higher strain rate during deformation. Hydraulic press has a longer dwell time before the deformation, which results in longer free resting heat transfer between the workpiece and the dies. Experimental work on the determination and comparison of the characteristics of forging presses was carried out on rings made from selected steel, titanium, and aluminum alloys [Douglas et al., 1971]. Finite element modeling was carried out to simulate the experiments and perform quantitative comparison of press speed and contact time on heat transfer in nonisothermal ring compression tests [Im et al., 1988a] and [Im et al., 1988b].

Ti-6242 and Al6061 ring compression tests were simulated using FEM package ALPID (a parent version of DEFORM(TM)). The process conditions used in the computer simulation such as the dimensions of the rings, the ram speeds of the forging equipment, and the reduction in height of the rings were all identical in the experiments [Douglas et al., 1971]. The shear friction factor used was the factor measured from the experimental ring tests. The interface heat transfer coefficient used was based on the experimental work [Semiatin et al., 1987]. Table 6.2 gives the conditions used in the finite element modeling. The hydraulic press used had a capacity of 700 metric tons. The ram velocity was assumed constant during deformation. Mechanical press used was a high-speed Erie press with scotch yoke design, rated at 500 metric tons at 0.25 in. (6.4 mm) above the bottom dead center. It had a stroke of 10 in. (250 mm) and a nominal speed of 90 strokes/min. As for the screw press, a Weingarten PSS 255 with a nominal rating of 400 metric tons, 2250 mkg (22 × 10^3 J) energy, was used. The starting speed of the mechanical press and screw press is shown in Table 6.2.

Table 6.1 Process conditions for nonisothermal 304 stainless steel ring tests

Lubricant	Deltaforge
Initial ring temperature, °F (°C)	2000 (1093)
Initial die temperature, °F (°C)	600 (316)
Ram speed, in./s (mm/s)	2.2 (56)

The other conditions used in the finite element modeling are:

Ring dimensions (OD:ID:height), in (mm)	3:1.5:0.5 (76:38:12.5) for Ti-6242 3:1.5:0.25 (76:38:6.4) for Ti-6242 6:3:2 (152:76:50.8) for Al6061 6:3:1 (152:76:25.4) for Al6061 6:3:0.5 (152:76:12.5) for Al6061
Die material	H-13 hot working tool steel
Billet temperature, °F (°C)	1750 °F (955 °C) for Ti-6242 800 °F (425 °C) for Al6061
Die temperature, °F (°C)	300 °F (150 °C) for both Ti-6242 and Al6061 ring tests

OD, outside diameter; ID, inside diameter

The contact time during deformation obtained from simulation is shown in Table 6.2. It is seen from Table 6.2 that the contact time during deformation is an order of magnitude longer in hydraulic press than in mechanical press and screw press. The temperature distribution obtained from the finite element modeling for the Ti-6242 3:1.5:0.5 in. (76:38:12.5 mm) ring compression in hydraulic press is presented in Fig. 6.10. The temperature distribution at the start of the deformation is shown in Fig. 6.10(a) where a heat loss to the die on the bottom surface of the ring was observed. The temperature loss was due to the dwell time before the deformation started for

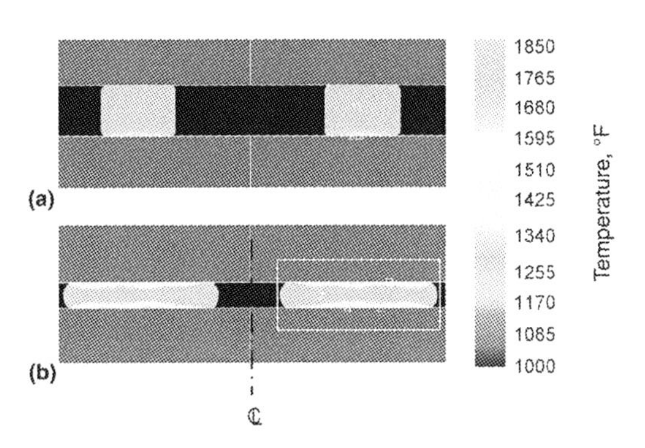

Fig. 6.10 The temperature distribution at (a) the beginning and (b) the end of a Ti-6242 3:1.5:0.5 in. (76:38:12.5 mm) ring test in a hydraulic press. The section inside the rectangle is used in Fig. 6.11.

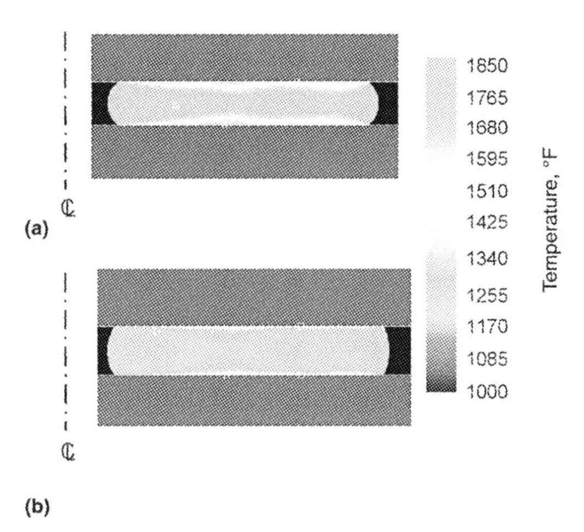

Fig. 6.11 The temperature distribution at the end of a Ti-6242 3:1.5:0.5 in. (76:38:12.5 mm) ring test in (a) a hydraulic press and (b) a mechanical press

Table 6.2 Process conditions used in ring compression of Ti and Al alloys

Press	Reduction, %	Shear friction (m)	Contact time during loading, s	Ram velocity in./s	Ram velocity mm/s
3:1.5:0.5 in. (76:38:12.5 mm) Ti-6242 ring					
Hydraulic	50	0.28	0.33	0.78	19.8
Mechanical	50	0.42	0.044	16	405
Screw	47	0.44	0.024	22	560
3:1.5:0.25 in. (76:38:6.4 mm) Ti-6242 ring					
Mechanical	30.8	0.42	0.033	15	380
Screw	37.6	0.2	0.019	22	560
6:3:2 in. (152:76:50.8 mm) Al6061 ring					
Hydraulic	51.2	0.65	0.83	1.23	31.2
Mechanical	51	0.53	0.079	26	660
Screw	34.9	0.49	0.051	22	560
6:3:1 in. (152:76:25.4 mm) Al6061 ring					
Hydraulic	51	0.42	0.53	1	25
Mechanical	49.8	0.31	0.047	19	480
Screw	47	0.35	0.031	22	560
6:3:0.5 in. (152:76:12.5 mm) Al6061 ring					
Mechanical	45.7	0.4	0.038	16	405
Screw	45.6	0.34	0.023	22	560

The interface heat transfer (h) was 0.0068 btu/s/in.^{2}/°F (20 kW/m^2 · K).

hydraulic press. Figures 6.11(a) and (b) show the temperature distribution at the end of the Ti-6242 3:1.5:0.5 in. (76:38:12.5 mm) ring test in the hydraulic press and the mechanical press. The hydraulic press has more die chilling due to longer contact time of the workpiece to the dies before and during deformation. The ring compressed in the mechanical press shows a lot of heat building up during deformation. There are zones inside the ring having temperatures above 1850 °F (1010 °C), which is around 25 °F (14 °C) higher than the beta transus of Ti-6242. However, the temperature gradient is less in the ring forged in the mechanical press and the deformation is more uniform and the bulge is less pronounced in this ring. The forgers may make use of the difference in press speed and contact time on heat transfer for different forging applications. For alpha-beta titanium forging, the temperature increase to above beta transus is not desired. Therefore, hydraulic press forging is beneficial. For beta titanium forging, steel forging, and selected superalloy forging, temperature increase during forging is not critical and high-speed forging machines such as mechanical press, screw press, and hammer can all be used.

REFERENCES

[Altan et al., 1970]: Altan, T., Gerds, A.F., "Temperature Effects in Closed-Die Forging," ASM Technical Report No. C70-30.1, Oct 1970.

[Altan et al., 1973]: Altan, T., et al., "Forging Equipment, Materials and Practices," *MCIC Handbook HB-03,* Battelle, Columbus, OH, 1973.

[Burte et al., 1989]: Burte, P., Semiatin, S.L., Altan, T., "Measurement and Analysis of Heat Transfer and Friction During Hot Forging," Report No. ERC/NSM-B-89-20, ERC for Net Shape Manufacturing, Ohio State University, June 1989.

[Douglas et al., 1971]: Douglas, J.R., Altan, T., "Characteristics of Forging Presses: Determination and Comparison," Proceedings of the 13th M.T.D.R. Conference, Birmingham, England, September, 1971.

[Farren et al., 1925]: Farren, W.S., Taylor, G.I., "The Heat Developed During Plastic Extrusion of Metals," *Proc. R. Soc., Ser. A,* Vol 107, 1925, p 422–451.

[Im et al., 1988a]: Im, Y.T., Shen, G., "A Study of the Influence of Press Speed, Contact Time and Heat Transfer in Nonisothermal Upset Forging of Ti and Al Rings," *16th North America Manufacturing Research Conference Proceedings,* 1988, p 91–98.

[Im et al., 1988b]: Im, Y.T., Vardan, O., Shen, G., Altan, T., "Investigation of Non-Isothermal Forging Using Ring and Spike Tests," *Ann. CIRP,* Vol 37/1, 1988, p 225–230.

[Lahoti et al., 1975]: Lahoti, G.D., Altan, T., "Prediction of Temperature Distribution in Axisymmetric Compression and Torsion," *ASME, J. Eng. Mater. Technol.,* Vol 97, 1975.

[Lahoti et al., 1978]: Lahoti, G.D., Altan, T., "Prediction of Metal Flow and Temperatures in Axisymmetric Deformation Process," *Advances in Deformation Processing,* J.J. Burke and V. Weiss, Ed., Plenum Publishing, 1978, p 113–120.

[Semiatin et al., 1987]: Semiatin, S.L., Collings, E.W., Wood, V.E., Altan, T., "Determination of the Interface Heat Transfer Coefficient for Non-Isothermal Bulk-Forming Processes," *Trans. ASME, J. Eng. Ind.,* Vol 109A, Aug 1987, p 49–57.

[SFTC, 2002]: Scientific Forming Technologies Corporation, *DEFORM 7.2 User Manual,* Columbus, OH, 2002.

[Vigor et al., 1961]: Vigor, C.W., Hornaday, J.W., "A Thermocouple for Measurement of Temperature Transients in Forging Dies," in *Temperature, Its Measurement and Control,* Vol 3, Part 2, Reinhold, 1961, p 625.

CHAPTER 7

Friction and Lubrication

Mark Gariety
Gracious Ngaile

7.1 Introduction

In forging, the flow of metal is caused by the pressure transmitted from the dies to the deforming workpiece. Therefore, the frictional conditions at the die/workpiece interface greatly influence metal flow, formation of surface and internal defects, stresses acting on the dies, and load and energy requirements [Altan et al., 1983]. Figure 7.1 illustrates this fundamental phenomenon as it applies to the upsetting of a cylindrical workpiece. As Fig. 7.1(a) shows, under frictionless conditions, the workpiece deforms uniformly and the resulting normal stress, σ_n, is constant across the diameter. However, Fig. 7.1(b) shows that under actual conditions, where some level of frictional stress, τ, is present, the deformation of the workpiece is not uniform (i.e., barreling). As a result, the normal stress, σ_n, increases from the outer diameter to the center of the workpiece and the total upsetting force is greater than for the frictionless conditions.

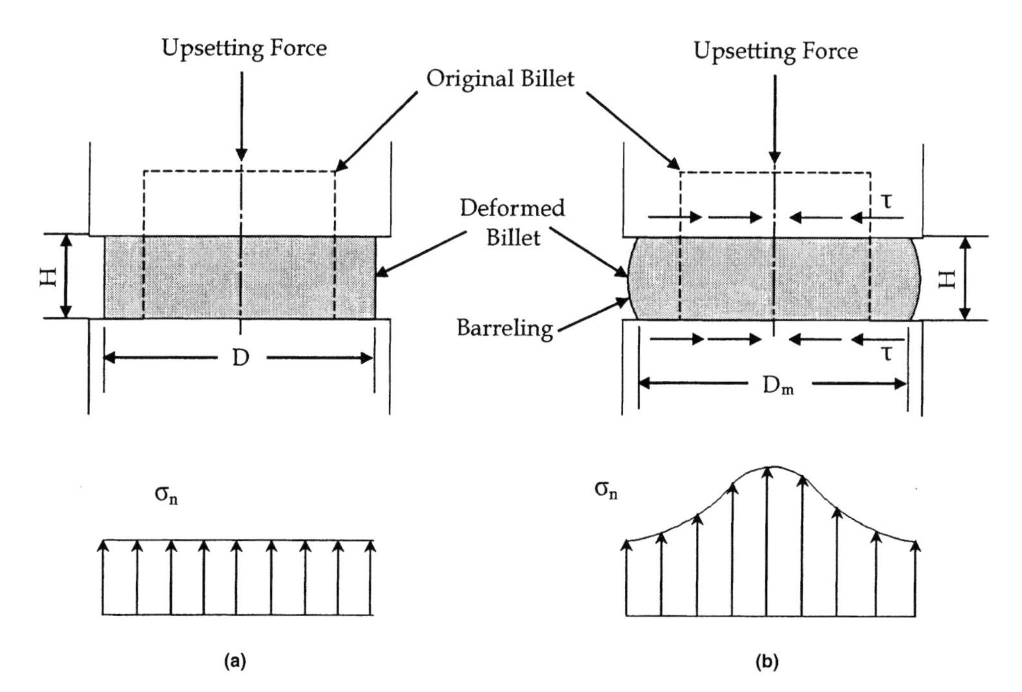

Fig. 7.1 Upsetting of cylindrical workpiece. (a) Frictionless. (b) With friction

7.2 Lubrication Mechanisms in Metal Forming

There are four basic types of lubrication that govern the frictional conditions in metal forming [Altan, 1970] [Schey, 1983]. The Stribeck curve shown in Fig. 7.2 illustrates the onset of these various types of lubrication as a function of the combination of lubricant viscosity, η, sliding velocity, v, and normal pressure, p.

Under dry conditions, no lubricant is present at the interface and only the oxide layers present on the die and workpiece materials may act as a "separating" layer. In this case, friction is high, and such a situation is desirable in only a few selected forming operations, such as hot rolling of plates and slabs and nonlubricated extrusion of aluminum alloys.

Boundary lubrication is governed by thin films (typically organic) physically adsorbed or chemically adhered to the metal surface. These films provide a barrier under conditions of large metal-to-metal contact where the properties of the bulk lubricant have no effect. As is the case with dry conditions, friction is high.

Full-film lubrication exists when a thick layer of solid lubricant/dry coating is present between the dies and the workpiece. In this case, the friction conditions are governed by the shear strength of the lubricant film.

Hydrodynamic conditions exist when a thick layer of liquid lubricant is present between the dies and the workpiece. In this case, the friction conditions are governed by the viscosity of the lubricant and by the relative velocity between the die and the workpiece. The viscosities of most lubricants decrease rapidly with increasing temperature. Consequently, in most practical high-speed forming operations, such as strip rolling and wiredrawing, the hydrodynamic conditions exist only within a certain regime of velocities, where the interface temperatures are relatively low [Altan, 1970]. As the Stribeck curve indicates, friction is relatively low.

Mixed-layer lubrication is the most widely encountered situation in metal forming. Because of the high pressures and low sliding velocities encountered in most metal forming operations, hydrodynamic conditions cannot be maintained. In this case, the peaks of the metal surface experience boundary lubrication conditions and the valleys of the metal surface become filled with the liquid lubricant. Thus, many liquid lubricants contain organics that will adsorb to or chemically react with the metal surface in order to help provide a barrier against metal-to-metal

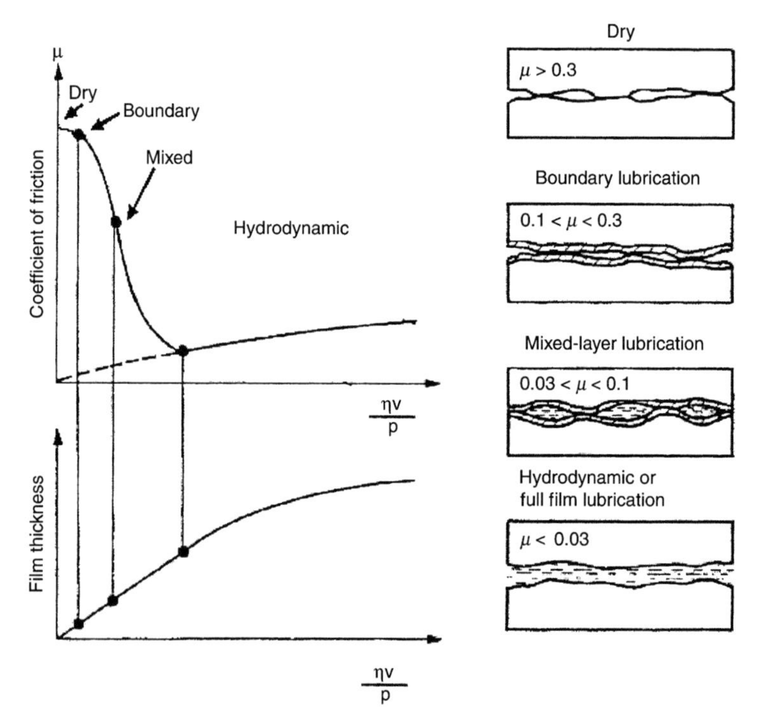

Fig. 7.2 Stribeck curve showing onset of various lubrication mechanisms. [Schey, 1983]

contact. If there is enough lubricant present, the lubricant in the valleys of the metal surface can act as a hydrostatic medium. In this case, both the contacting peaks of the metal surface and the hydrostatic pockets support the normal pressure. Thus, friction is moderate.

7.3 Friction Laws and Their Validity in Forging

In order to evaluate the performances (lubricity) of various lubricants under various material and process conditions and to be able to predict forming pressures, it is necessary to express the interface friction quantitatively. There are two laws that can be utilized for this purpose. Both of these laws quantify interface friction by lumping all of the interface phenomena into one nondimensional coefficient or factor. The Coulomb friction law uses a coefficient of friction, μ, to quantify the interface friction. Equation 7.1 shows that μ is simply the ratio of the frictional shear stress, τ, and the normal stress (pressure), σ_n.

$$\tau = \sigma_n \mu \quad \text{(Eq 7.1)}$$

As is illustrated in Fig. 7.3, the linear relationship defined by Coulomb's law is not valid at all normal stress (pressure) levels because the shear stress, τ, cannot exceed the shear strength, k, of the material. Thus, a second law named the interface shear friction law has been developed [Schey, 1983].

The interface shear friction law uses a friction factor, f, or a shear factor, m, to quantify the interface friction. Equation 7.2 shows that the frictional shear stress, τ, is dependent on the flow stress of the deforming material, $\bar{\sigma}$, and the friction factor, f, or the shear factor, m. Thus, for a frictionless condition, m = 0, and for a sticking friction condition, m = 1. Sticking friction is the case where sliding at the interface is preempted by shearing of the bulk material [Schey, 1983].

$$\tau = f\bar{\sigma} = \frac{m}{\sqrt{3}}\bar{\sigma} = mk \quad \text{(Eq 7.2)}$$

The shear factor, m, in Eq 7.2 is not to be confused with the exponent, m, in the simple exponential law, $\bar{\sigma} = C(\dot{\varepsilon})^m$, used to express the strain-rate dependency of flow stress, discussed in Chapter 4. Recent studies in forming mechanics indicate that Eq 7.2 adequately represents the frictional shear stress in forging, where the normal stresses are high, and offers advantages in evaluating friction and in performing stress and load calculations [Altan et al., 1983] [Schey, 1983] [Bhushan, 2001].

For various forming conditions, the shear factor values, m, vary as follows [Altan et al., 1983]:

- m = 0.05 to 0.15 in cold forming of steels, aluminum alloys, and copper, using conventional phosphate-soap lubricants or oils.
- m = 0.2 to 0.4 in hot forming of steels, copper, and aluminum alloys with graphite-based (graphite-water or graphite-oil) lubricants.
- m = 0.1 to 0.3 in hot forming of titanium and high-temperature alloys with glass lubricants.
- m = 0.7 to 1.0 when no lubricant is used, e.g., in hot rolling of plates or slabs and in nonlubricated extrusion of aluminum alloys.

7.4 Parameters Influencing Friction and Lubrication

There are numerous parameters that influence the friction and lubrication conditions present at the die/workpiece interface in a forging operation. These parameters can be outlined as follows [Schey, 1983] [Bay, 1995] [Ngaile et al., 1999] [Saiki et al., 1999] [Lenard, 2000]:

Tool/Workpiece Parameters

- The properties of the workpiece *material* (i.e., flow stress) influence how the work-

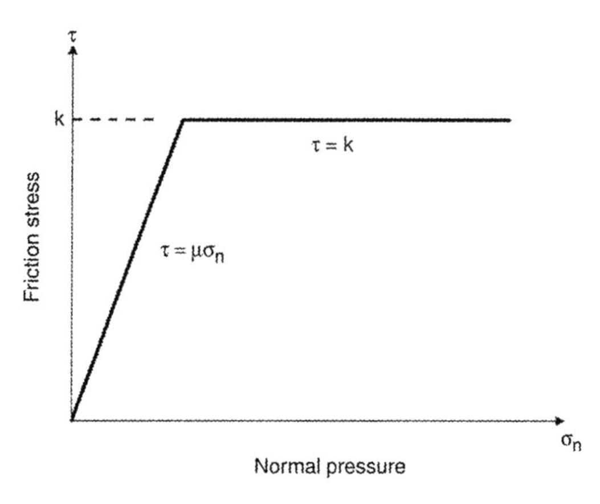

Fig. 7.3 Friction at high normal pressures. Courtesy of N. Bay

piece deforms and thus how the lubricant must flow. The properties of both the die and workpiece material influence how the lubricant reacts with the surfaces (i.e., the boundary lubrication).

- The *geometry* of the die influences how the workpiece deforms and thus how the lubricant must flow.
- The *surface finish* of both the tool and the workpiece influence how hydrostatic lubricant pockets (i.e., mixed-layer lubrication) are formed.
- In hot forging, the *scale* present on the workpiece surface influences the interface conditions. If the scale is soft and ductile, it may act as a lubricant. If it is hard and brittle, it may cause an abrasive wear mechanism.

Lubricant Parameters

- The *composition* of the lubricant influences the viscosity (i.e., hydrodynamic lubrication) and how the viscosity changes when subjected to extreme heat and pressure. The lubricant composition also influences how the lubricant reacts with both the die and the workpiece (i.e., boundary lubrication).
- The *viscosity* of the lubricant influences how it flows as the workpiece is deformed (i.e., hydrodynamic lubrication).
- The *amount* of lubricant influences how the lubricant spreads as the workpiece is deformed and how hydrostatic lubricant pockets (i.e., mixed-layer lubrication) are formed.

Process Parameters

- The *pressure* exerted by the die on the workpiece influences the viscosity of the lubricant (i.e., hydrodynamic lubrication) and the deformation of the surface asperities, which affects the formation of hydrostatic lubricant pockets (i.e., mixed-layer lubrication).
- The *sliding velocity* at which the die moves relative to the workpiece influences the heat generation at the die/workpiece interface. It also influences the onset of hydrodynamic lubrication.
- The *sliding length* at which the die moves over the workpiece influences the heat generation at the die/workpiece interface, the extent to which the lubricant must spread out, and the extent to which the lubricant will break down.
- The amount of *surface expansion* generated during the deformation process influences the extent to which the lubricant must spread out.
- The *heat* generated due to the deformation process and the machine operation influences the material properties (i.e., flow stress) of both the die and the workpiece and the viscosity of the lubricant.

7.5 Characteristics of Lubricants Used

In metal forming, friction is controlled by use of appropriate lubricants for given applications. The lubricant is expected to have certain characteristics and to perform some, if not most, of the following significant functions [Schey, 1983]:

- Reduce the sliding friction between the dies and the workpiece; this is achieved by using a lubricant of high lubricity
- Act as a parting agent and prevent sticking and galling of the workpiece to the dies
- Possess good insulating properties, especially in hot forming, so as to reduce heat losses from the workpiece to the dies
- Possess inertness to prevent or minimize reactions that will degrade the dies and the workpiece materials at the forming temperatures used
- Be nonabrasive so as to reduce erosion of the die surface and die wear
- Be free of polluting and poisonous components and not produce unpleasant or dangerous gases
- Be easily applicable to and removable from dies and workpiece
- Be commercially available at reasonable cost

7.6 Lubrication Systems for Cold Forging

The choice of which lubricant to use for a cold forging process depends on the severity of the operation (i.e., process parameters such as interface pressure and surface expansion) and the parameters associated with the billet itself (i.e., hardness). For example, upsetting to small diameter-to-height ratios does not require the same lubricant performance (lubricity) as backward extrusion processes. In addition, the lubrication systems used for steel may be very different from those used for aluminum or titanium [Schey, 1983]. Figure 7.4 shows an example of this lubricant selection process for cold forging of aluminum alloys [Bay, 1994].

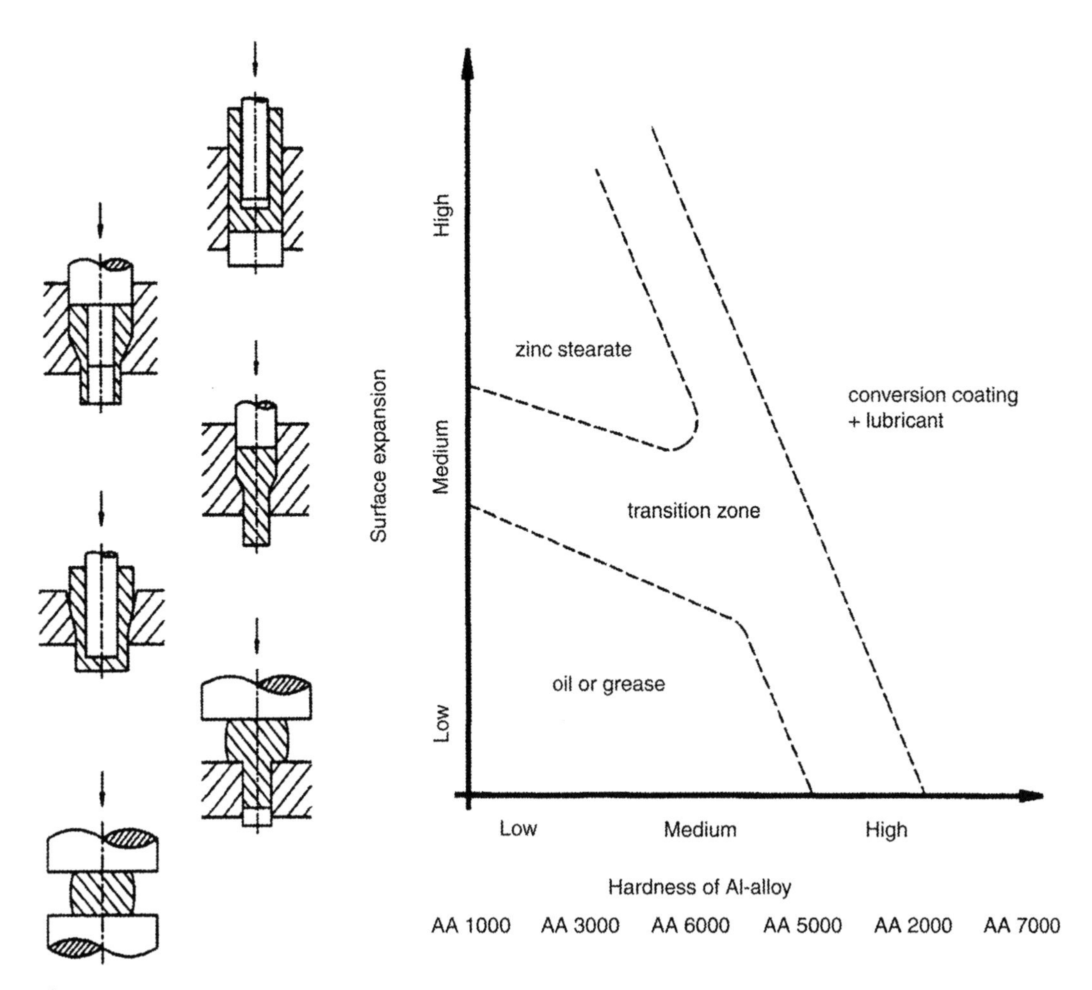

Fig. 7.4 Lubricant selection based on deformation severity. [Bay, 1994]

7.6.1 Ferrous Materials

Carbon Steels. Because of the severe deformation conditions typical of many cold forging operations, the most widely used lubrication system in the cold forging of carbon steels is a zinc phosphate coating and soaping system (Fig. 7.5). However, simple forging processes such as light upsetting may be completed without lubrication or with a simple mineral oil. In the zinc phosphate coating lubrication system, the basic workpiece material is first cleaned to remove grease and scales and then dipped in a zinc phosphate solution. Through chemical reaction, a zinc phosphate layer on the order of 5 to 20 μm is formed. The coated billets are then dipped in an alkaline solution (usually sodium or calcium soap), resulting in firmly adhered and adsorbed layers of alkaline soap, zinc soap, and phosphate crystals with very low shear strength. The typical procedure for applying this lubrication system to a billet is described in Table 7.1 [Altan et al., 1983] [Bay, 1994] [Manji, 1994] [ICFG, 1996].

Process parameters such as bath age and temperature, process time for phosphating and lubrication, and type of activators influence the properties of the zinc phosphate coating and soaping lubrication system. In general, thicker coating or soaping layers are obtained by allow-

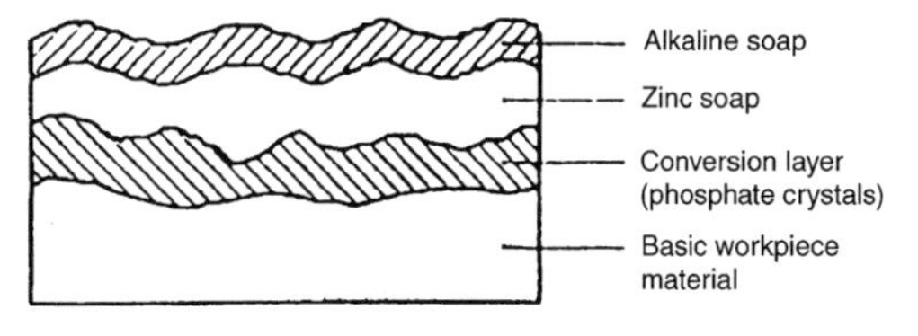

Fig. 7.5 Zinc phosphate coating and soaping lubrication system. [Bay, 1994]

Table 7.1. Treatment sequence for zinc phosphate coating of steel billets for cold forging

Operation	Bath temperature		Process time, min
	°F	°C	
Cleaning			
Degreasing in alkaline solution	140–205	60–95	5–15
Rinsing in cold water	N/A	N/A	N/A
Removing scale, usually by pickling but occasionally by shot blasting	104–160	40–70	1–5
Rinsing in cold water and neutralizing (if pickling used)	N/A	N/A	N/A
Dipping in warm water with activators	N/A	N/A	N/A
Phosphating			
Dipping in zinc phosphate solution	130–205	55–95	5–10
Rinsing in cold water and neutralizng	N/A	N/A	N/A
Lubrication			
Lubricating with sodium soap	160–175	70–80	0.5–5
Drying	N/A	N/A	N/A

Source: [Altan et al., 1983] [Bay, 1994] [Manji, 1994] [ICFG, 1996]

ing the billets to remain in the zinc phosphate solution or the sodium soap solution for longer periods of time. However, care should be taken because excessive amounts of lubricant could cause dimensional tolerance errors or unsatisfactory surface finish [Bay, 1994] [Lazzarotto et al., 1999].

Despite its success as a cold forging lubrication system, the zinc phosphate coating has several disadvantages. Hence, today, lubricant manufacturers are attempting to design replacements for this lubrication system [Ngaile et al., 2002]. The disadvantages are summarized as follows [Schmoeckel et al., 1997]:

Profitability

- The initial purchase, as well as the maintenance, of the zinc phosphate coating and soaping line is expensive.
- Removal of the zinc phosphate layer is difficult and thus expensive.

Productivity

- The zinc phosphate coating and soaping process is time consuming.

Energy Usage

- It is necessary to heat multiple baths to temperatures between 105 and 205 °F (40 and 95 °C).

Worker Environment

- Dust accumulates as a result of surface enlargement during forging. This dust is a health risk to the workers in the facility.
- The baths are a source of toxic chemicals and fumes, which lead to unhealthy working conditions.

Waste Removal

- The baths contain acids, ion of the basic metal, the alloying constituents, and phosphates. The wastewater contains organic compounds and emulsifying agents. After phosphating, the baths become polluted with heavy metals like lead and cadmium. The wastewater treatment and the baths result in solids, which contain metals, oils, and other pollutants. Most of this waste cannot be reused and thus becomes hazardous waste.

Mechanical Properties of the Billets

- Zinc phosphate can increase corrosion and diffuse into the workpiece material during heat treatment. This is a common cause of surface embrittlement.

Even though zinc phosphate coating based lubrication systems are the most widely used, cold forging of carbon steels involves a wide variety of processes and thus a wide variety of lubrication systems. These processes and lubrication systems are summarized in Table 7.2.

Table 7.2 Lubrication systems for cold forging of steel

Process	Deformation	Lubricant
Upsetting	Light	None
		Mi + EP + FA
	Severe	Ph + SP
Ironing and open-die extrusion	Light	Ph + Mi + EP + FA
	Severe	Ph + SP
Extrusion	Light	Ph + Mi + EP + FA
	Severe	Ph + SP
		Ph + MoS_2
		Ph + MoS_2 + SP

Mi, mineral oil; SP, soap; EP, extreme pressure additive; Ph, phosphate coating; FA, fatty additives. Source: [Bay, 1994]

Extreme-pressure additives include chlorine, sulfur, and phosphorus. These additives react with carbon steel surfaces to produce excellent barriers (boundary lubrication) against metal-to-metal contact. In severe extrusion operations, the bulk surface temperature may exceed the melting point of the soap. In these cases, molybdenum disulfide (MoS_2) is used in place of the soap. In the most severe operations, the entire phosphate coating is replaced by a thin copper coating [Schey, 1983].

Stainless Steels. For cold forging of stainless steels and other steels containing more than 5% Cr, an oxalate coating is used in place of the phosphate coating. This is done because it is difficult to phosphate these materials [Schey, 1983].

7.6.2 Nonferrous Materials

Aluminum. Lubrication in the cold forging of aluminum is especially important because of the high adhesion between the aluminum and the die material. The lubrication systems used for the cold forging of aluminum are given in Fig. 7.4. In general, one of three conversion coatings are used with aluminum; namely, zinc phosphate, calcium aluminate, or aluminum fluoride. The lubricants used with these conversion coatings include soaps and molybdenum disulfide. The general treatment sequence for these conversion coatings is the same as described in Table 7.1; however, the type of conversion coating determines the bath temperature and process time as shown in Table 7.3 [Bay, 1994] [ICFG, 1996].

Copper. Lubrication systems for cold forging of copper are summarized in Table 7.4. It should be noted that many EP additives are useless when used in a mineral oil for lubrication of a copper alloy because they do not react with the copper surface to create a barrier (boundary lubrication) to withstand metal-to-metal contact as they do with carbon steels. In addition, sulfur additives stain the copper [Schey, 1983] [Gariety et al., 2002].

Titanium. Cold forging of titanium has very limited application. However, in those limited applications, the lubrication systems can be summarized as shown in Table 7.4 [Schey, 1983].

Table 7.3 Alternative conversion coatings for aluminum billets for cold forging

Phosphating operation	Bath temperature °F	Bath temperature °C	Process time, min
Dipping in zinc phosphate solution	130–150	55–65	5–10
Dipping in calcium aluminate solution	140–175	60–80	5–15
Dipping in aluminum fluoride solution	185–195	85–90	5–10

Source: [Bay, 1994] [ICFG, 1996]

7.7 Lubrication Systems for Warm and Hot Forging

The main difference in lubrication conditions between cold and hot or warm forging is the temperature range in which the lubricant must function. Excessive die temperatures combined with high die stresses as the result of heat transfer from the billet to the dies and deformation stresses cause increased wear, plastic deformation, and heat checking in the dies [Saiki, 1997]. Thus, in order to increase tool life and part quality, a good lubrication system should be capable of minimizing both the heat transfer to the dies and the shear stresses at the tool/workpiece interface.

Unlike cold forging, the application of the lubricant in hot forging is constrained by the total forging cycle time, which is on the order of a few seconds. Figure 7.6 illustrates a typical die lubrication process [Doege et al., 1996].

It is therefore essential to apply the appropriate lubrication within the shortest amount of time. Thus, factors such as spray pressure, lubricant flow rate, spray angle, spray distance, and spray pattern are of great importance for a successful warm or hot forging operation.

Table 7.4 Lubrication systems for the cold forging of copper and titanium

Process	Deformation	Lubricant
Copper		
Upsetting	Light	Emulsion
	Severe	Mineral oil
Extrusion	Light	Emulsion
		Mineral oil
	Severe	Soaps
Titanium		
Upsetting	Light	Emulsion or mineral oil
	Severe	Copper coating + soap
		Fluoride-phosphate coating + soap
Extrusion	Light	Copper coating + soap
		Fluoride-phosphate coating + soap
	Severe	Copper coating + graphite
		Fluoride-phosphate coating + graphite

Source: [Schey, 1983]

The temperatures used in warm and hot forging do not readily facilitate the use of organic-based lubrication systems (i.e., mineral oils) or soaps. Organic-based lubricants will burn and soap-based lubricants will melt at these temperatures. Thus, the choices of lubrication systems are very limited [Schey, 1983].

The four most common lubrication systems are MoS_2, graphite, synthetics, and glass; however, MoS_2 is only useful at warm forging temperatures (up to 750 °F, or 400 °C). MoS_2 and graphite are solid lubricants. Because of their layered molecular structure, they demonstrate low frictional stresses. They are usually mixed into an aqueous solution and sprayed onto the dies. This serves two purposes. First, the aqueous solution evaporates upon contact with the dies, thus acting to cool the dies and protect them against increased wear due to thermal softening. Second, an MoS_2 or graphite layer remains on the dies following evaporation of the aqueous solution. This layer not only acts as a lubricant, but also as insulation against excessive die heating [Schey, 1983] [Manji, 1994]. Today, concerns over the environmental friendliness of graphite as well as the accumulation of graphite within the dies have led to the development of water-based synthetic lubricants [Manji, 1994].

If the lubricants were applied to the workpiece instead of the dies, they would be destroyed during the heating of the workpiece. There are some exceptions to this. In the hot forging of aluminum and magnesium, lower forging temperatures permit the use of a graphite/mineral oil combination applied directly to the workpiece. In steels, it is possible to use graphite-based coatings that are applied very rapidly to the billets prior to induction heating. Glass can also be used as a lubricant and as a protective coating in hot forging of titanium, nickel, and tungsten alloys for aerospace applications. When glass is used, it is applied to the workpiece from an aqueous slurry or to the preheated workpiece from a powder. The glass subsequently melts into a highly viscous liquid [Schey, 1983] [Manji, 1994]. Because the dies are at much lower temperatures than the workpiece, a sharp temperature gradient is created through the glass film, which produces a sharp viscosity gradient (molten glass near the workpiece and solid glass near the dies) vital for lubrication.

The lubrication systems used for the warm and hot forging of steels are summarized in Table 7.5 [Schey, 1983]. The lubrication systems used for the warm and hot forging of aluminum, magnesium, copper, titanium, nickel, and tungsten are summarized in Table 7.6 [Schey, 1983].

Table 7.5 Lubrication systems for the warm and hot forging of steels

Material	Process	Deformation	Lubricant
Carbon steel	Warm forging	Severe	MoS_2 in aqueous solution
			Graphite in aqueous solution
	Hot forging	Severe	Graphite in aqueous solution
Stainless steel	Hot forging	Severe	Glass in aqueous slurry or powder

Source: [Schey, 1983]

7.8 Methods for Evaluation of Lubricants

The cost of lubricants is small compared to the costs of items such as raw material, equipment, and labor. As a result, the economic incentive to evaluate or change lubricants is not always very significant. However, lubricant breakdown resulting in excessive die wear or die failure is one of the largest factors contributing to reduced production as a result of press downtime and part rejection. Therefore, it is essential to evaluate the lubricants in use and to compare

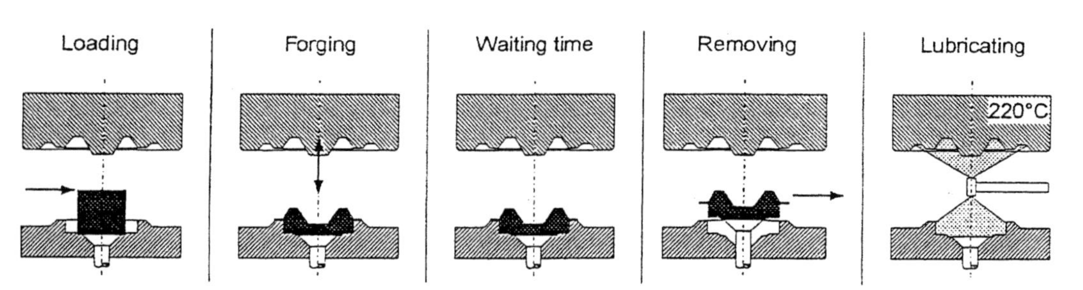

Fig. 7.6 Die lubrication process in warm and hot forging. [Doege et al., 1996]

them to alternative types of lubricants. Such an evaluation is necessary in order to utilize effectively the large investment required for installing a coating and lubrication line for cold forging [Shen et al., 1992].

There are many bench-type simulation tests designed to evaluate friction and lubrication in forging operations [Schey, 1983]. Here, however, only two of the most common tests are presented, i.e., the ring compression test and the

Table 7.6 Lubrication systems for the warm and hot forging of aluminum, magnesium, copper, titanium, nickel, and tungsten

Material	Process	Deformation	Lubricant
Aluminum	Warm and hot forging	Severe	None
			Graphite in mineral oil
Magnesium	Warm and hot forging	Severe	Graphite in mineral oil
Copper	Warm and hot forging	Severe	Graphite in aqueous solution
Titanium, nickel, and tungsten	Warm forging	Severe	MoS_2 compounds
			Graphite compounds
	Hot forging	Severe	Graphite compounds
		Most severe	Glass in aqueous slurry or powder

Source: [Schey, 1983]

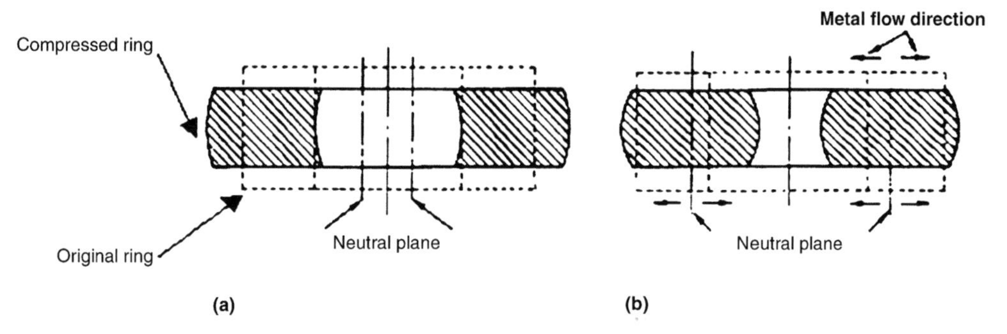

Fig. 7.7 Metal flow in ring compression test. (a) Low friction. (b) High friction

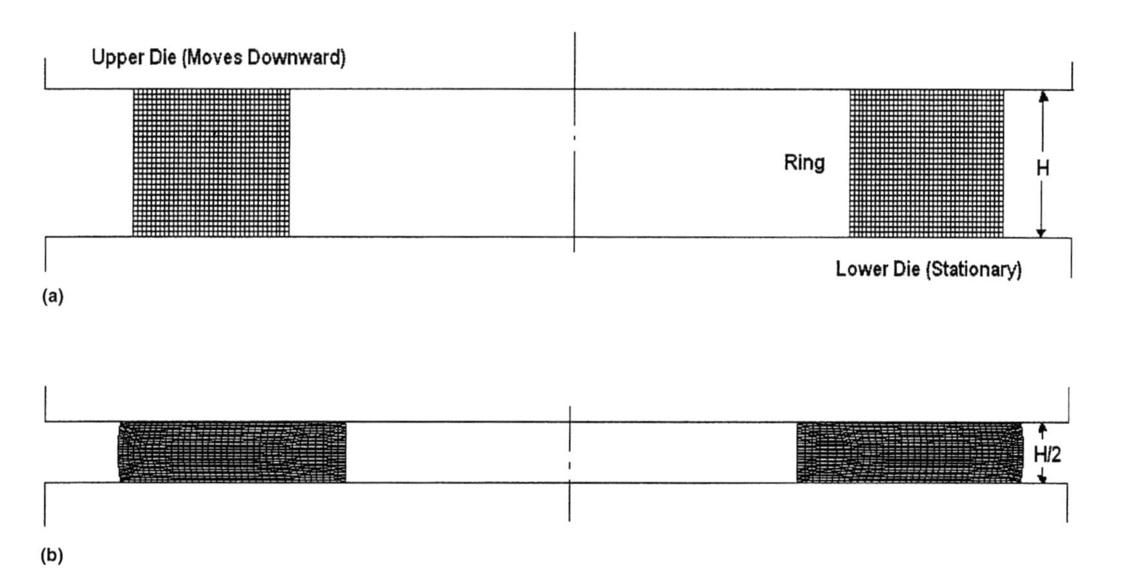

Fig. 7.8 Finite element model of ring compression test. (a) Initial ring. (b) Compressed ring (50% height reduction) (shear factor m = 0.1). (Gariety et al., 2003)

double cup backward extrusion test. The ring compression test best simulates forging applications with a moderate amount of deformation, where the surface expansion induced is on the order of only 100%, while the double cup backward extrusion test best simulates more severe forging applications, where the surface expansion and the interface pressure induced are over 500% and 290 ksi (2000 MPa), respectively.

In determining the friction factor, f, or the shear factor, m, for hot forming, in addition to lubrication effects, the effects of die chilling or heat transfer from the hot material to colder dies must be considered. Therefore, the lubrication tests used for determining friction factors must include both lubrication and die-chilling effects. Consequently, in hot forming, a good test must satisfy the following requirements [Altan et al., 1983]:

- The specimen and die temperatures must be approximately the same as those encountered in the actual hot forming operation.
- The contact time between specimen and tools under pressure must be approximately the same as in the forming operation of interest.
- The ratio of the new generated deformed surface area to the original surface area of the undeformed specimen (i.e., surface expansion) must be approximately the same as in the process investigated.
- The relative velocity between deforming metal and dies should have approximately the same magnitude and direction as in the forming process.

7.8.1 Ring Compression Test

Lubricity, as defined by the friction factor, f, or the shear factor, m, is commonly measured by using the ring test [Male et al., 1970] [Douglas et al., 1975]. In the ring test, a flat ring-shape specimen is compressed to a known reduction (Fig. 7.7). The change in internal and external diameters of the forged ring is very much de-

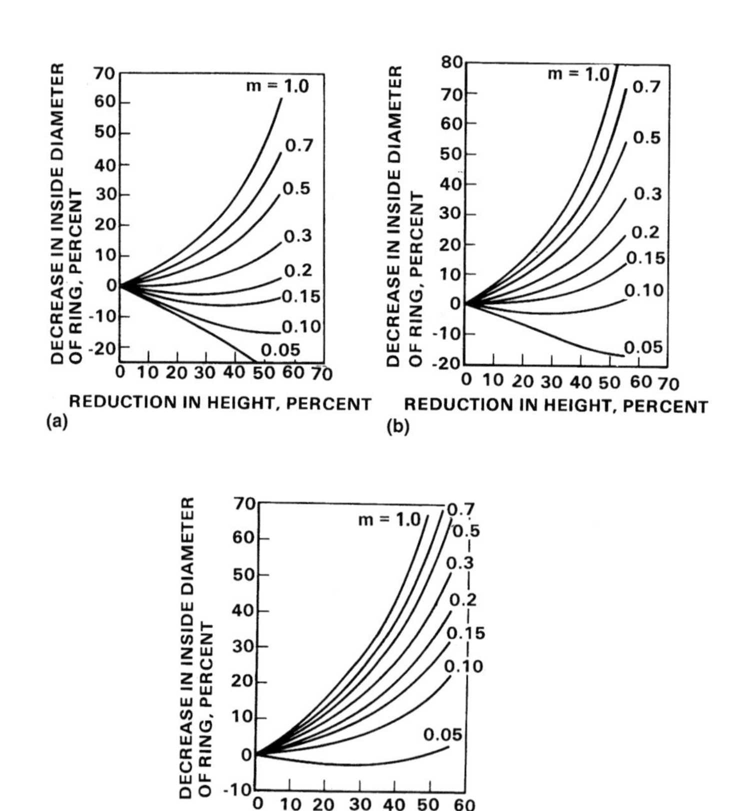

Fig. 7.9 Theoretical calibration curves for ring compression test having indicated OD: ID:thickness ratios. (a) 6:3:2 ratio. (b) 6:3:1 ratio. (c) 6:3:0.5 ratio. [Altan et al., 1983]

Table 7.7 Values of frictional shear factor, m, obtained from ring compression tests conducted in a hydraulic press

Material	Specimen/die temperatures °F	Specimen/die temperatures °C	Ring size OD:ID:h(a) in.	Ring size OD:ID:h(a) mm	Frictional shear factor (m)	Lubrication system
AISI 1018	200/200	95/95	1.75:1.13:0.5	44.5:28.7:12.7	0.040	Zinc phosphate coating + soap
					0.045	Metallic compounds + sulfur compounds(b)
					0.060	Mineral oil + EP additives
Copper	75/75	24/24	2:1:0.67	50.8:25.4:16.9	0.30	Emulsion
					0.27	Water-based synthetic

(a) OD, ring outside diameter; ID, ring inside diameter; h, ring height. (b) Environmentally friendly lubrication system developed to replace zinc phosphate coating based systems. Source: [Gariety et al., 2003] [Hannan et al., 2000]

pendent on the friction at the die/ring interface. If friction were equal to zero, the ring would deform in the same way as a solid disk, with each element flowing radially outward at a rate proportional to its distance from the center. With increasing deformation, the internal diameter of the ring is reduced if friction is high and is increased if friction is low. Thus, the change in the internal diameter represents a simple method for evaluating interface friction.

Simulation of Cold Forging Conditions. The ring test has an advantage when applied to the study of friction under cold forging conditions. In order to measure friction with this test, the force necessary to deform the ring and the flow stress of the specimen do not have to be known. Thus, evaluation of test results is greatly simplified. To obtain the magnitude of the friction factor, the internal diameter of the compressed ring must be compared with the values predicted by using various friction factors, f, or shear factors, m. Today, these values are most often predicted by the finite element method (FEM). Figure 7.8 shows an example of an FEM model used for this purpose. The results are plotted in the form of "theoretical calibration curves," as can be seen in Fig. 7.9, for rings having OD:ID:thickness ratios of 6:3:2, 6:3:1, and 6:3:0.5. The internal diameters used in this figure are the diameters at the internal bulge. Under cold forging conditions, these calibration curves may be considered as "universal" because changes in material properties (i.e., strain hardening) have little effect on the curves. In determining the value of the shear factor, m, for a given experimental condition, the measured dimensions (reduction in height and variation in internal diameter) are plotted on the appropriate calibration figure. From the position of that point with respect to theoretical curves given for various values of "m," the value of the shear factor, m, which existed in the experiment is obtained.

Some results obtained from ring compression tests conducted in a 160-ton hydraulic press with a ram velocity of 15 mm/s and a ring height reduction of 50% are shown in Table 7.7.

Simulation of Hot Forging Conditions. In contrast to the simulation of cold forging conditions, the simulation of hot forging conditions do not provide for a "universal" set of calibration curves. The friction calibration curves must be generated for the specific ring material under the specified ring and die temperatures and the ram speed conditions. Hence, knowledge of the flow stress of the material is required [Lee et al., 1972].

The results from some ring compression tests conducted under hot forging conditions have been compiled. The results from ring compression tests conducted for various materials in a 500 ton mechanical press with a nominal speed of 90 strokes/min and a total stroke of 10 in. (255 mm) are shown in Table 7.8.

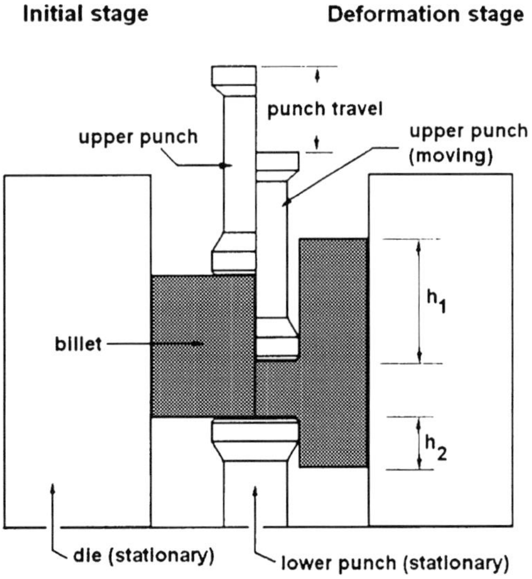

Fig. 7.10 Metal flow in double cup backward extrusion test

7.8.2 Double Cup Backward Extrusion Test

Lubricity, as defined by the friction factor, f, or the shear factor, m, is also measured by using the double cup backward extrusion test. As shown in Fig. 7.10, the test is a combination of the single cup forward and single cup backward extrusion processes. The ratio of the cup heights, H_1/H_2, is very dependent on the friction at the billet/die and billet/punch interfaces [Buschhausen et al., 1992] [Forcellese et al., 1994]. In par-

Table 7.8 Values of frictional shear factor, m, obtained from ring compression tests conducted in a mechanical press (die temperatures ≈ 300 °F, or 150 °C)

Material	Specimen temperature		Ring ratio OD:ID:t		Frictional shear factor (m)	Contact time, s	Lubrication system
	°F	°C	in.	mm			
6061 Al	800	425	6:3:0.5	150:75:13	0.40	0.038	(a)
			6:3:1	150:75:25	0.31	0.047	(a)
			6:3:2	150:75:51	0.53	0.079	(a)
Ti-7Al-4Mo	1750	955	3:1.5:0.25	75:38:6.5	0.42	0.033	(b)
			3:1.5:0.5	75:38:13	0.42	0.044	(b)
			3:1.5:1	75:38:25	0.42	0.056	(b)
403 SS	1800	980	3:1.5:0.25	75:38:6.5	0.23	0.029	(b)
			3:1.5:0.5	75:38:13	0.24	0.039	(b)
			3:1.5:1	75:38:25	0.34	0.047	(b)
	1950	1065	3:1.5:1	75:38:25	0.28	0.06	(b)
	2050	1120	3:1.5:1	75:38:25	0.35	0.06	(b)
Waspaloy	2100	1150	3:1.5:1	75:38:25	0.18	0.06	(b)
17-7PH SS	1950	1065	3:1.5:1	75:38:25	0.28	0.06	(b)
	2100	1150	3:1.5:1	75:38:25	0.35	0.06	(b)
Ti-6Al-4V	1700	925	3:1.5:1	75:38:25	0.30	0.06	(b)
	1750	955	3:1.5:1	75:38:25	0.46	0.06	(b)
Inconel 718	2000	1095	3:1.5:1	75:38:25	0.18	0.06	(b)
	2100	1150	3:1.5:1	75:38:25	0.33	0.06	(b)
Ti-8Al-1Mo-1V	1750	955	3:1.5:1	75:38:25	0.27	0.06	(b)
	1800	980	3:1.5:1	75:38:25	0.27	0.06	(b)
Udimet	2050	1120	3:1.5:1	75:38:25	0.40	0.06	(b)
7075 Al	700	370	5:3:1	125:75:25	0.37	0.06	(a)
	800	425	5:3:1	125:75:25	0.31	0.06	(a)

SS, stainless steel. (a) Caustic precoat + graphite coating Dag 137 (Acheson) on the specimens and graphite spray Deltaforge 43 (Acheson) on the dies. (b) Glass-based coating Deltaforge 347 (Acheson) on the specimens and graphite spray Deltaforge 43 (Acheson) on the dies. Source: [Douglas et al., 1975]

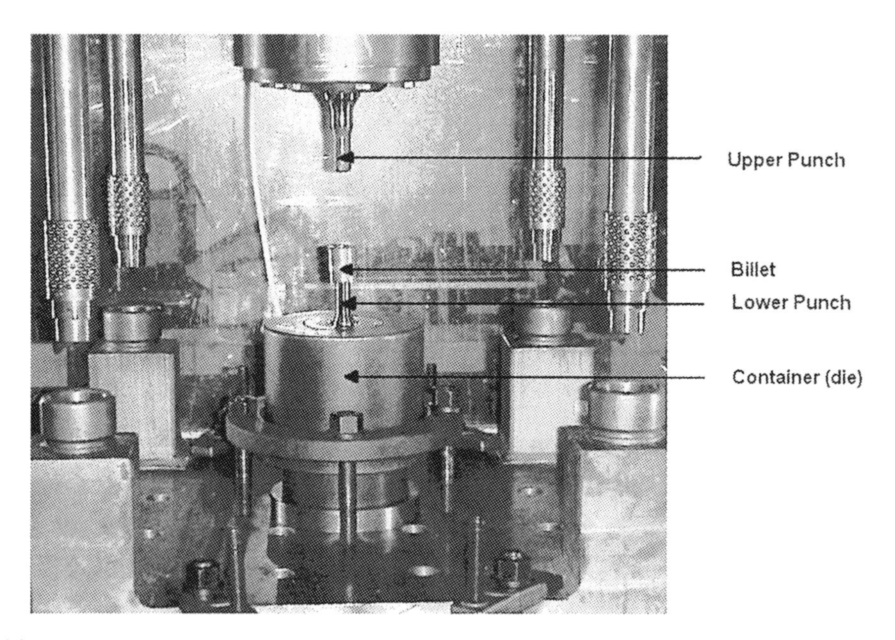

Fig. 7.11 Double cup backward extrusion test tooling at the ERC/NSM

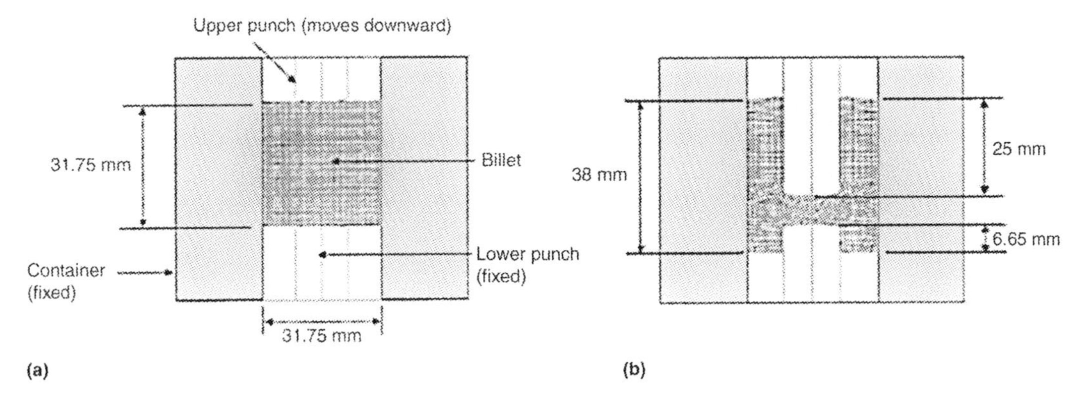

Fig. 7.12 FEM model of double cup backward extrusion test (shear factor m = 0.1) (dimensions in millimeters). (a) Initial. (b) Final. (Ngaile et al., 2002)

ticular, it has been found that the ratio of the cup heights increases as the friction factor, f, or the shear factor, m, increases. In other words, if there is no friction, the cup heights will be the same and the cup height ratio, H_1/H_2, will be equal to one. Thus, the ratio of the cup heights represents a simple method for evaluating interface friction.

Figure 7.11 shows the actual tooling used for the double cup backward extrusion test. It should be noted that the lower punch was raised out of the container for illustration purposes only. During the test, the container and lower punch are fixed on the bed of the press and held stationary with the lower punch located completely inside the container. In addition, the upper punch is fixed on the ram of the press and moves downward. Thus, there is a relative velocity between the container and the upper punch, but not between the container and the lower punch. Therefore, the material flow to the lower punch is more restricted in the presence of friction. This explains why the height of the upper cup is larger than the height of the lower cup.

To obtain the magnitude of the friction factor, f, or the shear factor, m, the ratio of the cup heights, H_1/H_2, must be compared with the values predicted by using various friction factors, f, or shear factors, m. Today, these values are most often predicted by the finite element method (FEM). Figure 7.12 shows an example of an FEM model used for this purpose. The results are plotted in the form of "theoretical calibration curves," as can be seen in Fig. 7.13. In determining the value of the shear factor, m, for a given experimental condition, the measured dimensions (cup height ratio and stroke) are plotted on the calibration figure. Figure 7.14 illustrates how the cup height ratio and stroke are measured. From the position of that point with respect to the theoretical curves given for various values of "m," the value of the shear factor, m, which existed in the experiment, is obtained. It should be noted that in addition to interface friction, the metal flow in this test is dependent on the billet material, the billet diameter, and the

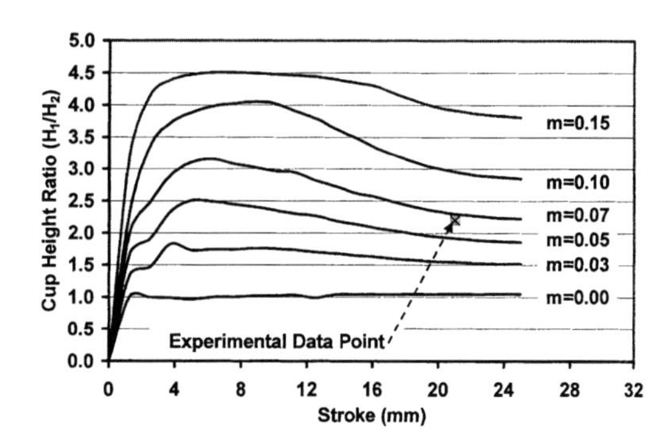

Fig. 7.13 Theoretical calibration curves for double cup backward extrusion test with experimental data point (shear factor m ≈ 0.065). [Ngaile et al., 2002]

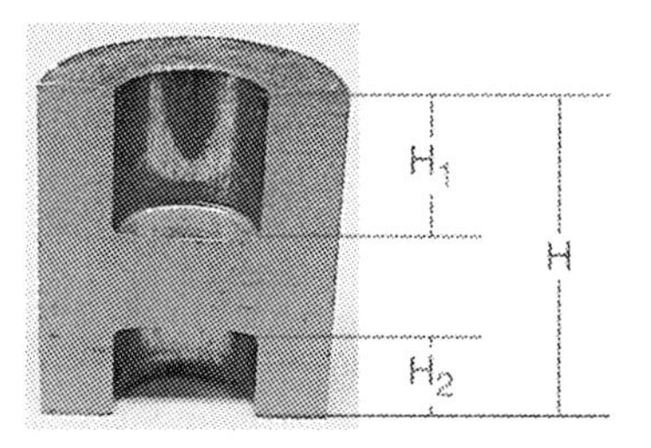

Fig. 7.14 Measurement of cup height ratio and stroke. Cup height ratio: $R_{ch} = H_1/H_2$. Stroke: S = initial height − $(H - H_1 - H_2)$. [Ngaile et al., 2002]

Table 7.9 Values of frictional shear factor, m, obtained from double cup backward extrusion tests conducted in a hydraulic press (punch/die temperatures ≈ 75 °F, or 24 °C)

Material	Specimen temperatures °F	Specimen temperatures °C	Frictional shear factor (m)	Lubrication system
AISI 8610	75	24	0.065	Zinc phosphate coating + soap
			0.035	Metallic compounds + sulfur compounds(a)
			0.075	Zinc-based dry film(b)
AISI 1038	75	24	0.050	Zinc phosphate coating + soap

(a) Billet size = 1.25 in. diam × 1.25 in. height (31.75 mm diameter × 31.75 mm height). (b) Environmentally friendly lubrication system developed for replacement of zinc phosphate coating based systems. Source: [Ngaile et al., 2002]

punch diameter. Thus, the flow stress of the material must be known and the appropriate theoretical curves should be used to quantify the interface friction. In other words, there is no "universal" set of calibration curves for this test.

Several double cup backward extrusion tests have been conducted for various materials in a 160 ton hydraulic press with a ram velocity of 15 mm/s and a punch stroke of 21 mm. The results of these tests are summarized in Table 7.9.

REFERENCES

[Altan, 1970]: Altan, T., "Heat Generation and Temperatures in Wire and Rod Drawing," *Wire J.,* March 1970, p 54.

[Altan et al., 1983]: Altan, T., Oh, S., Gegel, H., *Metal Forming Fundamentals and Applications,* American Society for Metals, 1983.

[Bay, 1994]: Bay, N., The State of the Art in Cold Forging Lubrication, *J. Mater. Process. Technol.,* Vol 46, 1994, p 19–40.

[Bay, 1995]: Bay, N., Aspects of Lubrication in Cold Forging of Aluminum and Steel, *Proceedings of the 9th International Cold Forging Congress,* Solihull, UK, May 1995, p 135–146.

[Bhushan, 2001]: Bhushan, B., *Modern Tribology Handbook—Vol 2: Materials, Coatings, and Industrial Applications,* CRC Press, 2000.

[Buschhausen et al., 1992]: Buschhausen, A., Lee, J.Y., Weinmann, K., Altan, T., "Evaluation of Lubrication and Friction in Cold Forging Using Double Backward Extrusion Process," *J. Mater. Process. Technol.,* Vol 33, 1992, p 95–108.

[Doege et al., 1996]: Doege, E., Seidel, R., Romanowski, C., "Increasing Tool Life Quantity in Die Forging: Chances and Limits of Tribological Measures," Technical Papers of the North American Manufacturing Research Institution of SME, 1996, p 89–94.

[Douglas et al., 1975]: Douglas, J.R., Altan, T., Flow Stress Determination for Metals at Forging Rates and Temperatures, *Trans. ASME, J. Eng. Ind.,* Feb 1975, p 66.

[Forcellese et al., 1994]: Forcellese, A., Gabrielli, F., Barcellona, A., Micari, F., "Evaluation of Friction in Cold Metal Forming," *J. Mater. Process. Technol.,* Vol 45, 1994, p 619–624.

[Gariety et al., 2002]: Gariety, M., Ngaile, G., Altan, T., "Identification of Lubricants and Enhancement of Lubricant Performance for Cold Heading—Progress Report 1—Identification of Lubricants Used for Cold Heading," Report No. PF/ERC/NSM-02-R-32A, 2002.

[Gariety et al., 2003]: Gariety, M., Padwad, S., Ngaile, G., Altan, T., "Identification of Lubricants and Enhancement of Lubricant Performance for Cold Heading—Progress Report 2—Preliminary Lubrication Tests for Cold Heading," Report No. PF/ERC/NSM-02-R-32B, 2003.

[Hannan et al., 2000]: Hannan, D., Ngaile, G., Altan, T., "Development of Forming Processes for Copper Components for Stanford Linear Accelerator," Report No. PF/ERC/NSM-B-00-20, 2000.

[ICFG, 1996]: International Cold Forging Group, "Lubrication Aspects in Cold Forging of Aluminum and Aluminum Alloys," Document No. 10/95, 1996.

[Lazzarotto et al., 1999]: Lazzarotto, L., Marechal, C., Dubar, L., Dubois, A., Oudin, J., "The Effects of Processing Bath Parameters on the Quality and Performance of Zinc Phosphate Stearate Coatings," *Surf. Coat. Technol.,* Vol 122, 1999, p 94–100.

[Lee et al., 1972]: Lee, C.H., Altan, T., "Influence of Flow Stress and Friction Upon Metal Flow in Upset Forging of Rings and Cylinders," *Trans. ASME, J. Eng. Ind.,* Aug 1972, p 775.

[Lenard, 2000]: Lenard, J., "Tribology in Metal Rolling," *Ann. CIRP,* Vol 49, 2000, p 1–24.

[Male et al., 1970]: Male, A.T., DePierre, V., "The Validity of Mathematical Solutions for Determining Friction from the Ring Compression Test," *Trans. ASME, J. Lubr. Technol.,* Vol 92, 1970, p 389.

[Manji, 1994]: Manji, J., "Die Lubricants," *Forging,* Spring 1994, p 39–44.

[Ngaile et al., 1999]: Ngaile, G., Saiki, H., "Cold Forging Tribo-Test Based on Variation of Deformation Patterns at the Tool-Workpiece Interface," *Lubr. Eng.: J. Soc. Tribol. Lubr. Eng.,* Feb 1999, p 23–31.

[Ngaile et al., 2002]: Ngaile, G., Schumacher, R., Gariety, M., Altan, T., Kolodziej, J., "Development of Replacements for Phoscoating Used in Forging Extrusion and Metal Forming Processes," 2002, Report No. PF/ERC/NSM-02-R-85.

[Saiki, 1997]: Saiki, H., "The Role of Tribology for Improvement of Tool Life in Hot Forging," *Proceedings of the 1st International Conference on Tribology in Manufacturing Processes,* Gifu, Japan, 1997, p 22–31.

[Saiki et al., 1999]: Saiki, H., Ngaile, G., Ruan, L., Marumo, Y., "Evaluation of Cold Forging Lubricants Under Realistic Forging Temperature Conditions," *Adv. Technol. Plast.: Ann. CIRP,* Vol 1, p 377–382.

[Schey, 1983]: Schey, J., *Tribology in Metalworking: Lubrication, Friction, and Wear,* American Society for Metals, 1983.

[Schmoeckel et al., 1997]: Schmoeckel, D., Rupp, M., "More Environment Friendly Cold Massive Forming—Production of Steel without Zinc Phosphate Layer," Symposium, *Latest Developments in Massive Forming,* Fellbach near Stuttgart, 1997, p 183–200.

[Shen et al., 1992]: Shen, G., Vedhanayagam, A., Kropp, E., Altan, T., "A Method for Evaluation of Friction Using a Backward Extrusion Type Forging," *J. Mater. Process. Technol.,* Vol 33, 1992, p 109–123.

CHAPTER 8

Inverse Analysis for Simultaneous Determination of Flow Stress and Friction

Hyunjoong Cho

8.1 Introduction

The finite element analysis (FEA) based simulation of metal forming processes has been widely used to predict metal flow and to optimize the manufacturing operations. In using user-friendly commercial FEA software, it is necessary to assign input parameters for the simulation. Among those inputs, the parameters in the flow stress equation, friction factor, and anisotropy coefficients of a material are usually obtained from the appropriate tests. The results of process simulation are extremely sensitive to the accuracy of flow stress and interface friction that are input to FEM programs. Therefore, it is essential that these input values are determined using (a) reliable material tests and (b) accurate evaluation methods. A test used to determine material properties should replicate processing conditions that exist in practical applications.

A common method for the determination of the flow stress data for forging simulation is the cylinder upset test as discussed in Chapter 4 because (a) during the test the deformation is done in a state of compressive stress, which represents well the true stress state of most forging processes and (b) the test can be done for a large strain. However, even in the simplest cylinder upset test, interface friction leads to an inevitable bulging of the sample and thereby to an inaccurate flow stress determination. Thus, the evaluation of the test results should be able to overcome difficulties introduced by friction and inhomogeneous deformation. It is desirable to consider the unavoidable friction at the tool/workpiece interface in the test and to identify the friction together with flow stress using an appropriate evaluation method. In this chapter, an inverse analysis technique for the accurate determination of the input data for FEM simulation is introduced to determine material parameters in the flow stress model and the friction at the tool/workpiece interface.

8.2 Inverse Analysis in Metal Forming

8.2.1 Direct and Inverse Problems

An FEA of metal forming process as illustrated in Fig. 8.1(a) is regarded as direct problem. The required input data for direct problem (i.e., FE simulation) are geometry, process conditions, flow stress, interface friction, etc. In the direct problem FEA predicts the metal flow, forming load, and energy by simulating the forming operation assuming that the flow stress and friction values are known. Compared with the direct problem, in the inverse problem the authors determine one or more of input data of the direct problem, leading to the best fit between experimental measurements and FEM prediction. With experimental measurements provided to the inverse problem, the input data are identified or calibrated (if initial guess is given). This inverse problem can be applied to any material test in which FEA can be done including the cylinder upset and ring compression tests, provided experimental measurements are accurate enough, Fig. 8.1(b). Therefore, an inverse problem is regarded as a parameter identification problem that can be formulated further as an optimization problem where the difference between measurement and FEM prediction is minimized by adjusting the input parameters.

8.2.2 Procedure for Parameter Identification

The basic concept of an inverse analysis in flow stress determination consists of a set of unknown parameters defined in flow stress equation. First, a finite element simulation of the selected material test with the assumed parameters of the flow stress equation is conducted and the computed load-stroke curve is compared with the experimentally measured curve. Then, the assumed parameters of the flow stress equation are adjusted in such a way that the difference in the calculated and measured load-stroke curves is reduced in the next comparison. This procedure is repeated until the difference between experimental measurements and computed data disappears. The result of inverse analysis is a set of the identified material parameters of the flow stress equation, which represents material properties. The procedure used to identify the material parameters includes:

1. Guess the material parameters in flow stress equation.
2. Start FEM simulation of the selected material test with given flow stress data.
3. Compare the computed forming load with experimentally measured one.
4. Obtain the amount of adjustments in material parameters by minimizing the difference between the computed and measured loads.
5. Improve the material parameters until the difference becomes within a desired tolerance.

The trial and error procedure is the simplest way to solve the above inverse problem, i.e., parameter identification. This method may be used to get some prior information of the parameter values and get some rough idea what is the most important parameter for a given problem. However, this method is time consuming and parameters cannot be identified accurately. Therefore, it is necessary to use a numerical optimization technique for robust determination of the parameters in the flow stress equation. Generally, the unknown parameters are determined by minimizing a least-square functional consisting of experimental data and FEM simulated data. The FEM is used to analyze the behavior of the material during the test, whereas the optimization technique allows for automatic adjustment of parameters until the calculated response matches the measured one within a specified tolerance. Derivation of inverse analysis based on rigid-plastic finite element formulation was developed at ERC/NSM [Cho et al., 2003].

8.2.3 Past Studies on the Inverse Analysis Used for Flow Stress Determination

Chenot et al. formulated an inverse problem in developing a methodology for automatic identification of rheological parameters. The inverse problem was formulated as finding a set of rheological parameters starting from a known constitutive equation. An optimization algorithm was coupled with the finite element simulation for computing the parameter vector that minimizes an objective function representing, in the least-square sense, the difference between experimental and numerical data. For sensitivity

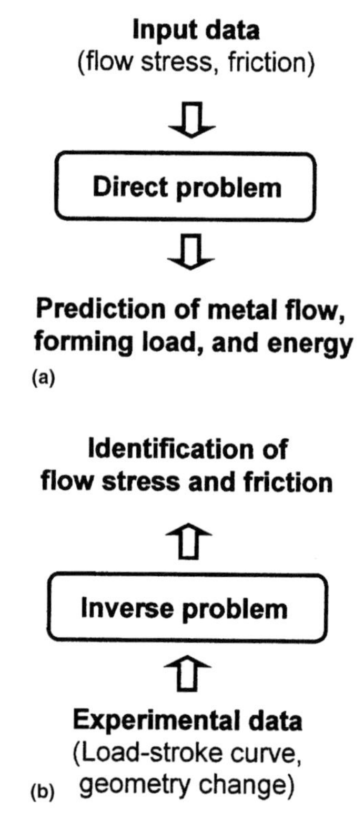

Fig. 8.1 Direct (a) and inverse (b) problems

analysis of the objective function with respect to the searching parameters during the optimization, Chenot differentiated the FEM code with respect to the searching parameters [Chenot et al., 1996].

Boyer and Massoni developed the semianalytical method for sensitivity analysis of inverse problem in material forming domain. This method compromises between computation time and effort of analytical code differentiation. This technique proved to be a good alternative to the finite difference method or analytical differentiation of the FEM code in conducting sensitivity analysis. As a result of research, the identification software CART (Computer Aided Rheology and Tribology) was introduced [Boyer & Massoni, 2001].

Pietrzyk et al. used inverse analysis technique to evaluate the coefficients in the friction and flow stress model for metal forming processes. He identified parameters in the conventional flow stress equation and in a dislocation density based internal variable model as well as friction factor from one set of ring compression test. He concluded that the determination of both rheological and frictional parameters from one combined test is ideal because the interpretation of tests to determine the flow stress depends on an assumed value of the friction factor [Pietrzyk et al., 2001].

Zhiliang et al. introduced a new method combining compression tests with FEM simulation (C-FEM) to determine flow stress from the compression tests where inhomogeneous deformation is present due to interface friction. In this method, the flow stress obtained from the compression test is improved by minimizing the target function defined in load-stroke curves [Zhiliang et al., 2002].

8.3 Flow Stress Determination in Forging by Inverse Analysis

In a large plastic deformation problem, usually encountered in most forging applications, the deformation behavior of the material can be assumed to be rigid-plastic by neglecting the elastic part. If the material shows strain hardening behavior in cold forging, the power-law type flow stress equation is used to describe a stress-strain relationship for plastic range. At elevated temperature, flow stress is sensitive to rate of deformation. Thus, for warm and hot forgings, flow stress is expressed in function of strain, strain-rate, and temperature. Examples of flow stress equations and material parameters are summarized in Table 8.1.

8.3.1 Material Parameters

In cold forging, K and n are the two material parameters (i.e., the material strength coefficient, K, and the strain hardening exponent, n) used to define the stress-strain relationship during plastic deformation. When the experimentally measured load-stroke curve is available, FEM simulations are made with initial guesses of K and n values, and then the two material parameters are to be identified. Therefore, the material parameters become design variables in the optimization problem.

8.3.2 Objective Function

The unknown material parameters are determined by minimizing an objective function, E, representing the difference between the experimental and the simulated loads in a least-square sense:

$$E = \frac{1}{N}\sum_{i=1}^{N}\left(\frac{F_{EXP} - F_{COM}(P_k)}{F_{EXP}}\right)^2 \qquad \text{(Eq 8.1)}$$

where F_{EXP} is the experimental load and F_{COM} is the computed load. N is the number of data sampling points selected from a load-stroke to construct the objective function. Figure 8.2 illustrates a definition of the objective function. The objective function, E, is a nonlinear implicit function of material parameters P_k. Therefore, the parameter identification problem is reduced to compute a set of the unknown parameter $P_k = \{K,n\}$, which leads to the best fit between experimental measurements and corresponding computed data.

For given material parameters P_k, the objective function $E = E(P_k)$ will be minimum at:

$$\frac{\partial E(P_k)}{\partial P_k} = 0 \text{ for } k = 1, 2 \qquad \text{(Eq 8.2)}$$

where P_k are the $P_1 = K$, $P_2 = n$.

Table 8.1 Flow stress models and parameters

Forging type	Flow stress equation	Parameters
Cold	$\bar{\sigma} = K\bar{\varepsilon}^n$	K,n
Warm	$\bar{\sigma} = K\bar{\varepsilon}^n\dot{\bar{\varepsilon}}^m$	K,n,m
Hot	$\bar{\sigma} = K\dot{\bar{\varepsilon}}^m$	K,m

The nonlinear Eq 8.2 is solved with respect to the parameters P_k using Newton-Raphson iterative procedure.

$$\frac{\partial^2 E}{\partial P_k \partial P_j} \Delta P_j = -\frac{\partial E}{\partial P_k} \text{ for j, k } = 1, 2 \qquad \text{(Eq 8.3)}$$

The first and second gradients of the objective function with respect to the parameters P_k are evaluated by taking the derivatives of the objective function $E = E(P_k)$ with respect to P_k:

$$\frac{\partial E}{\partial P_k} = -\frac{2}{N}\sum_{i=1}^{N}\left\{\frac{(F_{EXP} - F_{COM})}{F_{EXP}^2}\frac{\partial F_{COM}}{\partial P_k}\right\} \text{ for j, k } = 1, 2 \qquad \text{(Eq 8.4)}$$

$$\frac{\partial^2 E}{\partial P_k \partial P_j} = -\frac{2}{N}\sum_{i=1}^{N} \left\{-\frac{1}{F_{EXP}^2}\frac{\partial F_{COM}}{\partial P_k}\frac{\partial F_{COM}}{\partial P_j} + \frac{(F_{EXP} - F_{COM})}{F_{EXP}^2}\frac{\partial^2 F_{COM}}{\partial P_k \partial P_j}\right\} \text{ for j, k } = 1, 2 \qquad \text{(Eq 8.5)}$$

8.3.3 *Advantages*

In the inverse analysis, FEM simulation is used to describe the deformation behavior of material during the test and the optimization algorithm identifies material parameters using FEM simulation results. Any material test that can be simulated by FEM can be selected to determine the material property. Uniform strain-rate condition, which requires a sophisticated control of test machine, is not needed in the test because inverse analysis takes advantage of FEM simulation where complex stress and strain states can be handled. Therefore, inverse analysis technique gives flexibility in selecting material tests. For determining flow stress data for warm and hot forging, only two different test velocities are required to identify strain-rate sensitivity (m-value) instead of conducting the tests at several different constant strain-rates.

8.4 Inverse Analysis for Simultaneous Determination of Flow Stress and Friction

In the upset test, for determining bulk material property for a large strain, the main problem is the existing friction at the die/specimen interface. In order to overcome this problem, friction is minimized by using a lubricant together with geometry proposed by Rastegaev [Dahl et al., 1999]. This treatment will lead to a uniaxial stress state for a limited reduction in height during the test. However, the frictional force at large compression ratio (a) starts to bulge the sample (i.e., inhomogeneous deformation) regardless of the quality of the lubricant and (b) affects the measured load-stroke curve. In other words, the measured load-stroke curve has a force contribution from inhomogeneous deformation caused by frictional force, and this causes an error in flow stress calculation. Therefore, it is necessary to consider the inevitable interface friction in the test and then identify it together with the flow stress.

For simultaneous identification of both the material parameters P_k and the friction factor m_f, in addition to the measured load-stroke curve, one more measurable geometrical quantity in the test, namely the barreling, is used. Barreling reflects a degree of inhomogeneous deformation caused by friction. Therefore, it allows the identification of friction by measuring the barreling shape of the specimen. During the inverse analysis, the computed barreling shape is compared with the measured barreling shape and then the difference is minimized by adjusting the friction factor. After several iterations, the difference in barreling shape disappears and the friction factor is identified. Figure 8.3 shows a methodology for determining flow stress and interface friction simultaneously.

8.5 Example of Inverse Analysis

8.5.1 *Flow Stress Model*

The developed inverse analysis algorithm has been tested by using the real experimental data

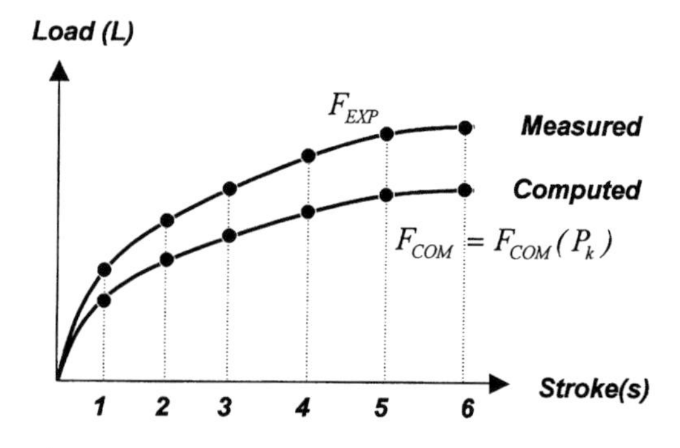

Fig. 8.2 The difference between the computed and the experimental load

obtained from the ring compression test. The investigated material was assumed to follow strain-hardening behavior and the following power-law-type flow stress equation was considered.

$$\bar{\sigma} = K\bar{\varepsilon}^n \qquad \text{(Eq 8.6)}$$

Therefore, a set of material parameters defined by $P_k = \{K,n\}$ and friction factor m_f are the unknown parameters that have to be identified. Two experimental quantities: (1) the measured load-stroke curve and (2) the maximum diameter of the specimen at the end of stroke were used as experimental values in the inverse analysis.

8.5.2 Experiment

Aluminum rings made from Aluminum 6061-T6 with 2.13 in. OD × 1.06 in. ID × 0.71 in. height (54 mm OD × 27 mm ID × 18 mm height) were compressed to various reductions

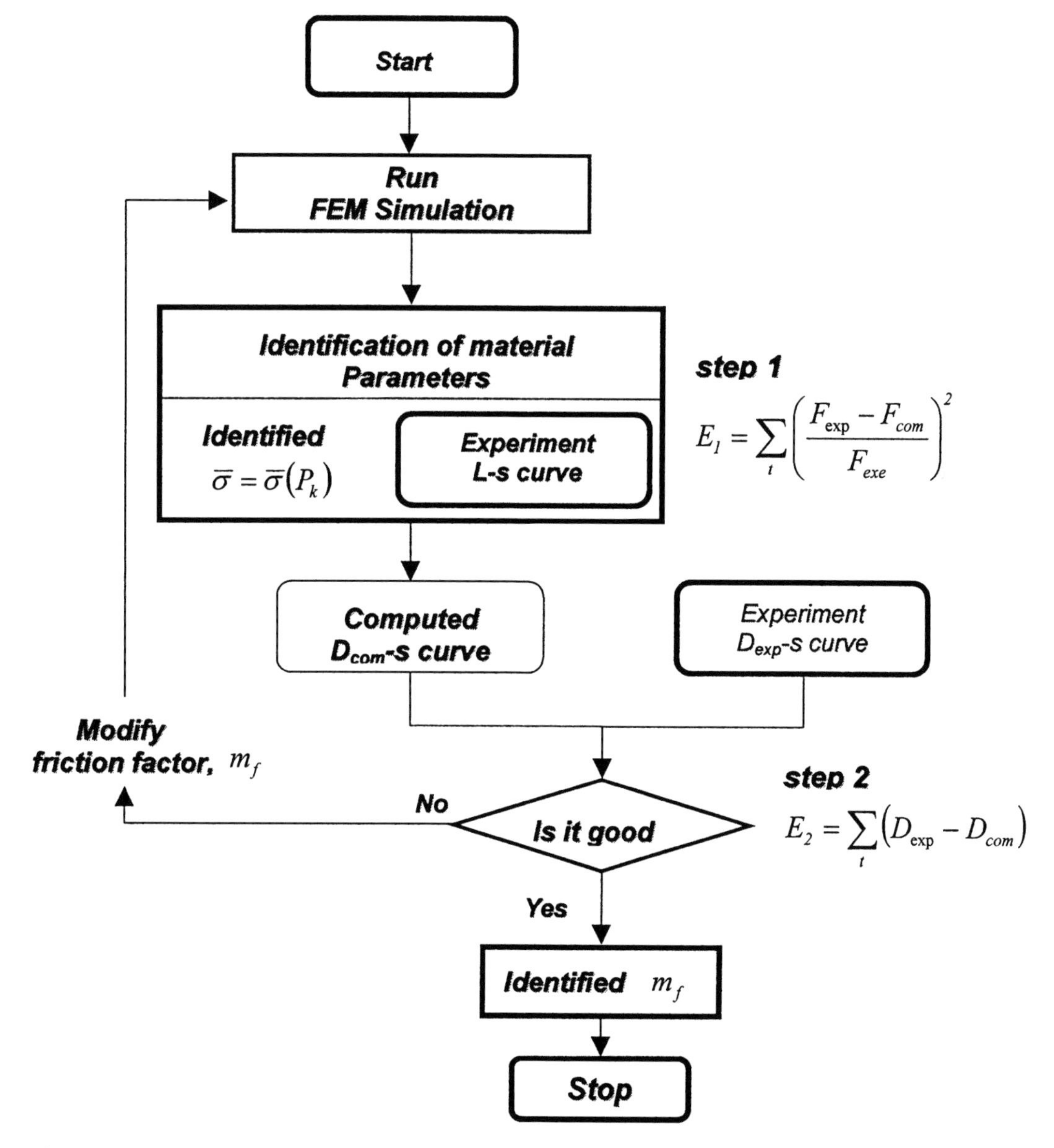

Fig. 8.3 Flow chart of simultaneous determination of flow stress and friction. L-s, load versus stroke; D-s, bulge diameter versus stroke; m_f, friction factor

in height. The rings were lubricated with Teflon spray on all surfaces of the samples and on the top and bottom dies. In order to observe the internal diameter variation the test was stopped at reductions of 7.2, 22.2, and 40%. Figure 8.4 shows the compressed ring samples, and Table 8.2 shows the decrease in ID of the ring at different reductions.

8.5.3 Determination of Flow Stress and Friction

The results of identifed parameters (K-value and n-value) in the flow stress equation and friction factor by the inverse analysis are summarized in Table 8.3. Three inverse analyses were conducted by varying the friction factor from 0.15 to 0.2. As initial guesses of material parameters, K = 430 (MPa) and n = 0.1 were used for every case. When the friction factor 0.175 was assumed, the inverse analysis prediction produced only 8.1% underestimation in ID comparison of the ring. Thus, a combination of friction factor m_f = 0.175 and flow stress $\bar{\sigma} = 452\bar{\varepsilon}^{0.074}$ (MPa) gives the best minimum for the objective function. As can be seen in Fig. 8.5, after four optimization iterations, computed and experimental loads are nearly identical.

Fig. 8.4 Compressed ring samples

Table 8.2 Percent decrease in ID of ring

Reduction in height, %	Decrease in ID of ring, %
7.2	−1.48
22.2	−2.96
40.0	−1.85

Table 8.3 Predicted inverse analysis results (ring test)

Friction (m_f)	Decrease in ID of ring, %	K-value, MPa ksi	K-value, MPa MPa	n-value
0.2	+0.5	65	446	0.073
0.15	−5.9	67	459	0.076
0.175	−1.7	66	452	0.074

8.5.4 Verification of the Determined Friction Factor

To verify the accuracy of determined friction factor, FEM simulations using the determined flow stress with friction factor of 0.175 were conducted for various friction factors. Thus, the ring calibration curves were generated as shown in Fig. 8.6. It is seen that measurements match well the curve obtained with the friction factor of 0.182, which is very close to 0.175.

8.5.5 Verification of the Determined Flow Stress

In order to verify the flow stress determined with friction factor by the inverse analysis, the aluminum cylinders with a 30 mm diam × 30 mm height were upset to 38% reduction in height. To minimize interface friction, the interface was lubricated with Ecoform lubricant made by Fuchs. As shown in Fig. 8.7, the lubrication nearly eliminated bulging in upsetting of cylinder. Using the measured load-stroke curve, flow stress $\bar{\sigma} = 437\bar{\varepsilon}^{0.067}$ (MPa) was obtained. The difference between the flow stress

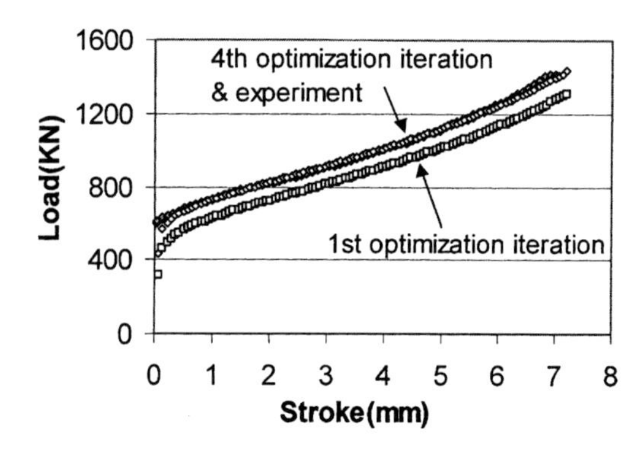

Fig. 8.5 Comparison of computed and experimental load-stroke curves (ring test)

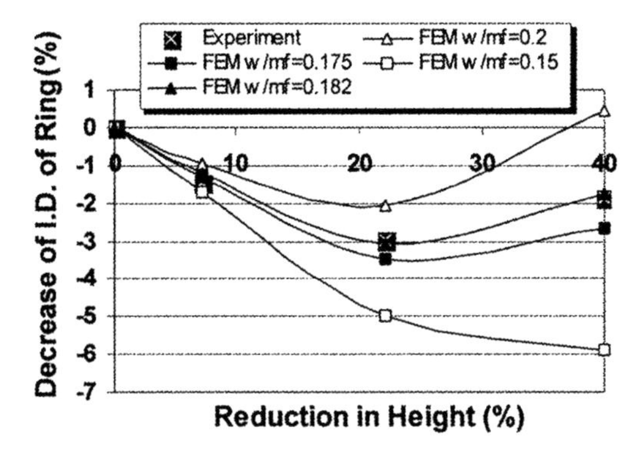

Fig. 8.6 Ring calibration curves obtained with $\bar{\sigma} = 452\bar{\varepsilon}^{0.074}$ (MPa)

Fig. 8.7 Compressed ring samples

data obtained in ring and cylinder compression tests is about 3.3% in K-value and 9.5% in n-value, respectively.

REFERENCES

[Boyer & Massoni, 2001]: Boyer, B., Massoni, E., "Inverse Analysis for Identification of Parameters During Thermo-Mechanical Tests," *Simulation of Materials Processing: Theory, Method and Applications,* K.-I. Mori, Ed., 2001, p 281–284.

[Chenot et al., 1996]: Chenot, J., Massoni, E., Fourment, L., "Inverse Problems in Finite Element Simulation of Metal Forming Processes," *Eng. Comput.,* Vol 13 (No. 2/3/4), 1996, p 190–225.

[Cho et al., 2003]: Cho, H., Ngaile, G., "Simultaneous Determination of Flow Stress and Interface Friction by Finite Element Based Inverse Analysis Technique," *Ann. CIRP,* Vol 52/1, 2003, p 221–224.

[Dahl et al., 1999]: Dahl, C., Vazquez, V., Altan, T., "Determination of Flow Stress of 1524 Steel at Room Temperature Using the Compression Test," Engineering Research Center for Net Shape Manufacturing, ERC/NSM-99-R-22.

[Pietrzyk et al., 2001]: Pietrzyk, M., Szyndler, D., Hodgson, P.D., "Identification of Parameters in the Internal Variable Constitutive Model and Friction Model for Hot Forming of Steels," *Simulation of Materials Processing: Theory, Method and Applications,* Mori, Ed., 2001, p 281–284.

[Zhiliang et al., 2002]: Zhiliang, Z., Xinbo, L., Fubao, Z., "Determination of Metal Material Flow Stress by the Method of C-FEM," *J. Mater. Process. Technol.,* Vol 120, 2002, p 144–150.

CHAPTER 9

Methods of Analysis for Forging Operations

Manas Shirgaokar

9.1 Introduction

The major process variables involved in forging can be summarized as: (a) the billet material properties, (b) the tooling/dies, (c) tool/workpiece interface conditions, (d) forging equipment, (e) mechanics of the deformation zone, and (f) the environmental conditions. The major objectives of analyzing any forging operation are:

- Establish the kinematic relationships (shape, velocities, strain rates, and strain) between the undeformed part (billet or preform) and the deformed part (final forged product), i.e., predict the metal flow during the forming operation.
- Establish the limits of formability or producibility, i.e., determine whether it is possible to perform the forming operation without causing any surface or internal failures (cracks or folds) in the deforming material.
- Predict the stresses, the forces, and the energy necessary to carry out the forming operation. This information is necessary for tool design and for selecting the appropriate equipment, with adequate force and energy capabilities, to perform the forming operation.

There are several different approximate methods, both analytical and numerical, for analyzing forging processes. None of these methods is perfect because of the assumptions made in developing the mathematical approach. In addition, every method of analysis requires as input:

- A description of the material behavior under the process conditions, i.e., the flow stress data
- A quantitative value to describe the friction, i.e., the friction factor, m, or the friction coefficient, μ.

These two quantities themselves (flow stress and friction) must be determined by experiment and are difficult to obtain accurately. Thus, any errors in flow stress measurement or uncertainties in the value of the friction factor are expected to influence the accuracy of the results of the analysis.

Forging processes can be analyzed by several methods including the slab method, the slip-line method, the visioplasticity method, the upper-bound method, finite difference method, and the finite element method. The capabilities and characteristics of these methods are summarized in Table 9.1 [Altan et al., 1979].

In the slab method, the workpiece being deformed is decomposed into several slabs. For each slab, simplifying assumptions are made mainly with respect to stress distributions. The resulting approximate equilibrium equations are solved with imposition of stress compatibility between slabs and boundary tractions. The final result is a reasonable load prediction with an approximate stress distribution [Kobayashi et al., 1989].

The slip-line field method is used in plane strain for perfectly plastic materials (constant yield stress) and uses the hyperbolic properties that the stress equations have in such cases. The construction of slip-line fields, although producing an "exact" stress distribution, is still quite limited in predicting results that give good correlations with experimental work. From the stress distributions, velocity fields can be calculated through plasticity equations [Kobayashi et al., 1989].

The visioplasticity method [Thomsen et al., 1954] combines experiment and analysis. A grid is imprinted on the metal or modeling substance before deformation starts. Pictures taken at small intervals during processing enable the investigator to construct a flow pattern. After the velocity vectors have been determined from an actual test, strain rates are calculated and the stress distributions are obtained from plasticity equations. The method can be used to obtain reliable solutions in detail for processes in which the experimental determination of the velocity vectors was possible.

The upper-bound method requires the "guessing" of admissible velocity fields (i.e., satisfying the boundary conditions), among which the best one is chosen by minimizing total potential energy. Information leading to a good selection of velocity fields comes from experimental evidence and experience. This method, with experience, can deliver fast and relatively accurate prediction of loads and velocity distributions [Kobayashi et al., 1989].

All of the above highlighted methods of analysis fail to consider temperature gradients, which are present in the deforming material during hot forming operations. As a result, the effect of temperatures on flow stress and metal flow during hot forming are often not considered adequately.

In the finite difference method, the derivatives in the governing partial differential equations are written in terms of difference equations. Therefore, for a two-dimensional domain, a grid of cells is placed inside the domain and the differencing approximation applied to each interior point. This results in a system of linear algebraic equations (with a banded solution matrix), which yields a unique solution provided the boundary conditions of the actual problem are satisfied. Though temperature gradients can be taken into account, this method is limited to problems with simple boundaries [Becker, 1992].

In the finite element method, the entire solution domain is divided into small finite segments (hence, the name "finite elements"). Over each element, the behavior is described by the differential equations. All these small elements are assembled together, and the requirements of continuity and equilibrium are satisfied between neighboring elements. Provided the boundary conditions of the actual problem are satisfied, a unique solution can be obtained to the overall system of linear algebraic equations (with a sparsely populated solution matrix) [Becker, 1992].

In recent years, the finite element method has gained wide acceptance in the industry and academia. This can be attributed to the rapid advancement in the computing technology, user-friendly commercial FE software and the detailed information FEM can provide as compared to other methods of analysis. The FE method allows the user to incorporate in the simulation: (a) the tool and workpiece temperatures, (b) the heat transfer during deformation, (c) strain-rate-dependent material properties, (d) strain hardening characteristics, and (e) capabilities for microstructure analysis. This results in a more accurate analysis of the forging process. Commercial FE software packages have been

Table 9.1 Characteristics of various methods of analysis

	Input		Output				
Method	Flow stress	Friction	Velocity field	Stress field	Temperature field	Stresses on tools	Comments
Slab	Average	(a)(b)	No	Yes	No	Yes	Ignores redundant work
Uniform energy	Average	(b)	No	No	No	Average	Redundant work can be included approximately
Slip line	Average	(a)(b)	Yes	Yes	No	Yes	Valid for plane-strain problems
Upper bound	Distribution	(b)	Yes	No	No	Average	Gives upper bound on loads, can determine free boundaries
Hill's	Distribution	(a)(b)	Yes	No	No	Average	Can treat 3-D problems
Finite difference	Distribution	(a)(b)	Yes	Yes	Yes	Yes	Requires considerable computer time
Finite element	Distribution	(a)(b)	Yes	Yes	Yes	Yes	Same as above
Matrix	Distribution	(a)(b)	Yes	Yes	Yes	Yes	Treats rigid/plastic material
Weighted residuals	Distribution	(a)(b)	Yes	Yes	Yes	Yes	Very general approach

(a) $\tau = \mu\sigma_n$. (b) $\tau = m\bar{\sigma}/\sqrt{3}$. Source: [Altan et al., 1979]

used successfully in simulating complex two-dimensional (2-D) and three-dimensional (3-D) forging operations.

This chapter briefly discusses the slab, upper-bound, and the finite element (FE) methods.

9.2 Slab Method of Analysis

The following assumptions are made in using the slab method of analysis:

- The deforming material is isotropic and incompressible.
- The elastic deformations of the deforming material and the tool are neglected.
- The inertial forces are small and are neglected.
- The frictional shear stress, τ, is constant at the die/material interface and is defined as $\tau = f\bar{\sigma} = m\bar{\sigma}/\sqrt{3}$.
- The material flows according to the von Mises rule.
- The flow stress and the temperature are constant within the analyzed portion of the deforming material.

The basic approach for the practical use of the slab method is as follows:

1. Estimate or assume a velocity or metal flow field.
2. For this velocity field, estimate the average strains, strain rates, and temperatures within each distinct time zone of the velocity field.
3. Thus, estimate an average value of the flow stress, $\bar{\sigma}$, within each distinct zone of deformation.
4. Knowing $\bar{\sigma}$ and friction, derive or apply the necessary equations for predicting the stress distribution and the forming load (in the slab method) or the forming load and the average forming pressure (in the upper-bound method).

9.2.1 Application of Slab Method to Plane-Strain Upsetting

The Velocity Field. In this case, deformation is homogeneous and takes place in the x-z plane (Fig. 9.1). The velocity field, with the velocities in the x, y, and z directions, is defined as:

$$v_z = -V_D z/h; \; v_x = V_D x/h; \; v_y = 0 \qquad \text{(Eq 9.1)}$$

where V_D is the velocity of the top die.

The strain rates are:

$$\dot{\varepsilon}_z = \frac{\partial v_z}{\partial z} = -\frac{V_D}{h} \qquad \text{(Eq 9.2a)}$$

$$\dot{\varepsilon}_x = \frac{\partial v_x}{\partial x} = \frac{V_D}{h} = -\dot{\varepsilon}_z \qquad \text{(Eq 9.2b)}$$

$$\dot{\varepsilon}_y = \frac{\partial v_y}{\partial y} = 0 \qquad \text{(Eq 9.2c)}$$

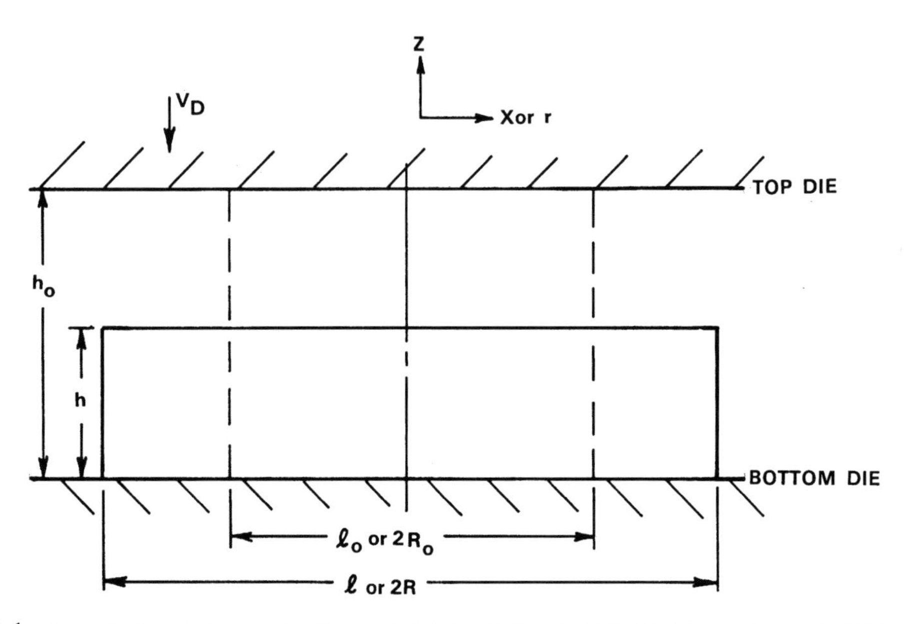

Fig. 9.1 Changes in shape during upsetting. Plane strain (initial width ℓ_o and initial height h_o) and axisymmetric (initial radius R_o)

It can be shown easily that the shear strain rates are $\dot{\gamma}_{xz} = \dot{\gamma}_{yz} = 0$.

The strains are:

$$\varepsilon_z = \ln \frac{h}{h_o}; \ \varepsilon_x = -\varepsilon_z; \ \varepsilon_y = 0 \qquad \text{(Eq 9.3)}$$

The effective strain rate is given by the equation:

$$\dot{\bar{\varepsilon}} = \sqrt{\frac{2}{3}(\dot{\varepsilon}_x^2 + \dot{\varepsilon}_y^2 + \dot{\varepsilon}_z^2)}$$

$$\dot{\bar{\varepsilon}} = \sqrt{2\left(\frac{\dot{\varepsilon}_x^2 + \dot{\varepsilon}_z^2}{3}\right)} = \frac{2}{\sqrt{3}}|\dot{\varepsilon}_x| = \frac{2}{\sqrt{3}}|\dot{\varepsilon}_z| \qquad \text{(Eq 9.4)}$$

The effective strain is:

$$\bar{\varepsilon} = \frac{2}{\sqrt{3}}|\varepsilon_z| \qquad \text{(Eq 9.5)}$$

The slab method of analysis assumes that the stresses in the metal flow direction and in the directions perpendicular to the metal flow direction are principal stresses, i.e.:

$$\sigma_z = \sigma_1, \ \sigma_x = \sigma_3; \ \sigma_y = \sigma_2 \qquad \text{(Eq 9.6)}$$

Plastic deformation/plastic flow starts when the stresses at a given point in the metal reach a certain level, as specified by a flow rule such as the Tresca or von Mises rule discussed in Chapter 5. Analysis of plastic deformation requires a certain relationship between the applied stresses and the velocity field (kinematics as described by velocity, strain (ε) and strain rate ($\dot{\varepsilon}$) fields). Such a relation between the stresses (in principal axes) and strain rates is given as follows:

$$\dot{\varepsilon}_1 = \lambda(\sigma_1 - \sigma_m)$$
$$\dot{\varepsilon}_2 = \lambda(\sigma_2 - \sigma_m)$$
$$\dot{\varepsilon}_3 = \lambda(\sigma_3 - \sigma_m)$$

where λ is a constant and σ_m is the hydrostatic stress.

These equations are called the plasticity equations. From these equations, for the plane strain case one obtains:

$$\dot{\varepsilon}_2 = \dot{\varepsilon}_y = \lambda(\sigma_2 - \sigma_m) = 0$$

or

$$\sigma_2 = \sigma_m \qquad \text{(Eq 9.7)}$$

Per definition:

$$\sigma_m = \frac{\sigma_1 + \sigma_2 + \sigma_3}{3}$$

or, with Eq 9.7:

$$\sigma_m = \sigma_2 = \frac{\sigma_1 + \sigma_3}{2}$$

For plane strain, i.e., $\sigma_2 = \sigma_m$, the von Mises rule gives:

$$3[(\sigma_1 - \sigma_m)^2 + (\sigma_3 - \sigma_m)^2 - 0] = 2\bar{\sigma}^2 \qquad \text{(Eq 9.8)}$$

After simplification, the flow rule is:

$$\sigma_1 - \sigma_3 = \sigma_z - \sigma_x = \left|\frac{2\bar{\sigma}}{\sqrt{3}}\right| \qquad \text{(Eq 9.9)}$$

Estimation of Stress Distribution. In applying slab analysis to plane strain upsetting, a slab of infinitesimal thickness is selected perpendicular to the direction of metal flow (Fig. 9.2). Assuming a depth of "1" or unit length, a force balance is made on this slab. Thus, a simple equation of static equilibrium is obtained [Thomsen et al., 1965] [Hoffman et al., 1953].

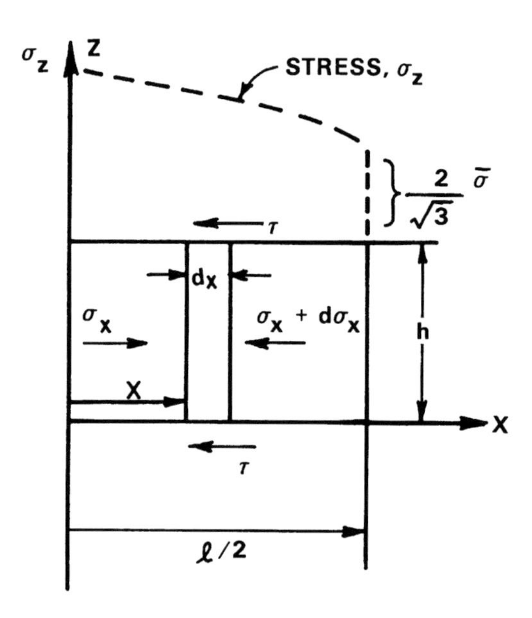

Fig. 9.2 Equilibrium of forces in plane strain homogeneous upsetting

Summation of forces in the X direction is zero or:

$$\sum F_x = \sigma_x h - (\sigma_x + d\sigma_x)h - 2\tau dx = 0$$

or

$$d\sigma_x = -2\tau dx/h$$

Thus, by integration one gets:

$$\sigma_x = -\frac{2\tau}{h} x + C$$

From the flow rule of plane strain, it follows that:

$$\sigma_z = -\frac{2\tau}{h} x + C + \left|\frac{2}{\sqrt{3}}\bar{\sigma}\right| \quad \text{(Eq 9.10)}$$

The constant C is determined from the boundary condition at $x = \ell/2$, where $\sigma_x = 0$, and, from Eq 9.9:

$$\sigma_z = \left|\frac{2}{\sqrt{3}}\bar{\sigma}\right|$$

Thus:

$$\sigma_z = -\frac{2\tau}{h}\left(\frac{\ell}{2} - x\right) - \frac{2}{\sqrt{3}}\bar{\sigma} \quad \text{(Eq 9.11)}$$

Equation 9.11 illustrates that the vertical stress increases linearly from the edge ($x = \ell/2$) of Fig. 9.2 toward the center ($x = 0$). The value of σ_z is negative, because z is considered to be positive acting upward and the upsetting stress is acting downward. Integration of Eq 9.11 gives the upsetting load.

In Eq 9.11, the frictional shear stress, τ, is equal to $m\bar{\sigma}/\sqrt{3}$. Thus, integration of Eq 9.11 over the entire width, ℓ, of the strip of unit depth gives the upsetting load per unit depth:

$$L = \frac{2\bar{\sigma}}{\sqrt{3}}\left(1 + \frac{m\ell}{4h}\right)\ell$$

9.2.2 *Application of the Slab Analysis Method to Axisymmetric Upsetting*

Figure 9.3 illustrates the notations used in the homogeneous axisymmetric upsetting. The analysis procedure is similar to that used in plane strain upsetting.

Velocity Field. The volume constancy holds; i.e., the volume of the material moved in the z direction is equal to that moved in the radial direction, or:

$$\pi r^2 V_D = 2\pi r v_r h, \text{ or } v_r = V_D r/2h$$

In the z direction, v_z can be considered to vary linearly while satisfying the boundary conditions at $z = 0$ and $z = h$. In the tangential di-

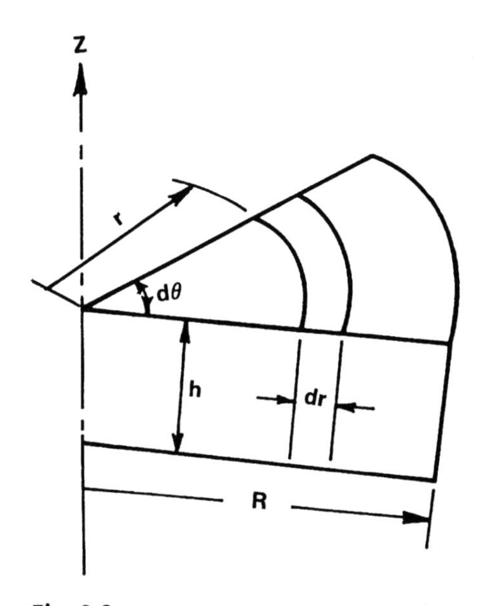

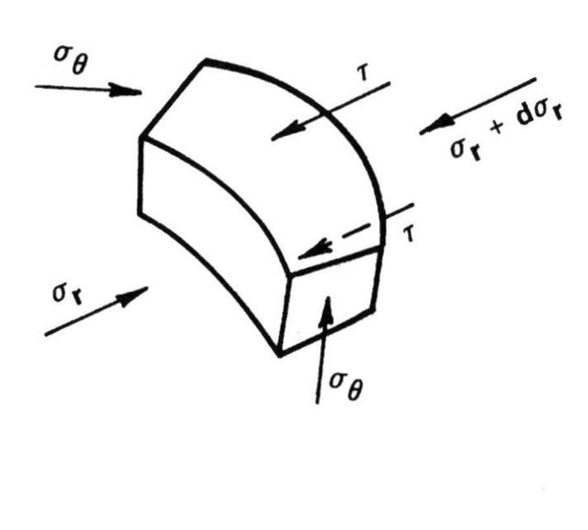

Fig. 9.3 Equilibrium of forces in axisymmetric homogeneous upsetting

rection, Θ, there is no metal flow. Thus, the velocities are:

$$v_r = V_D r/2h;\ v_z = -V_D z/h;\ v_\Theta = 0 \quad \text{(Eq 9.13)}$$

In order to obtain the strain rate in the tangential direction, it is necessary to consider the actual metal flow since $v_\Theta = 0$ and cannot be used for taking a partial derivative. Following Fig. 9.3, the increase in strain in the Θ direction, i.e., the length of the arc, is given by:

$$d\varepsilon_\Theta = \frac{(r + dr)d\Theta - rd\Theta}{rd\Theta} = \frac{dr}{r} \quad \text{(Eq 9.14)}$$

or the strain rate is

$$\dot{\varepsilon}_\Theta = \frac{d\varepsilon_\Theta}{dt} = \frac{dr}{dt}\frac{1}{r} = \frac{v_r}{r} = \frac{V_D}{2h} \quad \text{(Eq 9.14a)}$$

The other strain rates are:

$$\dot{\varepsilon}_z = \frac{\partial v_z}{\partial z} = -\frac{V_D}{h} \quad \text{(Eq 9.14b)}$$

$$\dot{\varepsilon}_r = \frac{\partial v_r}{\partial r} = \frac{V_D}{2h} = \dot{\varepsilon}_\Theta \quad \text{(Eq 9.14c)}$$

$$\dot{\gamma}_{rz} = \frac{1}{2}\left(\frac{\partial v_r}{\partial z} + \frac{\partial v}{\partial r}\right) = 0 \quad \text{(Eq 9.14d)}$$

$$\dot{\gamma}_{\Theta z} = \dot{\gamma}_{r\Theta} = 0 \quad \text{(Eq 9.14e)}$$

Thus, the effective strain rate is:

$$\dot{\bar{\varepsilon}} = \sqrt{\frac{2}{3}}\,(\dot{\varepsilon}_\Theta^2 + \dot{\varepsilon}_r^2 + \dot{\varepsilon}_z^2) = |\dot{\varepsilon}_z| \quad \text{(Eq 9.15)}$$

The strains can be obtained by integrating the strain rates with respect to time, i.e.:

$$\varepsilon_z = \int_{t_o}^{t} \dot{\varepsilon}_z dt = -\int_{t_o}^{t} \frac{V_D dt}{h}$$

or with $-dh = -V_D dt$:

$$\varepsilon_z = \int_{h_o}^{h} -\frac{dh}{h} = -\ln\frac{h}{h_o} \quad \text{(Eq 9.16a)}$$

Similarly, the other strains can be obtained as:

$$\varepsilon_\Theta = \varepsilon_r = \frac{1}{2}\ln\frac{h}{h_o} = -\frac{\varepsilon_z}{2} \quad \text{(Eq 9.16b)}$$

In analogy with Eq 9.15, the effective strain is:

$$\bar{\varepsilon} = |\varepsilon_z| \quad \text{(Eq 9.17)}$$

The flow rule for axisymmetric deformation is obtained by using a derivation similar to that used in plane strain deformation. Because $\dot{\varepsilon}_r = \dot{\varepsilon}_\theta$, the plasticity equations give:

$$\sigma_r = \sigma_\theta \text{ or } \sigma_2 = \sigma_3$$

Thus, the von Mises flow rule for axisymmetric upsetting is:

$$\sigma_1 - \sigma_2 = |\bar{\sigma}| \text{ or } \sigma_z - \sigma_r = |\bar{\sigma}| \quad \text{(Eq 9.18)}$$

Estimation of Stress Distribution. The equilibrium of forces in the r direction (Fig. 9.3) gives [Thomsen et al., 1965] [Hoffman et al., 1953]:

$$\sum F_r = \sigma_r(d\theta)rh - (\sigma_r + d\sigma_r)$$
$$(r + dr)hd\theta + 2\sigma_\theta \sin\frac{d\theta}{2}$$
$$hdr - 2\tau r d\theta dr = 0 \quad \text{(Eq 9.19)}$$

The angle $d\theta$ is very small. Thus, with $\sin d\theta/2 = d\theta/2$, and after canceling appropriate terms, Eq 9.19 reduces to:

$$-\sigma_r - \frac{d\sigma_r}{dr}r + \sigma_\theta - \frac{2\tau}{h}r = 0 \quad \text{(Eq 9.20)}$$

Since in axisymmetric deformation, $\dot{\varepsilon}_r = \dot{\varepsilon}_\theta$, the plasticity equations give:

$$\sigma_r = \sigma_\theta, \text{ or } \frac{d\sigma_r}{dr} + \frac{2\tau}{h} = 0 \quad \text{(Eq 9.21)}$$

Integration gives:

$$\sigma_r = -\frac{2\tau}{h}r + C$$

The constant C is determined from the condition that at the free boundary, $r = R$ in Fig. 9.3, and the radial stress $\sigma_r = 0$. Thus, integration of Eq 9.21 gives:

$$\sigma_r = \frac{2\tau}{h}(r - R) \quad \text{(Eq 9.22)}$$

With the flow rule, Eq 9.22 is transformed into:

$$\sigma_z = \frac{2\tau}{h}(r - R) - \bar{\sigma} \qquad \text{(Eq 9.23)}$$

Equation 9.23 illustrates that the stress increases linearly from the edge toward the center. The upsetting load can now be obtained by integrating the stress distribution over the circular surface of the cylindrical upset:

$$L = \int_0^R \sigma_z 2\pi r dr$$

Considering that $\tau = m\bar{\sigma}/\sqrt{3}$, integration gives:

$$L = \bar{\sigma}\pi R^2 \left(1 + \frac{2mR}{3h\sqrt{3}}\right) \qquad \text{(Eq 9.24)}$$

9.3 Upper Bound Method and Its Application to Axisymmetric Upsetting

9.3.1 Principles of the Method

This method can be used to estimate the deformation load and the average forming pressure. For describing metal flow with the upper-bound method, it is necessary to make the usual assumptions, discussed earlier in the slab method, as well as perform the following steps:

1. Describe a family of admissible velocity fields (use parameters to be determined later); these must satisfy the conditions of: incompressibility, continuity, and velocity boundaries.
2. Calculate the energy rates of deformation, internal shear, and friction shear.
3. Calculate the total energy rate and minimize it with respect to unknown parameters of velocity field formulation.

The load is then obtained by dividing the energy rate by the relative velocity between the die and the deforming material.

The total energy rate, $\dot{E}_T$, is given by $\dot{E}_T$ = load × die velocity, or [Avitzur, 1968]:

$$\dot{E}_T = LV_D = \dot{E}_D + \dot{E}_S + \dot{E}_F$$

or

$$\dot{E}_T = \int_V \bar{\sigma}\dot{\bar{\varepsilon}}dV + \int_{SS} \tau|\Delta v|ds + \int_{SF} \tau_i v_i ds \qquad \text{(Eq 9.25)}$$

where $\dot{E}_D$, $\dot{E}_S$, and $\dot{E}_F$ are the energy rates for deformation, internal shear, and friction, respectively; L is the forming load; V is the volume of deforming material; v is the relative velocity between two zones of material, when the velocity field has internal shear surfaces; S indicates surface (internal or at die/material interface); v_i is the die material interface velocity in the "i" portion of the deforming material; $\tau = \bar{\sigma}/\sqrt{3}$; and $\tau_i = m_i\bar{\sigma}/\sqrt{3}$ = interface shear stress at the "i" portion of the deforming material.

Based on limit theorems [Avitzur, 1968], the load calculated with Eq 9.25 is necessarily higher than the actual load and therefore represents an upper bound to the actual forming load. Thus, the lower this upper bound load is, the better the prediction. Often the velocity field considered includes one or more parameters that are determined by minimizing the total energy rate with respect to those parameters. Thus, a somewhat better upper-bound velocity field and solution are obtained. In general, with an increasing number of parameters in the velocity field, the solution improves while the computations become more complex. Consequently, in the practical use of the upper-bound method, practical compromises are made in selecting an admissible velocity field.

9.3.2 Application to Axisymmetric Homogeneous Upsetting

The velocity field for homogeneous upsetting is given by Eq 9.13 to 9.15. Thus, all the velocities and strain rates are known. Assuming a constant flow stress, $\bar{\sigma}$, the deformation energy rate is:

$$\dot{E}_D = \int_V \bar{\sigma}\dot{\bar{\varepsilon}}dV = h\pi R^2\bar{\sigma}\frac{V_D}{h} \qquad \text{(Eq 9.26)}$$

$\dot{E}_S$ (internal shear energy rate) = 0, because there are no internal velocity discontinuities in the present homogeneous velocity field.

The friction energy rate is:

$$\dot{E}_F = 2\int_{SF} \tau_i v_i ds$$

where v_i is the radial velocity, given in Eq 9.13, and $ds = 2\pi r dr$. $\dot{E}_F$ includes the friction energies on both the top and bottom surfaces of the deforming part. Thus,

$$\dot{E}_F = 2\int_0^R \tau_i \frac{V_D}{2h} r2\pi r dr = \frac{4\pi\tau_i V_D}{2h}\int_0^R r^2 dr$$

or, with $\tau = m\bar{\sigma}/\sqrt{3}$,

$$\dot{E}_F = \frac{2}{3}\pi m \frac{\bar{\sigma}}{\sqrt{3}}\frac{V_D}{h}R^3 \quad \text{(Eq 9.27)}$$

The total energy rate is:

$$\dot{E}_T = \dot{E}_D + \dot{E}_F$$

or

$$\dot{E}_T = \pi R^2 \bar{\sigma} V_D + \frac{2}{3}\pi m \frac{\bar{\sigma}}{\sqrt{3}}\frac{V_D}{h}R^3 \quad \text{(Eq 9.28)}$$

The load is:

$$L = \frac{\dot{E}_T}{V_D} = \pi R^2 \bar{\sigma}\left(1 + \frac{2}{3\sqrt{3}} m \frac{R}{h}\right) \quad \text{(Eq 9.29)}$$

Comparison of Eq 9.29 and 9.24 indicates that in axisymmetric homogeneous upsetting the loads calculated by the slab and upper-bound methods both give the same end result.

9.3.3 Application to Nonhomogeneous Upsetting

Homogeneous upsetting can only be achieved at low strains and with nearly perfect lubrication. In all practical upsetting operations, the friction at the die/material interface prevents the metal from flowing radially in a uniform fashion. As a result, bulging of the free surfaces occurs, and the radial and axial velocities are functions of z as well as of r. In this case, a velocity field may be given by [Lee et al., 1972]:

$$v_\theta = 0 \quad \text{(Eq 9.30a)}$$

$$v_z = -2Az(1 - \beta z^2/3) \quad \text{(Eq 9.30b)}$$

$$v_r = A(1 - \beta z^2)r \quad \text{(Eq 9.30c)}$$

where β is a parameter representing the severity of the bulge, and A is determined from the velocity boundary condition at $z = h$, to be:

$$A = \frac{V_D}{2h(1 - \beta h^2/3)}$$

The strain rates are:

$$\dot{\varepsilon}_r = \frac{v_r}{r} = A(1 - \beta z^2) \quad \text{(Eq 9.31a)}$$

$$\dot{\varepsilon}_\theta = \frac{v_r}{r} = A(1 - \beta z^2) \quad \text{(Eq 9.31b)}$$

$$\dot{\varepsilon}_z = \frac{v_z}{z} = -2A(1 - \beta z^2) \quad \text{(Eq 9.31c)}$$

$$\dot{\gamma}_{r\theta} = \dot{\gamma}_{\theta z} = 0;$$

$$\dot{\gamma}_{rz} = \frac{\partial v_r}{\partial z} + \frac{\partial v_z}{\partial r} = -2A\beta zr \quad \text{(Eq 9.31d)}$$

The effective strain rate is calculated from:

$$\dot{\bar{\varepsilon}} = \left[\frac{2}{3}\left(\dot{\varepsilon}_r^2 + \dot{\varepsilon}_\theta^2 + \dot{\varepsilon}_z^2 + \frac{1}{2}\dot{\gamma}_{rz}^2\right)\right]^{1/2}$$

or

$$\dot{\bar{\varepsilon}} = \frac{2A}{\sqrt{3}}[3(1 - \beta z^2)^2 + (\beta rz)^2]^{1/2} \quad \text{(Eq 9.32)}$$

The total energy dissipation, $\dot{E}_T$, given by Eq 9.25 can now be calculated analytically or numerically. The exact value of β is determined from the minimization condition, i.e., from:

$$\frac{\partial \dot{E}_T}{\partial \beta} = 0$$

The value of β, obtained from Eq 9.33, is used to calculate the velocities and strain rates that then give the minimum value of the energy rate, $\dot{E}_{min}$. The upsetting load is then given by:

$$L = \dot{E}_{min}/V_D$$

9.4 Finite Element Method in Metal Forming

The basic approach of the finite element (FE) method is one of discretization. The FE model is constructed in the following manner [Kobayashi et al., 1989]:

- A number of finite points are identified in the domain of the function and the values of the function and its derivatives, when appropriate, are specified at these points.

- The domain of the function is represented approximately by a finite collection of subdomains called finite elements.
- The domain is then an assemblage of elements connected together appropriately on their boundaries.
- The function is approximated locally within each element by continuous functions that are uniquely described in terms of the nodal point values associated with the particular element.

The path to the solution of a finite element problem consists of five specific steps: (1) identification of the problem, (2) definition of the element, (3) establishment of the element equation, (4) the assemblage of element equations, and (5) the numerical solution of the global equations. The formation of element equations is accomplished from one of four directions: (a) direct approach, (b) variational method, (c) method of weighted residuals, and (d) energy balance approach.

The main advantages of the FE method are:

- The capability of obtaining detailed solutions of the mechanics in a deforming body, namely, velocities, shapes, strains, stresses, temperatures, or contact pressure distributions
- The fact that a computer code, once written, can be used for a large variety of problems by simply changing the input data

9.4.1 *Basis for the Finite Element Formulation*

The basis for the finite element metal forming formulation, using the "variational approach," is to formulate the proper functional (function of functions) depending on specific constitutive relations. The variational approach is based on one of two variational principles. It requires that among admissible velocities u_i that satisfy the conditions of compatibility and incompressibility, as well as the velocity boundary conditions, the actual solution gives the following functional a stationary value [Kobayashi et al., 1989]:

$$\pi = \int_V \bar{\sigma}\dot{\bar{\varepsilon}}dV - \int_{S_F} F_i u_i dS \qquad \text{(Eq 9.35a)}$$

(for rigid-plastic materials) and:

$$\pi = \int_V E(\dot{\varepsilon}_{ij})dV - \int_{S_F} F_i u_i dS \qquad \text{(Eq 9.35b)}$$

(for rigid-viscoplastic materials) where $\bar{\sigma}$ is the effective stress, $\dot{\bar{\varepsilon}}$ is the effective strain rate, F_i represents surface tractions, $E(\dot{\varepsilon}_{ij})$ is the work function, and V and S the volume and surface of the deforming workpiece, respectively. The solution of the original boundary-value problem is then obtained from the solution of the dual-variational problem, where the first-order variation of the functional vanishes, namely:

$$\delta\pi = \int_V \bar{\sigma}\delta\dot{\bar{\varepsilon}}dV - \int_{S_F} F_i\delta u_i dS = 0 \qquad \text{(Eq 9.36)}$$

where $\bar{\sigma} = \bar{\sigma}(\bar{\varepsilon})$ and $\bar{\sigma} = \bar{\sigma}(\bar{\varepsilon},\dot{\bar{\varepsilon}})$ for the rigid-plastic and rigid-viscoplastic materials, respectively.

For an accurate finite element prediction of material flow in a forging process, the formulation must take into account the large plastic deformation, incompressibility, workpiece-tool contact, and, when necessary, temperature coupling.

The basic equations to be satisfied are the equilibrium equation, the incompressibility condition, and the constitutive relationship. When applying the penalty method, the velocity is the primary solution variable. The variational equation is in the form [Li et al., 2001]:

$$\delta\pi(v) = \int_V \bar{\sigma}\delta\dot{\bar{\varepsilon}}dV + K\int_V \dot{\varepsilon}_v\delta\dot{\varepsilon}_v dV - \int_{S_F} F_i\delta u_i dS = 0 \qquad \text{(Eq 9.37)}$$

where K, a penalty constant, is a very large positive constant.

An alternative method of removing the incompressibility constraint is to use a Lagrange multiplier [Washizu, 1968] and [Lee et al., 1973] and modifying the functional by adding the term $\int\lambda\dot{\varepsilon}_v dV$, where $\dot{\varepsilon}_v = \dot{\varepsilon}_{ii}$, is the volumetric strain rate. Then:

$$\delta\pi = \int_V \bar{\sigma}\delta\dot{\bar{\varepsilon}}dV + \int_V \lambda\delta\dot{\varepsilon}_v dV + \int_V \dot{\varepsilon}\delta\lambda dV - \int_{S_F} F_i\delta u_i dS = 0 \qquad \text{(Eq 9.38)}$$

In the mixed formulation, both the velocity and pressure are solution variables. They are solved by the following variational equation [Li et al., 2001],

$$\delta\pi(v,p) = \int_V \bar{\sigma}\delta\dot{\bar{\varepsilon}}dV + \int_V p\delta\dot{\varepsilon}_v dV + \int_V \dot{\varepsilon}_v \delta p - \int_{S_F} F_i \delta u_i dS = 0 \quad \text{(Eq 9.39)}$$

where p is the pressure.

Equations 9.37 to 9.39 can be converted into a set of algebraic equations by utilizing the standard FEM discretization procedures. Due to the nonlinearity involved in the material properties and frictional contact conditions, this solution is obtained iteratively.

The temperature distribution of the workpiece and/or dies can be obtained readily by solving the energy balance equation rewritten by using the weighted residual method as [Li et al., 2001]:

$$\int_V kT_{ij}\delta T_{ij} dV + \int_V \rho c\dot{T}\delta T dV - \int_V \alpha\bar{\sigma}\dot{\bar{\varepsilon}}\delta T dV = \int_S q_n \delta T dS \quad \text{(Eq 9.40)}$$

where k is the thermal conductivity, T the temperature, ρ the density, c the specific heat, α a fraction of deformation energy that converts into heat, and q_n the heat flux normal to the boundary, including heat loss to the environment and friction heat between two contacting objects. Using FEM discretization, Eq 9.40 can also be converted to a system of algebraic equations and solved by a standard method. In practice, the solutions of mechanical and thermal problems are coupled in a staggered manner. After the nodal velocities are solved at a given step, the deformed configuration can be obtained by updating the nodal coordinates [Li et al., 2001].

9.4.2 Computer Implementation of the Finite Element Method

Computer implementation of the basic steps in a standard finite element analysis consists of three distinct units, viz. the preprocessor, processor, and the postprocessor.

Preprocessor. This operation precedes the analysis operation. It takes in minimal information from the user (input) to generate all necessary problem parameters (output) required for a finite element analysis.

The input to the preprocessor includes information on the solid model, discretization requirements, material identification and parameters, and boundary conditions; whereas the output includes coordinates for nodes, element connectivity and element information, values of material parameters for each element, and boundary loading conditions on each node.

In the preprocessing stage, a continuum is divided into a finite number of subregions (or elements) of simple geometry (triangles, rectangles, tetrahedral, etc.). Key points are then selected on elements to serve as nodes where problem equations such as equilibrium and compatibility are satisfied.

In general, preprocessing involves the following steps to run a successful simulation using a commercial FE package:

1. Select the appropriate geometry/section for simulation based on the symmetry of the component to be analyzed.
2. Assign workpiece material properties in the form of the flow stress curves. These may be user defined or selected from the database provided in the FE package.
3. Mesh the workpiece using mesh density windows, if necessary, to refine mesh in critical areas.
4. Define the boundary conditions (velocity, pressure, force, etc.). Specify the movement and direction for the dies. Specify the interface boundary and friction conditions. This prevents penetration of the dies into the workpiece and also affects the metal flow depending on the friction specified. In massive forming processes such as forging, extrusion, etc., the shear friction factor "m" is specified.

Note: If die stress analysis or thermal analysis is required, one would need to mesh the dies and specify the material properties for them. Also die and workpiece temperatures and interface heat transfer coefficients would need to be specified. Material data at elevated temperatures would be required for the tools and the workpiece. The die stress analysis procedure is described in Chapter 16 on process modeling in impression die forging.

Processor. This is the main operation in the analysis. The output from the preprocessor (including nodal coordinates, element connectivity, material parameters, and boundary/loading conditions) serves as the input to this module. It establishes a set of algebraic equations that are to be solved, from the governing equations of the boundary value problem. The governing equations include conservation principles, kinematic relations, and constitutive relations. The FE engine/solver then solves the set of linear or nonlinear algebraic equations to obtain the state variables at the nodes. It also evaluates the flux quantities inside each element.

The following steps are pursued in this operation:

1. A suitable interpolation function is assumed for each of the dependent variables in terms of the nodal values.
2. Kinematic and constitutive relations are satisfied within each element.
3. Using work or energy principles, stiffness matrices and equivalent nodal loads are established.
4. Equations are solved for nodal values of the dependent variables.

Postprocessor. This operation prints and plots the values of state variables and fluxes in the meshed domain. Reactions may be evaluated. Output may be in the form of data tables or as contour plots.

Figure 9.4 shows a flowchart, which shows the sequence of finite element implementation in analysis of metal forming processes.

In interpreting FE results, careful attention is required since the idealization of the physical problem to a mathematical one involves certain assumptions that lead to the differential equation governing the mathematical model. As illustrated in Fig. 9.5, refinement of the mesh size, alteration of the boundary conditions, etc., may be needed to increase the accuracy of the solution.

9.4.3 *Analysis of Axisymmetric Upsetting by the FE Method*

The use of the FE method in analysis of forging processes is discussed with the help of a simple cylindrical compression simulation. Figure 9.6(a) shows the cutaway of a cylinder, which is to be upset. Since the cylinder is symmetric about the central axis, i.e., axisymmetric, a 2-D simulation is adequate to analyze the metal flow during deformation. Figure 9.6(b) shows the half model selected for simulation as a result of rotational symmetry, with the nodes along the axis of symmetry (centerline) restricted along the R-direction. This model can be further simplified to yield a quarter model (Fig. 9.6c) with velocity/displacement boundary conditions restricting the movement of the nodes along the Z-direction as shown. The boundary conditions necessary to be prescribed for this problem under isothermal conditions can be summarized as:

- Velocity/displacement boundary conditions depending on the symmetry of the workpiece, to restrict to movement of nodes along the planes of symmetry as required
- Die movement and direction according to the process being analyzed
- Interface friction conditions at the surfaces of contact between the tools and the workpiece

9.4.4 *Analysis of Plane Strain Deformation*

Figure 9.7 shows the setup of an FE model for a plane strain condition. In this case the part has a uniform cross section along the length and

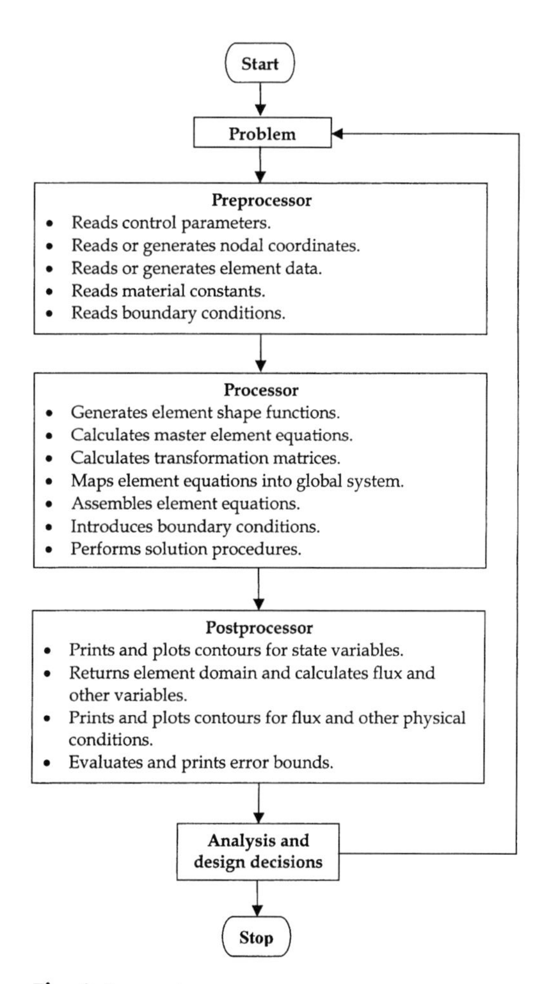

Fig. 9.4 Flowchart of finite element implementation

the strain along the Z or length direction is negligible; i.e., ε_z is negligible. Hence, only the two-dimensional cross section shown in Fig. 9.7(b) is selected for process simulation.

The procedure for simulation is the same as mentioned for the axisymmetric case. However, the velocity boundary conditions are different since the metal is free to flow in the X direction (Fig. 9.7b). If, however, the cross section is uniform, a quarter model can be used for simulation as shown in Fig. 9.7(c) with velocity boundary conditions similar to those used for the axisymmetric case discussed in the previous section. It should be noted, however, that the results obtained from the plane strain simulation have to be multiplied by the length of the billet used for the forging process to get the final results.

9.4.5 Three-Dimensional Nonisothermal FE Analysis

Figure 9.8 shows the FE model of the hot forging process of an automotive component. The starting preform geometry is not axisymmetric; hence a 2-D simulation could not be used to analyze the metal flow. The preform is, however, symmetric, and hence a quarter model could be used for modeling the process. Since the forging process is performed at elevated temperatures, a nonisothermal simulation is conducted. The simulation input differs from those of an isothermal simulation in the following ways:

- Tool and workpiece temperatures have to be considered, and thermal properties such as

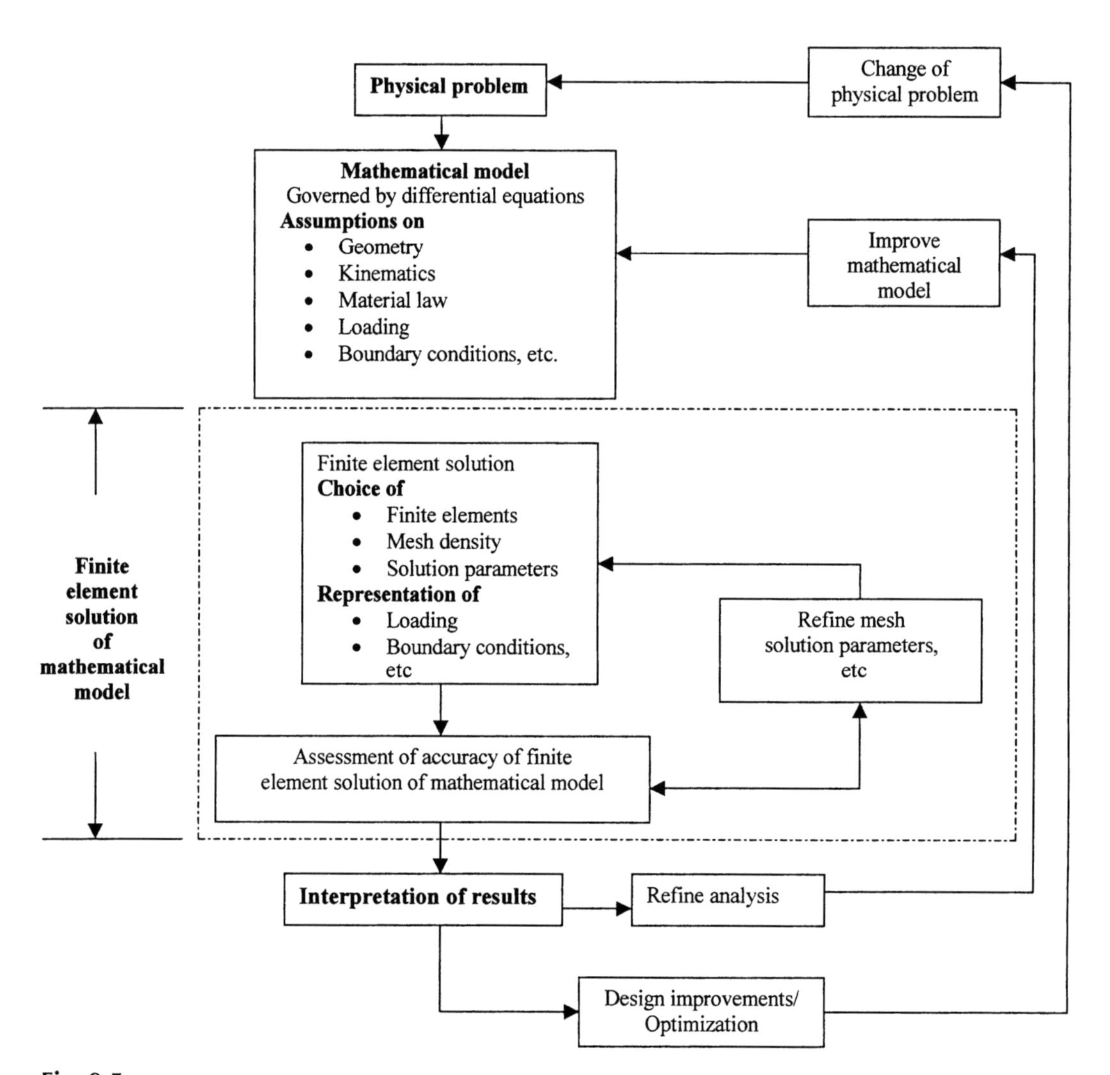

Fig. 9.5 The process of finite element analysis. [Bathe, 1996]

thermal conductivity, heat capacity, interface heat transfer coefficients, etc. have to be specified.
- Die and workpiece material properties are specified as a function of temperature and strain rate. Dies (still considered as rigid) have to be meshed since heat transfer between workpiece and tools is simulated.

The boundary conditions in a 3-D simulation are similar to those in a 2-D process:

- Velocity/displacement boundary conditions need to be specified depending on the workpiece geometry. In Fig. 9.8(b), movement of the boundary nodes is restricted along the planes of symmetry.
- Interface friction is specified depending on the type of lubrication at the tool/workpiece interface. In addition to the friction, the interface heat transfer coefficient is also a critical input to the simulation.

9.4.6 *Mesh Generation in FE Simulation of Forging Processes*

Bulk forming processes are generally characterized by large plastic deformation of the

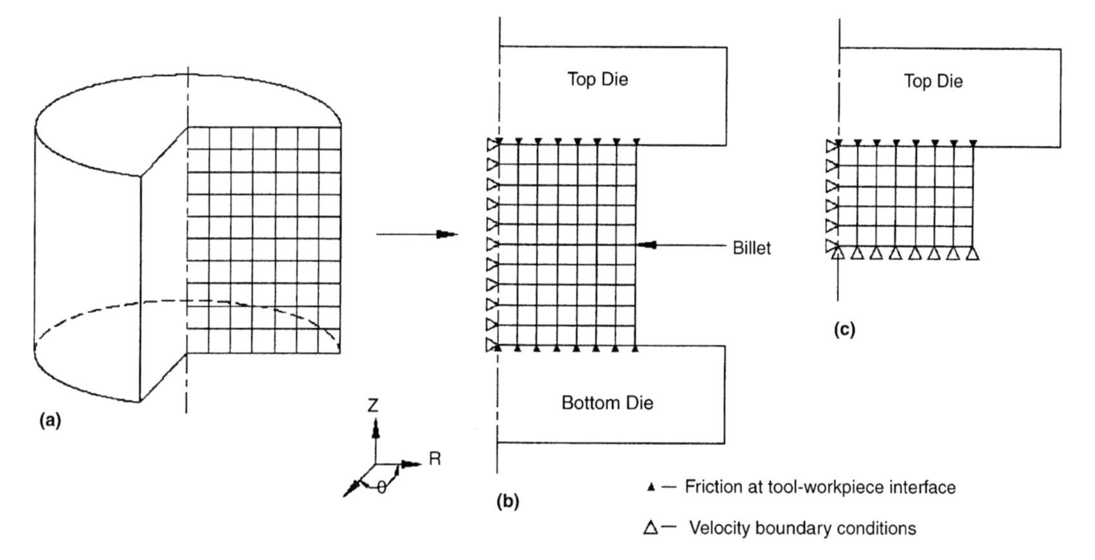

Fig. 9.6 Analysis of axisymmetric cylinder upsetting using the FE method. (a) Cutaway of the cylinder. (b) Half model. (c) Quarter model

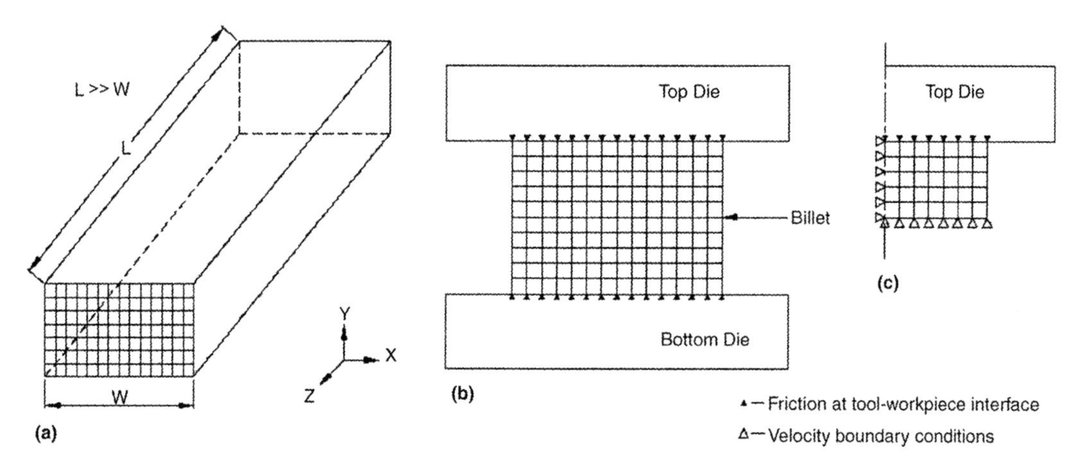

Fig. 9.7 Analysis of plane strain upsetting by the FE method. (a) Part for plane strain upsetting. (b) Full model. (c) Quarter model

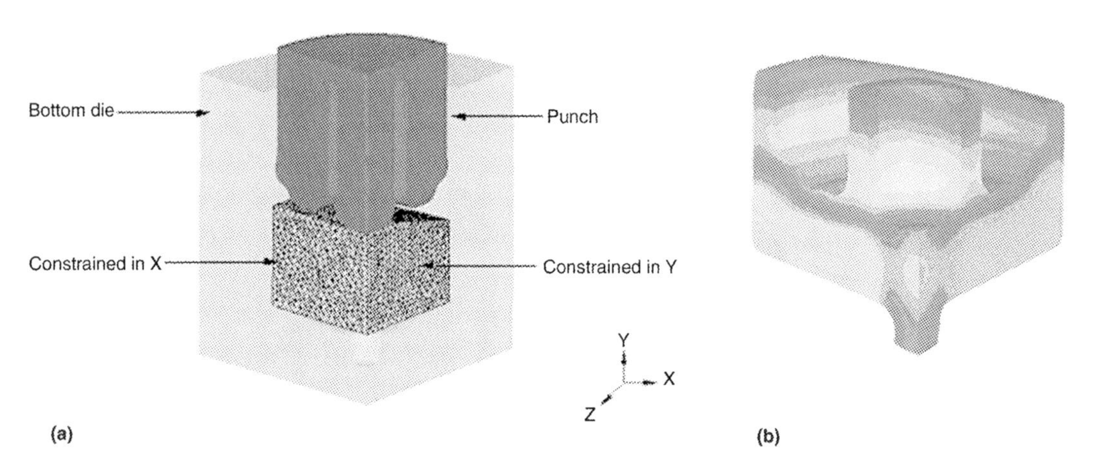

Fig. 9.8 Three-dimensional non-isothermal FE simulation of a forging process. (a) FE model. (b) Final part with temperature contour.

workpiece accompanied by significant relative motion between the deforming material and the tool surfaces. The starting mesh in the FE simulation is generally well defined with the help of mesh density windows to refine the mesh in critical areas. However, as the simulation progresses the mesh tends to get significantly distorted. This calls for (a) the generation of a new mesh taking into consideration the updated geometry of the workpiece and (b) interpolation of the deformation history from the old mesh to the new mesh. Current commercial FE codes, such as DEFORM™ accomplish this using automatic mesh generation (AMG) subroutines. AMG basically determines the optimal mesh density distribution and generates the mesh based on the given density. Once the mesh density was defined, the mesh generation procedure considers the following factors [Altan et al., 1979]:

- Geometry representation.
- Density representation. The density specification is accommodated in the geometry.
- Identification of critical points (like sharp corner points) for accurate representation of the geometry.
- Node generation. The number of nodes between critical points are generated and repositioned based on the density distribution.
- Shape improvement and bandwidth minimization. The shape of the elements after the mesh generation procedure has to be smoothed to yield a usable mesh. Also the bandwidth of the stiffness matrix has to be improved to obtain a computationally efficient mesh.

The FE modeling inputs and outputs are discussed in detail in Chapter 16 on process modeling.

FE simulation is widely used in the industry for the design of forging sequences, prediction of defects, optimization of flash dimensions, die stress analysis, etc. A number of commercial FE codes are available in the market for simulation of bulk forming processes, viz. DEFORM™, FORGE™, QFORM™, etc. Some practical applications of FE simulation in hot and cold forging processes are presented in Chapters 16 and 18.

REFERENCES

[Altan et al., 1979]: Altan, T., Lahoti, G.D., "Limitations, Applicability and Usefulness of Different Methods in Analyzing Forming Problems," *Ann. CIRP,* Vol 28 (No. 2), 1979, p 473.

[Avitzur, 1968]: Avitzur, B., *Metal Forming: Processes and Analysis,* McGraw-Hill, 1968.

[Bathe, 1996]: Bathe, K.J., *Finite Element Procedures,* Prentice Hall, 1996.

[Becker, 1992]: Becker, A.A., *The Boundary Element Method in Engineering,* McGraw-Hill International Editions, 1992.

[Hoffman et al., 1953]: Hoffman, O., Sachs, G., *Introduction to the Theory of Plasticity for Engineers,* McGraw-Hill, 1953.

[Kobayashi et al., 1989]: Kobayashi, S., Oh, S.I., Altan, T., *Metal Forming and the Finite Element Method,* Oxford University Press, 1989.

[Lee et al., 1972]: Lee, C.H., Altan, T., "Influence of Flow Stress and Friction upon Metal Flow in Upset Forging of Rings and Cylinders," *ASME Trans., J. Eng. Ind.,* Aug 1972, p 775.

[Lee et al., 1973]: Lee, C.H., Kobayashi, S., "New Solutions to Rigid-Plastic Deformation Problems using a Matrix Method," *Trans. ASME, J. Eng. Ind.,* Vol 95, p 865.

[Li et al., 2001]: Li, G., Jinn, J.T., Wu, W.T., Oh, S.I., "Recent Development and Applications of Three-Dimensional Finite Element Modeling in Bulk Forming Processes," *J. Mater. Process. Technol.,* Vol 113, 2001, p 40–45.

[Thomsen et al., 1954]: Thomsen, E.G., Yang, C.T., Bierbower, J.B., "An Experimental Investigation of the Mechanics of Plastic Deformation of Metals," *Univ. of California Pub. Eng.,* Vol 5, 1954.

[Thomsen et al., 1965]: Thomsen, E.G., Yang, C.T., Kobayashi, S., *Mechanics of Plastic Deformation in Metal Processing,* Macmillan, 1965.

[Washizu, 1968]: Washizu, K., *Variational Methods in Elasticity and Plasticity,* Pergamon Press, Oxford, 1968.

SELECTED REFERENCE

[Altan et al., 1983]: Altan, T., Oh, S.I., Gegel, H., *Metal Forming: Fundamentals and Applications,* ASM International, 1983.

CHAPTER 10

Principles of Forging Machines

Manas Shirgaokar

10.1 Introduction

In a practical sense, each forming process is associated with at least one type of forming machine (or "equipment," as it is sometimes called in practice). The forming machines vary in factors such as the rate at which energy is applied to the workpiece and the capability to control the energy. Each type has distinct advantages and disadvantages, depending on the number of forgings to be produced, dimensional precision, and the alloy being forged. The introduction of a new process invariably depends on the cost effectiveness and production rate of the machine associated with that process. Therefore, capabilities of the machine associated with the new process are of paramount consideration. The forming (industrial, mechanical, or metallurgical) engineer must have specific knowledge of forming machines so that he/she can:

- Use existing machinery more efficiently
- Define with accuracy the existing plant capacity
- Better communicate with, and at times request improved performance from, the machine builder
- If necessary, develop in-house proprietary machines and processes not available in the machine-tool market

10.2 Interaction between Process Requirements and Forming Machines

The behavior and characteristics of the forming machine influence:

- The flow stress and workability of the deforming material
- The temperatures in the material and in the tools, especially in hot forming
- The load and energy requirements for a given product geometry and material
- The "as-formed" tolerances of the parts
- The production rate

The interaction between the principal machine and process variables is illustrated in Fig. 10.1 for hot forming processes conducted in presses. As can be seen in Fig. 10.1, the flow stress, $\bar{\sigma}$, the interface friction conditions, and the forging geometry (dimensions, shape) determine (a) the load, L_P, at each position of the stroke and (b) the energy, E_P, required by the forming process.

The flow stress, $\bar{\sigma}$, increases with increasing deformation rate, $\dot{\bar{\varepsilon}}$, and with decreasing temperature, θ. The magnitudes of these variations depend on the specific forming material. The frictional conditions deteriorate with increasing die chilling.

As indicated by lines connected to the temperature block, for a given initial stock temperature, the temperature variations in the part are largely influenced by (a) the surface area of contact between the dies and the part, (b) the part thickness or volume, (c) the die temperature, (d) the amount of heat generated by deformation and friction, and (e) the contact time under pressure.

The velocity of the slide under pressure, V_p, determines mainly the contact time under pressure, t_p, and the deformation rate, $\dot{\bar{\varepsilon}}$. The number of strokes per minute under no-load conditions, n_o, the machine energy, E_M, and the deformation energy, E_p, required by the process influence the slide velocity under load, V_p, and the number of strokes under load, n_p; n_p determines the maximum number of parts formed per minute (i.e., the production rate) provided that feeding and unloading of the machine can be carried out at that speed.

The relationships illustrated in Fig. 10.1 apply directly to hot forming of discrete parts in hydraulic, mechanical, and screw presses, which are discussed later. However, in principle, most of the same relationships apply also in other hot forming processes such as hot extrusion and hot rolling.

10.3 Load and Energy Requirements in Forming

It is useful to consider forming load and energy as related to forming equipment. For a given material, a specific forming operation (such as closed-die forging with flash, forward, or backward extrusion, upset forging, bending, etc.) requires a certain variation of the forming load over the slide displacement (or stroke). This fact is illustrated qualitatively in Fig. 10.2, which shows load versus displacement curves characteristic of various forming operations.

For a given part geometry, the absolute load values will vary with the flow stress of the given material as well as with frictional conditions. In the forming operation, the equipment must supply the maximum load as well as the energy required by the process.

The load-displacement curves, in hot forging a steel part under different types of forging equipment, are shown in Fig. 10.3. These curves illustrate that, due to strain rate and temperature effects, for the same forging process, different forging loads and energies are required by different machines. For the hammer, the forging load is initially higher, due to strain-rate effects, but the maximum load is lower than for either hydraulic or screw presses. The reason is that the extruded flash cools rapidly in the presses, while in the hammer, the flash temperature remains nearly the same as the initial stock temperature.

Thus, in hot forging, not only the material and the forged shape, but also the rate of deformation

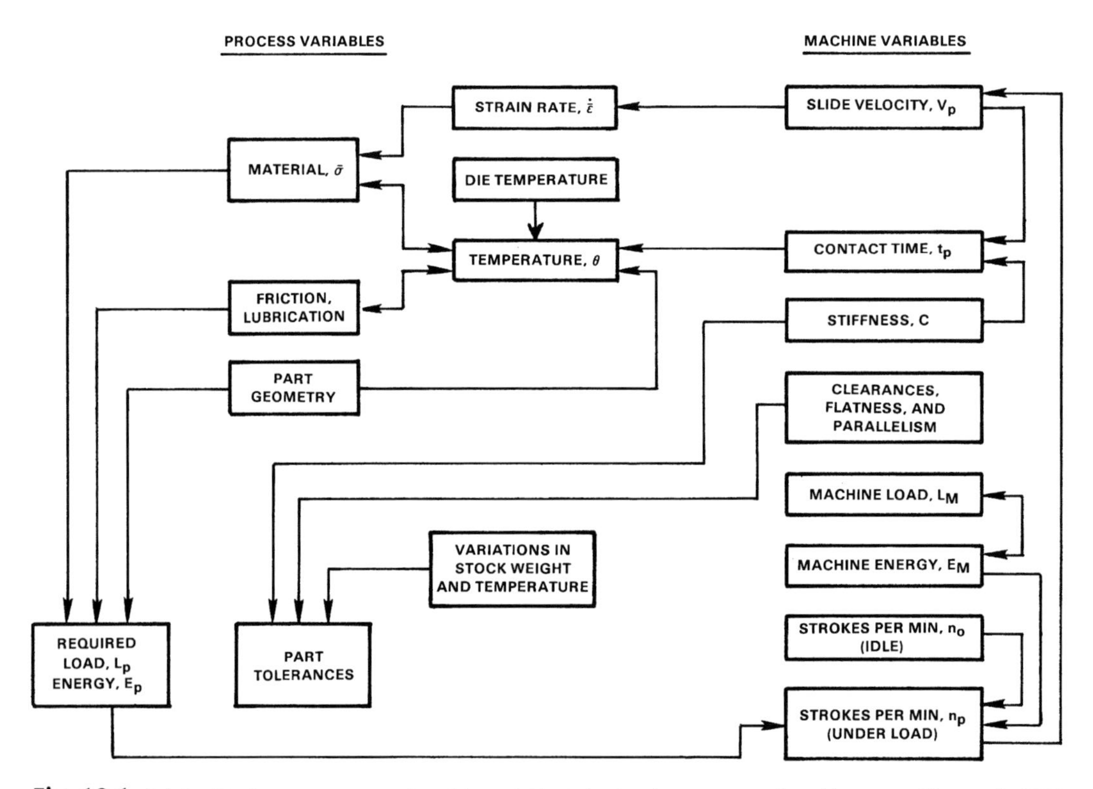

Fig. 10.1 Relationships between process and machine variables in hot forming process conducted in presses. [Altan et al., 1973]

and die-chilling effects and, therefore, the type of equipment used, determine the metal flow behavior and the forging load and energy required for the process. Surface tearing and cracking or development of shear bands in the forged material often can be explained by excessive chilling of the surface layers of the forged part near the die/material interface.

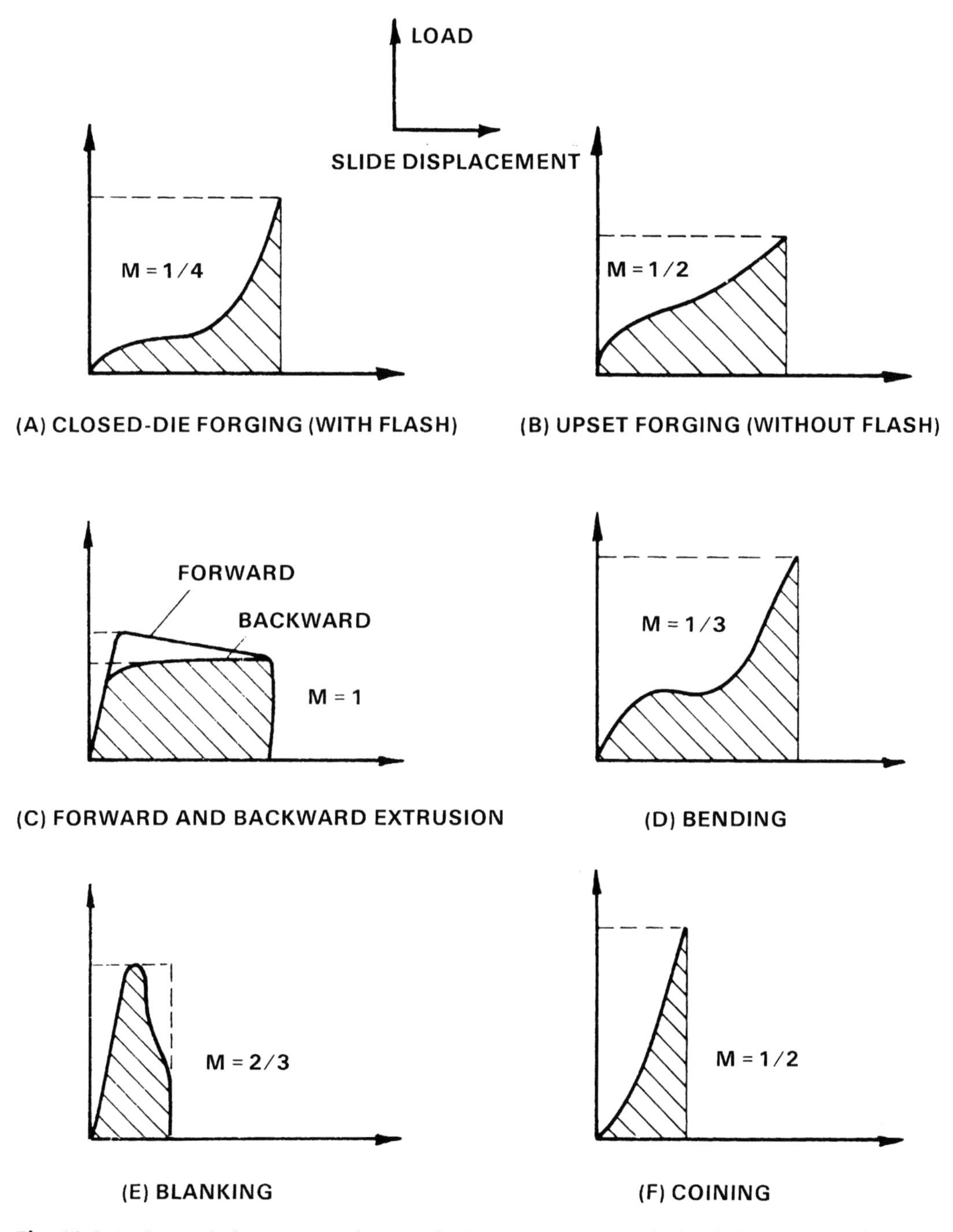

Fig. 10.2 Load versus displacement curves for various forming operations (energy = load × displacement × M, where M is a factor characteristic of the specific forming operation). [Altan et al., 1973]

10.4 Classification and Characteristics of Forming Machines

In metal forming processes, workpieces are generally fully or nearly fully formed by using two-piece tools. A metal forming machine tool is used to bring the two pieces together to form the workpiece. The machine also provides the necessary forces, energy, and torque for the process to be completed successfully, ensuring guidance of the two tool halves.

Based on the type of relative movement between the tools or the tool parts, the metal forming machine tools can be classified mainly into two groups:

- Machines with linear relative tool movement
- Machines with nonlinear relative tool movement

Machines in which the relative tool movements cannot be classified into either of the two groups are called special-purpose machines. The machines belonging to this category are those operated on working media and energy. The various forming processes, discussed in Chapter 2, are associated with a large number of forming machines. These include:

- Rolling mills for plate, strip and shapes
- Machines for profile rolling from strip
- Ring rolling machines
- Thread rolling and surface rolling machines
- Magnetic and explosive forming machines
- Draw benches for tube and rod; wire and rod drawing machines
- Machines for pressing-type operations, i.e., presses

Among those listed above, "pressing"-type machines are most widely used and applied for a variety of different purposes. These machines can be classified into three types [Kienzle, 1965] [Kienzle, 1953]:

- Load-restricted machines (hydraulic presses)
- Stroke-restricted machines (crank and eccentric presses)
- Energy-restricted machines (hammers and screw presses)

Hydraulic presses are essentially load-restricted machines; i.e., their capability for carrying out a forming operation is limited mainly by the maximum load capacity. Mechanical (eccentric or crank) presses are stroke-restricted machines, since the length of the press stroke and the available load at various stroke positions represent the capability of these machines. Hammers are energy-restricted machines, since the deformation results from dissipating the kinetic energy of the hammer ram. The hammer frame guides the ram, but is essentially not stressed during forging. The screw presses are also energy-restricted machines but they are similar to the hydraulic and mechanical presses since their frames are subject to loading during forging stroke. The speed range and the speed stroke behavior of different forging machines vary considerably according to machine design, as illustrated in Table 10.1.

The significant characteristics of these machines comprise all machine design and performance data, which are pertinent to the machine's economic use. These characteristics include:

- Characteristics for load and energy
- Time-related characteristics
- Characteristics for accuracy

In addition to these characteristic parameters, the geometric features of the machine such as the stroke in a press or hammer and the dimensions and features of the tool-mounting space (shut height) are also important. Other important values are the general machine data, space requirements, weight, and the associated power requirements.

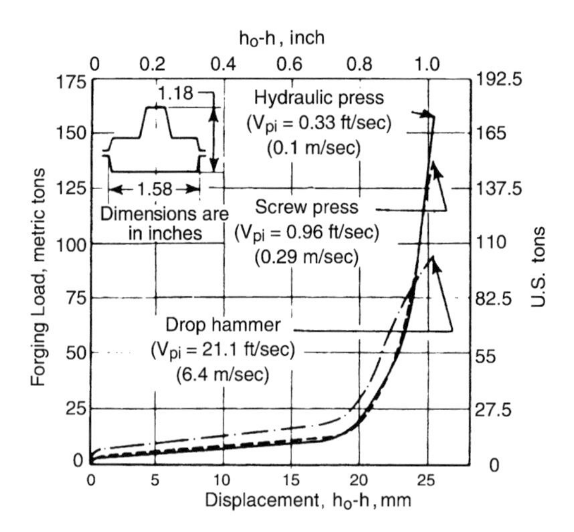

Fig. 10.3 Load versus displacement curves obtained in closed-die forging an axisymmetric steel part at 2012 °F (1100 °C) in three different machines with different initial velocities (V_{pi}). [Altan et al., 1973]

Apart from the features mentioned previously, some of the basic requirements that are expected of a good forging machine can be listed as:

- High tool pressure, which requires the stock to be tightly gripped and upsetting forces completely absorbed.
- Sufficient tool length to permit rigid bar reception apart from filling up the impression.
- The gripping tools must not open during the upsetting process.
- The device for moving the tools must be secured against overloading.
- The heading slide must be provided with long and accurate guides.
- The whole machine must be elastically secured against overloading.
- Design of a crankshaft of special rigidity.
- Readily interchangeable gripping and heading tools.
- The driving motor and the machine must be connected through a security coupling.
- The machine must have central lubrication.

10.5 Characteristic Data for Load and Energy

Available energy, E_M (in ft-lb or m-kg), is the energy supplied by the machine to carry out the deformation during an entire stroke. Available energy, E_M, does not include either E_f, the energy necessary to overcome the friction in the bearings and slides, or E_d, the energy lost because of elastic deflections in the frame and driving system.

Available load, L_M (in tons), is the load available at the slide to carry out the deformation process. This load can be essentially constant as in hydraulic presses, but it may vary with the slide position in respect to "bottom dead center" (BDC) as in mechanical presses.

Efficiency factor, η, is determined by dividing the energy available for deformation, E_M, by the total energy, E_T, supplied to the machine; i.e., $\eta = E_M/E_T$. The total energy, E_T, also includes in general: (a) the losses in the electric motor, E_e, (b) the friction losses in the gibs and in the driving system, E_f, and (c) the losses due to total elastic deflection of the machine, E_d.

The following two conditions must be satisfied to complete a forming operation: first, at any time during the forming operation,

$$L_M \geq L_P \quad \text{(Eq 10.1)}$$

where L_M is the available machine load and L_P is the load required by the process; and second, for an entire stroke,

$$E_M \geq E_P \quad \text{(Eq 10.2)}$$

where E_M is the available machine energy and E_P is the energy required by the process.

If the condition expressed by the former inequality above (Eq 10.1) is not fulfilled in a hydraulic press, the press will stall without accomplishing the required deformation. In a mechanical press, the friction clutch would slip and the press run would stop before reaching the bottom dead center position. If the condition expressed by the latter inequality (Eq 10.2) is not satisfied, either the flywheel will slow down to unacceptable speeds in a mechanical press or the part will not be formed completely in one blow in a screw press or hammer.

Table 10.1 Speed-range and speed-stroke behavior of forging equipment

Forging machine	Speed range, ft/s	Speed range, m/s	Speed-stroke behavior
Hydraulic press	0.2–1.0(a)	0.06–0.30(a)	
Mechanical press	0.2–5	0.06–1.5	
Screw press	2–4	0.6–1.2	Speed / Stroke
Gravity drop hammer	12–16	3.6–4.8	
Power drop hammer	10–30	3.0–9.0	
Counterblow hammer (total speed)	15–30	4.5–9.0	
HERF machines	20–80	6.0–24.0	
Low-speed Petroforge	8–20	2.4–6.0	

Source: [Altan et al., 1973]

10.6 Time-Dependent Characteristic Data

Number of strokes per minute, n, is the most important characteristic of any machine, because it determines the production rate. When a part is forged with multiple and successive blows (in hammers, open-die hydraulic presses, and screw presses), the number of strokes per minute of the machine greatly influences the ability to forge a part without reheating.

Contact time under pressure, t_p, is the time during which the part remains in the die under the deformation load. This value is especially important in hot forming. The heat transfer between the hotter formed part and the cooler dies is most significant under pressure. Extensive studies conducted on workpiece and die temperatures in hot forming clearly showed that the heat transfer coefficient is much larger under forming pressure than under free contact conditions. With increasing contact time under pressure, die wear increases. In addition, cooling of the workpiece results in higher forming load requirements.

Velocity under pressure, V_p, is the velocity of the slide under load. This is an important variable because it determines (a) the contact time under pressure and (b) the rate of deformation or the strain rate. The strain rate influences the flow stress of the formed material and consequently affects the load and energy required in hot forming.

10.7 Characteristic Data for Accuracy

Under unloaded conditions, the stationary surfaces and their relative positions are established by (a) clearances in the gibs, (b) parallelism of upper and lower beds, (c) flatness of upper and lower beds, (d) perpendicularity of slide motion with respect to lower bed, and (e) concentricity of tool holders. The machine characteristics influence the tolerances in formed parts. For instance, in backward extrusion a slight nonparallelism of the beds, or a slight deviation of the slide motion from ideal perpendicularity, would result in excessive bending stresses on the punch and in nonuniform dimensions in extruded products.

Under loaded conditions, the tilting of the ram and the ram and frame deflections, particularly under off-center loading, might result in excessive wear of the gibs, in thickness deviations in the formed part and in excessive tool wear. In multiple-operation processes, the tilting and deflections across the ram might determine the feasibility or the economics of forging a given part. In order to reduce off-center loading and ram tilting, the center of loading of a part, i.e., the point where the resultant total forming load vector is applied, should be placed under the center of loading of the forming machine.

In presses (mechanical, hydraulic, or screw), where the press frame and the drive mechanism are subject to loading, the stiffness, C, of the press is also a significant characteristic. The stiffness is the ratio of the load, L_M, to the total elastic deflection, d, between the upper and lower beds of the press, i.e.:

$$C = L_M/d \qquad \text{(Eq 10.3)}$$

In mechanical presses, the total elastic deflection, d, includes the deflection of the press frame (~25 to 35% of the total) and the deflection of the drive mechanism (~65 to 75% of the total). The main influences of stiffness, C, on the forming process can be summarized as follows:

- Under identical forming load, L_M, the deflection energy, E_d, i.e., the elastic energy stored in the press during buildup, is smaller for a stiffer press (larger C). The deflection energy is given by:

$$E_d = dL_M/2 = L_M^2/2C \qquad \text{(Eq 10.4)}$$

- The higher the stiffness, the lower the deflection of the press. Consequently, the variations in part thickness due to volume or temperature changes in the stock are also smaller in a stiffer press.
- Stiffness influences the velocity versus time curve under load. Since a less stiff machine takes more time to build up and remove pressure, the contact time under pressure, t_p, is longer. This fact contributes to the reduction of tool life in hot forming.

Using larger components in press design increases the stiffness of a press. Therefore, greater press stiffness is directly associated with increased costs, and it should not be specified unless it can be justified by expected gains in part tolerances or tool life.

REFERENCES

[Altan et al., 1973]: Altan, T., et al., *Forging Equipment, Materials and Practices,* Metal and Ceramics Information Center, HB03, 1973, p 4–7.

[Kienzle, 1965]: Kienzle, O., Characteristics of Data in Machine Tools for Closed Die Forging, (in German), *Werkstatttechnik,* Vol 55, 1965, p 509.

[Kienzle, 1953]: Kienzle, O., The Characteristic Data on Presses and Hammers (in German), *Werkst. Maschin.,* Vol 43, 1953, p 1.

SELECTED REFERENCES

[Altan et al., 1983]: Altan, T., Oh, S.I., Gegel, H., *Metal Forming: Fundamentals and Applications,* ASM International, 1983.

[FIA 1997]: Forging Industry Association, *Product Design Guide for Forging,* 1997.

[Geleji et al., 1967]: Geleji, A., et al., *Forge Equipment, Rolling Mills and Accessories* (in English), Akademiai Kiado, Budapest, 1967, p 168.

[Lange, 1972]: Lange, K., Ed., *Study Book of Forming Technology,* (in German), Vol 1, *Fundamentals,* Springer-Verlag, 1972.

CHAPTER 11

Presses and Hammers for Cold and Hot Forging

Manas Shirgaokar

11.1 Introduction

The continuous development of forging technology requires a sound and fundamental understanding of equipment capabilities and characteristics. The equipment, i.e., presses and hammers used in forging, influences the forging process, since it affects the deformation rate and temperature conditions, and it determines the rate of production. The requirements of a given forging process must be compatible with the load, energy, time, and accuracy characteristics of a given forging machine.

There are basically three types of presses: hydraulic, mechanical, and screw presses. These machines are used for hot and cold forging, cold extrusion trimming, and coining.

Developments in the forging industry are greatly influenced by the worldwide requirements for manufacturing ever-larger and more complex components for more difficult-to-forge materials. The present and future needs of the aerospace industry, the increase in demand for stationary power systems, jet engines, and aircrafts components, and the ever-increasing foreign technological competition require continuous upgrading of today's technology. Thus, the more efficient use of existing forging equipment and the installation of more sophisticated machinery have become unavoidable necessities. Development in all areas of forging has the objectives of (a) increasing the production rate, (b) improving forging tolerances, (c) reducing costs by minimizing scrap losses, by reducing preforming steps, and by increasing tool life, and (d) expanding capacity to forge larger and more intricate parts.

The purchase of new forging equipment requires a thorough understanding of the effect of equipment characteristics on the forging operations, load and energy requirements of the specific forging operation, and the capabilities and characteristics of the specific forging machine to be used for that operation. Increased knowledge on forging equipment would also specifically contribute to:

- More efficient and economical use of existing equipment
- More exact definition of the existing maximum plant capacity
- Better communication between the equipment user and the equipment builder
- Development of more refined processes such as precision forging of gears and of turbine and compressor blades

11.2 Hydraulic Presses

The operation of hydraulic presses is relatively simple and is based on the motion of a hydraulic piston guided in a cylinder [Geleji 1967], [Mueller 1969], [Peters 1969]. Hydraulic presses are essentially load-restricted machines; i.e., their capability for carrying out a forming operation is limited mainly by the maximum available load. The following important features are offered by hydraulic presses:

- In direct-driven hydraulic presses, the maximum press load is available at any point during the entire ram stroke. In accumulator-driven presses, the available load decreases slightly depending on the length of the stroke and the load-displacement characteristics of the forming process.

- Since the maximum load is available during the entire stroke, relatively large energies are available for deformation. This is why the hydraulic press is ideally suited for extrusion-type forming operations requiring a nearly constant load over a long stroke.
- Within the capacity of a hydraulic press, the maximum load can be limited to protect the tooling. It is not possible to exceed the set load, because a pressure-release valve limits the fluid pressure acting on the ram.
- Within the limits of the machine, the ram speed can be varied continuously at will during an entire stroke cycle. Adequate control systems can regulate the ram speed with respect to forming pressure or product temperature. This control feature can offer a considerable advantage in optimizing forming processes.

11.2.1 Drive Systems for Hydraulic Presses

The operational characteristics of a hydraulic press are essentially determined by the type and design of its hydraulic drive system. As shown in Fig. 11.1, two types of hydraulic drive systems give different time-dependent characteristic data [Hutson 1968] [Riemenschneider et al., 1959].

Direct-driven presses usually employ hydraulic oil or water emulsion as the working medium. In earlier vertical press designs, at the start of the downstroke the upper ram falls under gravity and oil is drawn from the reservoir into the ram cylinder through the suction of this fall. When the ram contacts the workpiece, the valve between the ram cylinder and the reservoir is closed and the pump builds up pressure in the ram cylinder. This mode of operation results in relatively long dwell times prior to the start of deformation. As illustrated in Fig. 11.1(b), during the downstroke in modern direct-driven presses a residual pressure is maintained in the return cylinders or in the return line by means of a pressure relief valve. Thus, the upper ram is forced down against pressure and the dwell inherent in the free fall is eliminated. When the pressure stroke is completed, i.e., when the upper ram reaches a predetermined position, or when the pressure reaches a certain value, the oil pressure on the ram cylinder is released and diverted to lift the ram.

Accumulator-driven presses usually employ a water-oil emulsion as the working me-

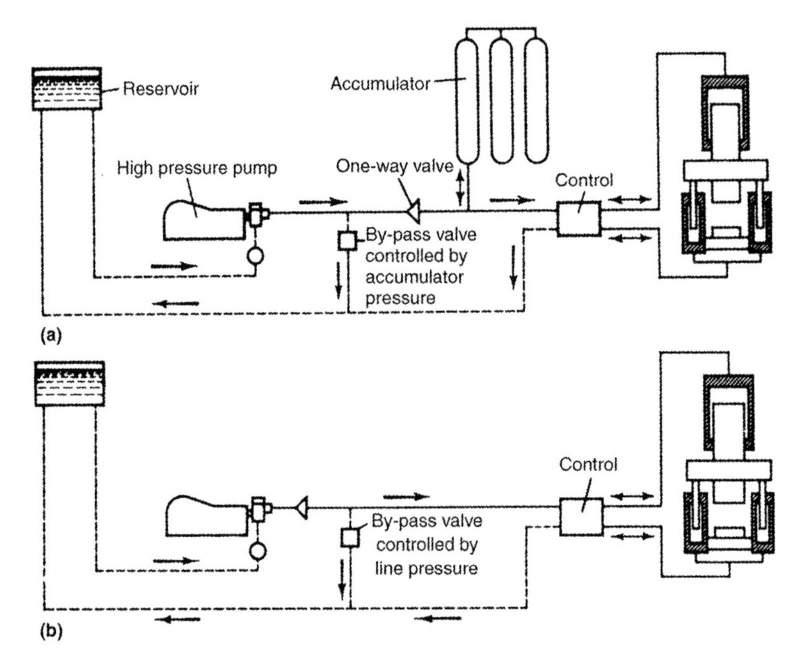

Fig. 11.1 Schematic illustration of drives for hydraulic presses. (a) Accumulator drive. (b) Direct drive. [Riemenschneider et al., 1959]

dium and use nitrogen, steam, or air-loaded accumulators to keep the medium under pressure (Fig. 11.1a). The sequence of operations is essentially similar to that for the direct-driven press except that the pressure is built up by means of the pressurized water-oil emulsion in the accumulators. Consequently, the rate of penetration, i.e., the ram speed under load, is not directly dependent on the pump characteristics and can vary depending on the pressure in the accumulator, the compressibility of the pressure medium, and the resistance of the workpiece to deformation. Toward the end of the forming stroke, as deformation progresses, the working medium expands, the force required to form the material increases, and the speed of penetration and the load available at the ram decrease.

In both direct and accumulator drives, as the pressure builds up and the working medium is compressed, a certain slowdown in penetration rate occurs. This slowdown is larger in direct oil-driven presses, mainly because oil is more compressible than a water emulsion.

The approach and initial deformation speeds are higher in accumulator-driven presses. This improves the hot forming conditions by reducing the contact times, but wear in hydraulic elements of the system also increases. Sealing problems are somewhat less severe in direct-oil drives, and control and accuracy are in general about the same for both types of drives.

From a practical point of view, in a new installation, the choice between direct or accumulator drive is decided by the economics of operation. Usually, the accumulator drive is more economical if several presses can use one accumulator system, or if very large press capacities (10,000 to 50,000 tons) are considered.

The frame of a hydraulic press must carry the full forming load exerted by the hydraulic cylinder on the press bed. The load-carrying capability of the frame is achieved by using various designs such as cast (or welded) structures prestressed by forged tie rods or laminated plates assembled through large transverse pins.

As can be seen in Fig. 11.2, the two principal types of press construction are designated as "pull-down" and "push-down" designs [Kirschbaum, 1968]. The conventional push-down design is often selected for four-column presses of

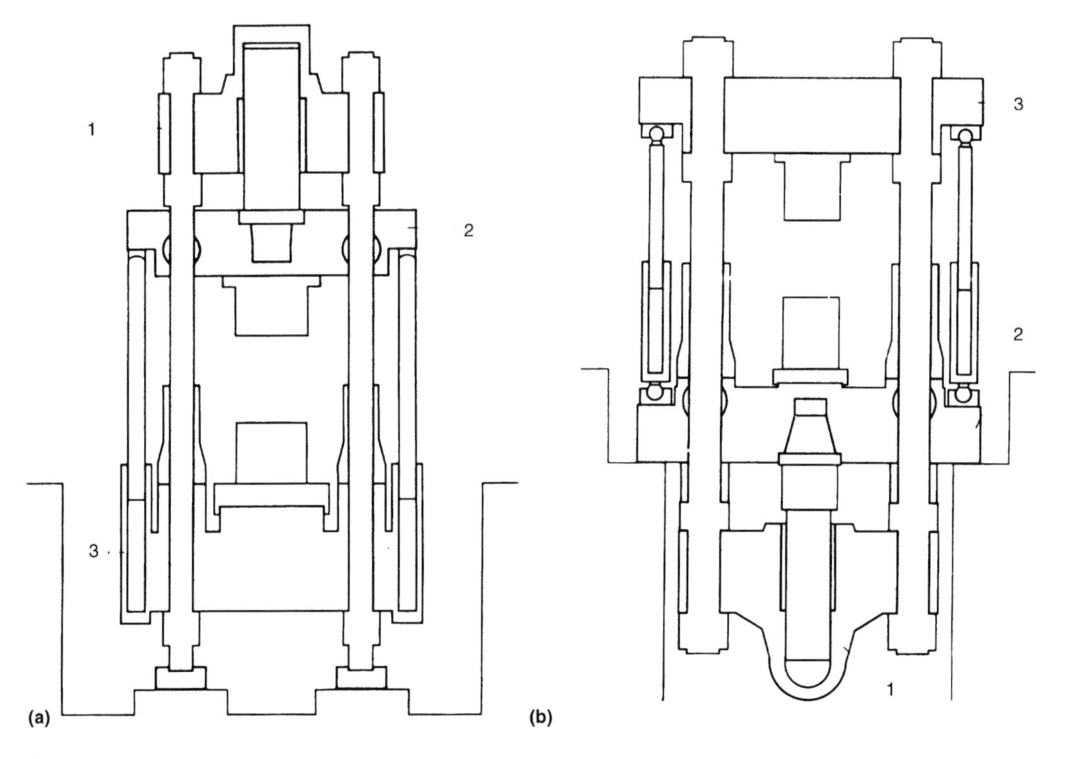

Fig. 11.2 Schematic illustration of two types of hydraulic press drives. (a) Push-down drive: 1, stationary cylinder cross head; 2, moving piston-ram assembly; 3, stationary press bed with return cylinders. (b) Pull-down drive: 1, movable cylinder-frame assembly; 2, press bed with return cylinders; 3, moving cross-head. [Kirschbaum, 1968]

all sizes. The cylinder cross head and base platen are rigidly connected by four columns that take up the press load and simultaneously guide the moving piston-ram assembly. Considerable elastic deflections are exhibited under off-center loading. This type of press requires a relatively tall shop building. In the pull-down design, the base platen rests on a foundation. The cylinder cross head, located below floor level, is rigidly connected to the press columns. This assembly is movable and is guided in the bed platen. The center of gravity of the press is low, at approximately floor level, and the overall static and dynamic stiffness of the press is increased accordingly. Pull-down presses are particularly suitable for installation in low buildings. Most of the hydraulic and auxiliary equipment may then be accommodated beneath floor level. This arrangement is particularly favorable for direct-oil drives since it minimizes fire hazard and reduces the length of piping between the pumping system and the press cylinder.

11.2.2 Characteristics of Hydraulic Presses

In direct-driven hydraulic presses, the maximum press load is established by the pressure capability of the pumping system and is available throughout the entire press stroke. Thus, hydraulic presses are ideally suited for extrusion-type operations requiring very large amounts of energy. With adequate dimensioning of the pressure system, an accumulator-driven press exhibits only a slight reduction in available press load as the forming operation proceeds.

In comparison with direct drive, the accumulator drive usually offers higher approach and penetration speeds and a short dwell time prior to forging. However, the dwell at the end of processing and prior to unloading is larger in accumulator drives. This is illustrated in Fig. 11.3 and 11.4, where the load and displacement variations are given for a forming process using a 2500 ton hydraulic press equipped with either accumulator or direct-drive systems.

Parallelism of the Slide. The capacity of the press frame to absorb eccentric loads plays a major role in forming a part with good dimensional accuracy. Eccentric forces occur during the forming process when the load of the resulting die force is not exerted centrally on the slide, causing it to tilt (Fig. 11.5). The standard press is able to absorb a maximum slide tilt of 0.8 mm/m. If a higher offcenter loading capa-

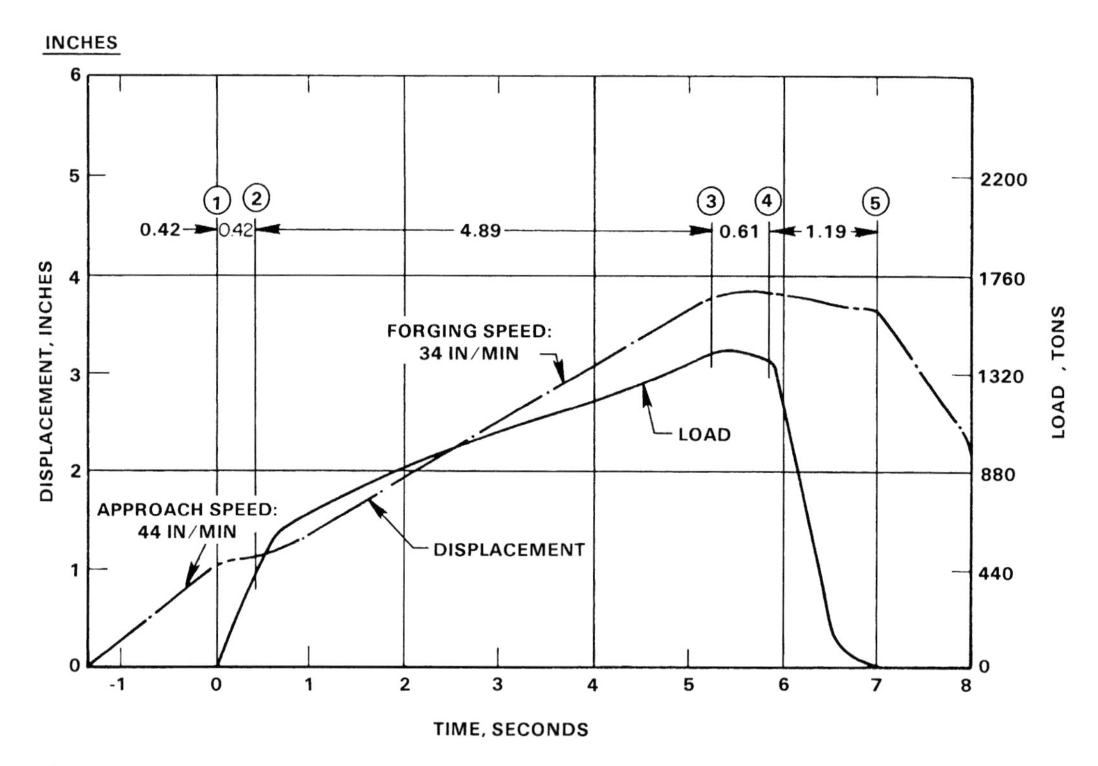

Fig. 11.3 Load and displacement versus time curves obtained on a 2500 ton hydraulic press in upsetting with direct drive. 1, start of deformation; 2, initial dwell; 3, end of deformations; 4, dwell before pressure release; 5, ram lift. [Altan et al., 1973)

bility is desired, then the press design must be more rigid. In this case, the slide gibs will have greater stability, the press frame will be more rigid, and the slide will be higher.

Often it is necessary to use hydraulic parallelism control systems, using electronic control technology (Fig. 11.5), for example, in the case of hydraulic transfer presses. The parallelism

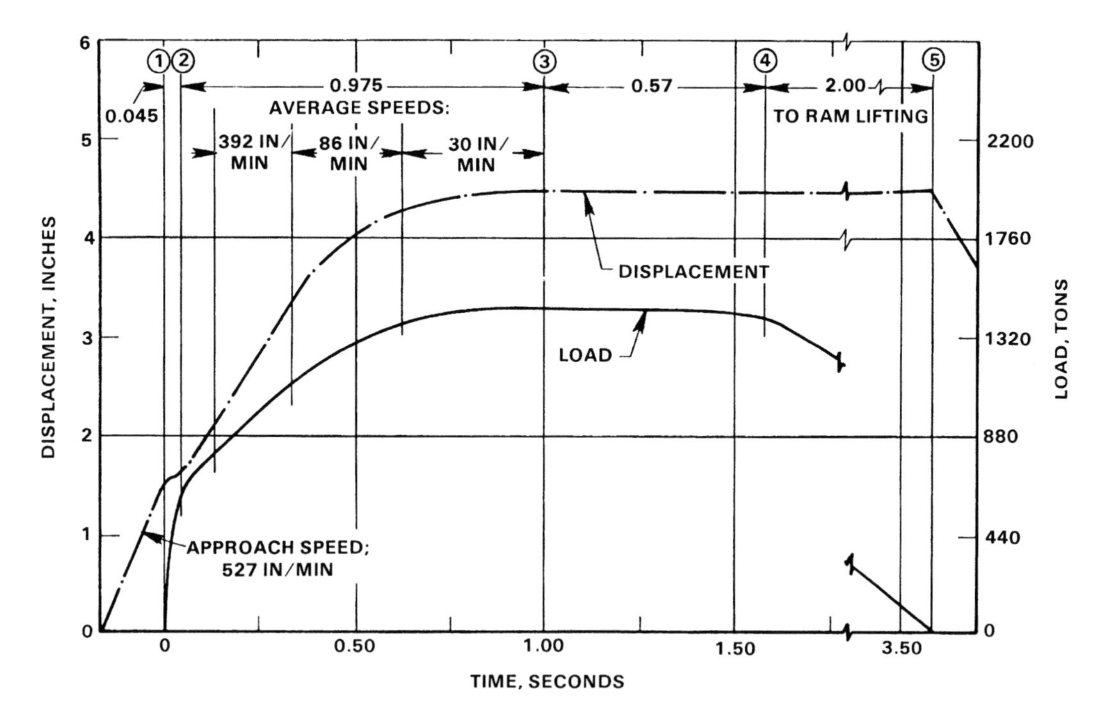

Fig. 11.4 Load and displacement versus time curves obtained on a 2500 ton hydraulic press in upsetting with accumulator drive. 1, start forming; 2, initial dwell; 3, end of forming; 4, dwell before pressure release; 5, ram lift. [Altan et al., 1973]

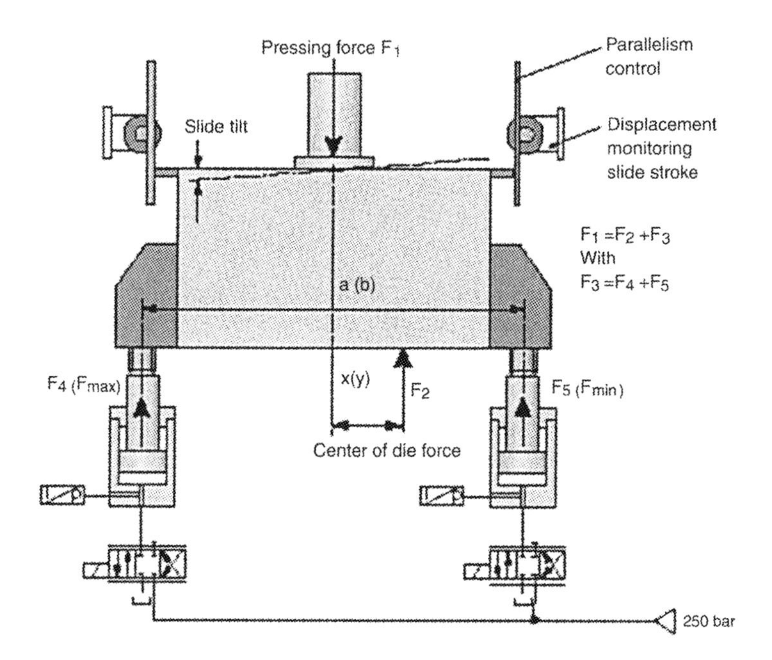

Fig. 11.5 Control system for maintaining slide parallelism. [Schuler Handbook, 1998]

control systems act in the die mounting area to counter slide tilt. Position measurement sensors monitor the position of the slide and activate the parallelism control system (Fig. 11.5). The parallelism controlling cylinders act on the corners of the slide plate, and they are pushed during the forming process against a centrally applied pressure. If the electronic parallelism monitor sensor detects a position error, the pressure on the leading side is increased by means of servo valves, and at the same time reduced on the opposite side to the same degree. The sum of exerted parallelism control forces remains constant, and the slide tilt balance is restored. Depending on the deformation speed, a slide parallelism of 0.05 to 0.2 mm/m is achieved. A central device adjusts the system to different die heights by means of spindles at the slide.

Full-stroke parallelism control involves the use of parallel control cylinders, with their pistons permanently connected to the slide (Fig. 11.6). These act over the entire stroke of the slide so that no setting spindles are required to adjust the working stroke. Two cylinders with the same surface area, arranged well outside the center of the press, are subjected to a mean pressure. The tensile and compressive forces are balanced out by means of diagonal pipe connections. The system is neutral in terms of force exerted on the slide. If an off-center load is exerted by the die on the slide, a tilt moment is generated. The slide position sensor detects a deviation from parallel and triggers the servo valve. The valve increases the pressure on the underside of the piston acting on the leading side of the slide and thus also on the opposite upper side of the piston. At the same time, the pressure in the other connecting pipe is reduced. The opposing supporting torques exerted on the two sides counteract the tilt moment.

11.2.3 Mechanical Crank and Eccentric Presses

All mechanical presses employ flywheel energy, which is transferred to the workpiece by a network of gears, cranks, eccentrics, or levers (Fig. 11.7). The ability of mechanical presses to deform the workpiece material is determined by the length of the press stroke and the available force at various stroke positions.

Two major groups of mechanical presses are:

- Presses with crank drive
- Presses with cam drive

Crank presses may have either simple or extended crank drives. Conventional crank presses (the total stroke cannot be varied) and eccentric presses (the total stroke is variable) belong to the simple drives. If either a knuckle or a lever is used to extend the crank drive, the designs are called knuckle joint or link drive presses. Multipoint presses are those in which two or more cranks are used to drive the same ram.

Other methods of classification are:

- Frame type: C or closed frame
- Number of useful motions: Single or more
- Location of drive: Top drive (connecting rod subjected to compression) and bottom drive (connecting rod subjected to tension)
- Position of drive shaft: Longitudinal or cross shaft
- Number of connecting rods: One-, two-, or four-point drive

The drive system used in most mechanical presses (crank or eccentric) is based on a slider-crank mechanism that translates rotary motion into reciprocating linear motion. The eccentric

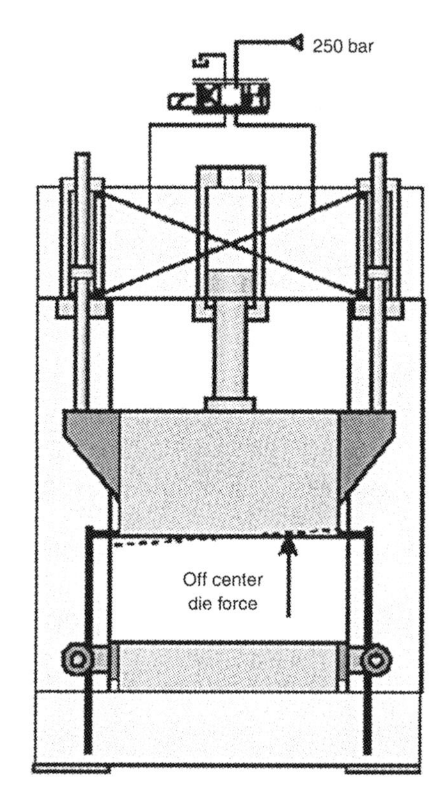

Fig. 11.6 Full-stroke parallelism control of the press slide. [Schuler Handbook, 1998]

shaft is connected through a clutch and brake system directly to the flywheel (Fig. 11.8). In designs for larger capacities, the flywheel is located on the pinion shaft, which drives the eccentric shaft (Fig. 11.9). The constant clutch torque, M, is available at the eccentric shaft, which transmits the torque and the flywheel energy to the slide through the pitman arm or connecting rod, as illustrated in Fig. 11.8. The flywheel, which is driven by an electric motor and "V" belts, stores energy that is used only during a small portion of the crank revolution, namely, during deformation of the formed material.

Figure 11.10 shows the basic slider-crank mechanism. The clutch at the flywheel transmits the constant torque, M, to the eccentric (or crank) shaft. The force diagram gives the relations between the torque, M, the force on the connecting rod, P, and the tangential force, T:

$$T = P \sin(\alpha + \beta) \qquad \text{(Eq 11.1)}$$

and

$$M = rT \qquad \text{(Eq 11.2)}$$

Usually, the ratio, λ, of crank radius r to connecting-rod length l is small, about:

$$\lambda = \frac{r}{l} = \frac{1}{10}$$

or

$$\frac{\sin \beta}{\sin \alpha} = \frac{1}{10} \qquad \text{(Eq 11.3)}$$

Using Eq 11.1 and considering that the total press stroke is $S = 2r$, the machine load, L_M, acting on the ram is:

$$L_M = P \cos \beta = \frac{T \cos \beta}{\sin(\alpha + \beta)} = \frac{2M \cos \beta}{S \sin(\alpha + \beta)} \qquad \text{(Eq 11.4)}$$

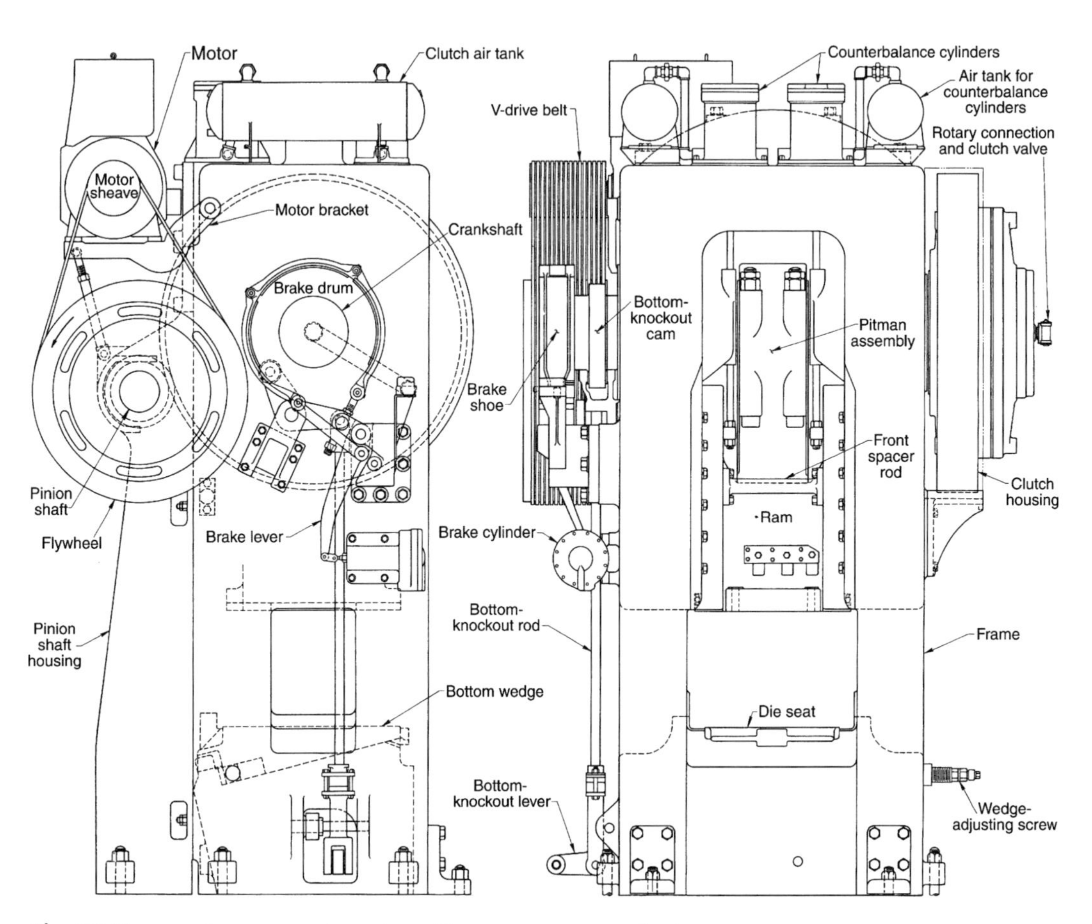

Fig. 11.7 Principal components of a mechanical forging press. [ASM Handbook, 1988]

When the angles α and β approach 0, i.e., toward bottom dead center (BDC), L_M may go to infinity for constant torque, M. This is illustrated in Fig. 11.11.

The stroke position, i.e., the distance, h, from BDC, as a function of the crank angle, α, can be derived from the geometric relationships illustrated in Fig. 11.10, to be:

$$h = (r + l) - (r \cos\alpha + \sqrt{l^2 - r^2 \sin^2\alpha})$$

or

$$h = r(1 - \cos\alpha) + l\left(1 - \sqrt{1 - \left(\frac{r}{l}\right)\sin^2\alpha}\right) \quad \text{(Eq 11.5)}$$

Using the binomial expansion, the term under the square root sign can be approximated as $1 - (r/l)^2 \sin^2 \alpha/2$.

Thus, Eq 11.5 can be transformed into:

$$h = r(1 - \cos\alpha) + \frac{r^2}{2l}\sin^2\alpha \quad \text{(Eq 11.6)}$$

For small values of α, i.e., near BDC, Eq 11.6 is approximated as:

$$h = r(1 - \cos\alpha) = \frac{S}{2}(1 - \cos\alpha) \quad \text{(Eq 11.7)}$$

The ram velocity, V, is obtained from Eq 11.6 by differentiation with respect to time, t:

$$V = \frac{dh}{dt} = \frac{dh}{d\alpha}\frac{d\alpha}{dt} = \frac{dh}{d\alpha}\omega$$

$$= \left(r \sin\alpha + \frac{r^2}{l}\sin\alpha\cos\alpha\right)\omega \quad \text{(Eq 11.8)}$$

where, with n being the rotational speed of the crank in revolutions per minute, the angular velocity $\omega = 2\pi n/60$ and the stroke $S = 2r$. Neglecting the second small term in Eq 11.8 gives:

$$V = \frac{S\pi n}{60}\sin\alpha \quad \text{(Eq 11.9)}$$

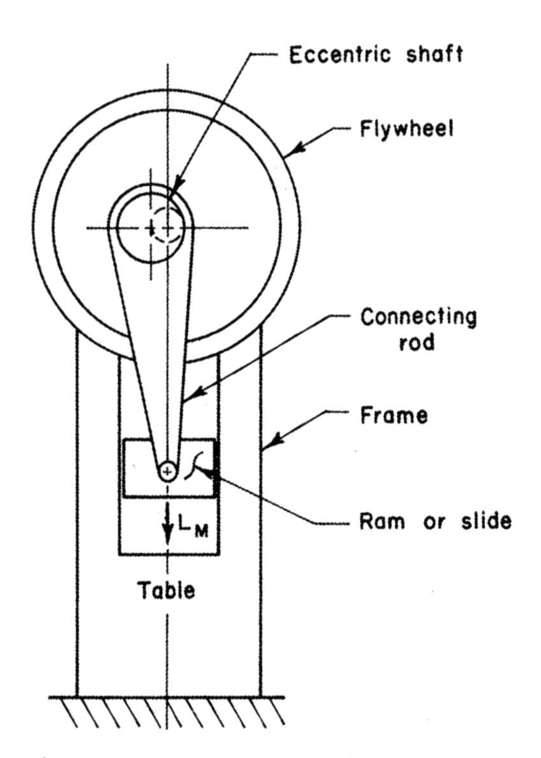

Fig. 11.8 Schematic of a mechanical press with eccentric drive (clutch and brake on eccentric shaft). [Altan et al., 1973]

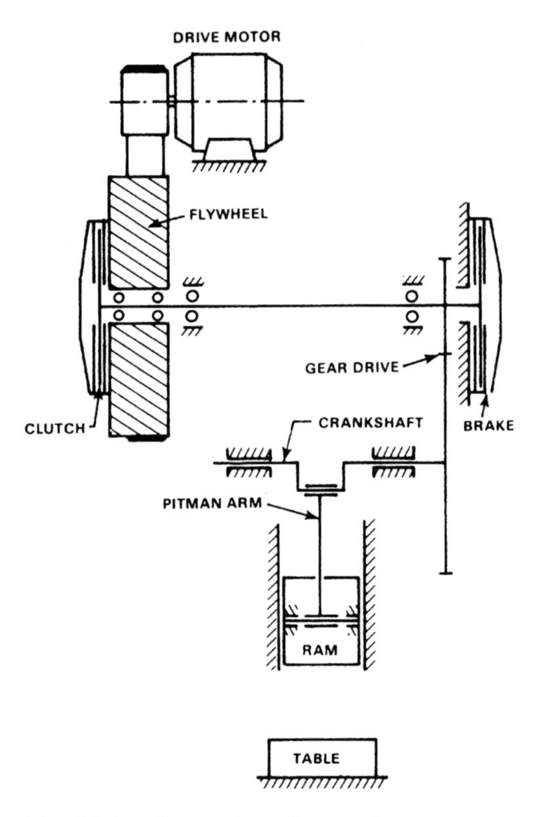

Fig. 11.9 Schematic of a crank press with pinion-gear drive (clutch and brake are on pinion shaft; for large capacities this design is more stable and provides high flywheel energy). [Altan et al., 1983]

Using the geometric relationships of Fig. 11.10, the ram velocity can be expressed as:

$$V = \frac{\pi n}{30} h \sqrt{\frac{S}{h} - 1} \qquad \text{(Eq 11.10)}$$

Thus with Eq 11.6 and 11.10 the displacement, h, and the velocity, V, at each point of the ram stroke can be calculated. Figure 11.12 illustrates the variation of these values with the crank angle α before BDC.

11.2.4 Load and Energy in Mechanical Presses

With the symbols used in Fig. 11.10, Eq 11.4 gives the ram or machine load, L_M. Considering that angle β is much smaller than angle α, L_M can be approximated as:

$$L_M = \frac{2M}{S \sin \alpha} \qquad \text{(Eq 11.11)}$$

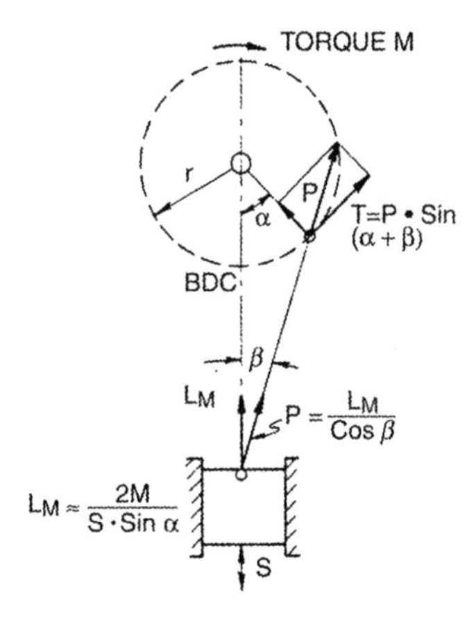

Fig. 11.10 The basic slider-crank mechanism used in crank presses. S, stroke; BDC, bottom dead center; α, crank angle before bottom dead center (BDC); L_M, machine load. [Altan et al., 1973]

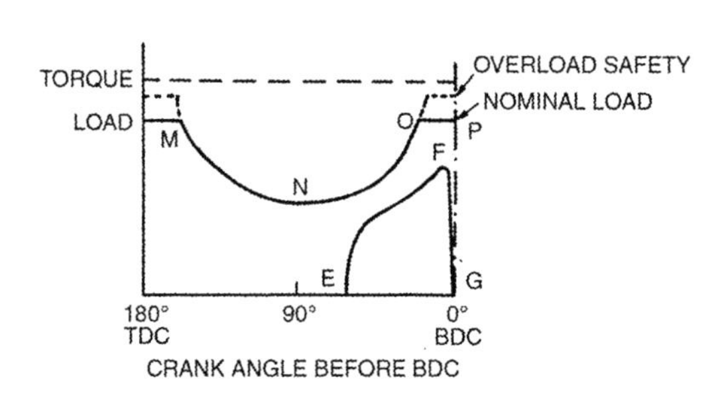

Fig. 11.11 Variations of clutch torque and machine load with crank angle in an eccentric or crank press. [Altan et al., 1973]

As shown in Fig. 11.11, Eq 11.11 illustrates the variation of the slide load, L_M, with the crank angle, α, before bottom dead center, BDC, for given values of torque, M, and stroke, S, of the press. The torque, M, at the clutch has a constant value for which the drive mechanism (i.e., eccentric shaft, pinion gear, clutch, brake, etc.) is designed. Thus, from Eq 11.11 it can be seen that, as the slide approaches the BDC, i.e., as angle α approaches zero, the available machine load L_M, may become infinitely large without exceeding the constant clutch torque, M, i.e., without causing the friction clutch to slip.

From the observations made so far, the following conclusions may be drawn:

- Crank and eccentric presses are displacement-restricted machines. The slide velocity, V, and the available slide load, L_M, vary according to the position of the slide before BDC. Most manufacturers in the United States rate their presses by specifying the nominal load at ¼ or ⅛ in. (6.4 or 3.2 mm) before BDC. For different applications, the nominal load may be specified at different positions before BDC according to the standards established by the American Joint Industry Conference.
- If the load required by the forming process is smaller than the load available at the press (i.e., if curve EFG in Fig. 11.11 remains below curve NOP), the process can be carried out provided that the flywheel can supply the necessary energy per stroke.
- For small angles α before BDC, within the OP portion of curve NOP in Fig. 11.11, the slide load, L_M, can become larger than the

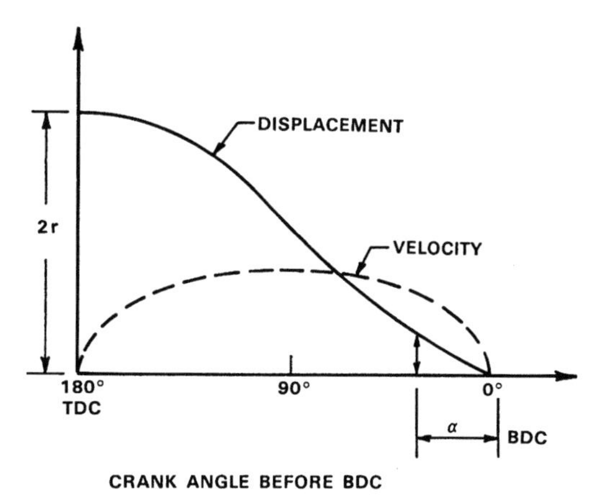

Fig. 11.12 Displacement and velocity in a simple slider-crank mechanism (stroke = 2r)

nominal press load if no overload safety (hydraulic or mechanical) is available on the press. In this case, the press stalls, the flywheel stops, and the entire flywheel energy is transformed into deflection energy by straining the press frame, the pitman arm, and the drive mechanism. Usually, the press can then be freed only by burning out the tooling.

- If the load curve EFG exceeds the press load NOP (Fig. 11.11) before point O is reached, then the friction clutch slides and the press slide stops, but the flywheel continues to turn. In this case, the press can be freed by increasing the air pressure on the clutch and by reversing the flywheel rotation if the slide has stopped before BDC.

The energy needed for the forming operation during each stroke is supplied by the flywheel, which slows down to a permissible percentage—usually 10 to 20%—of its idle speed. The total energy stored in a flywheel is:

$$E_{FT} = \frac{I\omega^2}{2} = \frac{I}{2}\left(\frac{\pi n}{30}\right)^2 \qquad \text{(Eq 11.12)}$$

where I is the moment of inertia of the flywheel, ω is the angular velocity in radians per second, and n is the rotational speed of the flywheel in revolutions per minute. The total energy, E_S, used during one stroke is:

$$E_S = \frac{1}{2} I(\omega_0^2 - \omega_1^2) = \frac{I}{2}\left(\frac{\pi}{30}\right)^2 (n_0^2 - n_1^2) \qquad \text{(Eq 11.13)}$$

where ω_0 is initial angular velocity, ω_1 is angular velocity after the work is done, n_0 is initial flywheel speed in revolutions per minute, and n_1 is flywheel speed after the work is done, also in rpm.

Note that the total energy, E_S, also includes the friction and elastic deflection losses. The electric motor must bring the flywheel from its lowered speed, n_1, to its idle speed, n_0, before the next forming stroke starts. The time available between two strokes depends on whether the mode of operation is continuous or intermittent. In a continuously operating mechanical press, less time is available to bring the flywheel to its idle speed, and consequently a higher-horsepower motor is necessary.

Very often the allowable slowdown of the flywheel is given as a percentage of the nominal speed. For instance, if a 13% slowdown is permissible, then:

$$\frac{n_0 - n_1}{n_0} = \frac{13}{100}, \text{ or } n_1 = 0.87\, n_0$$

The percentage energy supplied by the flywheel is obtained by using Eq 11.12 and 11.13 to arrive at:

$$\frac{E_S}{E_{FT}} = \frac{n_0^2 - n_1^2}{n_0^2} = 1 - (0.87)^2 = 0.25$$

The simple calculations given above illustrate that for a 13% slowdown of the flywheel, 25% of the flywheel energy will be used during one stroke.

As an example, the variation of load, displacement, and flywheel speed in upset forming of a copper sample under 1600 ton mechanical press is illustrated in Fig. 11.13. This press was instrumented with strain bars attached to the frame for measuring load, an inductive transducer (LVDT) for measuring ram displacement, and a dc tachometer for measuring flywheel speed. In Fig. 11.13 it can be seen that, due to frictional and inertial losses in the press drive, the flywheel slows down by about (5 rpm) before deformation begins. The flywheel requires 3.24 s to recover its idling speed; i.e., in forming this part the press can be operated at a maximum speed of 18 (60/3.24) strokes/min. For each mechanical press, there is a unique relationship between strokes per minute, or production rate, and the available energy per stroke. As shown in Fig. 11.14, the strokes per minute available on the machine decreases with increasing energy required per stroke. This relationship can be determined experimentally by upsetting samples, which require various amounts of deformation energy, and by measuring load, displacement, and flywheel recovery time. The energy consumed by each sample is obtained by calculating the surface area under the load-displacement curve.

11.2.5 Time-Dependent Characteristics of Mechanical Presses

The number of strokes per minute, n, was discussed as part of the energy considerations. As can be seen in Eq 11.9, the ram velocity is directly proportional to the number of strokes per minute, n, and to the press stroke, S. Thus, for a given press, i.e., a given stroke, the only way to increase ram velocity during deformation is to increase the stroking rate, n.

For a given idle-flywheel speed, the contact time under pressure, t_p, and the velocity under pressure, V_p, depend mainly on the dimensions of the slide-crank mechanism and on the total stiffness, C, of the press. The effect of press stiffness on contact time under pressure, t_p, is illus-

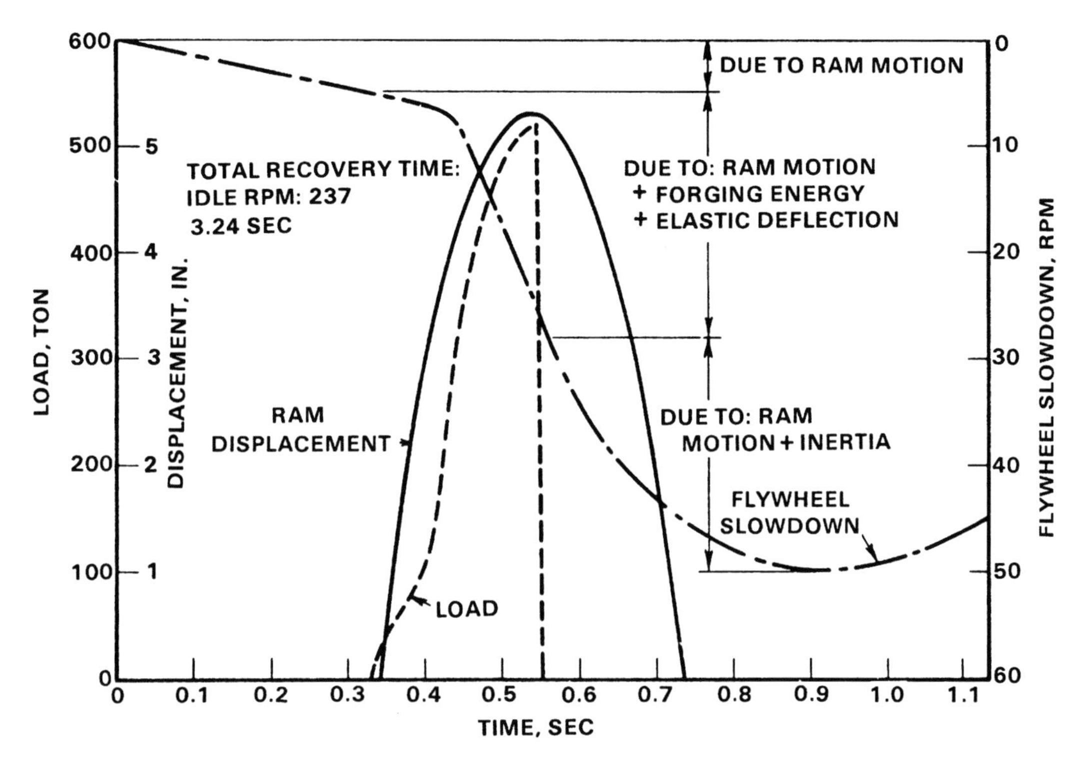

Fig. 11.13 Flywheel slowdown, ram displacement, and forming load in upsetting of copper samples in a 1600-ton mechanical press. [Altan et al., 1972]

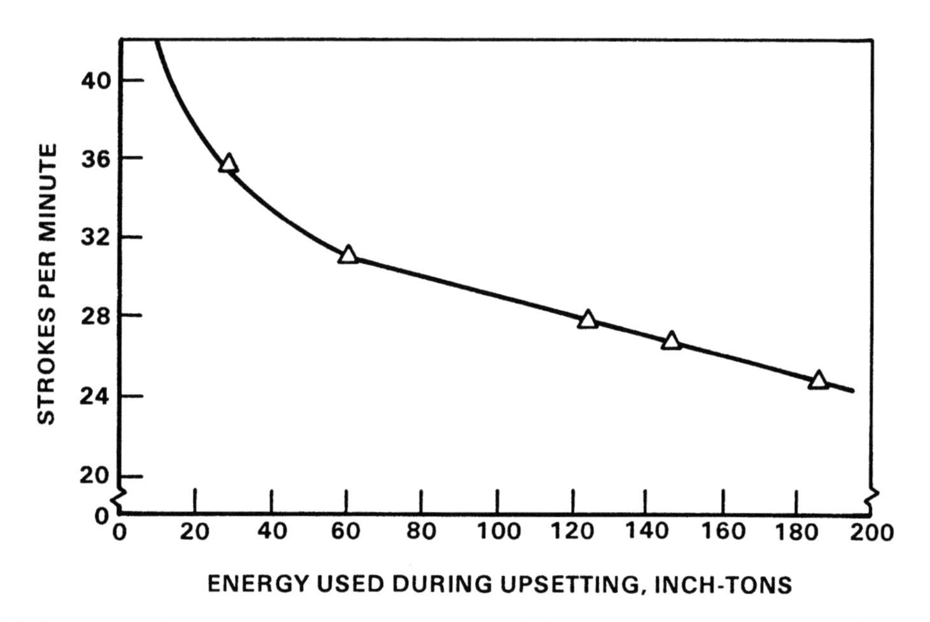

Fig. 11.14 Variation of strokes per minute with the energy available for forming in a 500 ton mechanical press. [Altan et al., 1972]

trated in Fig. 11.15. As the load builds up, the press deflects elastically. A stiffer press (larger C) requires less time, t_{p1}, for pressure buildup and also less time, t_{p2}, for pressure release as shown in Fig. 11.15(a). Consequently, the total contact time under pressure ($t_p = t_{p1} + t_{p2}$) is less for a stiffer press.

11.2.6 Accuracy of Mechanical Presses

The working accuracy of an eccentric press is substantially characterized by two features: the tilting angle of the ram under off-center loading and the total deflection under load or stiffness of the press. Tilting of the ram produces skewed surfaces and an offset on the part; stiffness influences the thickness tolerance [Rau, 1967]. Under off-center loading conditions, two- or four-point presses perform better than single-point presses because the tilting of the ram and the reduction forces into the gibways are minimized. Assuming the total deflection under load for a one-point eccentric press to be 100%, the distributions of total deflection shown in Table 11.1 were obtained from measurements under nominal load on one- and two-point presses of the same capacity [Rau, 1967]. It is interesting to note that a large percentage of the total deflection is in the drive mechanism, i.e., slide, pitman arm, drive shaft, and bearings.

Drive mechanisms that have considerable stiffness and off-center loading capability are provided by (a) the scotch-yoke design (Fig. 11.16) and (b) the wedge-type design (Fig. 11.17). Both these press drives have large bearing surfaces above the slide and can maintain larger off-center loads than the conventional eccentric drive forging presses.

Determination of the Dynamic Stiffness of a Mechanical Press. Unloaded machine con-

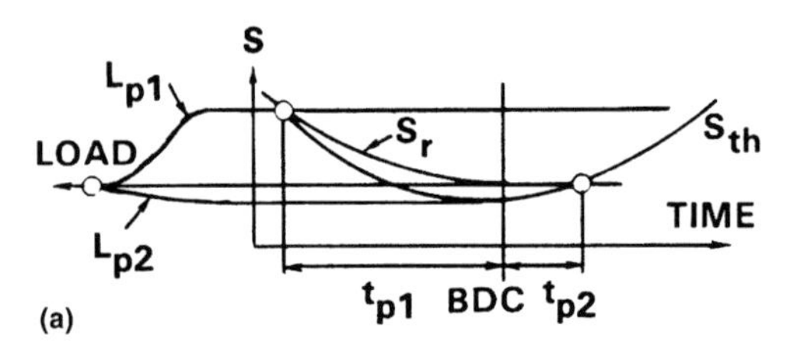

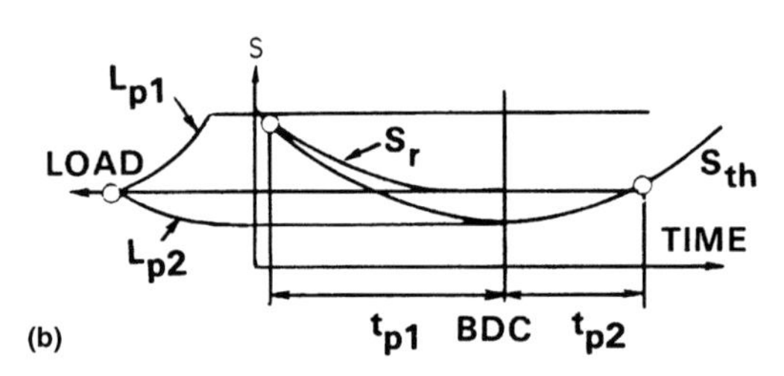

Fig. 11.15 Effect of press stiffness on contact time under pressure (S_{th} = theoretical displacement-time curve under load). (a) Stiffer press. (b) Less stiff press. [Kienzle, 1959]

Table 11.1 Total deflection under nominal load on one- and two-point presses of the same capacity

	Relative deflection	
	One-point eccentric press	Two-point eccentric press
Slide + pitman arm	30	21
Frame	33	31
Drive shaft + bearings	37	33
Total deflection	100	85

[Rau, 1967]

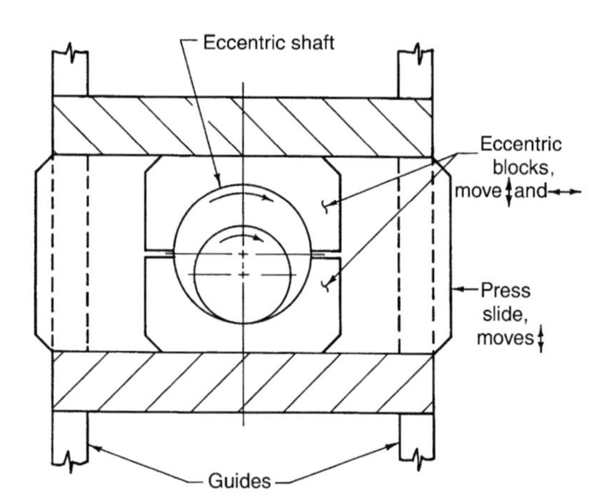

Fig. 11.16 Principle of the scotch-yoke type drive for mechanical presses. [Altan et al., 1973]

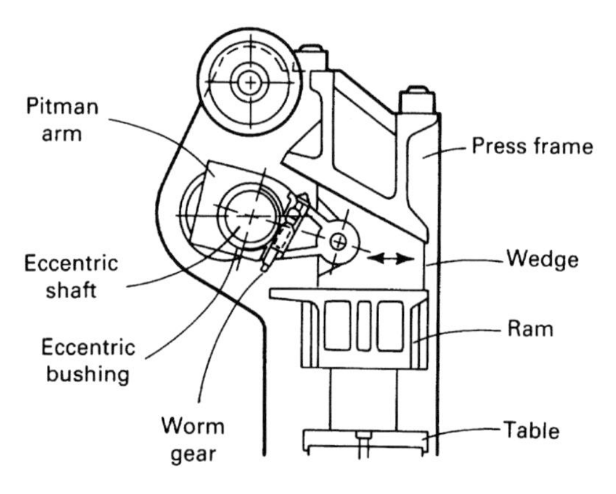

Fig. 11.17 Principle of the wedge-type mechanical press. [Rau, 1967]

ditions such as parallelism and flatness of upper and lower beds, perpendicularity of slide motion, etc. are important and affect the tolerances of the forged part. However, much more significant are the quantities obtained under load and under dynamic conditions. The stiffness of a press C (the ratio of the load to the total elastic deflection between the upper and lower dies) influences the energy lost in press deflection, the velocity versus time curve under load and the contact time. In mechanical presses, variations in forging thickness due to volume or temperature changes in the stock are also smaller in a stiffer press. Very often the stiffness of a press (ton/in.) is measured under static loading conditions, but such measurements are misleading. For practical purposes the stiffness has to be determined under dynamic loading conditions.

To obtain the dynamic stiffness of a mechanical press, copper samples of various diameters, but of the same height were forged under on-center conditions. A 500 ton Erie scotch-yoke type press was used for this study [Douglas et al., 1972]. The samples of wrought pure electrolytic copper were annealed for 1 h at 900 °F (480 °C). The press setup was not changed throughout the tests. Lead samples of about 1 in.^{2} (645 cm^2) and 1.5 in. (38 mm) height were placed near the forged copper sample, about 5 in. (125 mm) to the side. As indicated in Table 11.2, with increasing sample diameter the load required for forging increased as well. The press deflection is measured by the difference in heights of the lead samples forged with and without the copper at the same press setting. The variation of total press deflection versus forging load, obtained from these experiments, is illustrated in Fig. 11.18. During the initial nonlinear portion of the curve, the play in the press driving system is taken up. The linear portion represents the actual elastic deflection of the press components. The slope of the linear curve is the dynamic stiffness, which was determined as 5800 ton/in. for the 500 ton Erie forging press.

The method described above requires the measurement of load in forging annealed copper samples. If instrumentation for load and displacement would be impractical for forgeshop measurements, the flow stress of the copper can be used for estimating the load and energy for a given height reduction.

Ram Tilting in Off-Center Loading. Off-center loading conditions occur often in mechanical press forging when several operations are performed in the same press. Especially in automated mechanical presses, the finish blow (which requires the highest load) occurs on one side of the press. Consequently, the investigation of off-center forging is particularly significant in mechanical press forging.

The off-center loading characteristics of the 500 ton Erie press were evaluated using the following procedure [Douglas et al., 1972]. During each test, a copper specimen, which requires 220 ton to forge, was placed 5 in. (125 mm) from the press center in one of the four directions viz. left, right, front, or back. A lead specimen, which requires not more than 5 ton, was placed an equal distance on the opposite side of the center. On repeating the test for the remaining three directions, the comparison of the final height of the copper and lead forged during the same blow gave a good indication of the nonparallelity of the ram and bolster surfaces over a 10 in. (255 mm) span. In conducting this comparison, the local elastic deflection of the dies in forging copper must be considered. Therefore, the final thickness of the copper samples was corrected to counteract this local die deflection.

In off-center loading with 220 ton (or 44% or the nominal capacity), an average ram-bed nonparallelity of 0.038 in./ft was measured in both directions, front-to-back and left-to-right. In comparison, the nonparallelity under unloaded conditions was about 0.002 in./ft. Before conducting the experiments described above, the clearance in the press gibs was set to 0.010 in. (0.254 mm) [Douglas et al., 1972]. The nonpar-

Table 11.2 Copper samples forged under on-center conditions in the 500 ton mechanical press

	Sample size, in.					
Sample	Height	Diameter	Predicted load(a), tons	Measured load, tons	Predicted energy(b), tons	Measured energy, tons
1	2.00	1.102	48	45	24	29
2	2.00	1.560	96	106	48	60
3	2.00	2.241	197	210	98	120
4	2.00	2.510	247	253	124	140
5	2.00	2.715	289	290	144	163
6	2.00	2.995	352	350	176	175

(a) Based on an estimate of 50 ksi flowstress for copper at 50% reduction in height. (b) Estimated by assuming that the load-displacement curve has a triangular shape; that is, energy = 0.5 load × displacement. Source: [Douglas et al., 1972]

allelity in off-center forging would be expected to increase with increasing gib clearance.

11.2.7 Crank Presses with Modified Drives

For a long time, eccentric or crank drive systems were the only type of drive mechanisms used in mechanical presses. This section discusses some modified drives such as the knuckle-joint drive and the linkage drive. The sinusoidal slide displacement of an eccentric press is compared with those of a knuckle-joint and a linkage-driven press (Fig. 11.19). The relatively high impact speed on die closure and the reduction of slide speed during the forming processes are drawbacks that often preclude the use of eccentric or crank driven press for cold forging at high stroking rates. However, in presses with capacities up to a nominal force of 560 tonf (5000 kN), such as universal or blanking presses used for trimming, eccentric or crank drive is still the most effective drive system. This is especially true when using automated systems where the eccentric drive offers a good compromise between time necessary for processing and that required for part transport [Schuler Handbook, 1998].

Knuckle-Joint Drive Systems. The velocity versus stroke and the load versus stroke characteristics of crank presses can be modified by using different press drives. A well-known variation of the crank press is the knuckle-joint design (Fig. 11.20). This design is capable of generating high forces with a relatively small crank drive. In the knuckle-joint drive, the ram velocity slows down much more rapidly toward the BDC than in the regular crank drive. This machine is successfully used for cold forming and coining applications.

The knuckle-joint drive system consists of an eccentric or crank mechanism driving a knuckle

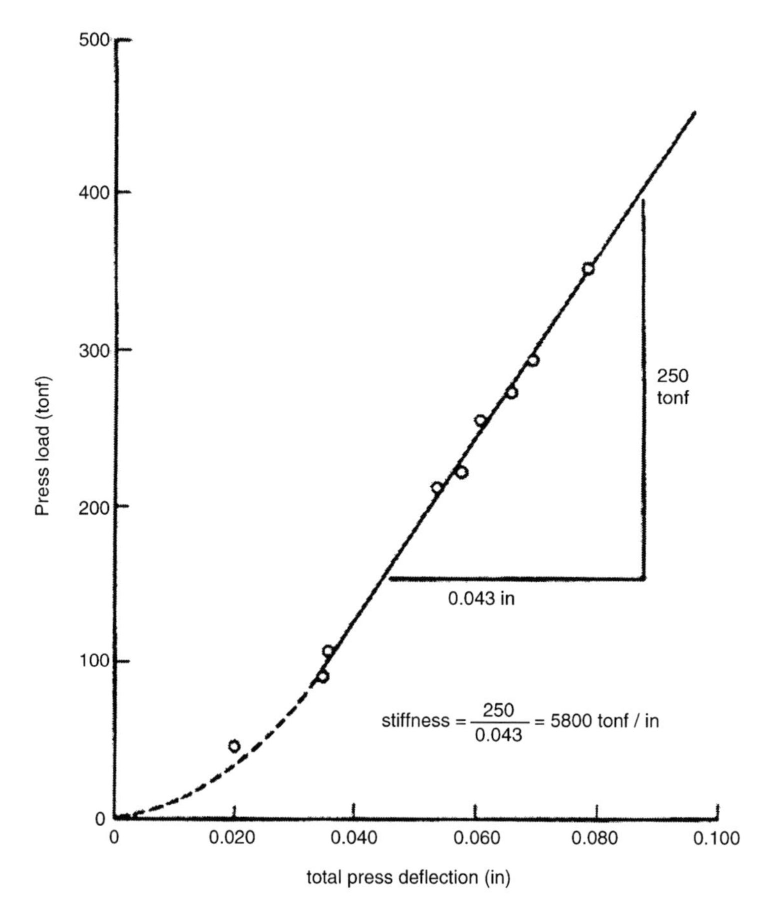

Fig. 11.18 Total press deflection versus press loading obtained under dynamic loading conditions for a 500 ton Erie scotch yoke type press. [Douglas et al., 1972]

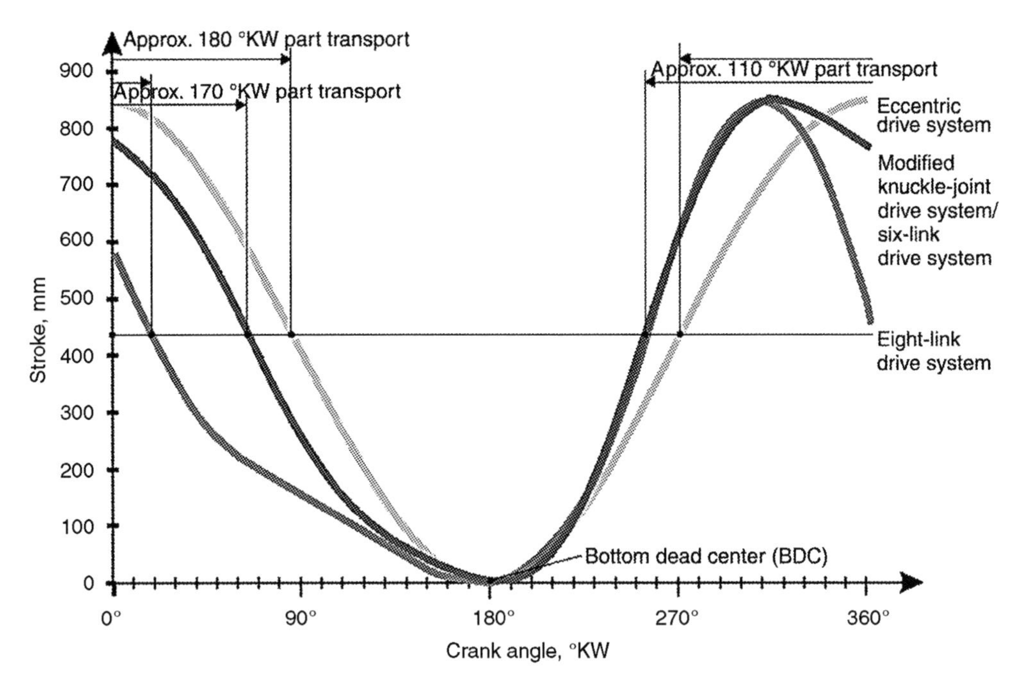

Fig. 11.19 Displacement-time diagram: comparison of the slide motion performed by an eccentric, a knuckle-joint, and a link-driven press. [Schuler Handbook, 1998]

joint. Figure 11.21 shows this concept used in a press with bottom drive [Schuler Handbook, 1998]. The fixed joint and bed plate form a compact unit. The lower joint moves the press frame. It acts as a slide and moves the attached top die up and down. Due to the optimum force flow and the favorable configuration possibilities offered by the force-transmitting elements, a highly rigid design with very low deflection characteristics is achieved. The knuckle joint, with a relatively small connecting rod force, generates a considerably larger pressing force. Thus, with the same drive moment, it is possible to reach around three to four times higher pressing forces as compared to eccentric presses. Furthermore, the slide speed in the region 30 to 40° above the bottom dead center is appreciably lower.

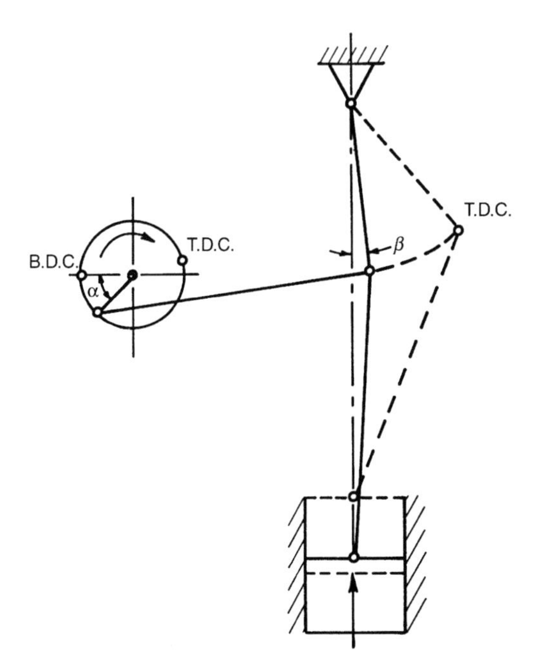

Fig. 11.20 Schematic of a toggle (or knuckle) joint mechanical press

By inserting an additional joint, the kinematic characteristics and the speed versus stroke of the slide can be modified. Knuckle-joint and modified knuckle-joint drive systems can be either top or bottom mounted. For cold forging, particularly, the modified top drive system is in popular use. Figure 11.22 illustrates the principle of a press configured according to this specification. The fixed point of the modified knuckle joint is mounted in the press crown. While the upper joint pivots around this fixed point, the lower joint describes a curve-shaped path. This results in a change of the stroke versus

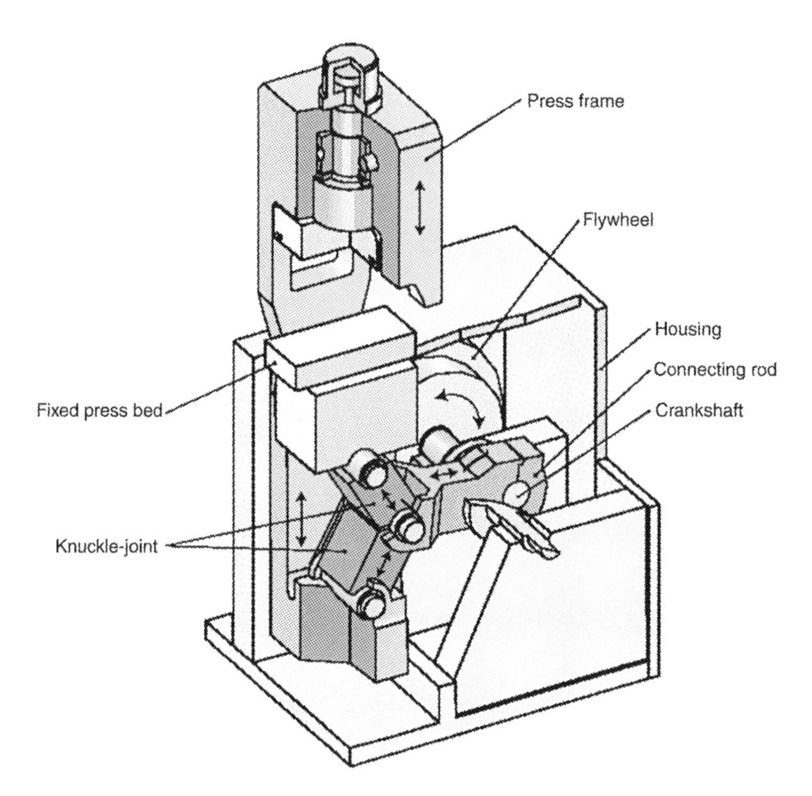

Fig. 11.21 Knuckle-joint press with bottom drive. [Schuler Handbook, 1998]

time characteristic of the slide, compared to the largely symmetrical stroke-time curve of the eccentric drive system (Fig. 11.19). This curve can be altered by modifying the arrangement of the joints (or possibly by integrating an additional joint).

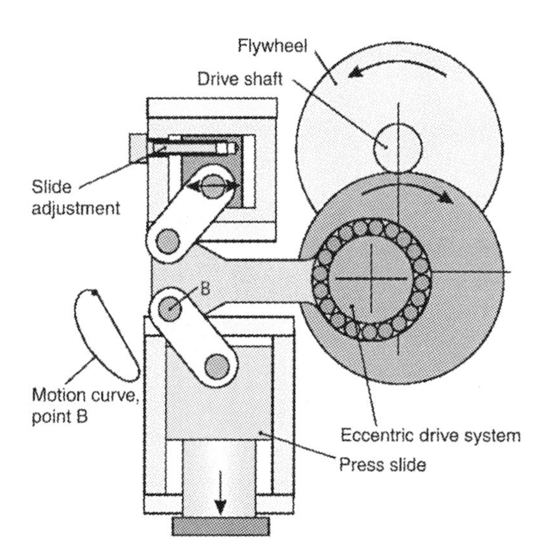

Fig. 11.22 Modified knuckle-joint drive system. [Schuler Handbook, 1998]

Linkage Drive. The mechanical press drive shown in Fig. 11.23 uses a four-bar linkage mechanism. In this mechanism, the load-stroke and velocity-stroke behavior of the slide can be established, at the design stage, by adjusting the length of one of the four links or by varying the

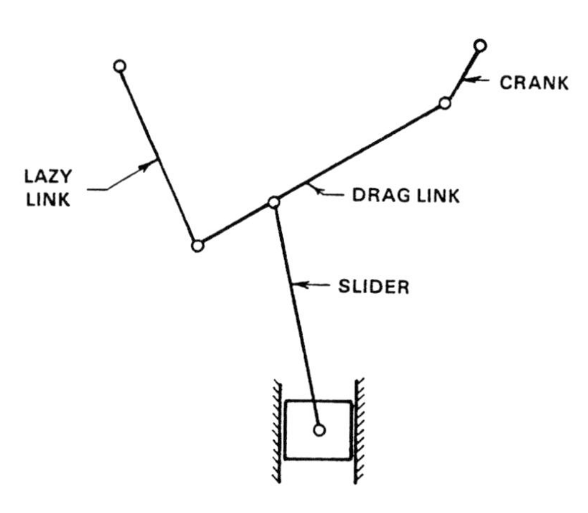

Fig. 11.23 Four-bar linkage mechanism for mechanical press drives

connection point of the slider link with the drag link. Thus, with this press it is possible to maintain the maximum load, as specified by press capacity, over a relatively long deformation stroke. Using a conventional slider-crank-type press, this capability can be achieved only by using a much larger-capacity press. A comparison is illustrated in Fig. 11.24, where the load-stroke curves for a four-bar linkage press and a conventional slider-crank press are shown. It can be seen that a slider-crank press equipped with a 1700 ton-in. torque drive can generate a force of about 1500 tons at 1/32 in. before BDC. The four-bar press equipped with a 600 ton-in. drive generates a force of about 750 tons at the same location. However, in both machines a 200 ton force is available at 6 in. before BDC. Thus, a 750 ton, four-bar press could perform the same forming operation, requiring 200 tons over 6 in., as a 1500 ton eccentric press. The four-bar press, which was originally developed for sheet metal forming and cold extrusion, is well suited for extrusion-type forming operations, where a nearly constant load is required over a long stroke.

11.3 Screw Presses

The screw press uses a friction, gear, electric, or hydraulic drive to accelerate the flywheel and the screw assembly, and it converts the angular kinetic energy into the linear energy of the slide or ram. Figure 11.25 shows two basic designs of screw presses [Bohringer et al., 1966].

In the friction-drive press, the driving disks are mounted on a horizontal shaft and are rotated continuously. For a downstroke, one of the driving disks is pressed against the flywheel by a servomotor. The flywheel, which is connected to the screw either positively or by a friction slip clutch, is accelerated by this driving disk through friction. The flywheel energy and the ram speed continue to increase until the ram hits the workpiece. Thus, the load necessary for forming is built up and transmitted through the slide, the screw, and the bed to the press frame. When the entire energy in the flywheel is used in deforming the workpiece and elastically deflecting the press, the flywheel, the screw, and the slide stop. At this moment, the servomotor activates the horizontal shaft and presses the upstroke driving disk wheel against the flywheel. Thus, the flywheel and the screw are accelerated in the reverse direction and the slide is lifted to its top position.

In the direct-electric-drive press, a reversible electric motor is built directly to the screw and on the frame, above the flywheel. The screw is threaded into the ram or the slide and does not move vertically. To reverse the direction of flywheel rotation, the electric motor is reversed after each downstroke and upstroke.

11.3.1 Load and Energy in Screw Presses

In a screw press the load is transmitted through the slide, screw, and bed to the press frame. The available load at a given stroke position is supplied by the energy stored in the flywheel. At the end of a stroke, the flywheel and the screw come to a standstill before reversing the direction of rotation. Thus, the following relationship holds:

$$E_T = E_P + E_F + E_d \quad \text{(Eq 11.14)}$$

where E_T is total flywheel energy, E_P is the energy consumed by the forming process, E_F is the energy required for overcoming machine friction, and E_d is the energy required for deflection of the press (bed + columns + screw).

If the total flywheel energy, E_T, is larger than necessary for overcoming machine losses and

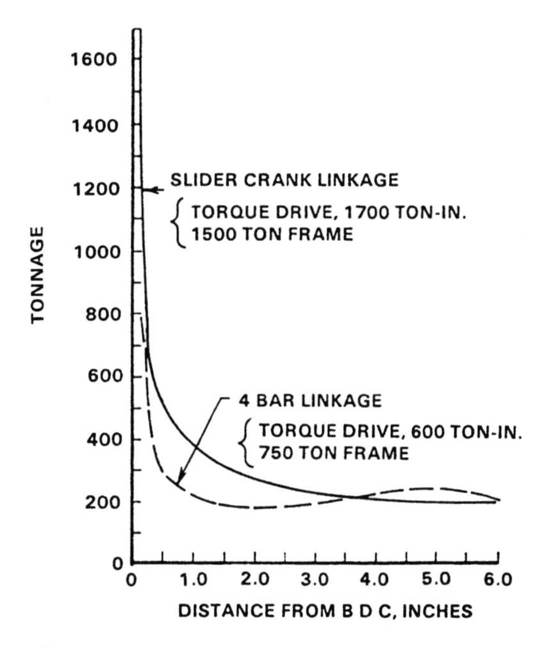

Fig. 11.24 Load-stroke curves for a 750 ton four-bar linkage press and a 1500 ton slider-crank press

for carrying out the forming process, the excess energy is transformed into additional deflection energy and both the die and the press are subjected to unnecessarily high loading. This is illustrated in Fig. 11.26. To annihilate the excess energy, which results in increased die wear and noise, the modern screw press is equipped with an energy-metering device that controls the flywheel velocity and regulates the total flywheel energy. The energy metering can also be programmed so that the machine supplies different amounts of energy during successive blows.

In a screw press, which is essentially an energy-bound machine (like a hammer), load and energy are in direct relation with each other. For a given press (i.e., for the same friction losses, elastic deflection properties, and available flywheel energy), the load available at the end of the stroke depends mainly on the deformation energy required by the process (i.e., on the shape, temperature, and material of the workpiece). Thus, for constant flywheel energy, low deformation energy, E_p, results in high end load, L_M, and high deformation energy, E_p, results in low end load, L_M. These relations are illustrated in the load-energy diagram of a screw press, as shown in Fig. 11.27.

The screw press can sustain maximum loads, L_{max}, up to 160 to 200% of its nominal load, L_M. In this sense, the nominal load of a screw press

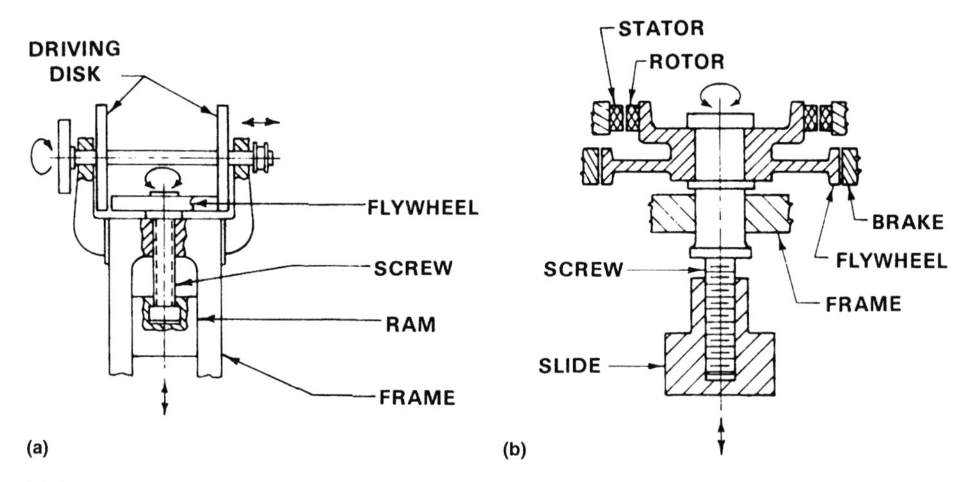

Fig. 11.25 Two widely used screw press drives. (a) Friction drive. (b) Direct electric drive. [Bohringer et al., 1966]

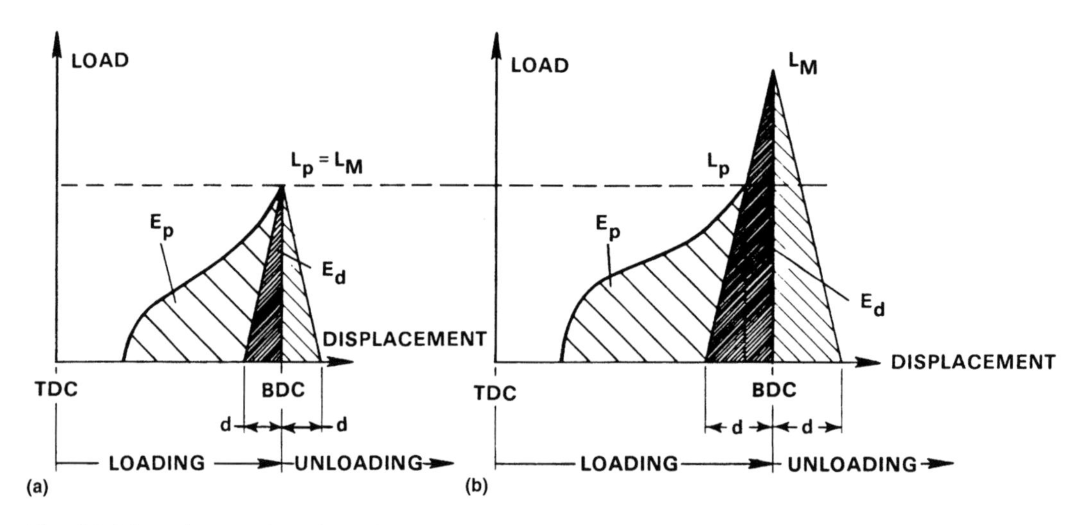

Fig. 11.26 Load-energy relationships in forming in a press. E_p, energy required by process; L_M, maximum machine load; E_d, elastic deflection energy; d, press deflection. (a) With energy or load metering. (b) Without energy or load metering

is set rather arbitrarily. The significant information about the press load is obtained from its load-energy diagram (see Fig. 11.27). The load-energy curve has a parabolic shape because the deflection energy, E_d, is given by a second-order equation:

$$E_d = L_M^2/2C \qquad \text{(Eq 11.15)}$$

where L_M is machine load and C is total press stiffness.

A screw press designed for a forming operation, where large energies, E_M, are needed, can also be used for operations where smaller energies are required. Here, however, a friction clutch is installed between the flywheel and the screw. When the ram load reaches the nominal load, this clutch starts slipping and uses up a part of the flywheel energy as frictional heat energy, E_c, at the clutch. Consequently, the maximum load at the end of the downstroke is reduced from L to L_{max} and the press is protected from overloading (Fig. 11.27). Screw presses used for coining are designed for hard blows (i.e., die-to-die blows without any workpiece) and do not have a friction slip clutch on the flywheel.

11.3.2 Time-Dependent Characteristics of Screw Presses

In a screw press, the number of strokes per minute under load, n_p, largely depends on the energy required by the specific forming process and on the capacity of the drive mechanism to accelerate the screw and the flywheel. In general, however, the production rate of a screw press is lower than that of a mechanical press, especially in automated high-volume operations.

During a downstroke, a velocity under pressure, V_p, increases until the slide hits the workpiece. In this respect, a screw press behaves like a hammer. After the actual deformation starts, the velocity of the slide decreases depending on the energy requirements of the process. Thus, the velocity, V_p, is greatly influenced by the geometry of the stock and of the part. As illustrated in Fig. 11.28, this is quite different from the conditions found in mechanical presses, where the ram velocity is established by the press kinematics and is not influenced significantly by the load and energy requirements of the process.

11.3.3 Accuracy in Screw Press Operation

In general, the dimensional accuracies of press components under unloaded conditions, such as parallelism of slide and bed surfaces, clearances in the gibs, etc., have basically the same significance in the operation of all presses—hydraulic, mechanical, and screw presses.

The off-center loading capacity of the press influences the parallelism of upset surfaces. This capacity is increased in modern presses by use of long gibs and by finish forming at the center, whenever possible. The off-center loading capacity of a screw press is less than that of a mechanical press or a hammer.

A screw press is operated like a hammer, i.e., the top and bottom dies "kiss" at each blow. Therefore, the stiffness of the press, which affects the load and energy characteristics, does not influence the thickness tolerances in the formed part.

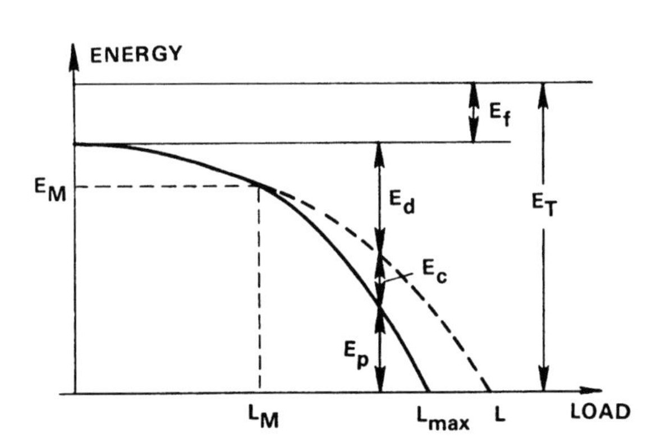

--- WITHOUT FRICTION CLUTCH AT FLYWHEEL
—— WITH SLIPPING FRICTION CLUTCH AT FLYWHEEL
E_T = TOTAL FLYWHEEL ENERGY
E_f = FRICTION ENERGY
E_d = DEFLECTION ENERGY
E_p = ENERGY REQUIRED BY PROCESS
E_c = ENERGY LOST IN SLIPPING CLUTCH
L_M = NOMINAL MACHINE LOAD
L_{max} = MAXIMUM LOAD
E_M = NOMINAL MACHINE ENERGY AVAILABLE FOR FORGING

Fig. 11.27 Schematic of load-energy relationship in a screw press. [Klaprodt, 1968]

11.3.4 Determination of Dynamic Stiffness of a Screw Press

The static stiffness of the screw press, as given by the manufacturer does not include the torsional stiffness of the screw that occurs under dynamic conditions [Bohringer et al., 1968]. As pointed out by Watermann [Watermann, 1963], who conducted an extensive study of the efficiency of screw presses, the torsional deflection of the screw may contribute up to 30% of the total losses at maximum load (about 2.5 times nominal load). Based on experiments conducted in a Weingarten press (Model P160, nominal load 180 metric ton, energy 800 kg-m), Waterman concluded that the dynamic stiffness was 0.7 times the static stiffness [Watermann, 1963]. Assuming that this ratio is approximately valid for the 400 ton press, the dynamic stiffness is $0.7 \times 8400 \approx 5900$ ton/in.

During the downstroke the total energy supplied by the screw press, E_T, is equal to the sum total of the machine energy used for the deformation process, E_P, the energy necessary to overcome friction in the press drive, E_F, and the energy necessary elastically to deflect the press, E_D (Eq 11.14). Expressing E_D in terms of the press stiffness, C, Eq 11.14 can be written as:

$$E_T - E_F = E_P + \frac{L_M^2}{2C} \quad \text{(Eq 11.16)}$$

In a forging test, the energy used for the process E_P (surface area under the load-displacement curve) and the maximum forging load L_P can be obtained from oscillograph recordings. By considering two tests simultaneously, and by assuming that E_F remains constant during tests, one equation with one unknown C can be derived from Eq 11.16. However, in order to obtain reasonable accuracy, it is necessary that in both tests considerable press deflection is obtained; that is, high loads L_P and low deformation energies E_P are measured. Thus, errors in calculating E_P do not impair the accuracy of the stiffness calculations.

11.3.5 Variations in Screw Press Drives

In addition to direct friction and electric drives (Fig. 11.25), there are several other types of mechanical, electric, and hydraulic drives that are commonly used in screw presses. A different design is shown in Fig. 11.29. A flywheel (1), supported on the press frame, is driven by one or several electric motors and rotates at a constant speed. When the stroke is initiated, an oil-operated clutch (2) engages the rotating flywheel against the stationary screw (3). This feature is similar to what is used to initiate the stroke of an eccentric forging press. On engagement of the clutch, the screw is rapidly accelerated and reaches the speed of the flywheel. As a result, the ram (4), which acts like a giant nut, moves

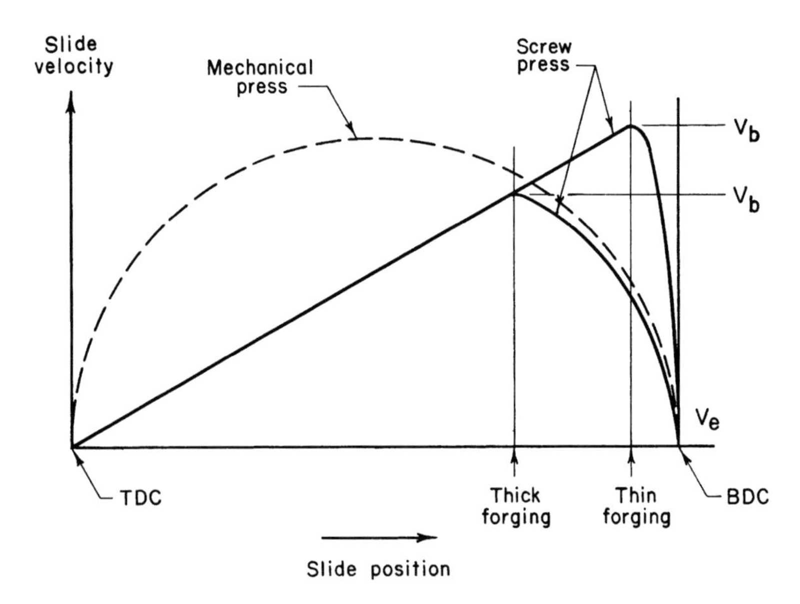

Fig. 11.28 Representation of slide velocities for mechanical and screw presses in forming a thick and a thin part. V_b, V_e = velocity at the beginning and end of forming, respectively. [Altan et al., 1973]

downward. During this downstroke, the oil is compressed in the hydraulic lift-up cylinders (5). The downstroke is terminated by controlling either the ram position, by means of a position switch, or the maximum load on the ram, by disengaging the clutch and the flywheel from the screw when the preset forming load is reached. The ram is then lifted by the lift-up cylinders, releasing the elastic energy stored in the press frame, the screw, and the lift-up cylinders. At the end of the downstroke, the ram is stopped and held in position by a hydraulic brake.

This press provides several distinct benefits: a high and nearly constant ram speed throughout the stroke, full press load at any position of the stroke, high deformation energy, overload protection, and short contact time between the workpiece and the tools. The press can also be equipped with variable-speed motors so that different flywheel and ram speeds are available. Thus, it offers considerable flexibility and can be used for hot as well as cold forming operations.

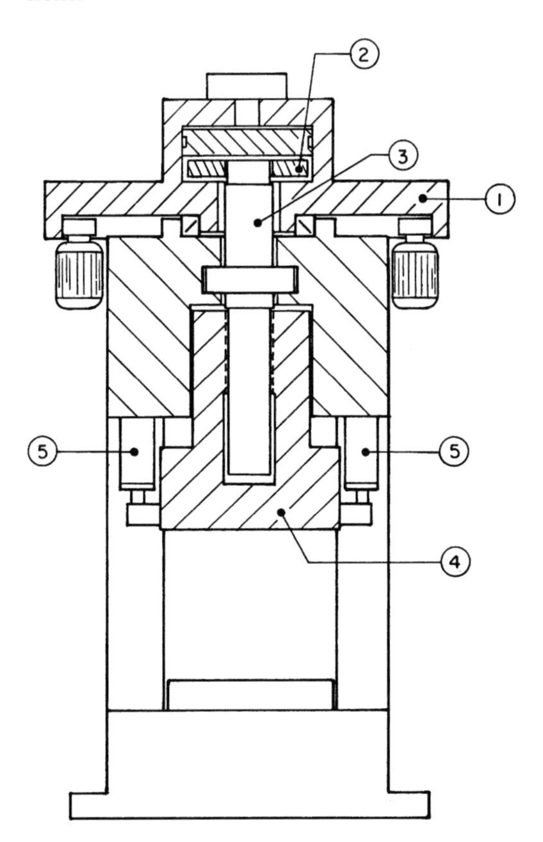

Fig. 11.29 A screw press drive that combines the characteristics of mechanical and screw presses. 1, flywheel; 2, oil-operated clutch; 3, screw; 4, ram; 5, lift-up cylinders. [Altan, 1978]

A wedge-screw press, similar in principle to the wedge mechanical press seen in Fig. 11.17, is shown in Fig. 11.30. The screw of this press drives a wedge that provides a large bearing surface on the top of the slide. As a result, this press can take larger off-center loads than regular screw presses. Thus, it can be used for multiple station forging operations.

11.4 Hammers

The hammer is the least expensive and most versatile type of equipment for generating load and energy to carry out a forming process. This technology is characterized by multiple impact blows between contoured dies. Hammers are primarily used for hot forging, for coining, and to a limited extent, for sheet metal forming of parts manufactured in small quantities—for example, in the aircraft/airframe industry. Hammer forging has a reputation as an excellent way to enhance the metallurgical properties of many materials, including high-performance materials such as Waspaloy, the nickel-base superalloy used for many turbine disk applications. The hammer is an energy-restricted machine. During a working stroke, the deformation proceeds until the total kinetic energy is dissipated by plastic deformation of the material and by elastic deformation of the ram and the anvil when the die

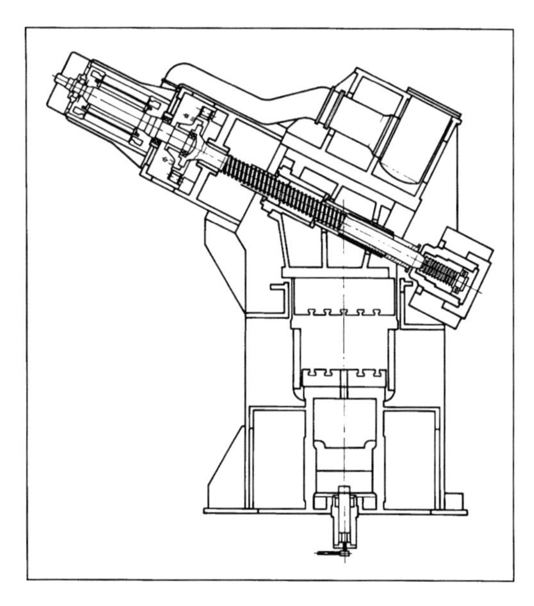

Fig. 11.30 Schematic of the wedge-screw press drive for hot forging

faces contact each other. Therefore, it is necessary to rate the capacities of these machines in terms of energy, i.e., foot-pounds, meter-kilograms, or meter-tons. The practice of specifying a hammer by its ram weight is not useful for the user. Ram weight can be regarded only as a model or specification number.

There are basically two types of anvil hammers: gravity-drop hammers and power-drop hammers. In a simple gravity-drop hammer, the upper ram is positively connected to a board (board-drop hammer), a belt (belt-drop hammer), a chain (chain-drop hammer), or a piston (oil-, air-, or steam-lift drop hammer); see Fig. 11.31. The ram is lifted to a certain height and then dropped on the stock placed on the anvil. During the downstroke, the ram is accelerated by gravity and builds up the blow energy. The upstroke takes place immediately after the blow; the force necessary to ensure quick lift-up of the ram can be three to five times the ram weight. The operation principle of a power-drop hammer is similar to that of an air-drop hammer (Fig. 11.31d). In the downstroke, in addition to gravity, the ram is accelerated by steam, cold air, or hot air pressure. In electrohydraulic gravity-drop hammers, the ram is lifted with oil pressure against an air cushion. The compressed air slows down the upstroke of the ram and contributes to its acceleration during the downstroke. Thus, electrohydraulic hammer also has a minor power hammer action.

In the power-drop hammer, the acceleration of the ram is enhanced with air pressure applied on the top side of the ram cylinder (Fig. 11.32). Today, most drop hammers are power or pressure drive. The air pressure and the ram height are measured and electronically controlled as a result. The energy per hammer blow can be metered automatically.

Counterblow hammers are widely used in Europe while their use in the United States is limited to a relatively small number of companies. The principles of two types of counterblow hammers are illustrated in Fig. 11.33. In both designs, the upper ram is accelerated downward by steam, cold air, or hot air [Kuhn, 1963]. At the same time, the lower ram is accelerated upward by a steel band (for smaller capacities) or by a hydraulic coupling system (for larger capacities). The lower ram, including the die assembly, is approximately 10% heavier than the upper ram. Therefore, after the blow, the lower ram accelerates downward and pulls the upper ram back up to its starting position. The combined speed of the rams is about 25 ft/s (7.5 m/s); both

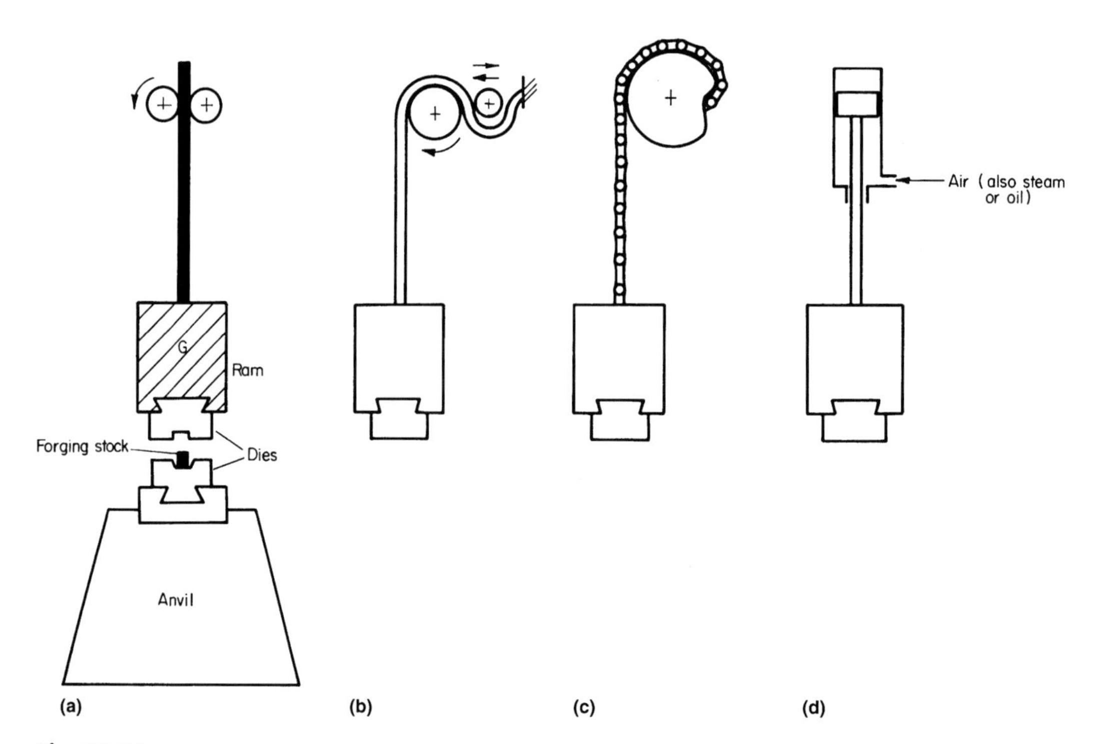

Fig. 11.31 Principles of various types of gravity-drop hammers. (a) Board drop. (b) Belt drop. (c) Chain drop. (d) Air drop

rams move with exactly half of the total closure speed. Due to the counterblow effect, relatively little energy is lost through vibration in the foundation and environment. Therefore, for comparable capacities, a counterblow hammer requires a smaller foundation than an anvil hammer.

The schematic of an electrohydraulically driven counterblow hammer is seen in Fig. 11.34. The upper ram, which weighs approximately ¼ to ⅕ as much as the lower ram, is guided in the latter. The ram speeds are inversely proportional to the ram weights so that, at the instant of blow, they both have the same momentum. Thus, the forging blow takes place at a lower plane than in conventional counterblow hammers. This facilitates the handling of the forging in the hammer.

Fig. 11.32 Schematic of a power-drop hammer

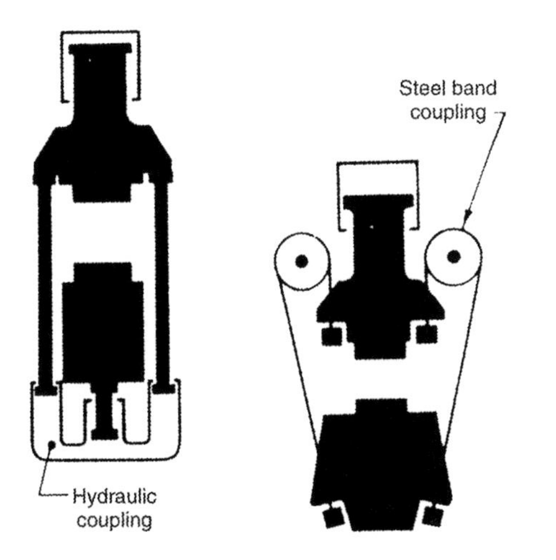

Fig. 11.33 Principles of operation of two types of counterblow hammers. [Altan et al., 1973]

11.4.1 Important Characteristics of Hammers

In a gravity-drop hammer, the total blow energy is equal to the kinetic energy of the ram and is generated solely through free-fall velocity, or:

$$E_T = \frac{1}{2} m_1 V_1^2 = \frac{1}{2} \frac{G_1}{g} V_1^2 = G_1 H \qquad \text{(Eq 11.17)}$$

where m_1 is the mass of the dropping ram, V_1 is the velocity of the ram at the start of the deformation, G_1 is the weight of the ram, g is the acceleration of gravity, and H is the height of the ram drop.

In a power-drop hammer, the total blow energy is generated by the free fall of the ram and by the pressure acting on the ram cylinder, or:

$$E_T = \frac{1}{2} m_1 V_1^2 = (G_1 + pA)H \qquad \text{(Eq 11.18)}$$

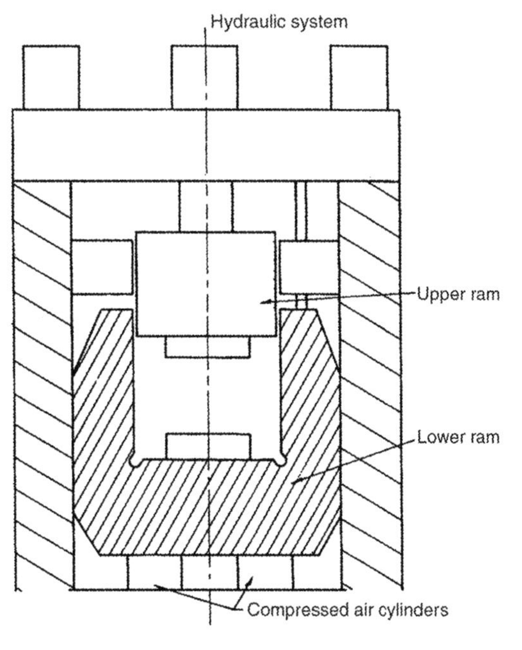

Fig. 11.34 Schematic of electrohydraulic counterblow hammer. [Altan et al., 1973]

where, in addition to the symbols given above, p is the air, steam, or oil pressure acting on the ram cylinder in the downstroke and A is the surface area of the ram cylinder.

In counterblow hammers, when both rams have approximately the same weight, the total energy per blow is given by:

$$E_T = 2\left(\frac{m_1 V_1^2}{2}\right) = \frac{m_1 V_t^2}{4} = \frac{G_1 V_t^2}{4g} \quad \text{(Eq 11.19)}$$

where m_1 is the mass of one ram, V_1 is the velocity of one ram, V_t is the actual velocity of the blow of two rams, which is equal to $2V_1$, and G_1 is the weight of one ram.

During a working stroke, the total nominal energy, E_T, of a hammer is not entirely transformed into useful energy available for deformation, E_A. Some small amount of energy is lost in overcoming friction of the guides, and a significant portion is lost in the form of noise and vibration to the environment. Thus, the blow efficiency, $\eta = E_A/E_T$, of hammers is always less than one. The blow efficiency varies from 0.8 to 0.9 for soft blows (small load and large displacement) and from 0.2 to 0.5 for hard blows (high load and small displacement).

The transformation of kinetic energy into deformation energy during a working blow can develop considerable forces. For instance, consider a deformation blow where the load, P, increases from P/3 at the start to P at the end of the stroke, h. The available energy, E_A, is the surface area under the curve in Fig. 11.35. Therefore:

$$E_A = \frac{P/3 + P}{2} h = \frac{4Ph}{6} \quad \text{(Eq 11.20)}$$

Consider a hammer with a total nominal energy, E_T, of 35,000 ft · lbf (47.5 kJ) and a blow efficiency, η, of 0.4; here, $E_A = E_T\eta = 14{,}000$ ft · lbf (19 kJ). With this value, for a working stroke, h, of 0.2 in., Eq 11.20 gives:

$$P = \frac{6E_A}{4h} = 1{,}260{,}000 \text{ lb} = 630 \text{ tons}$$

If the same energy were dissipated over a stroke, h, of 0.1 in., the load, P, would reach approximately double the calculated value. The simple hypothetical calculations given above illustrate the capabilities of relatively inexpensive hammers in exerting high forming loads.

11.4.2 Computer-Controlled Hammers

Up to now, only small forging hammers and impacters have benefited from sophisticated control systems that increase manufacturing speed, reduce cost, and increase quality. Because of the many benefits possible on large hammers, computerized controls are now being adapted to large forging hammers.

Adding computerized controls to a forging hammer significantly improves the hammer process in three critical areas: microstructural quality, operational uniformity and consistency, and cost containment for customers. However, transferring small-hammer control technology to large industrial forging hammers is not a trivial undertaking. It requires extensive equipment engineering, and many issues arise in the transfer of technology.

While hammer forging processes have always offered thermal-control-related advantages compared to press forging processes, these advantages are enhanced tremendously by the addition of microprocessor controls. These enhancements grow out of the ability to tailor the forging sequence of a computerized hammer, opening the door to process refinements that give engineers much greater control over the final microstructural and mechanical properties.

Computerized hammer controls allow unique processing schemes to be developed for optimum results through computer process modeling. Processing step combinations (blow energy, interblow dwell time, quantity of blows, etc.) can be engineered for best control of adiabatic heating, die chilling, strain, strain rate, recrystallization, and grain growth.

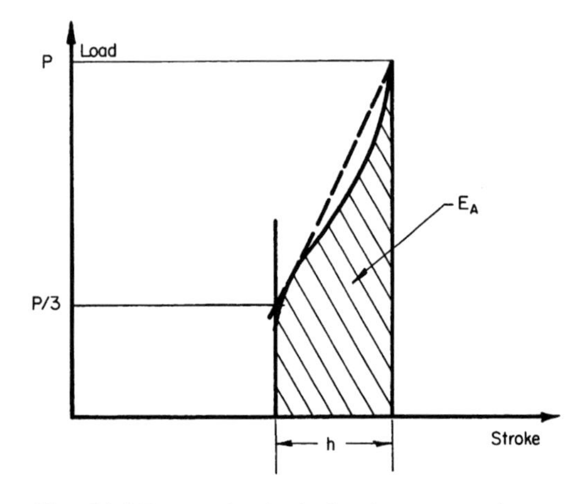

Fig. 11.35 Example of a load-stroke curve in a hammer blow

The engineered processing steps are precisely controlled by the computerized forging hammer controls, which also allow for greater understanding of process and equipment parameters. This leads to greater computer process modeling capability through precise boundary condition definitions and outstanding simulation accuracy. The capabilities of this process have dramatic implications for users of highly engineered forgings, and show why, metallurgically and economically, hammers are a tremendously valuable piece of equipment.

REFERENCES

[Altan et al., 1972]: Altan, T., Nichols, D.E., "Use of Standardized Copper Cylinders for Determining Load and Energy in Forging Equipment," *ASME Trans., J. Eng. Ind.,* Vol 94, Aug 1972, p 769.

[Altan et al., 1973]: Altan, T., et al., "Forging Equipment, Materials and Practices," HB03, Metal and Ceramics Information Center, 1973, p 4–7.

[Altan, 1978]: Altan, T., "Metalforming at 2.EMO," *Am. Mach.,* Jan 1978, p 132.

[Altan et al., 1983]: Altan, T., Oh, S.-I., Gegel, H., *Metal Forming: Fundamentals and Applications,* ASM International, 1983.

[ASM Handbook, 1988): Altan, T., "Hammer and Presses for Forging," *ASM Handbook,* Vol 14, *Forming and Forging,* ASM International, 1988, p 25–35.

[Bohringer et al., 1966]: Bohringer, H., Kilp, K.H., "Development of the Direct-Drive Percussion Press," *Sheet Metal Ind.,* Vol 43, Nov 1966, p 857.

[Bohringer et al., 1968]: Bohringer, H., Kilp, K.H., "The Significant Characteristics of Percussion Presses and Their Measurements," *Sheet Metal Ind.,* May 1968, p 335.

[Douglas et al., 1972]: Douglas, J.R., Altan, T., "Characteristics of Forging Presses: Determination and Comparison," *Proceedings of the 13th M.T.D.R. Conference,* Sept 1972, Birmingham, England, p. 536.

[Geleji, 1967]: Geleji, A., "Forge Equipment, Rolling Mills and Accessories" (in English), *Akademaii Kiado,* Budapest, 1967.

[Hutson, 1968]: Hutson, C., "An Investigation into Press Driving Systems," *Metal Forming,* Vol 35, March 1968, p 46.

[Kienzle, 1959]: Kienzle, O., "Development Trends in Forming Equipment" (in German), *Werkstattstechnik,* Vol 49, 1959, p 479.

[Kirschbaum, 1968]: Kirschbaum, Z., "A Comparative Study of the Stability and Economic Construction of Pushdown, Pulldown and Horizontal Double Opposed Forging Presses," *Iron Steel,* Feb 1968, p 46.

[Klaprodt, 1968]: Klaprodt, Th., "Comparison of Some Characteristics of Mechanical and Screw Presses for Die Forging" (in German), *Indust.-Anzeig.,* Vol 90, 1968, p 1423.

[Kuhn, 1963]: "Counterblow Hammers for Heavy Forgings" (in German), *Kleipzig Fachberichte,* No. 11, Dusseldorf, 1963.

[Mueller, 1969]: Mueller, E., *Hydraulic Forging Presses,* Springer Verlag, Berlin, 1969.

[Peters, 1969]: Peters, K.H., "Design Features of the Hydraulic Press and its Field of Application," *Sheet Metal Ind.,* Vol 46, March 1969, p 221–226.

[Rau, 1967]: Rau, G., A Die Forging Press with a New Drive, *Metal Form.,* July 1967, p 194–198.

[Riemenschneider et al., 1959]: Riemenschneider, F., Nickrawietz, K., "Drives for Forging Presses" (in German), *Stahl Eisen,* Vol 79, 1959, p 494.

[Schuler Handbook, 1998]: Schuler, *Metal Forging Handbook,* Springer, Goppingen, Germany, 1998.

[Watermann, 1963]: Watermann, H.D., "The Blow Efficiency in Hammers and Screw Presses" (in German), *Indust.-Anzeig.,* No. 77, 24 Sept 1963, p 53.

SELECTED REFERENCES

[Brauer, 1961]: Brauer, W., "The Development of Electrohydraulic Drop Hammers"(in German), *Werkstattstechnik,* 1961, p 105.

[Hamilton, 1960]: Hamilton, E., "Power Presses, Their Design and Characteristics," *Sheet Metal Ind.,* Vol 37, July 1960, p 501–513.

[Pahnke]: Pahnke, H.J., "Technical and Economic Limitations of Conventional Four Column Top Drive Forging Presses," SME Paper MF70-589.

[Spachner]: Spachner, S.A., "Use of a Four Bar Linkage as a Slide Drive for Mechanical Presses," SME Paper MF70-216.

CHAPTER 12

Special Machines for Forging

Pinak Barve

12.1 Introduction

Prior to forging in an impression die, billet stock must be often preformed to achieve adequate material distribution, especially in hot forging with flash. Several special machines are used for the purpose of preforming the incoming stock. Some of these machines may also be used for finish forging.

The principle of forging rolls, or reducer rolls, is illustrated schematically in Fig. 12.1. This machine is generally used for volume distribution in long and thin parts, prior to closed-die forging. A typical operation on reducer rolls, such as those of Fig. 12.2, is as follows.

The operator or a robot places the heated stock on a table in the front of the machine, grasps the stock with the tongs, and starts the machine with a foot pedal or automatic signal. During the portion of the roll rotation, when the rolls are in open position, the stock is placed between the rolls against a stock gage and in line with the first roll groove. As the rolls rotate, forging begins and the deformed stock is forced toward the front of the machine [Altan et al., 1973].

These sequences are repeated for the next grooves. Thus, the shape of the die segments on the forging rolls determines the rolled configuration. An example illustrating the application of preforming by roll forging, or reducer rolling, is shown in Fig. 12.3. Another example, illustrating preforming for a truck axle forging, is seen in Fig. 12.4.

In roll forging, the contact time between the workpiece and the roll segments is extremely short due to the high speed of the rolls. Therefore, even after the final rolling operation, the workpiece is still hot and can be finish forged under a hammer or press without reheating.

For high-volume production, the reducer rolling operation is automated. The stock is gripped, fed into the rolls, transferred from one die segment to the other, and released on a conveyor using a dedicated robot. In producing long parts,

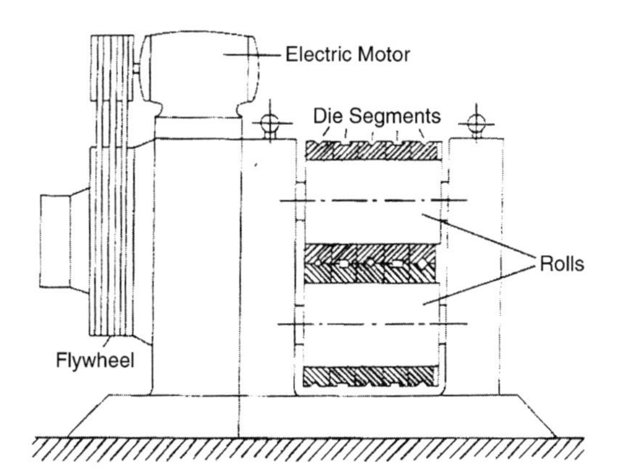

Fig. 12.1 Schematic of forging rolls for reducer rolling

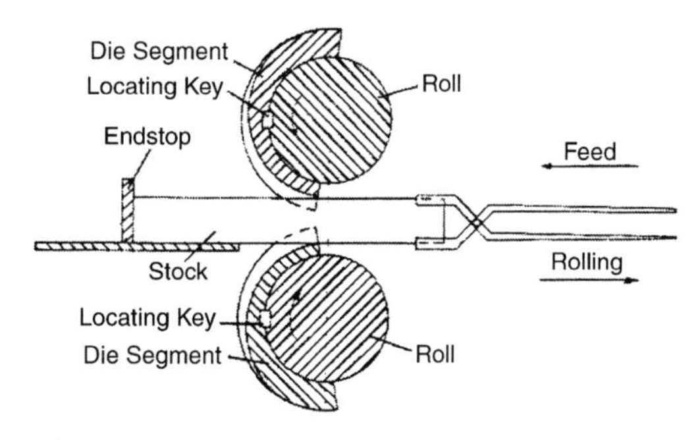

Fig. 12.2 Schematic of the reducer rolling operation

such as crankshafts and front axles, it is possible to use vertical reducer rollers, as seen in Fig. 12.5.

The design of the roll segments needs considerable experience. Recently, three-dimensional FEM codes have been used to (a) simulate the reducer rolling operation and (b) design and optimize the configuration of the die segments. Figure 12.6 shows, as an example, two roll passes, simulated by the commercial software DEFORM 3-D [SFTC, 2003]. The desired shape is produced by rolling the heated billet between two rotating dies having appropriately profiled grooves. After the first pass, the billet is fed into the roll segments of the second pass after 90° rotation.

12.2 Transverse or Cross-Rolling Machines

Transverse rolling is used for producing preforms or finish forgings from round billets. As seen in Fig 12.7, a round billet is inserted transversely between two or three rolls, which rotate in the same direction and drive the billet [Altan et al., 1973]. The rolls, which hold replaceable die segments with appropriate impressions, make one revolution while the workpiece rotates several times in the opposite direction. Thus, the transverse rolling method can form axially symmetrical shafts with complex geometry in one operation. As an example, a forging produced by this method is shown in Fig. 12.8. The transverse rolling machines are suitable for automatic production, using bar stock automatically fed to the rolls through an induction heating unit.

There are two main types of transverse rolling machines:

- The two- or-three-roll machine, (Fig. 12.7)
- A transverse rolling machine that uses two straight wedge-shaped tools (Fig. 12.9)

12.3 Electric Upsetters

Electric upsetters are used mostly in preform preparation for gathering a large amount of material at one end of a round bar. The principle of operation is illustrated in Fig. 12.10. A bar of circular cross section (d) is gripped between the tools (b) of the electrode (c) and is pushed by the hydraulically or pneumatically operated upsetting head against the anvil plate (f) on which the other electrode (e) is secured. On switching on the current, the rod section contained between the electrodes heats rapidly and the formation of the head begins. The cold bar is con-

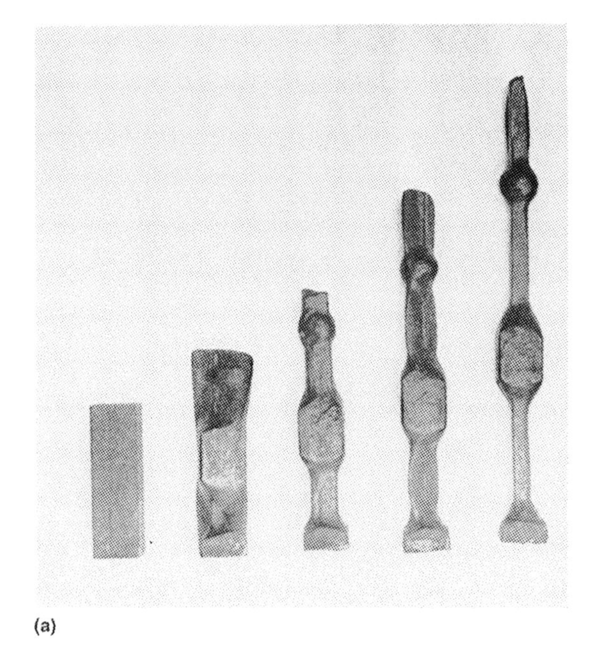
(a)

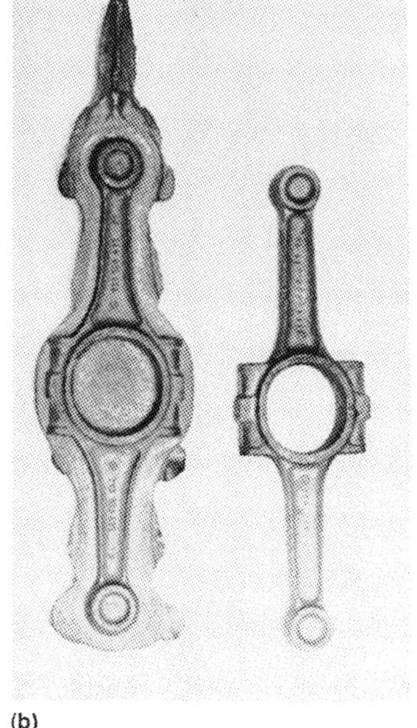
(b)

Fig. 12.3 Example of preforming by reducer rolling in forging of connecting rods. (a) Preforms prepared in reducer rolls. (b) Finish forging before and after trimming. [Altan et al., 1973]

tinuously fed between the gripping electrodes (b), thus the metal accumulates continuously in the head. The anvil electrode is gradually retracted to give enough space for the formation of the head. As soon as sufficient quantity of metal is gathered, the machine switches off and the product can be removed by its cold end. Normally, the head is formed to final shape in a mechanical or screw-type press in the same heat. Thus, the process is suitable for manufacturing components such as automotive exhaust valves, or steam turbine blades [Altan et al., 1973].

The flat anvil electrode can be replaced by a water-cooled copper mold into which material is gathered and formed to close shape and dimensions. Material can be gathered at any point on the length of the bar by placing a sheath around one end. The only limitation on size is the availability of electric current. Commercially available equipment is capable of upsetting 0.5 to 5 in. (12.5 to 125 mm) diameter bars. As time for upsetting a head of average size is 2 to 5 min, several units are required for achieving high volume production.

12.4 Ring-Rolling Mills

The principle of operation of a horizontal ring-rolling mill is illustrated in Fig. 12.11. The vertical mills operate essentially in the same way. The doughnut-shaped blank is placed over a mandrel with a diameter smaller than the inside diameter of the blank. The mandrel, in moving laterally toward the main roll, applies pressure on the blank. The main roll, which is driven, rotates the blank and the mandrel as the cross section of the blank is reduced. The axial rolls provide support to the deforming ring and control its width and its squareness. As seen in Fig. 12.12, by modifying the configurations of the mandrel and the main roll, it is possible to roll rings with internal and external profiles [Beseler, 1969].

For components required in large quantities, such as bearing races, completely automated ring-rolling installations are available. Such an installation may consist of a billet shear, a heating furnace, a forging press, and a ring-rolling mill. The principles of an automatic horizontal ring-rolling mill for manufacturing bearing races

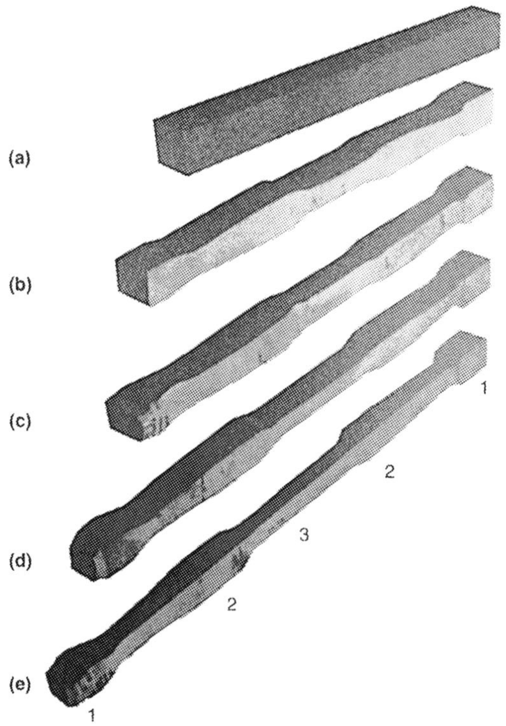

(a)
(b)
(c)
(d)
(e)
1
2
3
2
1

Fig. 12.4 Deformation stages in reducer rolling of a forging to produce a truck axle. (a) Starting billet. (b)–(e) Several reducer roll passes. (1)–(3) Locations where more material needed in the final forging. [Haller, 1982]

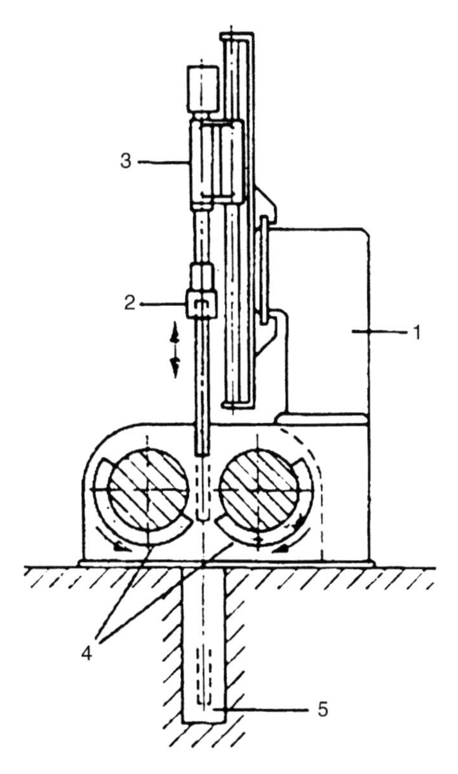

Fig. 12.5 Schematic of a vertical reducer roller. 1, roll stand; 2, holder on manipulator; 3, slide; 4, rolls; 5, part transfer conveyor. [Haller, 1982]

are illustrated in Fig. 12.13. Four mandrels are mounted in a rotating table, and the main roll, eccentrically located within the table, is driven independently by a variable-speed drive. The blank placed over the mandrel, in position 1, is rolled out into a ring as the clearance between mandrel and main roll decreases. After having passed the rolling zone, the hinged table segment is lifted by a cam operation, and the finished ring is discharged from the machine.

Rolling direction
in the simulation

(a)

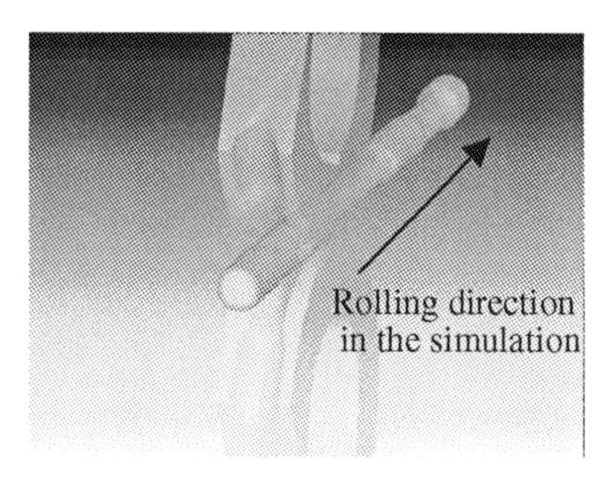

(b)

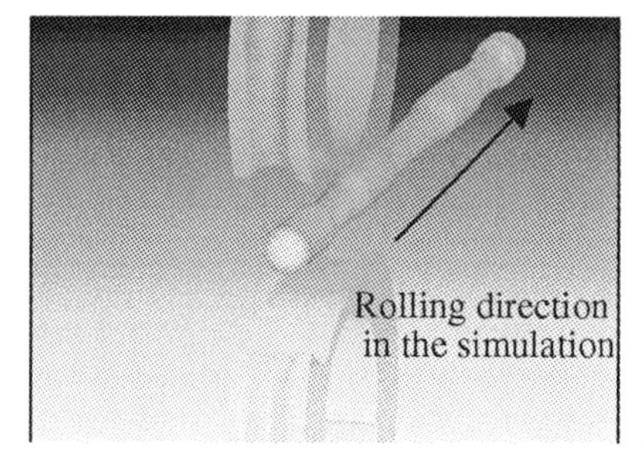

(c)

Fig. 12.6 Computer simulation of reducer rolling operation using DEFORM—3D (only two passes are shown). (a) First pass. (b) Second pass after rotating 90°. (c) At the end of the second pass. [SFTC, 2003]

12.5 Horizontal Forging Machines or Upsetters

The horizontal forging machines are essentially horizontal mechanical presses. As seen in Fig. 12.14, these machines employ two gripper dies, one stationary and the other movable. The dies are closed side-to-side by a toggle mecha-

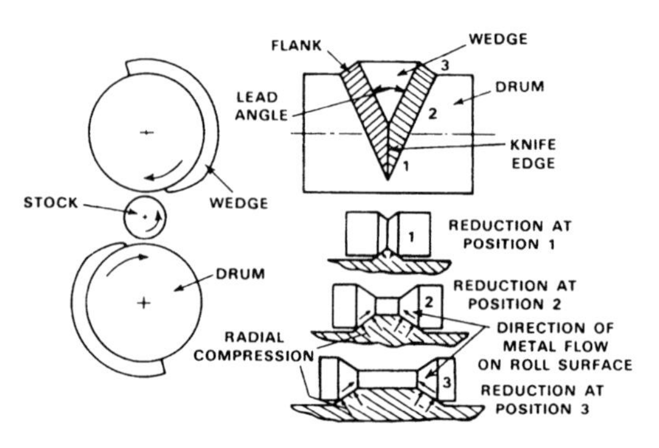

Fig. 12.7 Principle of operation of transverse rolling machines. [Neuberger et al., 1968]

Fig. 12.8 Forging produced in a transverse rolling machine. [Neuberger et al., 1968]

nism operated by a cam or eccentric located on the eccentric shaft. Several matching die inserts are placed in the gripper dies, while the slide carrying the punches is moved by an eccentric-pitman mechanism. The operational sequence of a horizontal forging machine is illustrated in Fig. 12.15 for the upsetting process: (a) the hot end of the bar is placed into the stationary gripper die against a stop, (b) the moving gripper die closes and the stop retracts, (c) the heading tool begins to deform the bar, and (d) completes the upsetting at the end of its stroke.

The upsetters are used for upset forging, piercing, and reducing of bars and tubes. In automatic operation, the parts are transported by finger-type cam-operated devices or walking-beam-type transfer devices from one die cavity to the next. Most horizontal forging machines are designed such that the gripper dies are oriented vertically, i.e., during closing action, the movable gripping die moves horizontally [Lange, 1958].

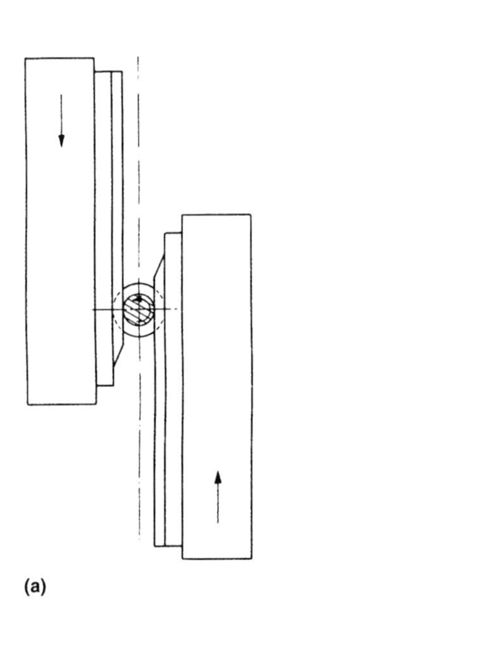

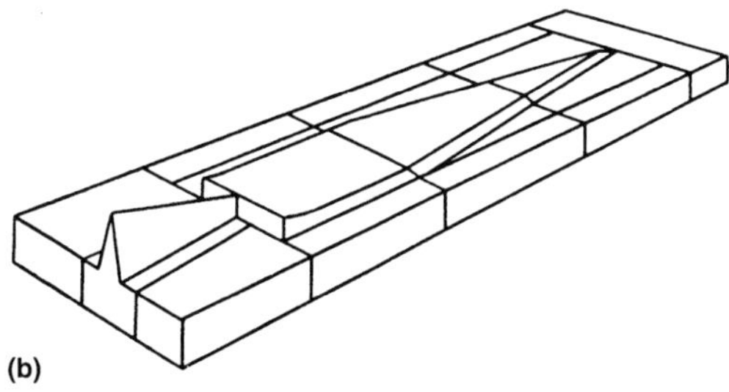

Fig. 12.9 Principles of and tooling for transverse rolling machine with straight dies. (a) Operation. (b) Assembly of simple die. [Altan et al., 1973]

12.6 Rotary or Orbital Forging Machines

The principles of rotary forging machines are illustrated in Fig. 12.16 for a simple upsetting operation.

Instead of the direct pressing action between two flat platens, the workpiece is subjected to a combined rolling and pressing action between a flat bottom platen and a swiveling upper die with a conical working face. The cone axis is inclined so that the narrow sector in contact with the workpiece is parallel to the lower platen. As the cone rotates about the cone apex, the contact zone also rotates. At the same time, the platens are pressed toward each other so that the workpiece is progressively compressed by the rolling action. Press loading is appreciably less than that of conventional upsetting because of relatively small area of instantaneous contact. The application of a rotary forging machine to a closed-die forging operation is illustrated in Fig. 12.17.

12.7 Radial Forging Machines

In many applications, it is necessary to forge, as final product or as a preforming step, solid

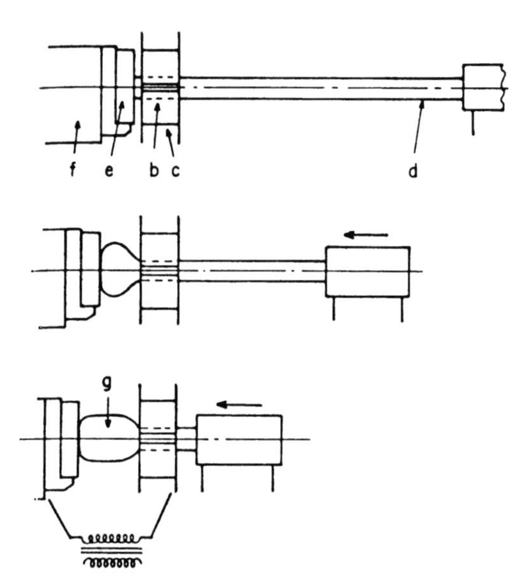

Fig. 12.10 Principle of operation of the electric upsetter. See text for details. [Altan et al., 1973]

shafts with varying or constant diameter along the length and tubes with internal or external varying diameter. All these parts require symmetrical reduction of cross sections. For some of these applications, open-die forging presses cannot be used economically because they are limited in number of strokes per minute and in the speed of feed and manipulation of the stock during forging. In addition, large reductions between two flat dies may cause excessive chilling at the billet corners and cracking at the center of the billet. Although these latter problems may be reduced by using V-dies, each V-die set can be used only for a certain dimension range and frequent die changes during forging increase the production costs. Figure 12.18 indicates that the best solution for high-production symmetrical reduction of cross sections is a forging system that squeezes the material from all sizes simultaneously. There are several machine types designed specifically for this purpose of radial or draw-forging of axisymmetric parts [Haller, 1971].

Automated and computer-numerical-controlled (CNC) radial forging machines can be used for hot or cold forging, with two, three, or four dies to produce solid or hollow, round, square, rectangular, or profiled sections. Some of the principal applications are production of stepped shafts or tubes, sizing of solid bars such

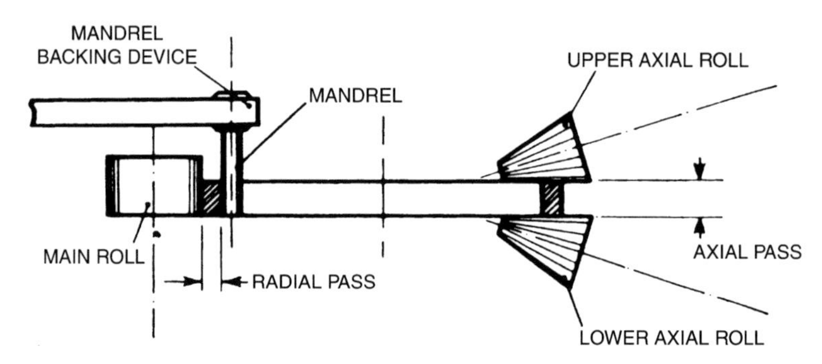

Fig. 12.11 Operational principles of a horizontal ring-rolling mill. [Beseler, 1969]

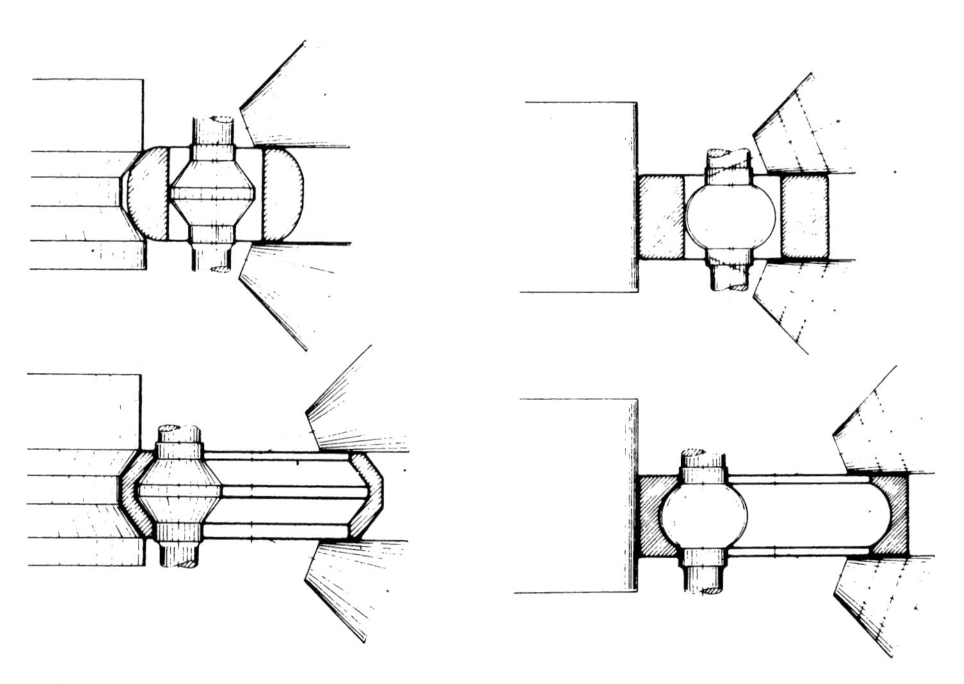

Fig. 12.12 Horizontal ring-rolling mill for producing rings with internal and external profiles. [Beseler, 1969]

as pilger mandrels, and sizing of the bores of tubes to exact round or profiled shapes. The tolerances in hot forged tubes are about ±0.004 in. (±0.1 mm) on the inside diameter (ID). In cold forging the outside diameter (OD) and ID tolerances are about ±0.004 (±0.1 mm) and ±0.001 in. (0.025 mm), respectively [Walter, 1965].

There are two types of radial precision forging machines: vertical and horizontal. Both models use essentially the same design principles. The vertical models are suitable only for relatively short components and are difficult to automate. The horizontal models are built in several variations depending on the application. The horizontal machines consist of a forging box with gear drive, one or two chuck heads to manipulate the workpiece, centering devices, and necessary hydraulic and electronic control components (Fig. 12.19). The core of the machine is a robust cast steel forging box that absorbs all

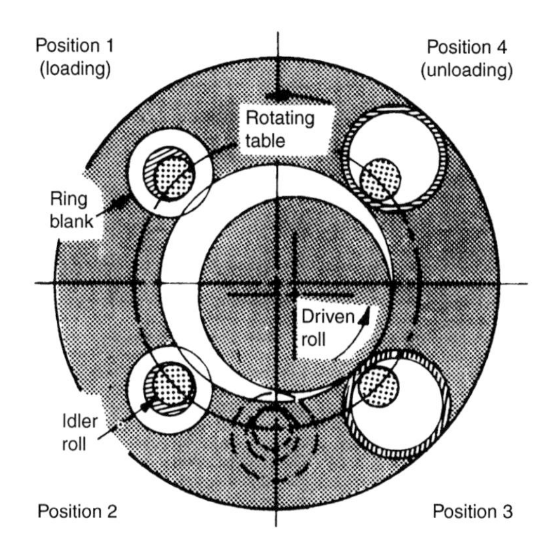

Fig. 12.13 Principle of semiautomatic ring-rolling machine for manufacturing of bearing races. [Beseler, 1969]

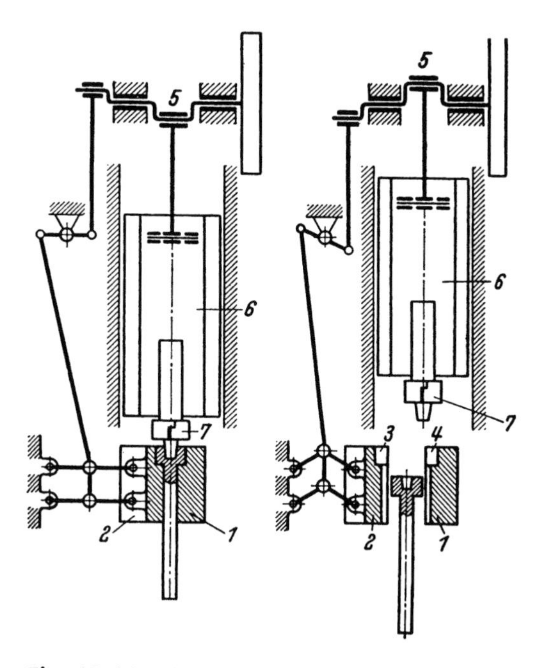

Fig. 12.14 Schematic of a horizontal forging machine. 1, stationary gripping die; 2, movable gripping die; 3 and 4, end-die cavities; 5, eccentric shaft; 6, slide carrying the punches; 7, upsetting and piercing punch. [Lange, 1958]

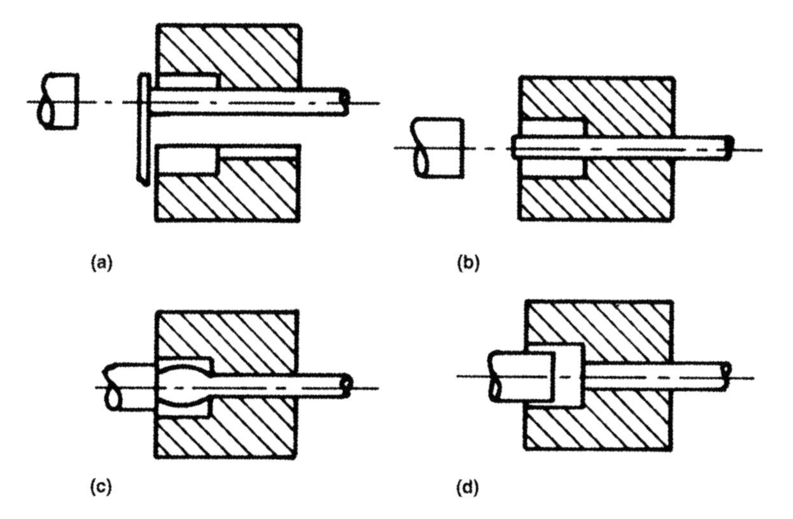

Fig. 12.15 Operating sequence in upsetting on a horizontal forging machine

forging forces (Fig. 12.20). It is mounted with the gear box in a support bolted to the foundation frame. The forging box contains four rotatable adjustment housings in which the eccentric shafts are mounted. The eccentric shafts, which are driven by an electric motor through a gear system, actuate the connecting rods and the forging dies at a rate of 250 to 1800 strokes/min. The stroke position of the connecting rods (or dies) is adjusted in pairs independently (for rectangles or special shapes) or in unison (for rounds and squares) by rotating each adjustment housing through a link, screw, adjustment nut, and worm gear drive powered by one or two hydraulic motors. Each adjustment nut rests, through a piston, on an oil cushion of a hydraulic cylinder. During operation, the forging pressure generates in this oil cushion a pressure proportional (about 20%) to the forging pressure. The pressure in this oil cushion is continuously monitored, and if it exceeds a certain limit the adjustment housings are immediately rotated to bring the dies to open position while the movement of the chuck heads is stopped simultaneously. This system protects the machine from overloading [Altan et al., 1973].

Depending on the machine type, one or two hydraulically driven chuck heads are provided

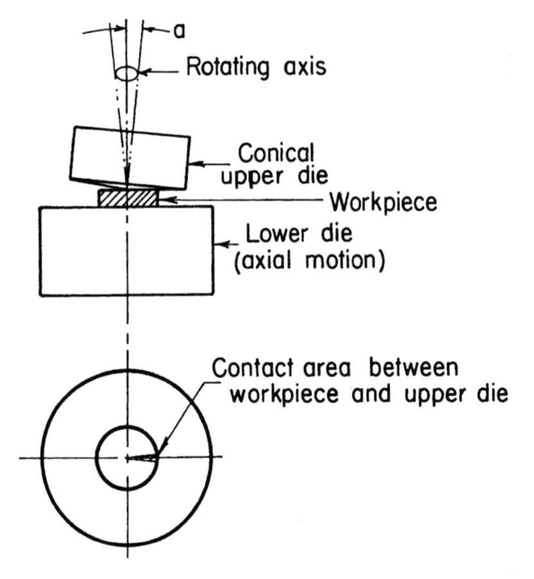

Fig. 12.16 Principle of rotary or orbital forging machines. [Altan et al., 1973]

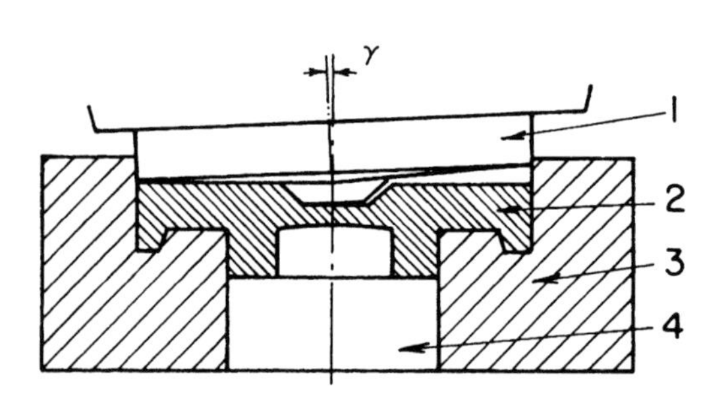

Fig. 12.17 Illustration of closed-die forging with a rotary forging machine. 1, rotating upper platen; 2, workpiece; 3, lower die; 4, ejector

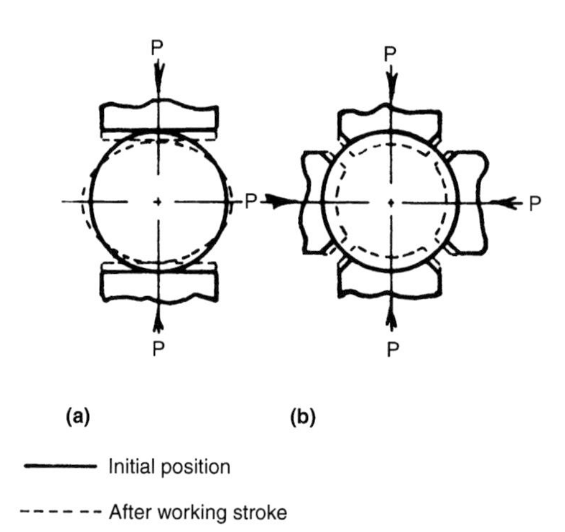

Fig. 12.18 Deformation of a round cross section in stretch forging. P, load. (a) Between flat anvils. (b) Between four curves of a radial forging machine. [Haller, 1971]

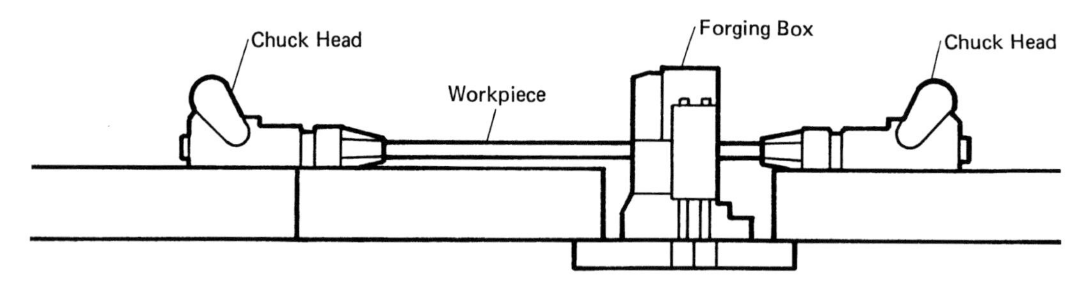

Fig. 12.19 Schematic of a GFM radial precision forging machine with two chuck heads. [Walter, 1965]

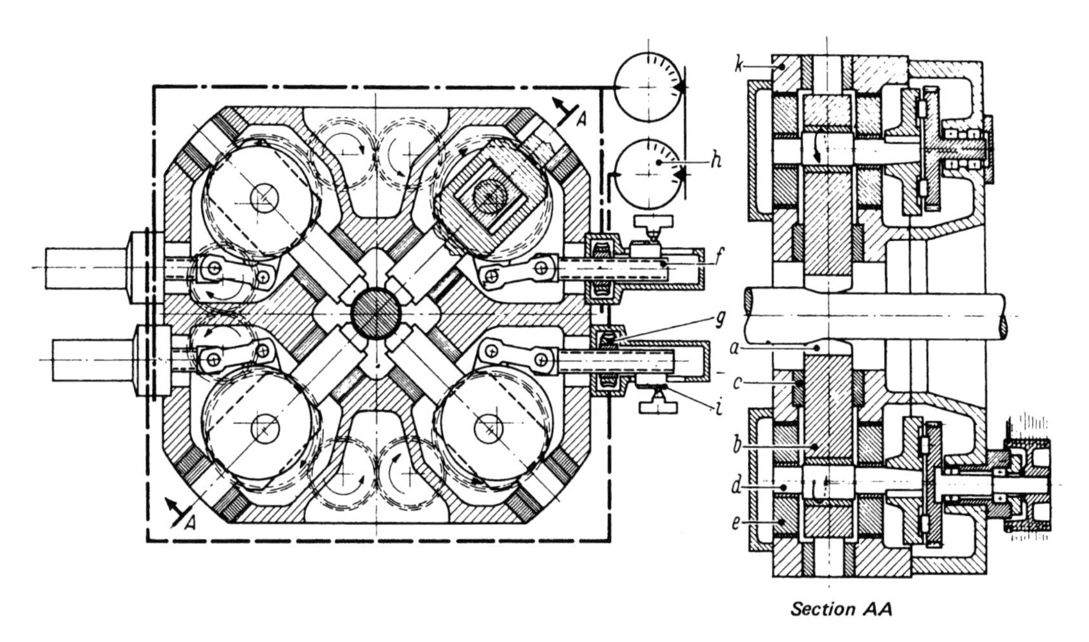

Fig. 12.20 Forging box of a radial precision forging machine illustrating the tool function and adjustment. (a) Dies. (b) Pitman arm. (c) Guides. (d) Eccentric shaft. (e) Adjustment housing. (f) Adjustment screw. (g) Worm gear drive. (h) Adjustment input. (i) Adjustable cam. (k) Forging box. [Altan et al., 1973]

for holding and manipulating the workpiece during forging. Two chuck heads are used when the workpiece must be forged over its entire length, including the chucked ends, in one heat. During forging, round components are rotated while square, rectangular, and profiled sections are forged at fixed positions. The movement of the chuck heads and the variation of the forging stroke in forging stepped components can be controlled by numerical control. For the forging of tubular parts, the chuck heads are provided with stationary or movable mandrels, which are cooled internally in hot forging applications. Central water cooling and lubrication of critical machine components are carried out automatically during the operation of the machine. The radial precision forging machines are capable of producing parts similar to those shown in Fig. 12.21.

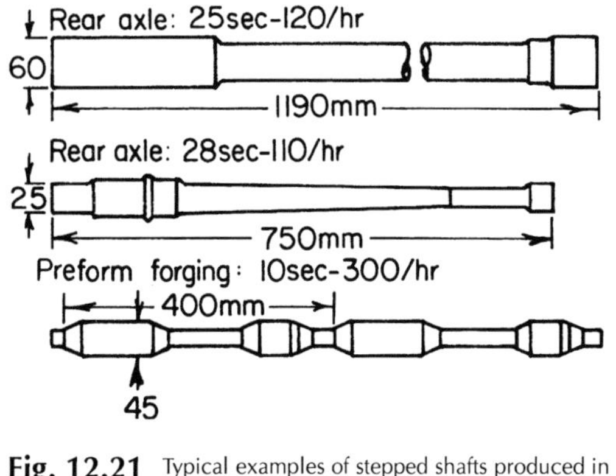

Driveshaft: 15 sec-180/hr
60
310 mm
Rear axle: 22 sec-140/hr
40
785 mm
Rear axle: 25 sec-120/hr
60
1190 mm
Rear axle: 28 sec-110/hr
25
750 mm
Preform forging: 10 sec-300/hr
400 mm
45

Fig. 12.21 Typical examples of stepped shafts produced in precision radial forging machines. [Altan et al., 1973]

REFERENCES

[Altan et al., 1973]: Altan, T., Boulger, F.W., Becker, J.R., Akgermon, N., Henning, H.J., "Forging Equipment, Materials, and Practices," Metals and Ceramics Information Center, Battelle Columbus Laboratories, 1973.

[Beseler, 1969]: Beseler, K.H., "Modern Ring-Rolling Practice," *Met. Form.,* Vol 36, Feb 1969, p 1.

[Haller, 1971]: Haller, H.W., *Handbook of Forging* (in German), Hanser Verlag, 1971.

[Haller, 1982]: Haller, H.W., *Practice of Impression Die Forging* (in German), Hanser Verlag, 1982.

[Lange, 1958]: Lange, K., *Closed-Die Forging of Steel* (in German), Springer-Verlag, Berlin, 1958.

[Neuberger et al., 1968]: Neuberger, F., et al., "Transverse Rolling," *Met. Form.,* Vol 35, Oct 1968, p 1.

[SFTC, 2003]: Scientific Forming Technologies Corp., DEFORM 2D and 3D Software, Columbus, OH, 2003.

[Walter, 1965]: Walter, L., "Use of Precision Forging Machines," *Met. Treat.,* Aug 1965, p 296.

CHAPTER 13

Billet Separation and Shearing

Serdar Isbir
Pinak Barve

13.1 Introduction

Forging stock must be cut from the initial mill products, usually rolled round or round-cornered square bars, into billets of exact lengths and volumes prior to forging. The method of cutting off bars is determined by the edge condition required for subsequent operations. Sawing usually produces a uniform cut edge with little or no damage to the microstructure in the immediate area. Gas cutting produces an edge that resembles a sawed edge in smoothness and squareness. However, the cut edge of some steels becomes hardened during gas cutting, thus making subsequent processing difficult. Separation of billets by shearing is a process without material loss and with considerably higher output with respect to sawing, abrasive cutting, or flame cutting.

Billets and bar sections are sheared between the lower and upper blades of a machine in which only the upper blade is movable. There are also shears, such as impact cutoff machines, that utilize a horizontal knife movement to shear the bar sections. The shear blade plastically deforms the material until its deformation limit in the shearing zone has been exhausted, shearing cracks appear, and fracture occurs. Figure 13.1 shows the appearance of a hot sheared round bar. The burnished area, or depth of shear action by the blade, is usually one-fifth to one-fourth the diameter of the bar. In visual examination of a sheared edge, the burnished portion appears smooth, while the fractured portion is relatively rough.

Many materials cannot be cut by simple shearing into billets with exact lengths and volumes. In general, high-strength steels having tensile strength above 60,000 psi (414 MPa) are heated to between 600 and 750 °F (315 and 400 °C) prior to shearing in order to eliminate the danger of cracking. Aluminum, magnesium, and copper alloys require sawing or cutting with a friction wheel. Nickel-base alloys, superalloys, and titanium alloys also require sawing or abrasive wheel cutting. Metal-cutting saws of various types, usually carbide tipped, are available for billet separation.

13.2 Billet and Sheared Surface Quality

Straight blades can be used to shear bars and bar sections, but in this case, a considerable amount of distortion occurs, as seen in Fig. 13.1.

Fig. 13.1 Sheared surface of a billet [Duvari et al., 2003]

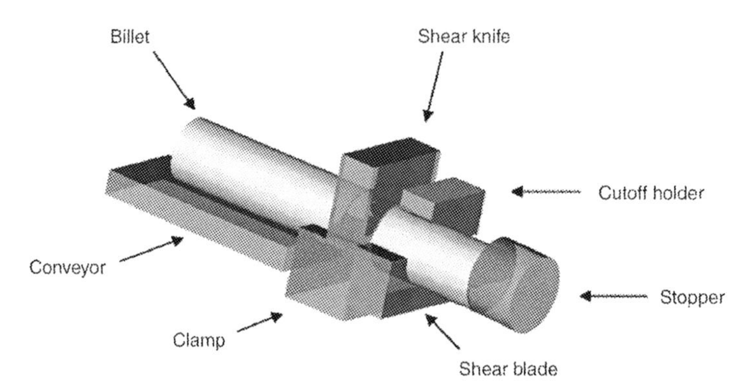

Fig. 13.2 Schematic display of a typical billet shearing system [Duvari et al., 2003]

In addition, the shock on the blades is high when shearing with straight blades, particularly when shearing round bars. Preferred practice is to use blades that conform to the shape of the work metal, as seen in Fig. 13.2. The quality of sheared edges usually increases as thickness or diameter of the billet decreases.

Depending on the initial geometry of the bar, the sheared surface may have different zones and defects (Fig. 13.3). Distortions such as ears, burrs, and scars are undesirable, because they reduce the quality of the billet.

As with every process, there is a load-stroke curve correlating to the different phases described previously. With the penetration of the blade into the bar, the load increases continuously. The load-stroke curve of the process can be divided into the following steps (Fig. 13.4): (1) The bar is deformed elastically. (2) The plastic deformation starts. The material flows along the cutting edges in the direction of the blade penetration and into the gap between the two blades. The material flow causes strain hardening, which results in an increase of the shearing force up to the maximum load. At this time, the cross section is not reduced and shearing has not appeared. (3) Once the pressure at the cutting edges increases sufficiently, the material stops deforming and shearing starts. Due to a decreasing cross section, the cutting force decreases despite the strain hardening of the material. (4) Fracture starts after the shear strength of the material is exceeded. Depending on the process parameters, the incipient cracks will run toward each other, separating the bar and the billet. The shearing force decreases rapidly during this phase.

It should be mentioned that this is an ideal (theoretical) load-stroke curve that does not take friction forces, inconsistent material properties, and tool and machine inaccuracies into account. This means that experimentally obtained load-stroke curves may be slightly different.

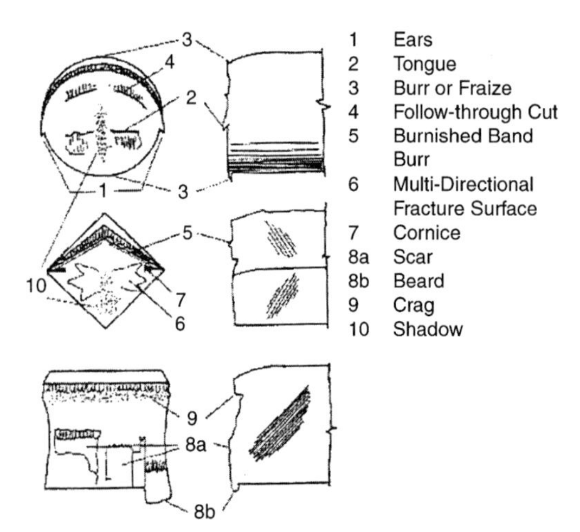

Fig. 13.3 Different zones of the sheared surface [Breitling et al., 1997]

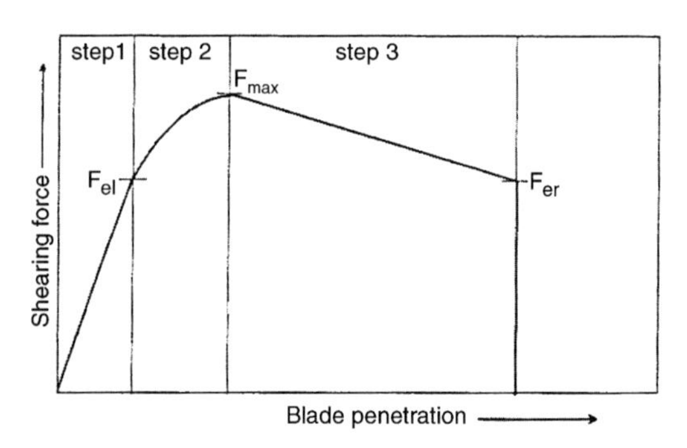

Fig. 13.4 Theoretical load-stroke curve [Breitling et al., 1997]

There are four ways a billet can be sheared: without bar and cutoff holder, with bar holder, with bar and cutoff holder, and with axial pressure application [Schuler, 1998]. In all the aforementioned methods, plastic flow lateral to the shearing direction is increasingly prevented, while compressive stress increases during the shearing operation. Both tendencies exercise a positive influence over the geometry (ovality, tolerance) of the sheared surface. Most accurate billets are produced using the shearing principle with bar and cutoff holder (Fig. 13.5). For cold forging, sheared billets should have the greatest possible straightness, volume control, and little plastic deformation. The sheared surfaces should be free of shearing defects and exhibit only a moderate amount of strain hardening. The appearance of the sheared surfaces is the result of interactions between workpiece characteristics, tool, machine, and friction.

Billets are usually supported on both sides of the shear blades by a roller conveyor table and placed squarely against a gage stop securely bolted to the exit side of the machine. In such a setup, billets can ordinarily be cut to lengths accurate to +1/8, −0 in. (+3.2, −0 mm) on shears that can cut bars up to 4 in. (102 mm) in diameter. When larger shears are used, the breakaway of the metal can cause a variation of ±3/16 in. (±4.8 mm) [Wick et al., 1984]. Fairly consistent accuracy in the shearing of the slugs can be obtained by careful adjustment of the gage setting, especially if the slugs are produced on a weight-per-piece basis. Supporting the free end of the material on a spring-supported table will minimize bending during the shearing operation, thus providing better control over the length of the cut. Shearing clearance, cutting speed, knife and blade edge radii, gap clearance, draft angle, and billet temperature are the parameters affecting sheared billet quality [Duvari et al., 2003, Camille et al., 1998].

Clearance. The shearing clearance exercises a major influence in the surface quality of sheared billets. The greater the strength of the steel, the smaller is the shearing clearance. The following values may be taken as guidelines for shearing clearance of steel; the given values are percentage of the starting material diameter in millimeters:

- Soft steel types, 5–10%
- Hard steel types, 3–5%
- Brittle steel types, 1–3%

Rough fractured surfaces, tears, and seams indicate an excessively large shearing tool clearance. Cross-fractured surfaces and material tongues indicate an insufficient tool clearance. With increasing shearing velocity, the deformation zone reduces, the hardness distribution becomes more uniform, and the hardness increase in the sheared surface becomes less pronounced, i.e. the material becomes more "brittle."

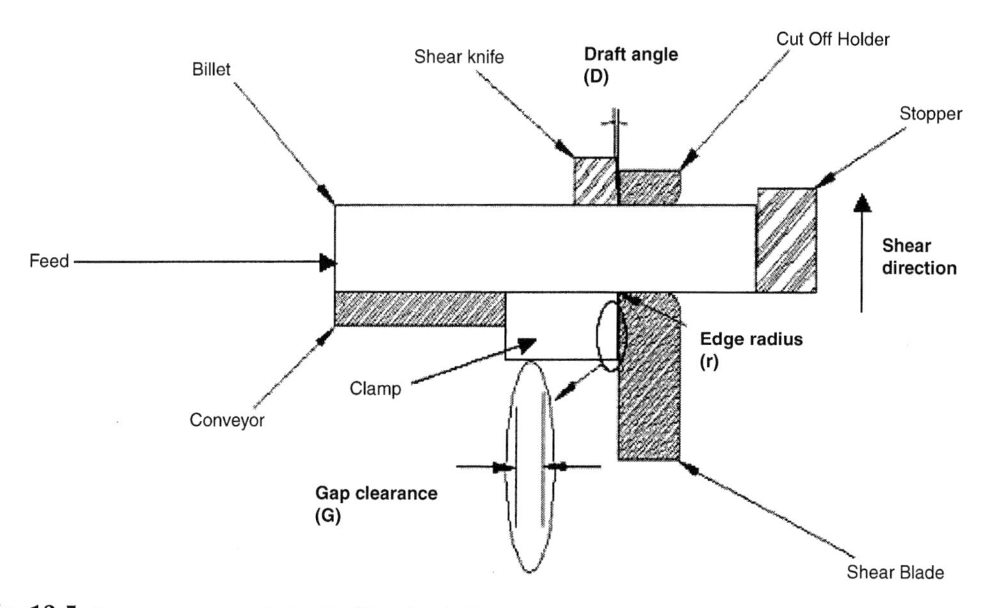

Fig. 13.5 Important parameters in shearing [Duvari et al., 2003]

Cutting Speed. The speed at which material is sheared without adverse effect can range from almost zero to 70 or 80 ft (21 or 24 m) per minute. However, as speed increases above 20 to 25 ft (6.1 to 7.6 m) per minute, problems are encountered in holding the workpiece securely at the blade without the far end whipping, especially with material ¼ in. (6.4 mm) thick or more. When bars harder than 30 HRC are cut at speeds of 40 to 50 ft (12 to 15 m) per minute or higher, chipping of the blade may occur.

Draft Angle. Draft angle and load on the tooling are inversely proportional. Load on the tooling reduces as the draft angle is increased. However, fracture length and burr length also increase with the increase in draft angle. Increase in both of these parameters, i.e., fracture length and burr length, reduce the sheared billet quality. It is also observed that rollover length, which is another indication of billet quality, increases with increasing draft angle. Therefore a compromise should be made between the tooling force and billet quality. Especially with ductile materials, draft angle should be kept at minimum, even if this increases tool forces.

13.3 Shearing Force, Work, and Power

The shearing force, F_s, and shearing work, W_s, can be calculated approximately using the following formula when separating round material with the diameter *d:*

$$F_s = A_s \cdot k_s \qquad W_s = x \cdot F_s \cdot s \qquad \text{(Eq 13.1)}$$

whereby A_s is the sectional surface to be sheared, k_s is the shearing resistance of the billet material, and s is an approximate portion of the shearing stroke, i.e., of the bar diameter. The factor s varies between 20% (hard, brittle material) to 40% (soft, ductile material). The correction factor x indicates the extent to which the increase in force deviates from a rectangular force-stroke curve. In general, x is taken to be between 0.4 and 0.7. The shearing resistance, k_s, amounts to approximately 70 to 80% of tensile strength of the material.

For low-carbon steels, the net horsepower required for shearing can be estimated from the following formula:

$$hp = \frac{A \cdot V \cdot S}{33{,}000} \qquad \text{(Eq 13.2)}$$

where A is the cross-sectional area of the workpiece (in square inches), V is the speed of the shear blade (in feet per minute), and S is the shear strength of the work metal (in pounds per square inch). The 33,000 is foot-pounds per minute per horsepower. For metric use, the power in English units (hp) should be multiplied by 0.746 to obtain kilowatts. It may be necessary to increase the calculated value as much as 25% to compensate for machine efficiency.

13.4 Shearing Equipment

The parameters influencing the quality of the sheared billet can be divided into workpiece-tooling and shear-related parameters. Machine- and tooling-related parameters, such as a weak frame design or inconsistent blade alignment, reduce the shearing quality and should be avoided [Breitling et al., 1997].

There are two ways to shear billets: A shearing tool may be mounted in a mechanical press, or, alternatively, a regular hydraulic billet shear, solely designed for shearing, may be used. The second option is more desirable, because it provides more accuracy and productivity. The latest shear designs incorporate the following features:

- Rugged frame construction and precise guidance of all moving parts in order to eliminate deflection under load
- Adjustable hydraulic billet support and bar holder in order to minimize billet bending

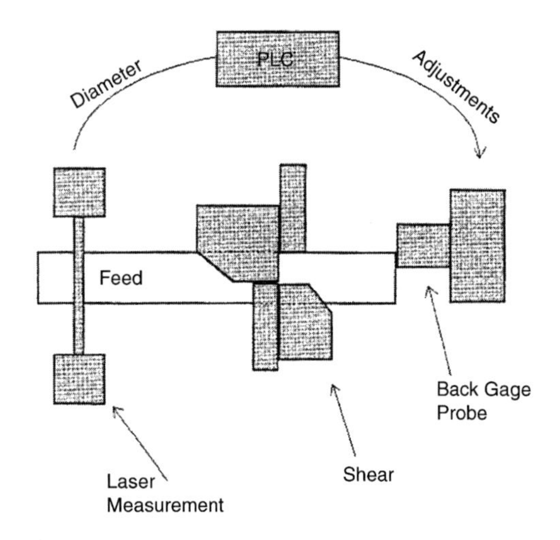

Fig. 13.6 Stock volume monitoring system [Breitling et al., 1997]

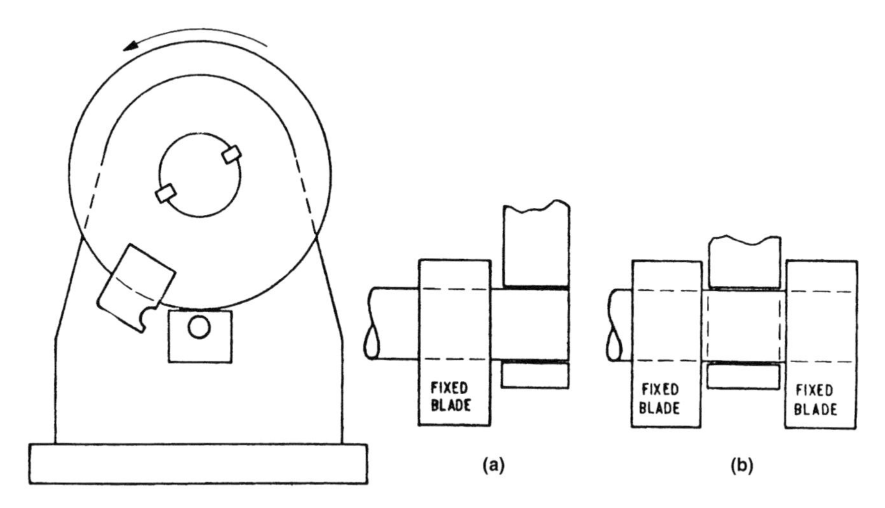

Fig. 13.7 High-velocity rotary-type shear. (a) Tooling for one billet per revolution. (b) Tooling for two billets per revolution [Altan et al., 1973]

- Fast blade clearance adjustment to reduce the lead time for new setups
- Hydraulic knife clamps for fast blade changes
- A tiltable shear base that can be used for inclining the bar when blanking soft materials
- A high shearing speed and shear rate in order to improve billet quality and process productivity
- Automated billet quality control (for example, continuous stock volume monitoring)

The last point becomes increasingly important, because it is not sufficient to control the billet weight and geometrical accuracy only manually and intermittently. Modern shearing machines use a stock volume monitoring system, which ensures the maintenance of a constant billet weight despite changes in bar diameter.

The system measures the bar diameter, by means of laser sensors, which is then sent to a programmable logic controller (PLC) that computes the adjustments and moves the back gage accordingly (Fig. 13.6). The billet is then sheared and ready for further processing.

Most conventional shears are mechanical and their operation is based on the eccentric slide principle, as in mechanical forging presses. The holddown mechanism is necessary for obtaining good sheared surfaces. It operates mechanically, through an additional linkage from the eccentric, or hydraulically. The use of an outboard support also improves the quality of the billets. In that device, the billet is supported during the entire operation.

A radically different design, a high-velocity rotary-type shear, is seen schematically in Fig. 13.7. In this machine, the energy is provided by a flywheel that carries an open-type moving blade. According to tooling arrangement, one or two billets per revolution can be obtained. With this shear, production rates of 300 billets/min are feasible. In the design seen in Fig. 13.8, the material to be sheared is confined by a close-fitting closed blade and held against a stop by an axial

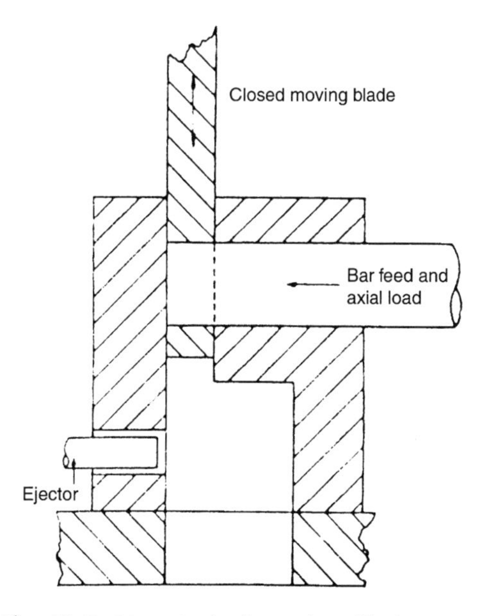

Fig. 13.8 Schematic of a shear with axial load to improve shear quality [Altan et al., 1973]

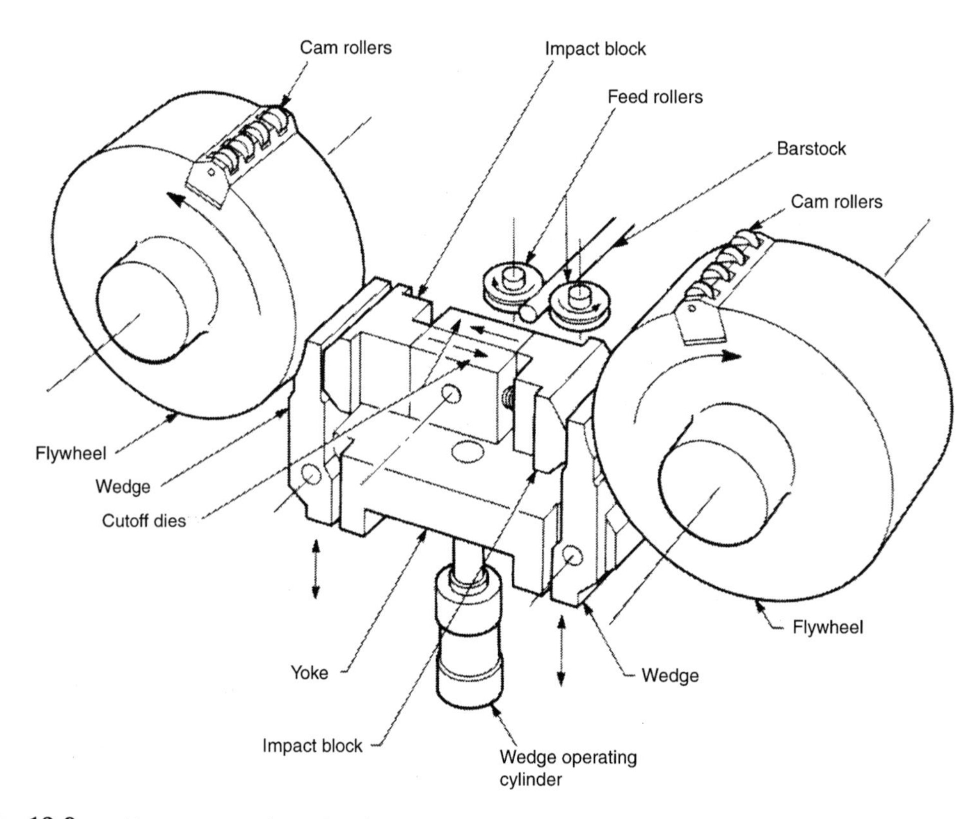

Fig. 13.9 Double-cutting principle [Wick et al., 1984]

load. The axial load ensures squareness and inhibits crack propagation.

In general, any metal that can be machined can be sheared, but power requirements increase as the strength of the work metal increases. Further, blade design is more critical and blade life decreases as the strength of the work metal increases. Equipment is available for shearing round, hexagonal, or octagonal bars up to 6 in. (152 mm) in diameter or thickness, rectangular bars and billets up to 3 × 12 in. (75 × 305 mm) in cross section, and angles up to 8 × 8 × 1½ in. (203 × 203 × 38 mm).

Cutoff-type shearing machines are used for cutting round, square, flat, or special-shaped bars into blanks or slugs. This process can be performed on a machine specifically designed for slug cutoff, or it can be performed using a box-type shearing die in conjunction with a press [Wick et al., 1984].

One manufacturer of cutoff machines utilizes a double-cutting principle to shear the blanks or slugs. The dies (Fig. 13.9) are actuated with short strokes by two flywheel-cam assemblies that rotate at a constant speed. The capacity of the machine is a 2½ in. (63.5 mm) diam bar having a maximum length of 36 in. (914 mm).

This method is fast, efficient, and economical when billets in large quantities are required. Some machines are capable of maintaining the length to within ±0.005 in. (±0.13 mm) as well as maintaining square cuts and ends that are free of burrs, distortion, and rollover. Production can be as high as 150 pieces per minute.

REFERENCES

[Altan et al., 1973]: Altan, T., Boulger, F., Becker, J., Akgerman, N., Henning, H., *Forging Equipment, Materials, and Practices,* Metal and Ceramics Information Center, Battelle Columbus Laboratories, HB03, p 4–7.

[Breitling et al., 1997]: Breitling, J., Chernauskas, V., Taupin, E., Altan, T., "Precision Shearing of Billets—Special Equipment and Process Simulation," *Journal of Materials*

Processing Technology, Vol. 71, 1997, p 119–125.

[Camille et al., 1998]: Santiago-Vega, C., Vasquez, V., Altan, T., "Simulation of Bar Shearing Process," ERC/NSM-97-27, Engineering Research Center for Net Shape Manufacturing.

[Duvari et al., 2003]: Duvari, S., Isbir, S., Ngaile, G., Altan, T., "Optimization of Tool Design in Hot Shearing of Billets for Forging," ERC/NSM-03-R-09, Engineering Research Center for Net Shape Manufacturing.

[Schuler, 1998]: Schuler, H., Hoffman, H., Frontzek, H., *Metal Forging Handbook,* Schuler Group, Springer, Goppingen, Germany, p 457–459.

[Wick et al., 1984]: Wick, C., Benedict, J.T., Veilleux, R., *Tool and Manufacturing Engineers Handbook,* Society of Manufacturing Engineers, Dearborn, MI, p 11-1–11-21.

SELECTED REFERENCES

- **[ASM International, 1999]:** Davis, J.R., Ed., *Forming and Forging,* Vol 14, *ASM Handbook,* p 714–719.
- **[Geleji et al., 1967]:** Geleji, A., *Forge Equipment, Rolling Mills, and Accessories* (in English), Akademiai Kiado, Budapest, p 168.
- **[Stotmann, 1968]:** Stotmann, W., "Evolution of Machines and Automation in the Drop Forging Industry," *Metal Forming,* May, p 136.

WEB SITES

- www.bemcor.com, Bemcor, Inc.
- www.ficep.it, Ficep Corp.
- www.sms-eumuco.de, SMS Eumuco GmbH

CHAPTER 14

Process Design in Impression-Die Forging

Manas Shirgaokar

14.1 Introduction

In impression-die forging, two or more dies are moved toward each other to form a metal billet that has a relatively simple geometry to obtain a more complex shape. Usually, the billet is heated to an appropriate forging temperature and the dies allow the excess billet material to flow outside of the die cavity to form a flash that is later trimmed and discarded. This process is capable of producing components of high quality at moderate cost. Forgings offer a high strength-to-weight ratio, toughness, and resistance to impact and fatigue. Forged components find application in the automobile/automotive industry, aircraft, railroad, and mining equipment.

Some parts can be forged in a single set of dies while others, due to shape complexity and material flow limitations, must be shaped in multiple sets of dies. In a common multistage forging process, the part is first forged in a set of busting dies, then moved to one or more sets of blocking dies, and finally forged in finisher dies. Finisher dies are used to enhance geometrical details without significant material flow. The quality of the finished part depends greatly on the design of the previous stages. If the material has been improperly distributed during the blocking stage, defects may appear in the finishing stage. In a good-quality forging, all sections of the die cavity must be filled, and the part must not contain flow defects such as laps, cold shuts, or folds.

Before being used in production, forging dies are tested to verify proper filling of the die cavities. The most commonly used method of process verification is die tryout in which full-scale dies are manufactured and prototype parts are forged to determine metal flow patterns and the possible occurrence of defects. This often takes several iterations and is very costly in terms of time, materials, facilities, and labor. Alternatively, two other methods for modeling metal flow, namely, physical modeling and process simulation using finite-element method (FEM)-based software, can be used to obtain information about the effects of die design and process variables on the forging process.

The design of any forging process begins with the geometry of the finished part (Fig. 14.1). Consideration is given to the shape of the part, the material to be forged, the type of forging equipment to be used, the number of parts to be forged, the application of the part, and the overall economy of the process being designed. The finisher die is then designed with allowances added for flash, draft, shrinkage, fillet and corner radii, and positioning of the parting line. When using multistage forging, the shapes of the preforms are selected, the blocker dies are designed, and the initial billet geometry is determined. In making these selections, the forging designer considers design parameters such as grain flow, parting line, flash dimensions, draft angles, and fillet and corner radii.

The terminology used to describe the flash zone in impression- and closed-die forging can be seen in Fig. 14.2. The flash dimensions and billet dimensions influence:

- The flash allowance, that is, the material that flows into the flash zone

- The forging load
- The forging energy
- The die life

The overall design of a forging process requires the prediction of:

- Shape complexity and volume of the forging
- Number and configurations of the preforms or blockers
- The flash dimensions in the dies and the additional flash volume required in the stock for preforming and finishing operations
- The forging load, energy, and center of loading for each of the forging operations

14.2 Forging Process Variables

The interaction of the most significant variables in forging is shown in a simplified manner in Fig. 14.3. It is seen that for a given billet material and part geometry, the ram speed of the forging machine influences the strain rate and flow stress. Ram speed, part geometry, and die temperature influence the temperature distribution in the forged part. Finally, flow stress, friction, and part geometry determine metal flow, forging load, and forging energy and consequently influence the loading and the design of the dies. Thus, in summary, the following three groups of factors influence the forging process:

- Characteristics of the stock or preform to be forged, flow stress and the workability at various strain rates and deformation conditions, stock temperature, preform shape, etc.
- Variables associated with the tooling and lubrication: tool materials, temperature, design of drafts and radii, configuration, flash de-

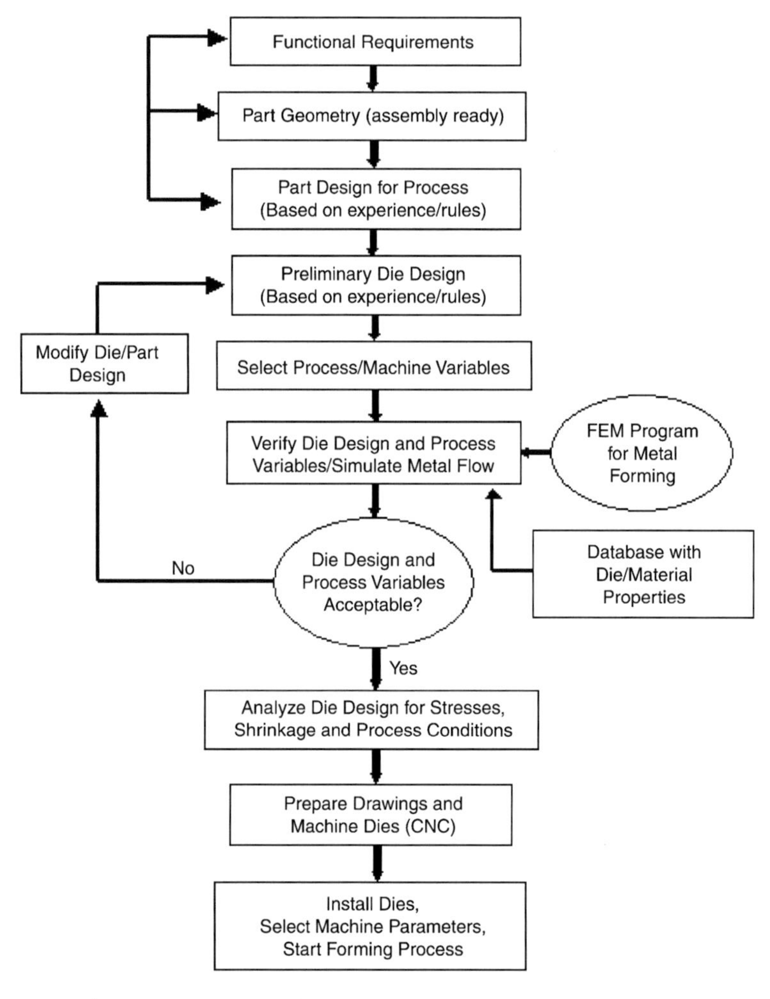

Fig. 14.1 A flow chart illustrating forging process design [Vasquez et al., 2000]

sign, friction conditions, forging stresses, etc.
- Characteristics of the available equipment: load and energy capacities, single or multiblow availability, stiffness, ram velocity under load, production rate, availability of ejectors, etc.

14.2.1 Forging Materials

Table 14.1 lists different metals and alloys in order of their respective forging difficulty [Sabroff et al., 1968]. The forging material influences the design of the forging itself as well as the details of the entire forging process. For example, Fig. 14.4 shows that owing to difficulties in forging, nickel alloys allow for less shape definition than do aluminum alloys.

In most practical hot forging operations, the temperature of the workpiece material is higher than that of the dies. Metal flow and die filling are largely determined by:

- The forging material resistance to flow and ability to flow, i.e., its flow stress and forgeability
- The friction and cooling effects at the die/material interface
- The complexity of the forging shape

For a given metal, both the flow stress and forgeability are influenced by the metallurgical characteristics of the billet material and the temperatures, strain, strain rates, and stresses that occur in the deforming material. The flow stress determines the resistance to deformation, i.e., the load, stress, and energy requirements. Forgeability has been used vaguely in the literature to denote a combination of both resistance to deformation and ability to deform without fracture. A diagram illustrating this type of information is presented in Fig. 14.5.

In general, the forgeabilities of metals increase with increasing temperature. However, as temperature increases, grain growth occurs, and in some alloy systems, forgeability decreases with increasing grain size. The forgeabilities of metals at various deformation rates and temperatures can be evaluated by using various tests such as torsion, tension, and compression tests. In all these tests, the amount of deformation prior to failure of the specimen is an indication of forgeability at the temperature and deformation rates used during that particular test.

14.2.2 Forging Equipment

In hot and warm forging, the behavior and the characteristics of the forging press influence:

- The contact time between the material and the dies, under load. This depends on the ram velocity and the stiffness of a given press. The contact time is extremely important, because it determines the heat transfer between the hot or warm material and the colder dies. Consequently, the contact time also influences the temperatures of the forging and that of the dies. When the contact time is large, the material cools down excessively during deformation, the flow stress increases, and the metal flow and die filling are reduced. Thus, in conventional forging operations, i.e., nonisothermal, it is desired to have short contact times.
- The rate of deformation, i.e., the strain rate. In certain cases, for example, in isothermal and hot-die forging of titanium and nickel alloys, that are highly rate dependent, the large rate of deformation would lead to an increase in flow stress and excessive die stresses.
- The production rate. With increasing stroke rate, the potential production rate increases, provided the machine can be loaded and unloaded with billet or preforms at these increased rates.

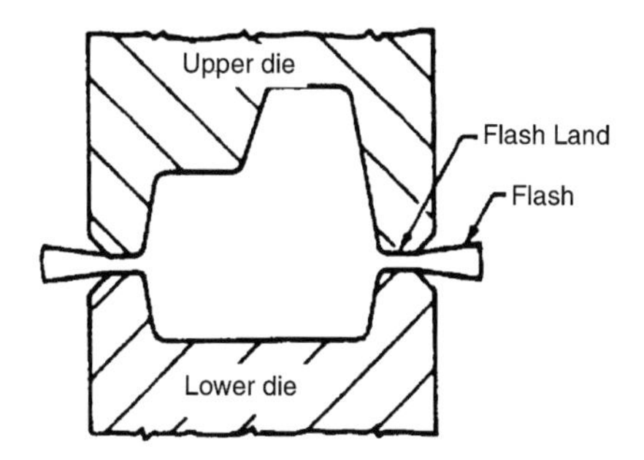

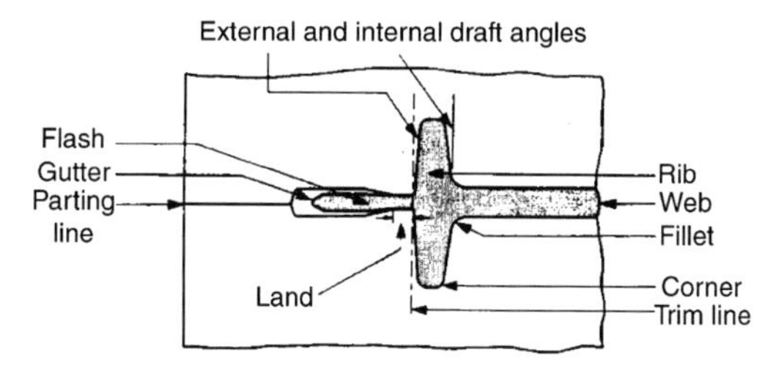

Fig. 14.2 Schematic of a die set and the terminology used in impressed-die forging with flash

- The part tolerances. Hydraulic and screw presses, for example, operate with kissing dies, i.e., the dies have flat surfaces that contact each other at the end of each working stroke of the forging press. This allows very close control of the thickness tolerances even if the flow stress and friction conditions change during a production run. Ram guiding, stiffness of the press frame, and drive also contribute to tolerances that can be achieved in forging.

14.2.3 Friction and Lubrication

The flow of metal in forging is caused by the pressure transmitted from the dies to the deforming material; therefore, the friction conditions at the die/material interface are extremely important and influence the die stresses and the forging load as well as the wear of the dies. In order to evaluate the performances of various lubricants and to be able to predict forming pressures, it is necessary to express the interface friction quantitatively, in terms of a factor or coefficient. In forging, the frictional shear stress, τ, is most commonly expressed as:

$$\tau = f\bar{\sigma} = \frac{m}{\sqrt{3}}\bar{\sigma} \qquad \text{(Eq 14.1)}$$

where τ is the frictional shear stress, f is the friction factor, and m is the shear friction factor ($0 \leqslant m \leqslant 1$).

For various forming conditions, the values of m vary as follows:

- m = 0.05 to 0.15 in cold forging of steels, aluminum alloys, and copper, using conventional phosphate soap lubricants or oils
- m = 0.2 to 0.4 in hot forging of steels, copper, and aluminum alloys with graphite-based lubricants
- m = 0.1 to 0.3 in hot forging of titanium and high-temperature alloys with glass lubricants
- m = 0.7 to 1 when no lubricant is used, e.g., in hot rolling of plates or slabs and in nonlubricated extrusion of aluminum alloys

14.2.4 Heat Transfer and Temperatures

Heat transfer between the forged material and the dies influences the lubrication conditions, die life, properties of the forged product, and die fill. Often, temperatures that exist in the material during forging are the most significant variables influencing the success and economics of a given forging operation. In forging, the magnitudes and distribution of temperatures depend mainly on:

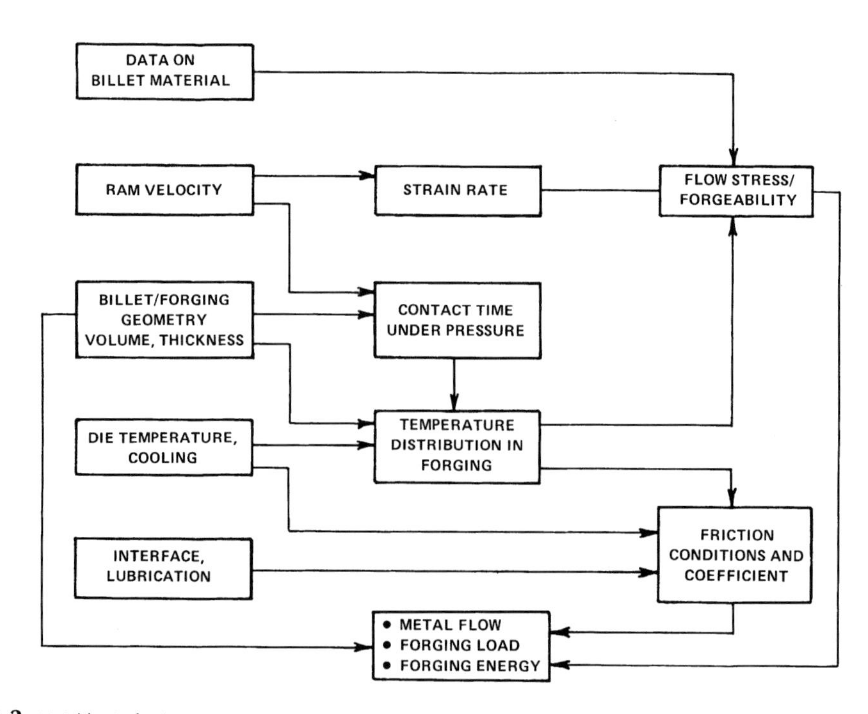

Fig. 14.3 Variables in forging

Table 14.1 Hot forging temperatures of different metals and alloys [Sabroff et al., 1968]

Metal or alloy	Approximate range of forging temperature °F	°C
Aluminum alloys (least difficult)	750–1020	400–550
Magnesium alloys	480–660	250–350
Copper alloys	1110–1650	600–900
Carbon and low-alloy steels	1560–2100	850–1150
Martensitic stainless steels	2010–2280	1100–1250
Maraging steels	2010–2280	1100–1250
Austenitic stainless steels	2010–2280	1100–1250
Nickel alloys	1830–2100	1000–1150
Semiaustenitic precipitation-hardenable stainless steels	2010–2280	1100–1250
Titanium alloys	1290–1740	700–950
Iron-base superalloys	1920–2160	1050–1180
Cobalt-base superalloys	2160–2280	1180–1250
Niobium alloys	1740–2100	950–1150
Tantalum alloys	1920–2460	1050–1350
Molybdenum alloys	2100–2460	1150–1350
Nickel-base superalloys	1920–2190	1050–1200
Tungsten alloys (most difficult)	2190–2370	1200–1300

- The initial material and die temperatures
- Heat generated due to plastic deformation and friction at the die/material interface
- Heat transfer between the deforming material and the dies as well as between the dies and the environment (air, coolant, lubricant)

The effect of contact time on temperatures and forging load is illustrated in Fig. 14.6, where the load-displacement curves are given for hot forging of a steel part using different types of forging equipment. These curves illustrate that, due to strain rate and temperature effects, for the same forging process, different forging loads and energies are required by different presses. For the hammer, the forging load is initially higher due to strain-rate effects, but the maximum load is lower than for either hydraulic or screw presses. The reason for this is that in the presses, the extruded flash cools rapidly,

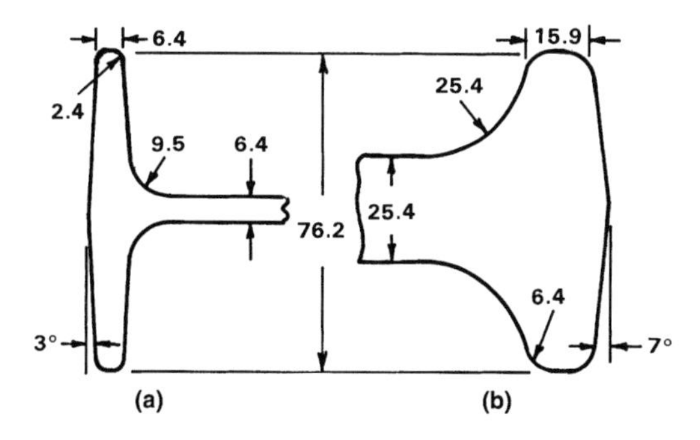

Fig. 14.4 Comparison of typical design limits for rib-web-type structural forgings of (a) aluminum alloys and (b) nickel-base superalloys (all dimensions are in mm) [Sabroff et al., 1968]

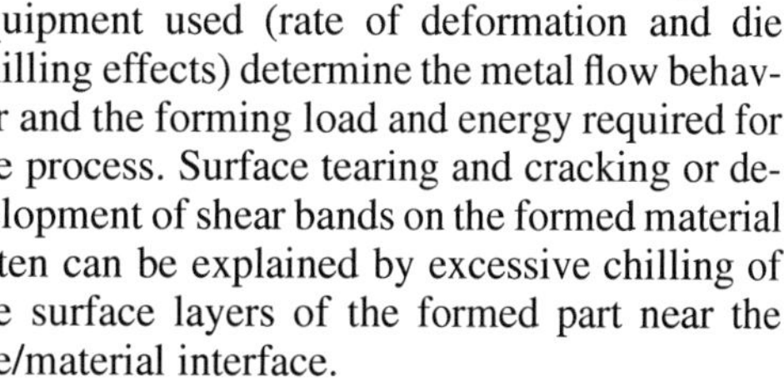
whereas in the hammer, the flash temperature remains nearly the same as the initial stock temperature. Thus, in hot forming, not only the material and the formed shape but also the type of equipment used (rate of deformation and die chilling effects) determine the metal flow behavior and the forming load and energy required for the process. Surface tearing and cracking or development of shear bands on the formed material often can be explained by excessive chilling of the surface layers of the formed part near the die/material interface.

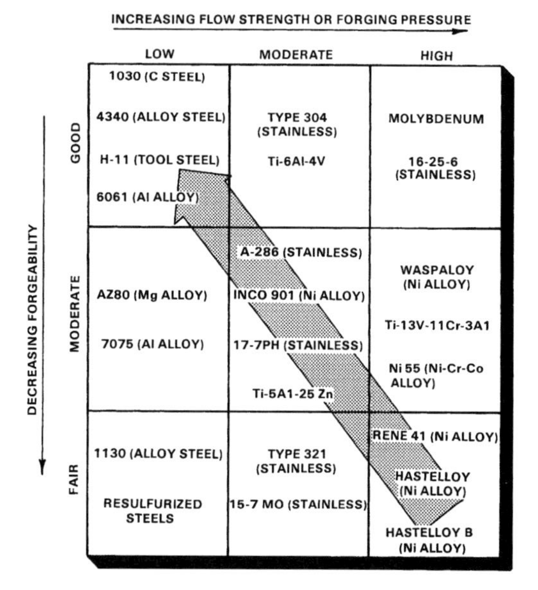

Fig. 14.5 Generalized diagram illustrating the influence of forgeability and flow stress on die filling [Sabroff et al., 1968]

14.2.5 Production Lot Size and Tolerances

As is the case in all manufacturing operations, these two factors have a significant influence on die design in forging. If the production lot size is large, the main reason for changing the dies would be die wear. In this case, die materials and their hardnesses would be selected to be especially wear resistant even if they are made from somewhat expensive alloys. The preforming and the finishing dies are designed such that relatively little material movement is allowed in the finisher dies; thus, the finisher dies, which determine the final part dimensions, will not wear out easily.

If the production lot size is small, as is the case in the aerospace forging industry, die wear is not a major problem, but die costs are very significant because these costs must be amortized over a smaller number of parts. As a result, some of the preforming or blocker dies may be omitted even if this would cause the use of more billet material. Also, in this case, the dies must be changed more often than in large-scale production. Therefore, quick die-changing and automatic die-holding mechanisms are required, for economic production.

Forging tolerances are very important in designing the die holders and die inserts, because they depend considerably on the manufacturing tolerances and elastic deflections of the dies during forging. Precision forging of gears and blades, for example, require not only very close manufacturing accuracies on the dies but also close control of die temperatures. In addition, often it is necessary to estimate the changes in die dimensions under forging conditions so that corrections can be made while designing and manufacturing these dies. Die dimensions vary during the forging operation because of thermal expansion, mechanical loading during assembling of the dies in a holder, and mechanical loading during the forging process itself.

14.3 Shape Complexity in Forging

The main objective of forging process design is to ensure adequate metal flow in the dies so that the desired finish part geometry can be obtained without any external or internal defects. Metal flow is greatly influenced by part or die geometry. Often, several operations (preforming or blocking) are needed to achieve gradual flow of the metal from an initially simple shape (cylinder or round-cornered square billet) into the more complex shape of the final forging. In a general sense, spherical and blocklike shapes are the easiest to forge in impression or closed dies. Parts with long thin sections or projections (webs and ribs) are more difficult to forge, because they have more surface area per unit volume. Such variations in shape maximize the effects of friction and temperature changes and hence influence the final pressure required to fill the die cavities. There is a direct relationship

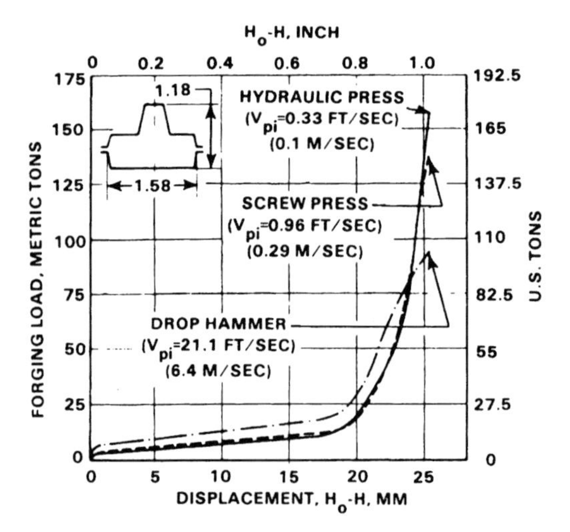

Fig. 14.6 Load-displacement curves for the same part forged in three different machines with three different ram speeds (dimensions of the part in inches, initial temperature = 2012 °F, or 1100 °C) [Altan et al., 1973]

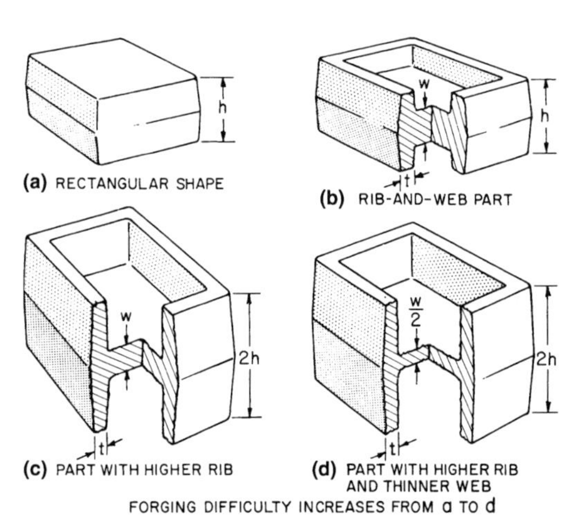

Fig. 14.7 Rectangular shape and three modifications showing increasing forging difficulty with increasing rib height and decreasing web thickness [Sabroff et al., 1968]

between the surface-to-volume ratio of a forging and the difficulty of producing the forging.

The ease of forging more complex shapes depends on the relative proportions of vertical and horizontal projections on the part. Figure 14.7 is a schematic representation of the effects of shape on forging difficulties. Parts "C" and "D" would require not only higher forging loads but also at least one more forging operation than parts "A" and "B" to ensure die filling.

As shown in Fig. 14.8, the majority of forgings can be classified into three main groups [Spies, 1959]:

- Compact shape, spherical and cubical shape (class 1)
- Disc shape (class 2)
- Oblong shape (class 3)

The first group of compact shapes has the three major dimensions, namely, the length (l), width (w), and height (h) approximately equal. The number of parts that fall into this group is rather small.

The second group consists of disk shapes, for which two of the three dimensions (length and width) are approximately equal and are larger than the height (h). All the round forgings belong to this group, which includes approximately 30% of all the commonly used forgings.

The third group of forgings consists of long shapes, which have one dimension significantly larger than the other two ($l > b \geq h$).

These three basic groups are further subdivided into subgroups depending on the presence and type of elements subsidiary to the basic shape. This "shape classification" is useful for practical purposes, such as for estimating costs and for predicting preforming steps. This method is, however, not entirely quantitative and requires some subjective evaluation based on past experience.

A quantitative value called the "shape difficulty factor" has been suggested by Teterin et al., 1968, for expressing the geometrical complexity of round forgings (having one axis of rotational symmetry). A "longitudinal shape factor," α, is defined as:

$$\alpha = \frac{X_f}{X_c} \qquad \text{(Eq 14.1)}$$

with

$$X_f = \frac{P^2}{F} \qquad \text{(Eq 14.2)}$$

and

$$X_c = \frac{P_c^2}{F_c} \qquad \text{(Eq 14.3)}$$

where P is the perimeter of the axial cross section of the forging, F is the surface area of the axial cross section of the forging (surface that includes the entire axis of symmetry), P_c is the perimeter of the axial cross section of the cylinder that circumscribes the forging, and F_c is the surface area of the axial cross section of the cylinder that circumscribes the forging. Because the circumscribing cylinder has the maximum diameter and the maximum height of the forging, the factor α represents a comparison of the shape of the forging with that of the cylinder.

On round forgings, bosses and rims placed farther from the center are increasingly more difficult to forge. Therefore, a "lateral shape factor," β, is defined as:

$$\beta = \frac{2R_g}{R_c} \qquad \text{(Eq 14.4)}$$

where R_g is the radial distance from the symmetry axis to the center of gravity of half of the cross section, and R_c is the maximum radius of the forged piece, which is equal to the radius of the circumscribing cylinder.

A "shape difficulty factor," S, incorporating both the longitudinal and lateral factors is defined as:

$$S = \alpha\beta \qquad \text{(Eq 14.5)}$$

The factor S expresses the complexity of a half cross section of a round forging with respect to that of the circumscribing cylinder. In round forgings, during the forging operation, the material is moved laterally down (toward the ends of the cylinder) from the center, which is considered to be at the "neutral axis." In a nonsymmetric forging the material is still moved out laterally from the "neutral surface." Thus, once this neutral surface is defined, a "shape difficulty factor" can also be calculated in nonsymmetric forgings.

14.4 Design of Finisher Dies

Using the shape complexity and the forging material as guidelines, the forging process engineer establishes the forging sequence (number of forging operations) and designs the dies for each operation, starting with the finisher dies. The most critical information necessary for forging die design is the geometry and the material of the forging to be produced. The forging geometry in turn is obtained from the machined part drawing by modifying this part to facilitate forging. Starting with the forging geometry, the die designer first designs the finisher dies by:

- Selecting the appropriate die block size and the flash dimensions
- Estimating the forging load and stresses to ascertain that the dies are not subjected to excessive loading

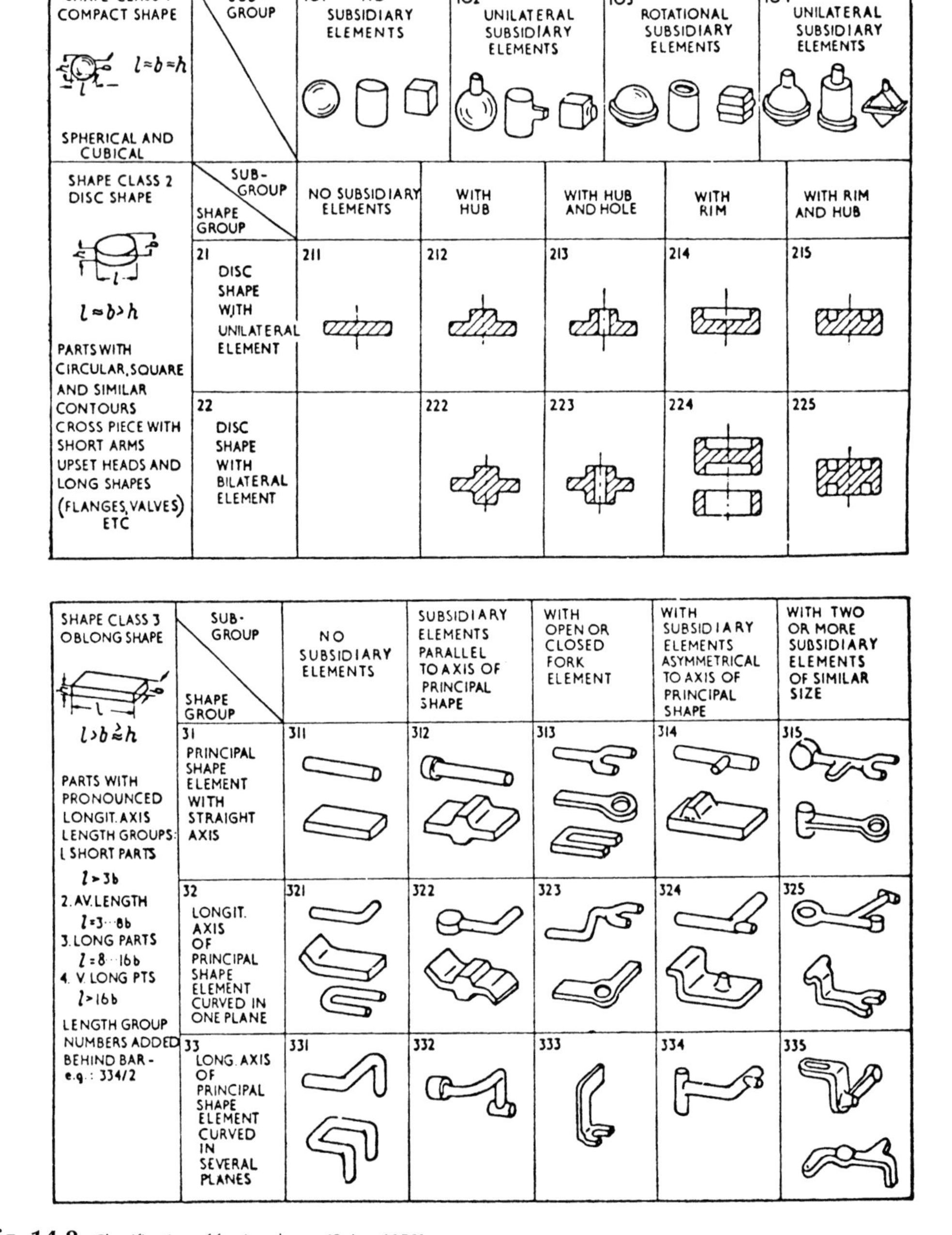

Fig. 14.8 Classification of forging shapes [Spies, 1959]

The geometry of the finisher die is essentially that of the finish forging augmented by flash configuration. In designing finisher dies, the dimensions of the flash should be optimized. The designer must make a compromise: on the one hand, to fill the die cavity it is desirable to increase the die stresses by restricting the flash dimensions (thinner and wider flash on the dies); but, on the other hand, the designer should not allow the forging pressure to reach a high value, which may cause die breakage due to mechanical fatigue. To analyze stresses, "slab method of analysis" or process simulation using finite-element method (FEM)-based computer codes is generally used. The FEM approach is discussed later.

By modifying the flash dimensions, the die and material temperatures, the press speed, and the friction factor, the die designer is able to evaluate the influence of these factors on the forging stresses and loads. Thus, conditions that appear most favorable can be selected. In addition, the calculated forging stress distribution can be utilized for estimating the local die stresses in the dies by means of elastic FEM analysis. After these forging stresses and loads are estimated, it is possible to determine the center of loading for the forging in order to locate the die cavities in the press, such that off-center loading is reduced.

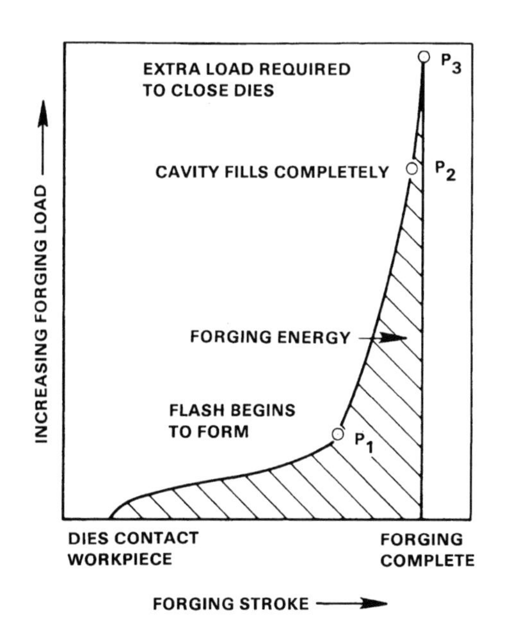

Fig. 14.9 Typical load-stroke curve for closed-die forging [Altan et al., 1983]

14.4.1 Flash Design and Forging Load

The flash dimensions and the billet dimensions influence the flash allowance, forging load, forging energy, and the die life. The selection of these variables influences the quality of the forged part and the magnitude of flash allowance, forging load, and the die life. The influence of flash thickness and flash-land width on the forging pressure is reasonably well understood from a qualitative point of view. The forging pressure increases with:

- Decreasing flash thickness
- Increasing flash-land width because of the combinations of increasing restriction, increasing frictional forces, and decreasing metal temperatures at the flash gap

A typical load-versus-stroke curve from an impression-die forging operation is shown in Fig. 14.9. Loads are relatively low until the more difficult details are partly filled and the metal reached the flash opening (Fig. 14.10). This stage corresponds to point P_1 in Fig. 14.9. For successful forging, two conditions must be fulfilled when this point is reached: A sufficient volume of metal must be trapped within the confines of the die to fill the remaining cavities, and extrusion of metal through the narrowing gap of the flash opening must be more difficult than filling of the more intricate detail in the die.

As the dies continue to close, the load increases sharply to point P_2, the stage at which the die cavity is filled completely. Ideally, at this point the cavity pressure provided by the flash geometry should be just sufficient to fill the en-

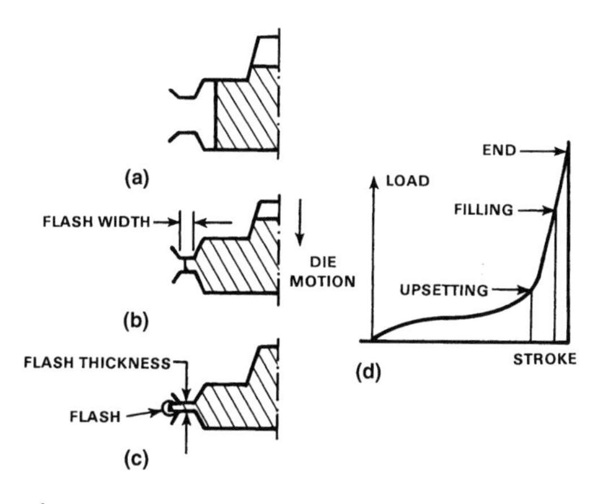

Fig. 14.10 Metal flow and the corresponding load-stroke curve. (a) Upsetting. (b) Filling. (c) End. (d) Load-stroke curve [Altan et al., 1983]

tire cavity, and the forging should be completed. However, P_3 represents the final load reached in normal practice for ensuring that the cavity is completely filled and that the forging has the proper dimensions. During the stroke from P_2 to P_3, all the metal flow occurs near or in the flash gap, which in turn becomes more restrictive as the dies close. In that respect, the detail most difficult to fill determines the minimum load for producing a fully filled forging. Thus, the dimensions of the flash determine the final load required for closing the dies. Formation of the flash, however, is greatly influenced by the amount of excess material available in the cavity, because that amount determines the instantaneous height of the extruded flash and, therefore, the die stresses.

The effect of excess metal volume in flash formation was studied extensively [Vieregge, 1968]. It was found that a cavity can be filled with various flash geometries provided that there is always a sufficient supply of material in the die. Thus, it is possible to fill the same cavity by using a less restrictive, i.e., thicker, flash and to do this at a lower total forging load if the necessary excess material is available (in this case, the advantages of lower forging load and lower cavity stress are offset by increased scrap loss) or if the workpiece is properly preformed (in which case low stresses and material losses are obtained by extra preforming). These relationships are illustrated in Fig. 14.11 and 14.12.

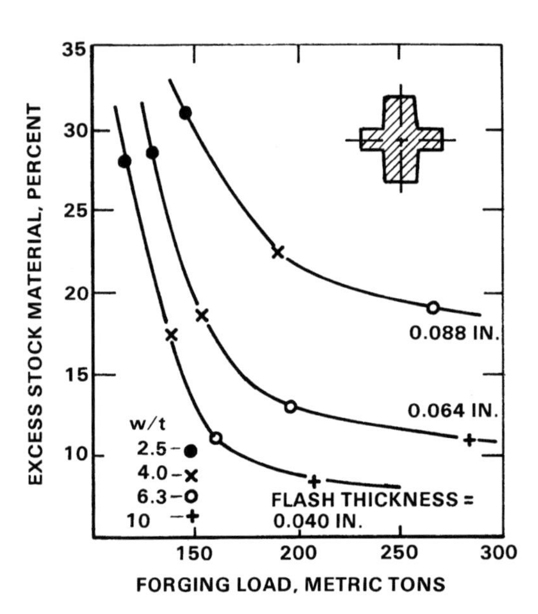

Fig. 14.11 Relationships among excess stock material, flash thickness, flash width/thickness ratio, and forging load for mechanical press forging of a round part approximately 3 in. (7.6 cm) in diameter by 3.5 in. (8.9 cm) high [Vieregge, 1968]

14.4.2 *Empirical Methods for Flash Design*

The "shape classification" (Fig. 14.8) has been utilized in systematic evaluation of flash dimensions in steel forgings. For this purpose, 1500 forgings from eight different forging companies were classified into shape groups, as shown in Fig. 14.8. By evaluating the flash designs suggested for these forgings, an attempt was made to establish a relationship between forging weight and flash dimensions. The results for group 224 are presented in Fig. 14.13 as an example [Altan et al., 1973]. This figure can be used for selecting the flash thickness based on the forging weight, Q, of the forging. This graph also shows the relationship between the flash width/thickness (w/t) ratio and the forging weight. Thus, knowing the weight of the part to be forged, it is possible to find the corresponding flash thickness and w/t ratio. Thus, the user can obtain the flash dimensions based on the weight of the forging.

There is no unique choice of the flash dimensions for a forging operation. The choice is variable within a range of values where the flash allowance and the forging load are not too high. There has to be a compromise between the two. In general, the flash thickness is shown to increase with increasing forging weight, while the ratio of flash width to flash thickness (w/t) decreases to a limiting value. In order to investi-

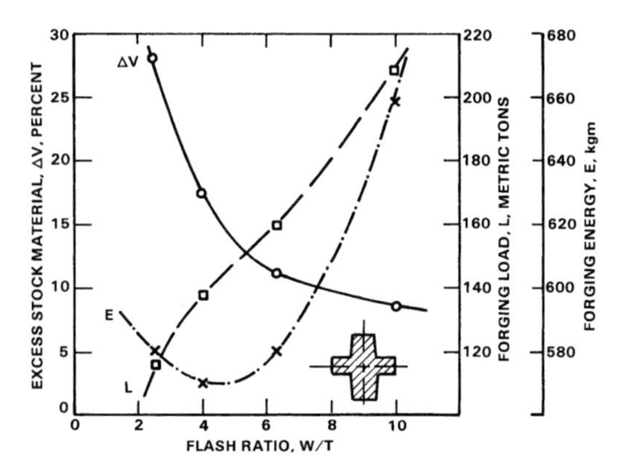

Fig. 14.12 Relationships among flash width/thickness ratio, excess stock material, forging load, and energy for a constant flash thickness, t, of 0.04 in. (1.0 mm) (same forging as that shown in Fig. 14.11) [Vieregge, 1968]

gate the effect of forging shape on flash dimensions, other subgroups were studied, and it was concluded that the influence of shape is not as significant as that of forging weight [Altan et al., 1973].

It is also possible to determine the flash dimensions for round forgings, using the billet dimensions. Thus, it is possible to obtain little flash allowance and minimize the forging energy. For round forgings, Eq 14.6 and 14.7 predict the flash dimensions that are a good compromise between the flash allowance and the forging load [Vieregge, 1968]:

$$t = [0.017 \cdot D] + \left[\frac{1}{\sqrt{D + 5}}\right] \quad \text{(Eq 14.6)}$$

$$\frac{W}{t} = \frac{30}{\sqrt[3]{D\left[1 + \left(\frac{2 \cdot D^2}{H(2R_h + D)}\right)\right]}} \quad \text{(Eq 14.7)}$$

where w is the flash width (mm), t is the flash thickness (mm), H is the height of the ribs or shaft (mm), D is the outside diameter of the forging (mm), and R_h is the radial distance of the center of a rib from the axis of symmetry of the forging.

14.5 Prediction of Forging Stresses and Loads

Prediction of forging load and pressure in closed- and impression-die forging operations is difficult. Most forging operations are of nonsteady-state type in terms of metal flow, stresses, and temperatures, i.e. all these variables vary continuously during the process. In addition, forgings comprise an enormously large number of geometrical shapes and materials, which require different, even though similar, techniques of engineering analysis. Because of these difficulties encountered in practice, forging loads are usually estimated on the basis of empirical procedures using empirically developed formulae.

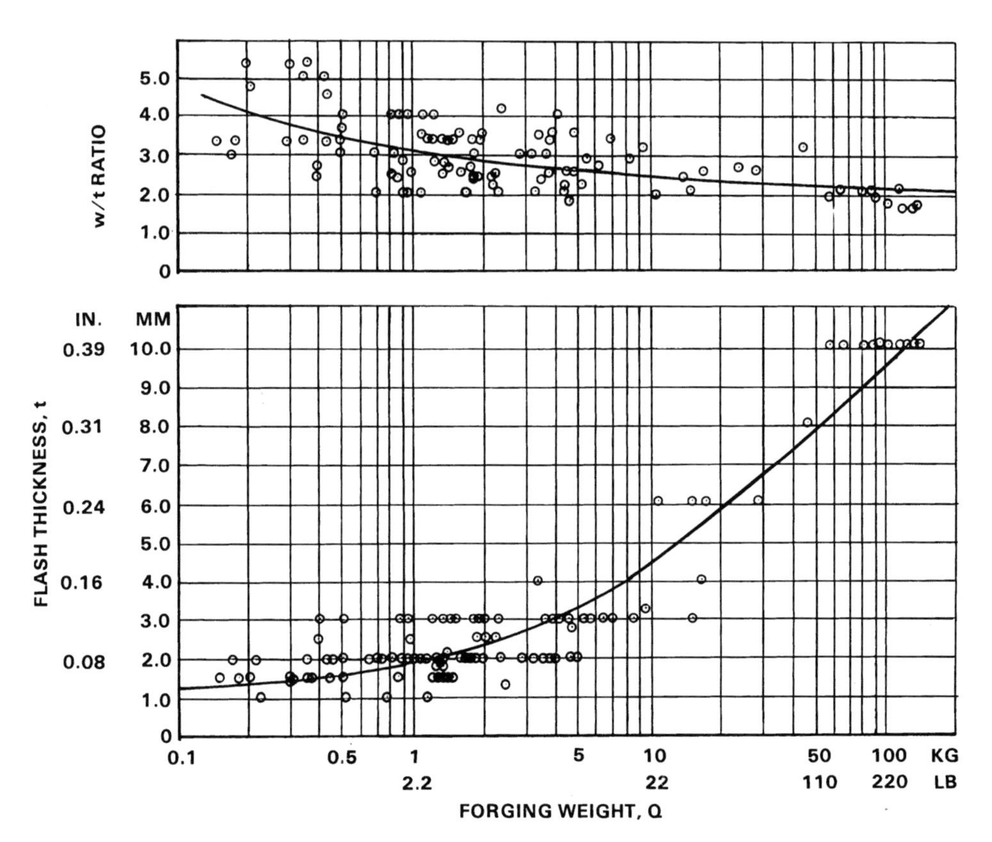

Fig. 14.13 Variations in flash-land-to-thickness ratio and in flash thickness, t, with weight, Q, of forgings of group 224 (materials: carbon and alloy steels)

14.5.1 Empirical Methods for Estimation of Forging Pressure and Load

In estimating the forging load empirically, the surface area of the forging, including the flash zone, is multiplied by an average forging pressure known from experience. The forging pressures encountered in practice vary from 20 to 70 tons/in.^{2} depending on the material and the geometry of the part. Neuberger and Pannasch [Neuberger et al., 1962] conducted forging experiments with various carbon steels (up to 0.6% C) and with low-alloy steels using flash ratios, w/t (where w is flash-land width and t is the flash thickness), from 2 to 4 (Fig. 14.14). They found that the variable that most influences the forging pressure, P_a, is the average height, H_a, of the forging. The lower curve relates to relatively simple parts, whereas the upper curve to slightly difficult ones [Neuberger et al., 1962].

Most empirical methods, summarized in terms of simple formulae or nomograms, are not sufficiently general to predict forging loads for a variety of parts and materials. Lacking a suitable empirical formula, one may use suitable analytical techniques of varying degrees of complexity for calculating forging load and stresses. Among these techniques, the relatively simple slab method has been proven to be very practical for predicting forging loads.

14.5.2 Simplified Slab Method to Estimate Forging Load

The slab method has been successfully used for predicting forging loads and stresses with acceptable engineering accuracy. For this purpose, a forging is divided into various plane strain and axisymmetric sections, and then simplified equations are used to predict the average pressure and load for each section before all these load components are added together. This method used in the practical prediction of forging loads is shown in Fig. 14.15 [Subramanian et al., 1980]. In this analysis, it is assumed that the cavity has a rectangular shape and the flash geometry illustrated in Fig. 14.15. In actual practice, where the cavity is not rectangular, the cross section is simplified to conform to this model.

As seen in Fig. 14.15, the cavity height is denoted by H, the radius (or half width of the cavity) by r, the flash thickness by t, and the flash width by w. The stresses at various locations of the cross section and hence the load acting on the cross section can be estimated as follows.

With the flow stress in the flash region denoted by σ_f and the frictional shear factor by m, the stress at the entrance from the cavity into the flash of an axisymmetric cross section, σ_{ea}, is given by:

$$\sigma_{ea} = \left(\frac{2}{\sqrt{3}} m \frac{w}{t} + 1\right)\sigma_f \qquad \text{(Eq 14.8)}$$

Because of rapid chilling and a high deformation rate, the flow stress in the flash region is considered to be different from the flow stress in the cavity. Hence, two different flow stresses are used for the flash and cavity regions. The total load (P_{ta}) on the cross section is the summation of the load acting on the flash region and the load acting on the die cavity:

$$P_{ta} = 2\pi\sigma_f\left[-\frac{2}{3}\frac{m}{\sqrt{3}}\frac{1}{t}(R^3 - r^3) + \left(1 + 2\frac{m}{\sqrt{3}}\frac{R}{t}\right)\left(\frac{R^2 - r^2}{2}\right)\right] + 2\pi r^2\left[\frac{m}{\sqrt{3}}\frac{\sigma_c}{3}\frac{r}{H} + \frac{\sigma_{ea}}{2}\right] \qquad \text{(Eq 14.9)}$$

where $R = r + w$, σ_f is the flow stress in the flash region, and σ_c is the flow stress in the cavity.

For the plane-strain cross sections, the equations corresponding to Eq 14.8 and 14.9 are:

$$\sigma_{ep} = \frac{2}{\sqrt{3}}\sigma_f\left(1 + m\frac{w}{t}\right) \qquad \text{(Eq 14.10)}$$

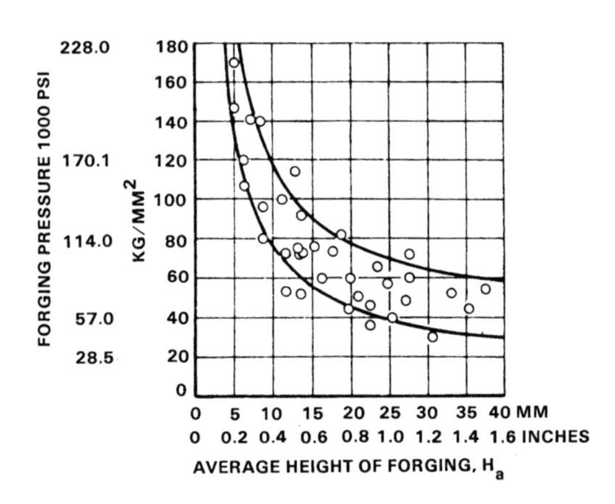

Fig. 14.14 Forging pressure versus average forging height (H_a) for forging of carbon and low-alloy steels at flash ratios, w/t, from 2 to 4 [Neuberger et al., 1962]

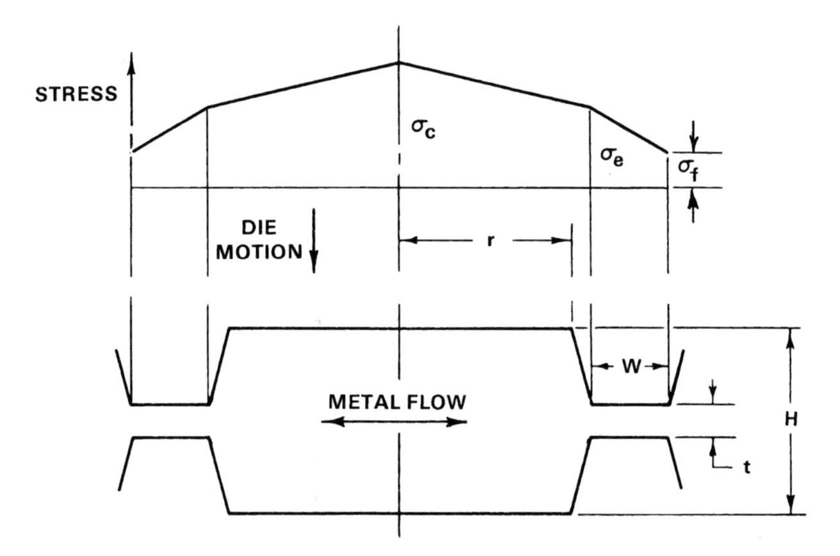

Fig. 14.15 Schematic of a simple closed-die forging and forging stress distribution [Subramanian et al., 1980]

$$P_{tp} = \frac{2}{\sqrt{3}} w\sigma_f\left(2 + \frac{mw}{t}\right) + \left(\sigma_{ep} + \frac{L}{2H}\frac{m}{\sqrt{3}}\sigma_c\right)L \quad \text{(Eq 14.11)}$$

where L is the cavity width, i.e., L = 2r in Fig. 14.15. The above equations are relatively simple and can be programmed for practical use. The following information is required to perform these calculations:

- The geometry of the part
- The flow stresses in the cavity and the flash during the final stages of the forging operation
- The friction at the die/forging interface

Appendix A gives an example of estimation of the load required for forging connecting rods. A comparison between the theoretical prediction and the actual data from forging trials is also provided.

14.6 Design of Blocker (Preform) Dies

One of the most important aspects of closed-die forging is the design of preforms or blockers to achieve adequate metal distribution [Altan et al., 1973]. Thus, in the finish forging operation, defect-free metal flow and complete die filling can be achieved, and metal losses into the flash can be minimized. In preforming, round or round-cornered square stock with constant cross section is deformed in such a manner that a desired volume distribution is achieved prior to impression-die forging. In blocking, the preform is forged in a blocker cavity prior to finish forging.

The determination of the preform configuration is an especially difficult task and an art in itself, requiring skills achieved only by years of extensive experience. Designing a correct pre-

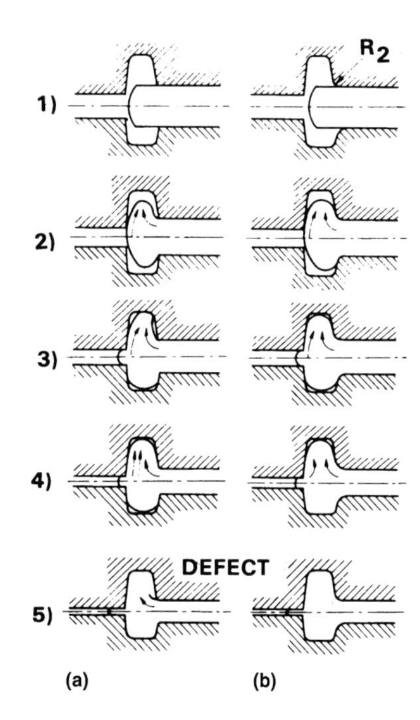

Fig. 14.16 Defect formation in forging when fillet radii are too small [Haller, 1971]

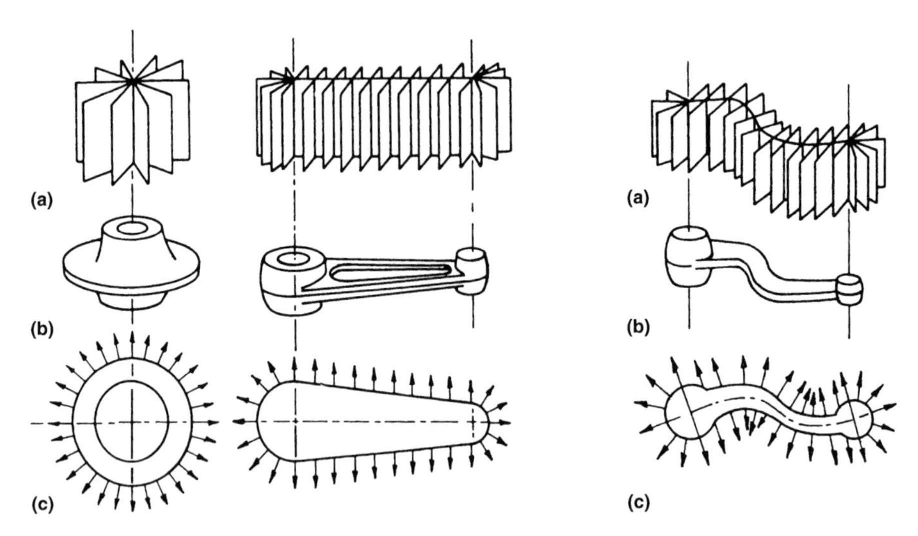

Fig. 14.17 Planes of metal flow. (a) Planes of flow. (b) Finished forged shapes. (c) Directions of flow [Altan et al., 1973]

form allows the control of the volume distribution of the part during forging as well as control over the material flow. The main objective of preform design is to distribute the metal in the preform in order to:

- Ensure defect-free metal flow and adequate die filling. Figure 14.16 shows how a defect can form with insufficient volume distribution in an H-shaped cross section [Haller, 1971]
- Minimize the amount of material lost as flash
- Minimize die wear in the finish-forging cavity by reducing the metal movement in this operation
- Achieve desired grain flow and control mechanical properties

The common practice in preform design is to consider planes of metal flow, i.e., selected cross sections of the forging (Fig. 14.17) [Altan et al., 1973]. Understanding the principles of the material flow during the forging operation can help attain a better understanding of the design rules. Any complex shape can be divided into axisymmetric or plane-strain flow regions, depending on the geometry in order to simplify the analysis.

The example steel forging presented in Fig. 14.18 illustrates the various preforming operations necessary to forge the part shown [Haller, 1971]. The round bar from rolled stock is rolled in a special machine called a reducer roller for volume distribution, bent in a die to provide the appropriate shape, blocked in a blocker die cavity, and finish forged.

In determining the forging steps for any part, it is first necessary to obtain the volume of the forging based on the areas of successive cross sections throughout the forging. The volume distribution can be obtained in the following manner [Haller, 1971]:

1. Lay out a dimensioned drawing of the finish configuration, complete with flash.

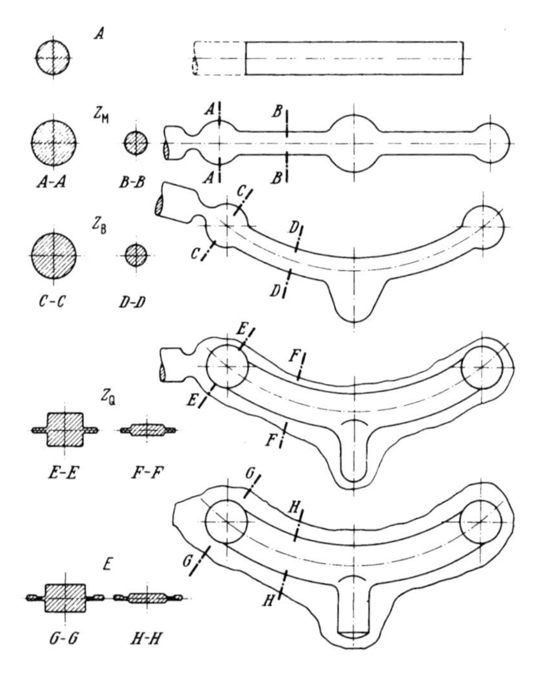

Fig. 14.18 Preforming, blocking, and finish forging operations for an example steel forging [Haller, 1971]

2. Construct a baseline for area determination parallel to the centerline of the part.
3. Determine the maximum and minimum cross-sectional areas perpendicular to the centerline of the part.
4. Plot these area values at proportional distances from the baseline.
5. Connect these points with a smooth curve (in instances where it is not clear how the curve would best show the changing cross-sectional areas, additional points should be plotted to assist in determining a smooth representation curve).
6. Above this curve, add the approximate area of the flash at each cross section, giving consideration to those sections where the flash should be widest. The flash will generally be of constant thickness but will be widest at the narrower sections and smallest at the wider sections (the proportional allowance for flash is illustrated by the examples in Fig. 14.19).
7. Convert the minimum and maximum area values to rounds or rectangular shapes having the same cross-sectional area.

Figure 14.19 shows two examples of obtaining a volume distribution through the above procedure. In both examples, (a) is the forging, (b) is the cross-sectional area versus length, (c) and (d) are the ideal preform, V_E and q_E are the volume and cross section of the finish forging, and V_G and q_G are the volume and cross section of the flash.

There are various methods of preforming i.e., for distributing the metal prior to die forging in a blocker or finisher die [Altan et al., 1973, Haller, 1971, and Lange et al., 1977]. In forging steel parts, a correct preform can be designed by using the following three general design rules (these rules do not apply to forging nonferrous materials) [Lange et al., 1977]:

- The area of cross section of the preform = the area of cross section of the finished product + the flash allowance (metal flowing into flash). Thus, the initial stock distribution is obtained by determining the areas of cross sections along the main axis of the forging.
- All the concave radii, including the fillet radii, on the preform must be greater than the corresponding radii on the finished part.
- In the forging direction, the thickness of the preform should be greater than that of the finished part so that the metal flow is mostly by upsetting rather than extrusion. During the finishing stage, the material will then be squeezed laterally toward the die cavity without additional shear at the die/material interface. Such conditions minimize friction and forging load and reduce wear along the die surfaces.

The application of these three design rules for preforming of steel forgings is illustrated by examples shown in Fig. 14.20 for H-shaped cross

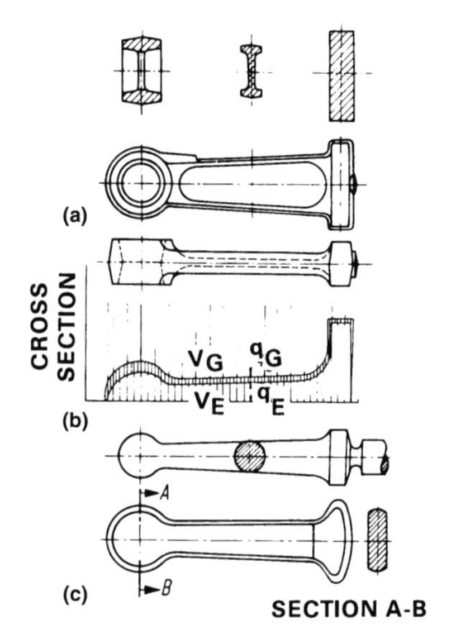

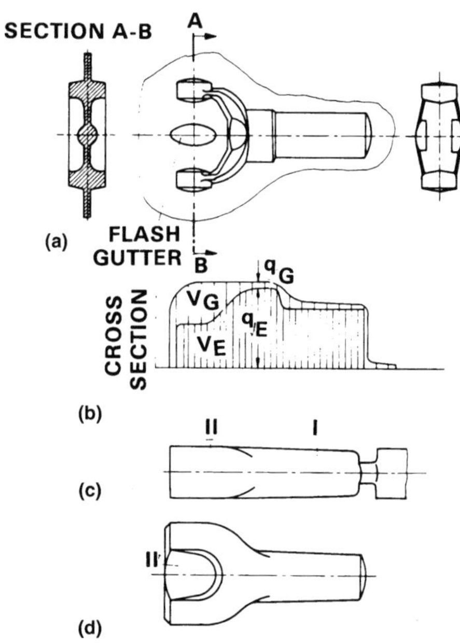

Fig. 14.19 Preform designs for two example parts. In both examples, (a) forging, (b) cross-sectional area vs. length, (c) and (d) ideal preform, V_E and q_E, volume and cross section of the finish forging, and V_G and q_G, volume and cross section of the flash [Haller 1971]

sections of various rib heights and in Fig. 14.21 for some solid cross sections [Lange et al., 1977].

The preform is the shape of the billet before the finish operation. In certain cases, depending on the ratio of the height of the preform to its width, there might be more than one preform operation involved.

Preform design guidelines differ from material to material. They are basically categorized into the following three categories.

- Carbon and low-alloy steel parts
- Aluminum alloy rib-web-type parts
- Titanium alloy rib-web-type parts

14.6.1 Guidelines for Carbon and Low-Alloy Steels

In hammer forging of carbon or low-alloy steels, the preform usually does not have flash. The blend-in radius (R_P) of the preform at the parting line is influenced by the adjacent cavity depth (Table 14.2). In the preform, the fillet radius (R_{PF}) between the web to a rib is larger than that in the finish forging (R_{FF}), especially when the height of the rib over the web is larger than the rib width, i.e., $D_F > W_F$ (Fig. 14.22).

14.6.2 Guidelines for Aluminum Parts

For rib-web-type aluminum alloy parts, the recommended preform dimensions fall into the following ranges given in Table 14.3. The preform is usually designed to have the same draft angles as the finish part. However, when very deep cavities are present in the finisher die, larger draft angles are provided in the preform. A greater web thickness in the preform is selected when the web area is relatively small and when the height of the adjoining ribs is very large. A comparison of the preform and the finished part is illustrated in Fig. 14.23.

14.6.3 Guidelines for Titanium Alloys

The guidelines for designing the titanium alloy preforms (Table 14.4) are similar to those for the aluminum alloys.

14.6.4 Guidelines for H Cross Sections

Most of the parts that are closed-die forged are H cross-section-type parts or they can be decomposed into H cross sections. For this reason it is good to have specific rules for these types of cross sections. These rules are different for different proportions in an H cross-section-type part. The design guidelines for preform design for rib-web-type sections must consider:

- Different rib-height (D_F) to rib-width (W_F) ratios
- Different distances between the ribs (W_D)

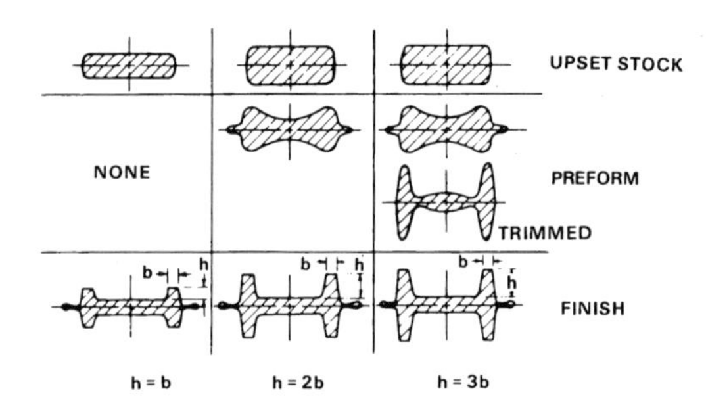

Fig. 14.20 Preforms for different H-shaped forgings [Lange et al., 1977]

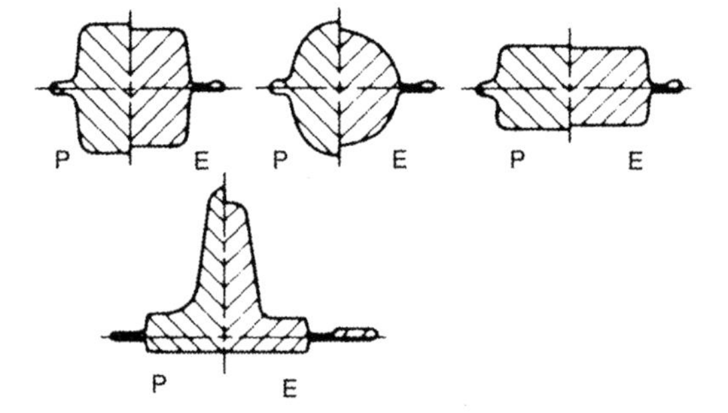

Fig. 14.21 The blocker and finish cross sections for various shapes (P, preform; E, end) [Lange et al., 1977]

Table 14.2 Preform dimensions for carbon or low-alloy steels [Altan et al., 1973]

Dimensions of the finish forgings	Dimensions of the preforms			
Flash	No flash			
Blend-in radii (R_F)	$R_P \cong R_F + C$ or $H_R/6 < R_P < H_R/4$			
Fillet radii (R_{FF})	$R_{PF} \cong 1.2\ R_{FF} + 0.125$ in. (3.175 mm)			
Depth of the cavity (H_R), in. (mm)	<0.4 (10)	0.4 to 1 (10 to 25)	1 to 2 (25 to 51)	>2 (51)
Value of C, in. (mm)	0.08 (2.0)	0.12 (3.0)	0.16 (4.0)	0.2 (5.1)

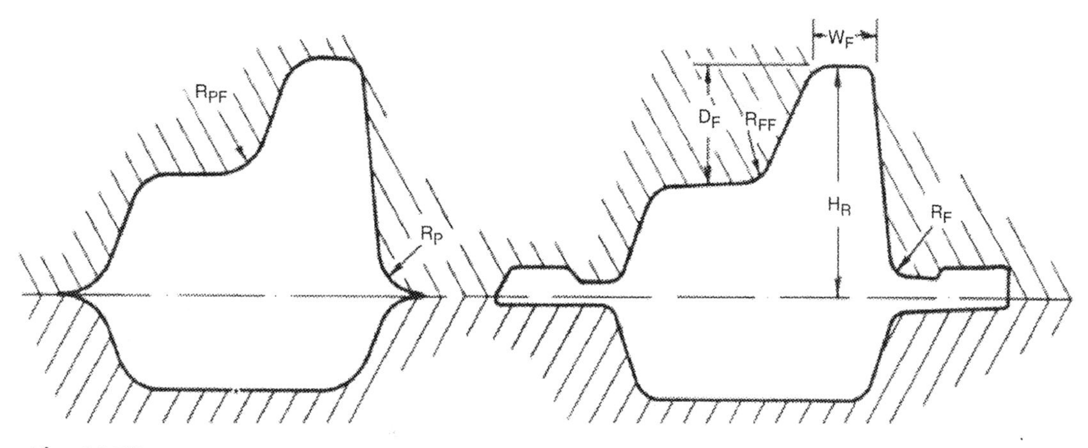

Fig. 14.22 Preform and finish shape [Altan et al., 1973]

Height of the Rib (D_F) < 2 · Width of the Rib (W_F). For an H cross-section part in which the rib height is smaller than two times the rib width, the preform will have a rectangular shape (Fig. 14.24). All one has to know is how to calculate the overall width of the preform, B_P, and its height, H_P. All the other parameters, such as radii and the draft angle, will be set according to the generic guidelines for different materials:

- The overall width of the preform (B_P) is determined in terms of the finish cavity width (B_F). It is expressed as:

$$B_P = B_F - (0.08 \text{ to } 0.4)$$

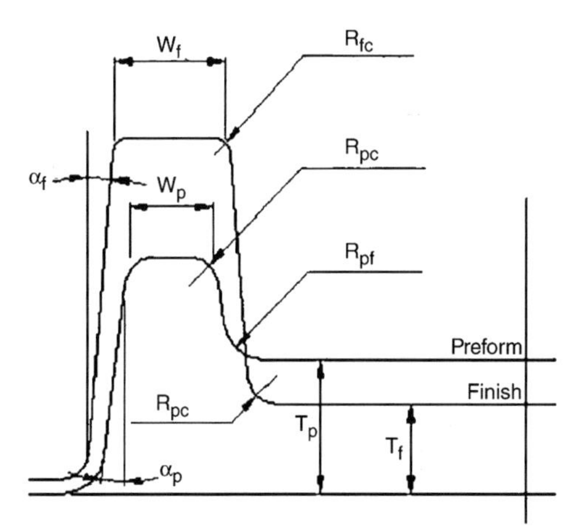

Fig. 14.23 Comparison of the preform and finished part for a quarter of an H cross section [Altan et al., 1973]

- The height of the preform (H_P) is obtained after calculating the preform width by dividing the surface area (SA) of the finish cross section by the preform width:

$$H_P = (\text{SA} + \text{flash})/B_P$$

$$\text{Flash} = 5 \text{ to } 15\% \text{ of SA}$$

Height of the Rib (D_F) > 2 · Width of the Rib (W_F). For an H cross-section part in which the rib height is larger than two times the rib

Table 14.3 Preform dimensions for aluminum alloys [Altan et al., 1973]

Dimensions of the finish forgings	Dimensions of the preforms
Web thickness (t_f)	$T_p \cong (1 \text{ to } 1.5) \cdot t_f$
Fillet radii (R_{ff})	$R_{pf} \cong (1.2 \text{ to } 2) \cdot R_{ff}$
Corner radii (R_{fc})	$R_{pc} \cong (1.2 \text{ to } 2) \cdot R_{fc}$
Draft angle (α_f)	$\alpha_p \cong \alpha_f + (2 \text{ to } 5°)$
Width of the rib (W_f)	$W_p \cong W_f - 1/32$ in. (0.8 mm)

Table 14.4 Preform dimensions for titanium alloys [Altan et al., 1973]

Dimensions of the finish forgings	Dimensions of the preforms
Web thickness (t_f)	$T_p \cong (1.5 \text{ to } 2.2) \cdot t_f$
Fillet radii (R_{ff})	$R_{pf} \cong (2 \text{ to } 3) \cdot R_{ff}$
Corner radii (R_{fc})	$R_{pc} \cong (2) \cdot R_{fc}$
Draft angle (α_f)	$\alpha_p \cong \alpha_f + (3 \text{ to } 5°)$
Width of the rib (W_f)	$W_p \cong W_f - 1/16$ to $1/8$ in. (1.6 to 3.2 mm)

width, the preform has a trapezoidal shape. The parameters to set are overall width of the preform (B_P), height of the preform (H_P), additional rib height (x), and thinning of the web portion (y). All the other parameters, such as radii and the draft angle, will be set according to the generic guidelines for different materials:

- The overall width of the preform (B_P) is determined in terms of the finish cavity width (B_F). It is expressed as:

$$B_P = B_F - (0.08 \text{ to } 0.4)$$

- The height of the preform (H_P) is obtained after calculating the preform width by dividing the surface area of the finish cross section by the preform width:

$$H_P = (SA + \text{flash})/B_P$$

$$\text{Flash} = 5 \text{ to } 15\% \text{ of SA}$$

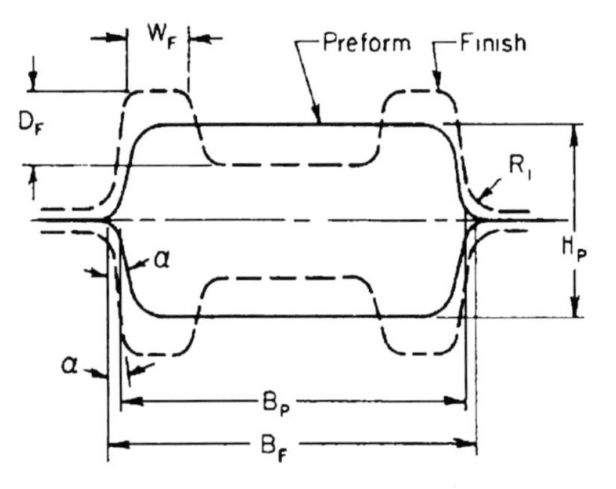

Fig. 14.24 Preform shape when $D_F < 2 \cdot W_F$ [Bruchanov et al., 1955]

- Additional rib height (x): The preform for this cross section is considered to have a trapezoidal form (Fig. 14.25). The necessary additional rib height (x) is determined by:

$$x = 0.25\ (H_F - H_P)$$

- Thinning of the web portion (y): Once the preform height and the additional rib height are determined, the thinning of the web portion is calculated by setting the areas $f_1 = f_2$ (Fig. 14.25), where the web and the rib are blended together with large radii.

Distance between Ribs (W_D) Very Large. For an H cross-section part in which the distance between the ribs is very large, the preform has a trapezoidal shape (Fig. 14.26). The parameters to be set are overall width and height of the preform (B_P and H_P), rib height (x), fillet and flash radius (R_{PF} and R_P), and preform thickness (H_P). All the other parameters, such as radii and the draft angle, will be set according to the generic guidelines for different materials:

- Overall width and height of the preform:

$$B_P = B_F - (0.08 \text{ to } 0.4)$$

$$H_P = (SA + \text{flash})/B_P$$

$$\text{Flash} = 5 \text{ to } 15\% \text{ of SA}$$

- Rib height (x): The preform for this cross section is assumed to have a trapezoidal form. The surface area f_2 can be significantly larger than the surface area f_1 (Fig. 14.26). The required rib height is determined by the relation:

$$x = (0.6 \text{ to } 0.8)\ D_F$$

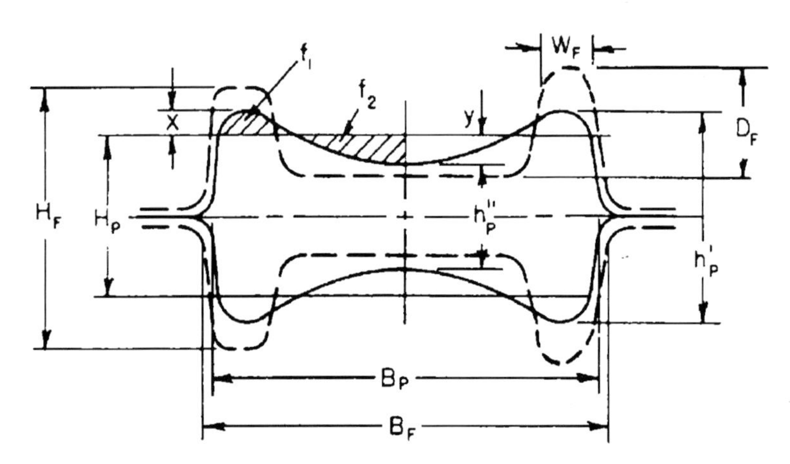

Fig. 14.25 Preform shape when $D_F > 2 \cdot W_F$ [Bruchanov et al., 1955]

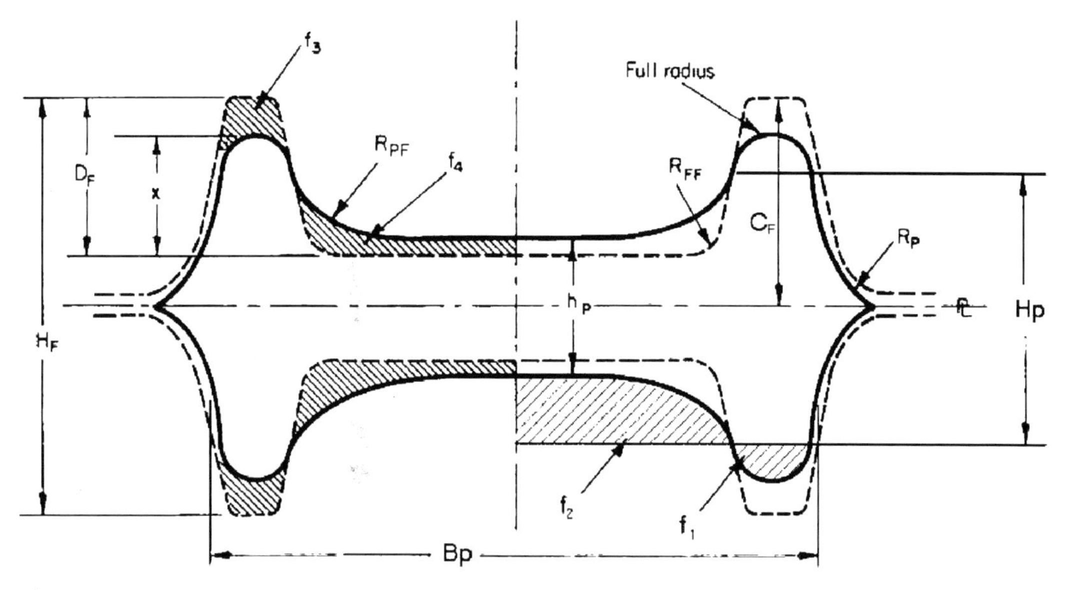

Fig. 14.26 Preform design when the distance between the ribs is very large [Bruchanov et al., 1955]

- Fillet and flash radii (R_{PF}, R_P):

  $$R_{PF} = 1.2\ R_{FF} + 0.125 \text{ in. } (3.175 \text{ mm})$$

  $$R_P \cong R_F + C$$

- Preform thickness (H_P): The preform thickness is determined by using the condition that the volume represented by f_4 should be larger than the volume represented by f_3 by the excess of the flash material (Fig. 14.26).

APPENDIX A Example: Prediction of Load for Forging of a Connecting Rod

A.1 Introduction

A connecting rod (Fig. A.1) was selected as a component to be used for practical evaluation of the calculations described in section 14.5.2, "Simplified Slab Method to Estimate Forging Load," in this chapter. The part was forged in a 500 ton mechanical press, and the measured loads were compared with the results obtained from computer-aided analysis as well as with results from the simplified slab method.

Three representative cross sections of the connecting rod were chosen for the estimation of the forging load (Fig. A.1). All the cross sections had the same flash dimensions, namely, blocker: 0.1 in. (2.5 mm) thick by 0.31 in. (7.9 mm) wide; finish: 0.06 in. (1.5 mm) thick by 0.31 in. (7.9 mm) wide. These dimensions were the dimensions of the flash lands in the dies. The values of flash thickness used in the forging load estimations were those actually measured on the forged parts. All the cavity cross sections were approximated as rectangles, as shown in Fig. 14.15 and A.1. The dimensions of the rectangles (in inches) used for load estimation were:

- Section A-A:
  Blocker: r = 0.96, H = 0.32
  Finish: L = 0.2, H = 0.18
- Section B-B:
  Blocker: L = 0.93, H = 0.32
  Finish: L = 0.95, H = 0.18
- Section C-C:
  Blocker: r = 0.53, H = 0.64
  Finish: r = 0.55, H = 0.59

During finish forging, in order to reduce the excessive load resulting from forging a very thin web, the central portion of cross-section A-A was relieved. Hence, to estimate the load in this cross section, only one-half of the section was considered and was treated as a plane-strain cross section of a length equal to the average circumference of the lower boss of the connecting rod.

A.2 Estimation of the Flow Stress

The forging trials were conducted in a mechanical press with a stroke of 10 in. (25.4 cm) and a speed of 90 strokes per minute (Fig. A.2). The blocker and finish dies were mounted side by side on the press bolster (Fig. A.3). Both dies were heated to approximately 350 °F (175 °C). AISI type 1016 steel billets were heated to 2100 °F (1150 °C) prior to blocker forging. The temperature of the billet prior to finish forging, as measured during the trial runs, was approximately 1950 °F (1065 °C). Thus, some cooling occurred during forging in the blocker cavity and transfer into the finisher.

The flow stress is a function of the strain, strain rate, and temperature that exist at a given

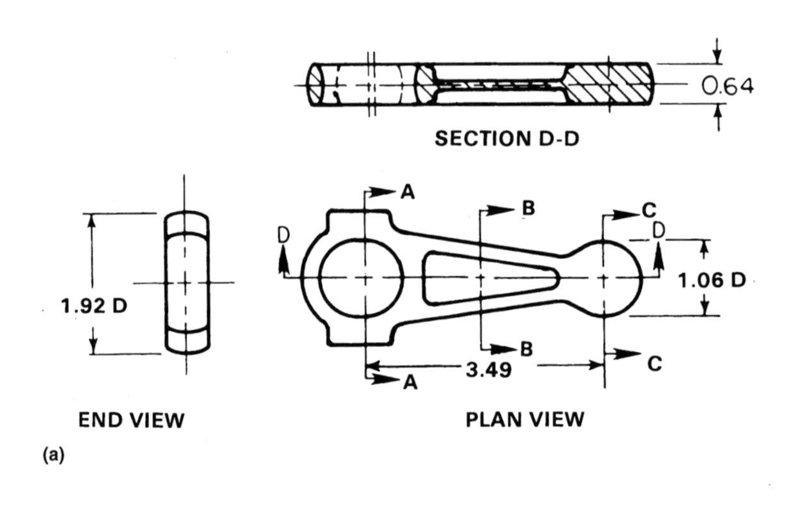

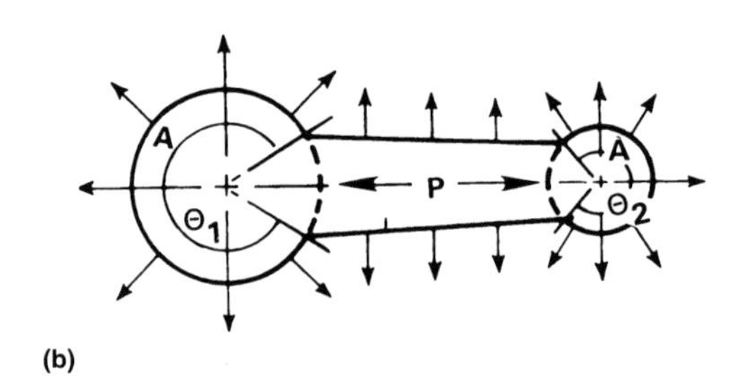

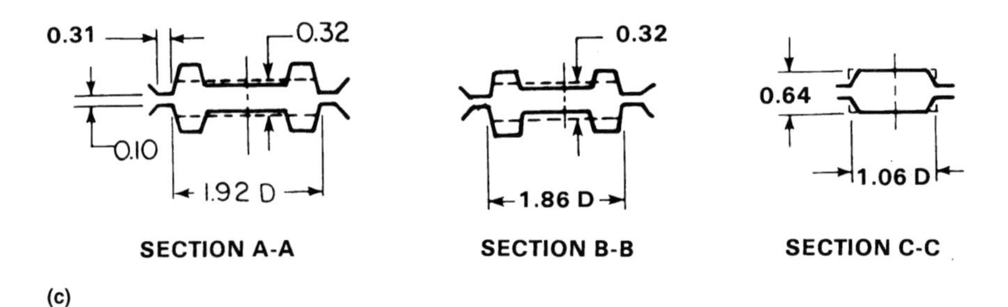

Fig. A.1 (a) Geometry of the connecting rod. (b) Directions of metal flow. (c) Representative sections and their simplification

time during the deformation process and can be expressed approximately as:

$$\sigma = C.\dot{\varepsilon}^{m} \quad \text{(Eq A.1)}$$

The values of C and m for the material used for the trials are given in Table A.1. These values vary significantly with temperature. Hence, in order to estimate the flow stress accurately, the temperature of the deforming material should be known. The temperature of the stock at the end of the forging stroke depends on the stock temperature, die temperature, speed of deformation, and frictional conditions. Further, the temperature varies across the forging due to die chilling and due to heat generation by friction and deformation. However, if the temperature gradient is neglected and the forging is considered to be a thin plate of uniform temperature cooled symmetrically from both sides, the average temperature of the forging in the cavity or in the flash can be expressed as follows:

$$\theta = \theta_1 + (\theta_s - \theta_1)\exp\left(-\frac{\alpha T}{c\rho t}\right) \quad \text{(Eq A.2)}$$

As an example, consider cross-section A-A of the blocker dies in Fig. A.1:

- θ_1 (initial die temperature) = 350 °F (175 °C)
- θ_s (initial stock temperature) = 2100 °F (1150 °C)
- α (heat-transfer coefficient) = 0.0039 Btu/in.^{2}/°F (3.1889 W/m^2-°K) (estimated from values obtained from the forging of steel)
- t (average forging or plate thickness) = 0.32 in. (8.1 mm)
- c (specific heat of the billet material) = 0.108 Btu/lb/°F (452 J/kg-°K)
- ρ (density of the billet material) = 0.285 lb/in.^{3} (7.89 g/cm^3)

To estimate the duration of contact, the average ram velocity during forging should be known. This velocity is half of the velocity of the ram when it touches the billet. In this case, the billet has round sections with an average diameter of 0.75 in. (19 mm). The average thickness of the blocker is 0.32 in. (8.1 mm). Hence, the average distance of the ram from the bottom dead center (BDC) position during forging is:

$$w = \frac{0.75 - 0.32}{2} = 0.125 \text{ in. (3.175 mm)} \quad \text{(Eq A.3)}$$

Thus, the ram velocity with respect to the ram location (w) before BDC obtained from the kinematics of the crank slider mechanism is:

$$V = w\frac{\pi n}{30}\sqrt{\frac{S}{w} - 1}$$

The mechanical press used for these trials has a stroke of 10 in. (25.4 cm) and a speed of 90 rpm. Hence:

$$V = 0.215\frac{\pi \times 90}{30}\sqrt{\frac{10}{0.215} - 1}$$

$$V = 13.67 \text{ in./s (0.35 m/s)}$$

The duration of contact, T, is:

$$T = \frac{\text{(Average billet thickness)} - \text{(Average forging thickness)}}{\text{Average ram velocity}}$$

$$= \frac{0.75 - 0.32}{13.67} = 0.0315 \text{ s}$$

The values of T can also be obtained from the load or stroke versus time curve similar to the one in Fig. A.4. The instantaneous average forging temperature is:

Fig. A.2 The 500 ton mechanical forging press used for forging trials

$$\theta = 350 + (2100 - 350) \exp\left(-\frac{(0.0039)(0.0315)}{(0.108)(0.285)(0.32)}\right) = 2078.3\ °F\ (1136.8\ °C)$$

The temperature increase due to deformation is given by:

$$\theta_d = \frac{A\bar{\sigma}_a\bar{\varepsilon}_a}{c\rho}$$

where, in addition to the symbols previously defined, θ_d is the temperature increase due to deformation; A is a factor used to convert mechanical energy to heat energy ($A = 1.07 \times 10^{-7}$ Btu/in.-lb) (9.798 J/m-kg); $\bar{\sigma}_a$ is the average flow stress in the material (18,000 psi, or 124 MPa, assumed); and $\bar{\varepsilon}_a$ is the average strain, estimated from the initial and final thickness:

$$\bar{\varepsilon}_a = \ln\left(\frac{\text{Initial thickness}}{\text{Final thickness}}\right) = \ln\frac{0.75}{0.32} = 0.85$$

Hence:

$$\theta_d = 53.2\ °F\ (11.8\ °C)$$

Ignoring the heat gain due to friction, the average temperature in the cavity of the blocker die is:

$$\theta_{cb} = \theta + \theta_d = 2078.3 + 53.2 = 2131.5\ °F\ (1166.4\ °C)$$

Fig. A.3 Blocker and finish forging dies as mounted on the bolster of the mechanical press

Table A.1 Summary of C (ksi) and m values describing the flow stress relation $\sigma = C.\dot{\varepsilon}^m$ for AISI 1016 steel at various temperatures

	Value of C or m at a temperature of, °F (°C)							
	1650 (900)		1830 (1000)		2010 (1100)		2190 (1200)	
Strain	C	m	C	m	C	m	C	m
0.05	11.8	0.133	10.7	0.124	9.0	0.117	6.4	0.150
0.1	16.5	0.099	13.7	0.099	9.7	0.130	7.1	0.157
0.2	20.8	0.082	16.5	0.090	12.1	0.119	9.1	0.140
0.3	22.8	0.085	18.2	0.088	13.4	0.109	9.5	0.148
0.4	23.0	0.084	18.2	0.098	12.9	0.126	9.1	0.164
0.5	23.9	0.088	18.1	0.109	12.5	0.141	8.2	0.189
0.6	23.3	0.097	16.9	0.127	12.1	0.156	7.8	0.205
0.7	22.8	0.104	17.1	0.127	12.4	0.151	8.1	0.196

AISI type 1016 steel (composition: 0.15 C, 0.12 Si, 0.68 Mn, 0.034 S, 0.025 P) in the hot rolled and annealed condition

From Table A.1, the values of C and m for a temperature of 2130 °F (1165 °C) and strain of 0.85 are calculated by linear interpolation as $C = 9.8 \times 10^3$ and $m = 0.187$.

The average strain rate in the deforming material is given by:

$$\dot{\bar{\varepsilon}} = \frac{\text{Velocity}}{\text{Average thickness}} = \frac{13.67}{(0.75 + 0.32)/2}$$

$$= 25.55/\text{s}$$

Using Eq A.1, the average flow stress is:

$$\bar{\sigma} = 9.8 \times 10^3 \times 25.55^{0.187}$$

$$= 17{,}964 \text{ psi (124 MPa)}$$

This is the value of the flow stress in the cavity of the section A-A in the blocker shape. Similarly, the corresponding values of flow stress for other sections are given in Table A.2.

A.3 Estimation of the Friction Factor

The frictional shear stress is given by:

$$\tau = \frac{m\bar{\sigma}}{\sqrt{3}}$$

The value of m varies between 0.25 and 0.4 for most hot steel forging operations. The simplified model assumes that metal flow occurs by sliding the entire die/material interface. In reality, however, the metal deforms by sliding along the web surfaces, but internal shearing is inevitable in the rib regions. Hence, to assume a nominal average m value of 0.4, which is usual for sliding surfaces in forging, would be unrealistic. In the present case, the length of the rib sections is almost equal to that of the web regions in cross-sections A-A and B-B. Hence, considering a weighted average, the values for m are chosen to be 0.7 for sections A-A and B-B and 0.4 for section C-C of both blocker and finish forgings.

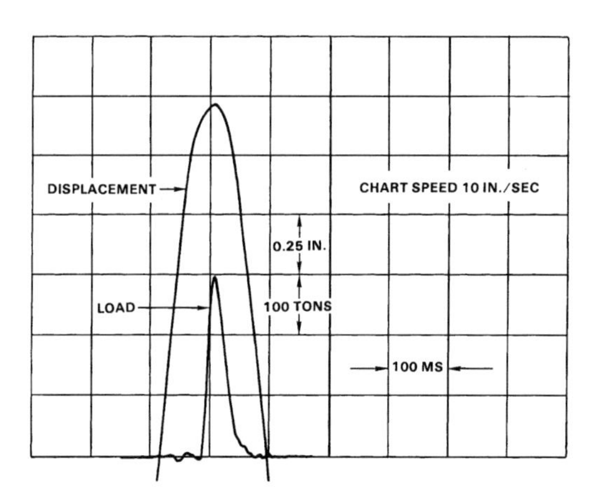

Fig. A.4 Load and displacement versus time for a forging operation

A.4 Estimation of the Forging Load

The average length of plane-strain cross-section B-B for both blocker and finish forgings is 2 in. (5 cm). Substituting the appropriate values in Eq 14.8 to 14.11, the loads were estimated for the cross sections as shown in Table A.3.

A.5 Comparison of Predictions with Data from Actual Forging Trials

To evaluate the accuracy of the simplified forging load estimation procedure, forging trials were conducted using a 500 ton Erie scotch-

Table A.2 Estimated flow stresses in different cross sections of the connecting rod forging shown in Fig. A.1

Forging	Stock/blocker temperature		Flow stress in section, psi (MPa)		
	°F	°C	A-A	B-B	C-C
Blocker	2100	1150	18,000 (124)	17,000 (117)	16,300 (112)
Finish	1950	1065	22,000 (152)	21,600 (149)	20,200 (139)

Table A.3 Estimated loads in different cross sections of the connecting rod forging

Forging	Loads in section, lb			Total	
	A-A	B-B	C-C	lb	Tons
Blocker	337,214	196,998	89,285	623,497	311.7
Finish	407,234	285,509	123,947	816,690	408.3

Table A.4 Summary and comparison of forging loads (tons)

Forging	Simple analysis	Experimental results
Blocker	311.7	320.67
Finish	408.3	425.00

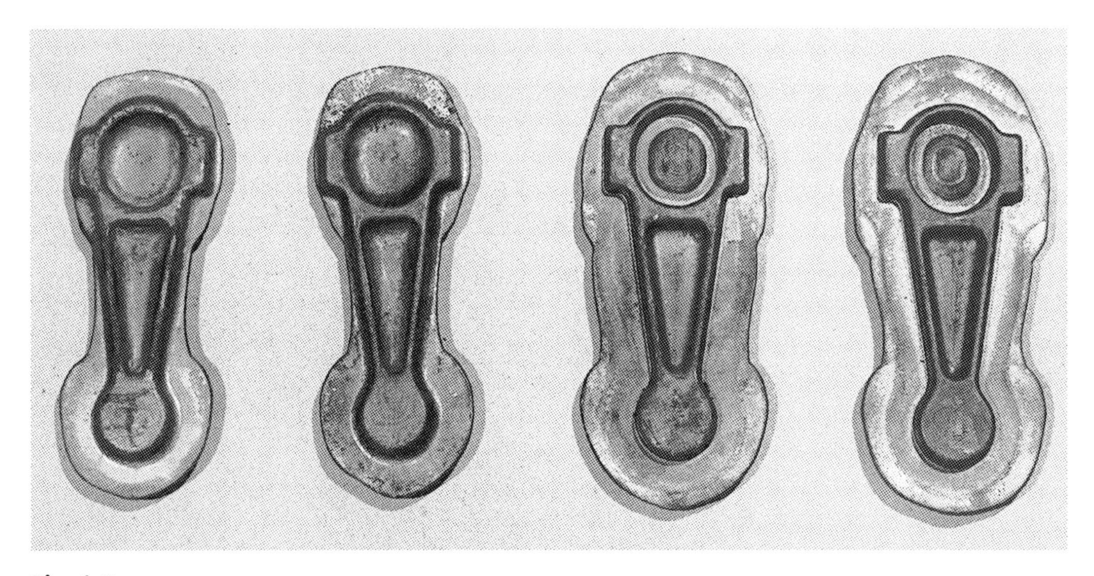

Fig. A.5 Parts that were blocker and finish forged in forging trials

yoke-type mechanical press. Both the blocker and finisher dies were mounted side by side on the press bolster. The dies were lubricated by spraying with Acheson's Delta-forge 105 (Acheson Colloids Co.). The billets were heated in an induction coil to 2100 °F (1150 °C). The dies were heated to 350 °F (175 °C) by infrared gas-fired burners. A typical load and displacement recording is shown in Fig. A.4. The displacement curve shows the position of the ram as a function of time. It is observed that the forging load starts increasing when the upper die contacts the workpiece. The entire forging operation takes place in less than 100 ms. Example forgings are shown in Fig. A.5.

The forging loads, measured in experiments and predicted by the simplified slab method, are compared in Table A.4. The experimental values represent averages of several measurements. It can be seen that the results of the simplified analysis are within practical engineering accuracy. For simple to moderately complex forgings, this analysis can be used effectively for die material selection and for press selection. However, it should be noted that the accuracy of the final results depends largely on proper estimation of the flow stress and frictional shear factor. Some experience and knowledge of forging analyses is necessary to make these estimates with acceptable accuracy. If the capabilities of a high-speed computer are available, then detailed calculations of flow stress, forging stress, and forging load can be made accurately.

REFERENCES

[Altan et al., 1973]: Altan, T., Boulger, F.W., Becker, J.R., Akgerman, N., Henning, H.J., *Forging Equipment, Materials and Practices,* Batelle, 1973.

[Altan et al., 1983]: Altan, T., Oh, S.-I., Gegel, H.L., *Metal Forming Fundamentals and Applications,* American Society for Metals, 1983.

[Bruchanov et al., 1955]: Bruchanov, A.N., and Rebelski, A.V., *Closed-Die Forging and Warm Forging,* Verlag Technik, 1955 (German translation from Russian).

[Haller, 1971]: Haller, H.W., *Handbook of Forging,* Carl Hanser Verlag, 1971 (in German).

[Lange et al., 1977]: Lange, K., and Meyer-Nolkemper, H., *Closed-Die Forging,* Springer-Verlag, 1977 (in German).

[Neuberger et al., 1962]: Neuberger, F., and Pannasch, S., "Material Consumption in Die Forging of Steel," *Fertiegungstechnik und Betrieb,* Vol. 12, 1962, p 775–779 (in German).

[Sabroff et al., 1968]: Sabroff, A.M. et al., *Forging Materials and Practices,* Reinhold, 1968.

[Spies, 1959]: Spies, K., "Preforming in Forging and Preparation of Reducer Rolling," Doctoral dissertation, University of Hannover, 1959.

[Subramanian et al., 1980]: Subramanian, T.L., and Altan, T., "Practical Method for Estimating Forging Loads with the Use of a Programmable Calculator," *Journal of Applied Metalworking,* Vol. 1, No. 2, Jan. 1980, p 60.

[Teterin et al., 1968]: Teterin, G.P., and Tarnovskij, I.J., "Calculation of Plastic Dimensions in Forging Axisymmetric Parts in Hammers," *Kuznechno-Stampovochnoe Proizvodstvo,* Vol 5, 1968, p 6 (in Russian).

[Vasquez et al., 2000]: Vasquez, V., and Altan, T., "New Concepts in Die Design—Physical and Computer Modeling Applications," *Journal of Materials Processing Technology,* Vol. 98, 2000, p 212–223.

[Vieregge, 1968]: Vieregge, K., "Contribution to Flash Design in Closed Die Forging," Doctoral dissertation, Technical University of Hannover, 1969.

SELECTED REFERENCES

- **[Brucelle et al., 1999]:** Brucelle, O., and Bernhart, G., "Methodology for Service Life Increase of Hot Forging Tools," *Journal of Material, Processing Technology,* Vol. 87, 1999, p 237.
- **[Feldman, 1977]:** Feldman, H.D., *Cold Extrusion of Steel,* Merkblatt 201, Düseldorf, 1977.
- **[Jenkins et al., 1989]:** Jenkins, B.L., Oh, S.I., Altan, T., "Investigation of Defect Formation in a 3-Station Closed Die Forging Operation," *CIRP Annals,* Vol. 381, p 243.
- **[Sagemuller, 1968]:** Sagemuller, F., "Cold Impact Extrusion of Large Formed Parts," *Wire,* No. 95, June, 1968, p. 2.
- **[Shirgaokar et al., 2002]:** Shirgaokar, M., Ngaile, G., Altan, T., "Multi-Stage Forging Simulations of Aircraft Components," ERC/NSM-02-R-84, Engineering Research Center for Net Shape Manufacturing, 2002.
- **[Snaith et al., 1986]:** Snaith, B., Probert, S.D., O'Callaghan, P.W., "Thermal Resistances of Pressed Contacts," *Appl. Energy,* Vol. 22, 1986, p 31–84.

CHAPTER 15

A Simplified Method to Estimate Forging Load in Impression-Die Forging

Hyunjoong Cho

15.1 Introduction

In hot impression-die forging, forging load and die stresses are important variables that affect die life and determine the selection of press capacity. During the die design and process planning stage, it is necessary to estimate these variables to avoid unexpected die failure and provide for necessary forging load to fill the die cavity.

Forging load may be estimated by experience-based values, i.e., by multiplying the plan area of the forging with an empirical pressure value, for example, 60 to 100 ksi (415 to 690 MPa) for forging steels and 20 to 30 ksi (140 to 205 MPa) for forging aluminum alloys, where higher values are used for thinner forgings and stronger alloys.

The forging load and die stresses may be calculated using finite-element analysis that requires a rather sophisticated software package, elaborate input data preparation, and considerable computer time. Therefore, it is often desirable to use a simple method for making quick estimates by using the so called "slab method" of analysis. This method takes into account the effects of material properties, friction and heat transfer at die/material interface, press ram speed, flash dimensions, forging geometry, and billet and die temperatures.

This method may help the designer to understand the effects of several forging parameters, such as flash dimensions, forging temperature, and friction (lubrication) on the forging load. For example, if the effect of flash dimensions on the forging load is known, then it is possible to reduce the forging pressure, within limits, by increasing the flash thickness or reducing the flash width. The effect of stock temperature on the forging load, if it is quantitatively known, can assist in optimizing the forging conditions. In some instances, it may be possible to forge at lower stock temperatures. As a result, heating costs and scale formation can be reduced.

15.2 Effect of Process Parameters on Forging Load

During the forging process, metal flow, die fill, and forging load are largely determined by the flow stress of the forging materials, the friction and cooling effect at the die/material interface, and the complexity of the forging shape. The interrelationships of the most significant forging variables are illustrated in the block diagram in Fig. 15.1.

For a given part, the factors that affect the forging load, briefly discussed in Fig. 15.1, were discussed in detail in Chapter 14, "Process Design in Impression-Die Forging." The factors include:

- The flow stress of the forged material as function of strain (or amount of deformation), strain rate (rate of deformation), and forging temperature

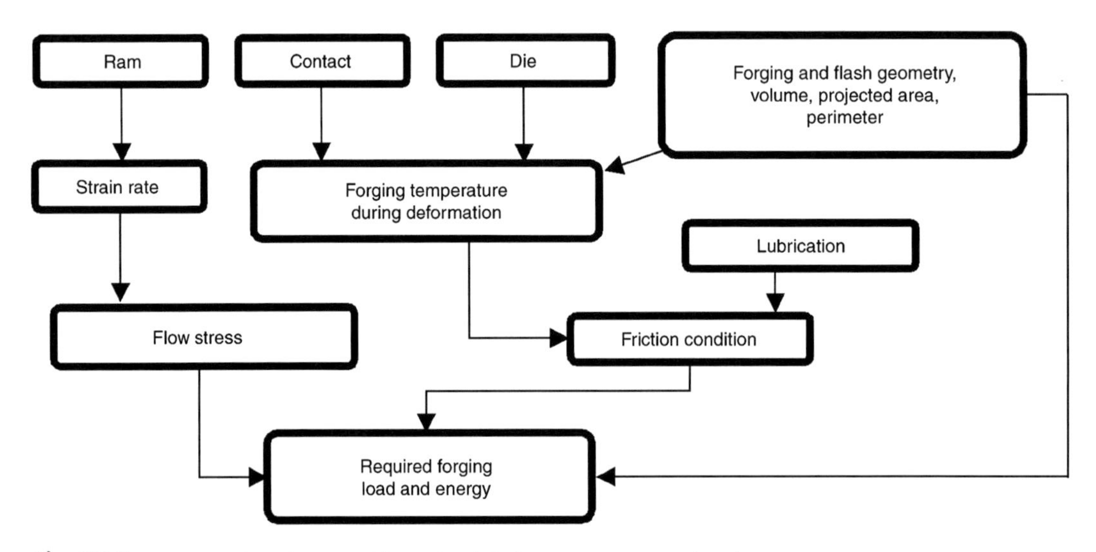

Fig. 15.1 Interaction of significant variables in closed-die forging process [Nagpal et al., 1975]

- Friction and heat transfer at the part/die interface
- Geometric complexity of the part and the number of forging operations used (preblocker, blocker, and finisher)
- Flash design, flash thickness and width (or flash land)

15.3 Methods for Load Estimation

In impression-die forging, the forging load is the maximum at the end of the forging stroke.

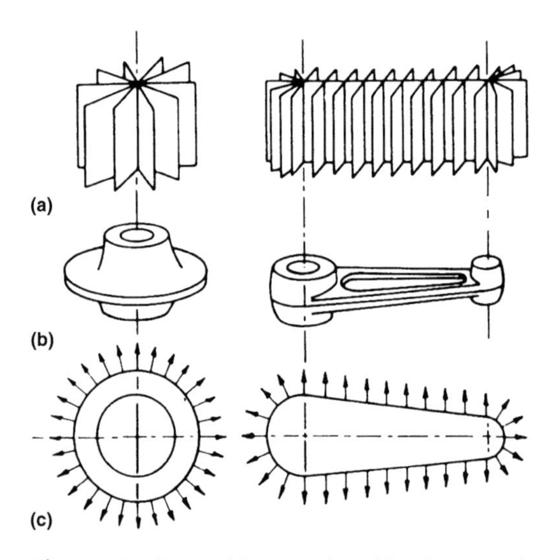

Fig. 15.2 Planes and directions of metal flow for two simple shapes. (a) Planes of flow. (b) Finish forging. (c) Directions of flow [Altan et al., 1983]

As discussed in Chapter 14, "Process Design in Impression-Die Forging," four broadly defined methods are used in estimating this maximum load in hot forging.

- Applied experience: The load is estimated based on available data from previous forging of similar parts. These estimations are very conservative and lead to significant errors.
- Empirical methods: Use simple empirical formulas and graphs or monographs to estimate the load for simple forging operations. These methods are quick but limited in their accuracy and are not sufficiently general to predict forging load for a variety of parts and materials.
- Analytical methods: A forging is viewed as being composed of several unit components.

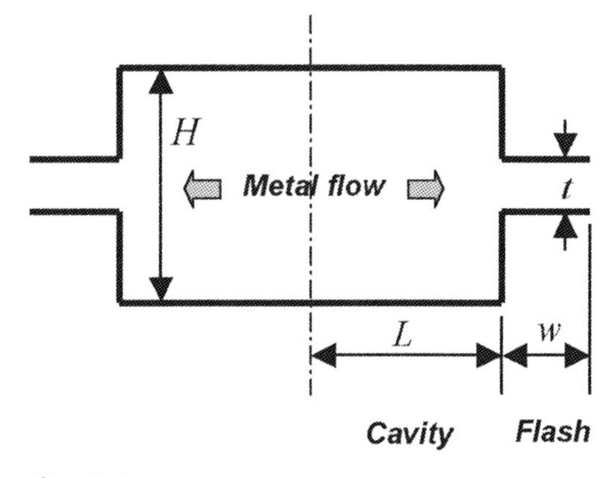

Fig. 15.3 Schematic of simple impression-die forging

Forces and stresses are calculated for every unit and then added together to calculate the total forging load and stresses. The slab method is the most widely known technique to perform this type of analysis. It is successfully used for predicting forging loads and stresses with acceptable engineering accuracy.

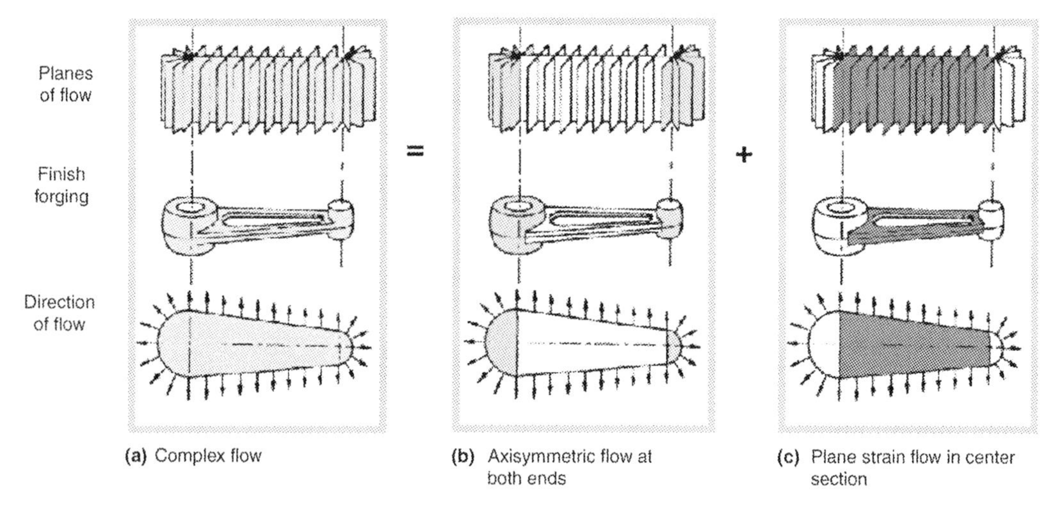

Fig. 15.4 Approximate load calculation for a complex part

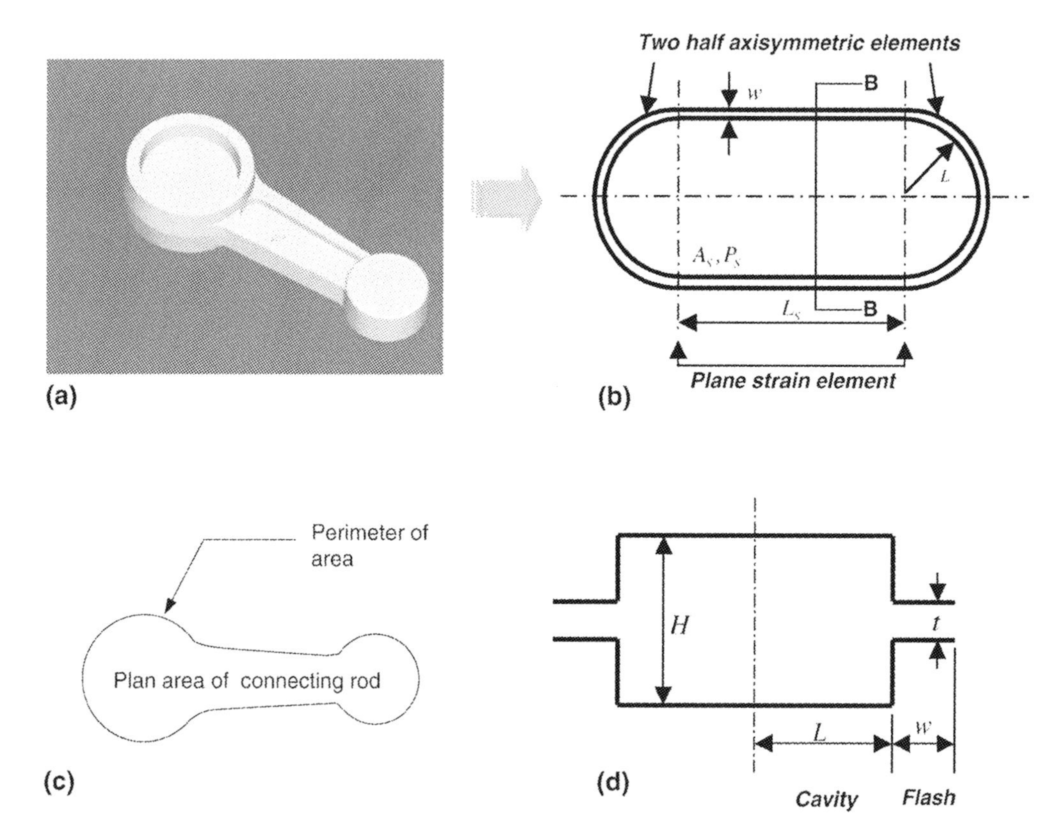

Fig. 15.5 Transformation of a complex forging part into a simplified model. (a) Connecting rod (example of complex forging). (b) Simplified model of the actual forging for forging load estimation [Mohammed et al., 1999]. (c) Plan area of connecting rod and perimeter of plan area. (d) Cross section of simplified model (section B-B)

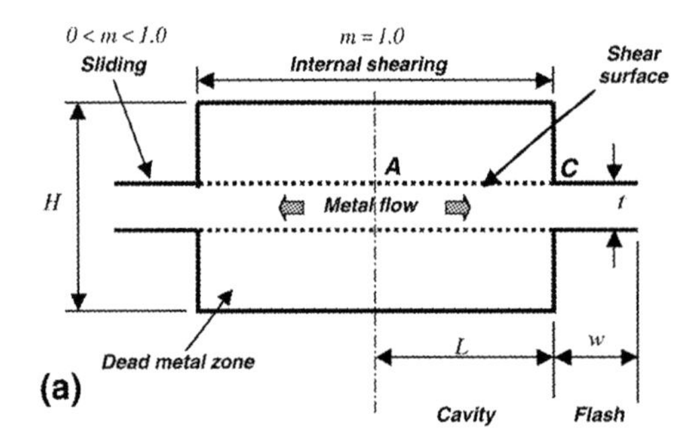

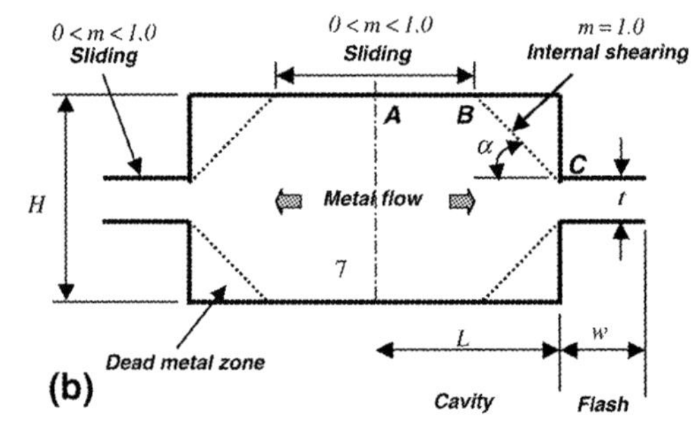

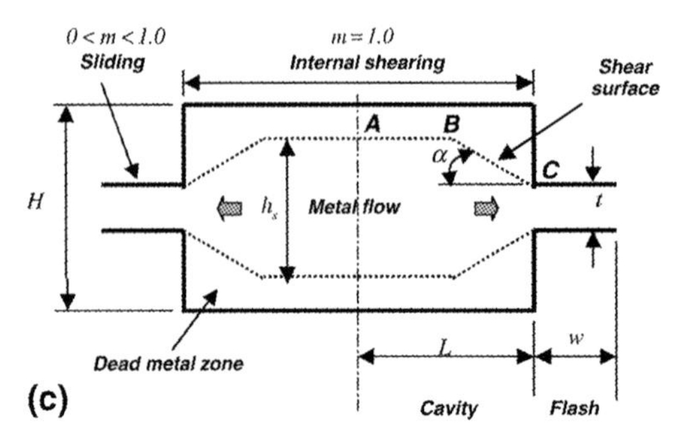

Fig. 15.6 Possible modes of metal flow at the end of forging stroke in impression-die forging. (a) Fictitious disk shearing. (b) Sliding in the central portion of the cavity. (c) Complete shearing in the cavity

Table 15.1 Derived equations for load calculation

Flash load	Cavity load	Stress at the cavity entrance
Fictitious disk shearing		
Plane strain		
$P_f = 2\left(\frac{2}{\sqrt{3}} + \frac{m}{\sqrt{3}}\frac{w}{t}\right) w\bar{\sigma}_f$	$P_c = 2\left(\frac{\bar{\sigma}_c}{\sqrt{3}}\frac{L}{t} + \sigma_{ce}\right)L$	$\sigma_{ce} = \left(\frac{2}{\sqrt{3}} + \frac{2m}{\sqrt{3}}\frac{w}{t}\right)\bar{\sigma}_f$
Axisymmetric		
$P_f = 2\pi\bar{\sigma}_f\left(-\frac{2m}{\sqrt{3}}\left(\frac{R^3 - L^3}{3t}\right) + \left(1 + \frac{2m}{\sqrt{3}}\frac{R}{t}\right)\left(\frac{R^2 - L^2}{2}\right)\right)$	$P_c = 2\pi L^2\left(\frac{\bar{\sigma}_c}{3\sqrt{3}}\frac{L}{t} + \frac{\sigma_{ce}}{2}\right)$	$\sigma_{ce} = \left(1 + \frac{2m}{\sqrt{3}}\frac{w}{t}\right)\bar{\sigma}_f$
Sliding in the cavity center		
Plane strain		
$P_f = 2\left(\frac{2}{\sqrt{3}} + \frac{m}{\sqrt{3}}\frac{w}{t}\right) w\bar{\sigma}_f$	$P_c = 2(K_p\bar{\sigma}_c t + \sigma_{ce}L)$	$\sigma_{ce} = \left(\frac{2}{\sqrt{3}} + \frac{2m}{\sqrt{3}}\frac{w}{t}\right)\bar{\sigma}_f$
Axisymmetric		
$P_f = 2\pi\bar{\sigma}_f\left(-\frac{2m}{\sqrt{3}}\left(\frac{R^3 - L^3}{3t}\right) + \left(1 + \frac{2m}{\sqrt{3}}\frac{R}{t}\right)\left(\frac{R^2 - L^2}{2}\right)\right)$	$P_c = (K_a\bar{\sigma}_c t^2 + \pi\sigma_{ce}L^2)$	$\sigma_{ce} = \left(1 + \frac{2m}{\sqrt{3}}\frac{w}{t}\right)\bar{\sigma}_f$
Complete shearing in the cavity		
Plane strain		
$P_f = 2\left(\frac{2}{\sqrt{3}} + \frac{m}{\sqrt{3}}\frac{w}{t}\right) w\bar{\sigma}_f$	$P_c = 2(K_{ps}\bar{\sigma}_c t + \sigma_{ce}L)$	$\sigma_{ce} = \left(\frac{2}{\sqrt{3}} + \frac{2m}{\sqrt{3}}\frac{w}{t}\right)\bar{\sigma}_f$
Axisymmetric		
$P_f = 2\pi\bar{\sigma}_f\left(-\frac{2m}{\sqrt{3}}\left(\frac{R^3 - L^3}{3t}\right) + \left(1 + \frac{2m}{\sqrt{3}}\frac{R}{t}\right)\left(\frac{R^2 - L^2}{2}\right)\right)$	$P_c = (K_{as}\bar{\sigma}_c t^2 + \pi\sigma_{ce}L^2)$	$\sigma_{ce} = \left(1 + \frac{2m}{\sqrt{3}}\frac{w}{t}\right)\bar{\sigma}_f$

Note: The factors K_p, K_a, K_{ps}, and K_{as} are determined from L/t and H/t ratios.

- Numerical methods: The finite-element method (FEM) is the most widely used method in this field. The major advantage of this method is its ability to generalize its applicability to various problems with little restriction on workpiece geometry. The FEM is able to analyze metal flow during hot forging and is able to predict instantaneous strains, stresses, and temperatures within the deforming metal. The disadvantage of the FEM is a large amount of computation time and expensive system requirements, depending on the problem.

The advantage of the slab method is that a complex forging can be divided into basic deformation units or blocks, and these could be analyzed separately. Then, by putting the deformation units together, loads for any complex shape forging can be determined. Metal flow for the deformation units is assumed to be either axisymmetric or plane strain. The plane-strain flow occurs in relatively long forgings where the deformation along the length is relatively small and can be neglected. Axisymmetric flow is encountered in round forgings and at the end of long forging where the metal flows radially toward the flash. Figure 15.2 shows examples of plane-strain and axisymmetric flow in complex forgings.

Even though the slab method cannot be as accurate as the FEM because of the assumptions made in developing the mathematical approach, it is still attractive because it does not require considerable computation time and does not require training for the user. In addition, the slab

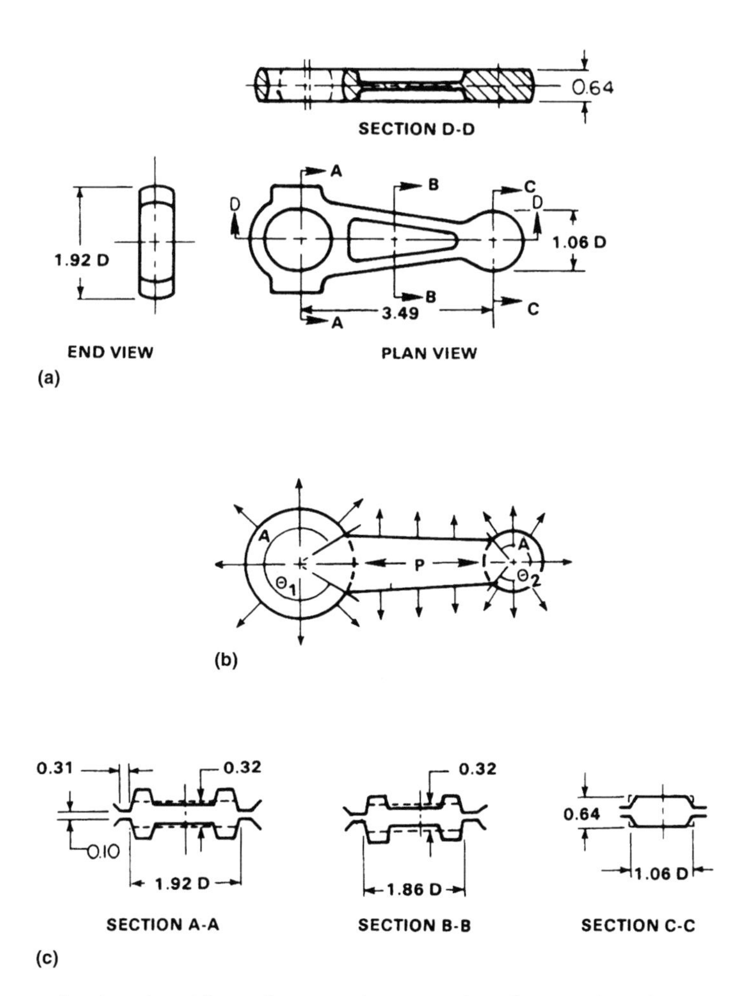

Fig. 15.7 Geometry, directions of metal flow and representative cross sections of a connecting rod: (a) cross-sectional views of the connecting rod, (b) directions of metal flow (A = axisymmetric, P = plane strain), (c) representative sections and their simplification

Table 15.2 Inputs for the load estimation of connecting rod

Material data	
Flow stress	SS 304
Specific heat, Btu/lb·°F (J/kg·K)	0.116 (486)
Density, lb/in.^{3} (g/cm^3)	0.285 (7.8)
Equipment data	
Press type	Mechanical
Press speed, rpm	50
Press stroke, in. (cm)	14 (36)
Geometry data	
Initial billet height, in. (mm)	1.12 (28)
Flash thickness, in. (mm)	0.1 (2.5)
Flash width, in. (mm)	0.3 (7.6)
Perimeter, in. (cm)	13.67 (34.7)
Projected area, in. (cm)	6.55 (16.6)
Cavity height, in. (mm)	0.56 (14.2)
Interface data	
Friction factor	0.3
Heat-transfer coefficient	0.0039 Btu/in.^{2}/s/°F (3.1889 W/m^2·°K)
Initial billet temperature, °F (°C)	2050 (1120)
Initial die temperature, °F (°C)	400 (205)

method can show the trend in the calculation of forging load while the FEM only provides the final calculated results.

15.4 A Simplified Method for Load Estimation

The present method, used in practical prediction of forging load, works on a simplified model of impression-die forging and uses the slab analysis technique. It is assumed that for all cross sections of forging, the cavity is rectangular and has the flash geometry, as illustrated in Fig. 15.3. In actual practice, where the cavity is not rectangular, the cross section is simplified to conform to this model. The load calculation is simplified by dividing a complex forging in simpler components where metal flow is either axisymmetric or plane strain, as shown in Fig. 15.4. Then the load is estimated by adding the loads calculated for each component. In load estimation, to take into account the change of flow

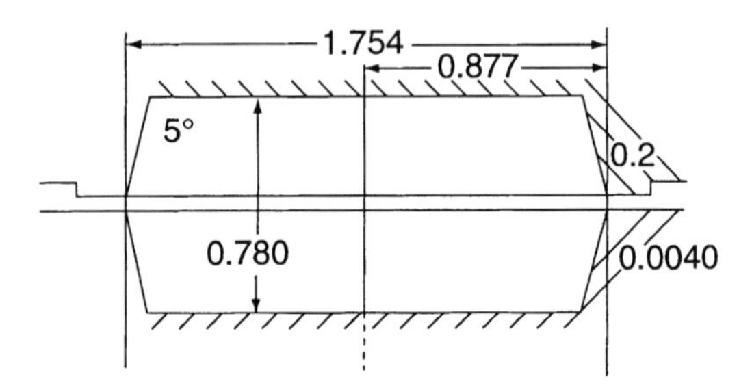

Fig. 15.8 Lead disk forging

stress due to temperature change during the hot forging operation, heat-transfer analysis is conducted. The advantage of this approach is that the effect of certain process parameters on the forging load can be investigated quickly. This enables an industrial designer to know the trend of design variables by conducting a parametric study when he is designing a new forging operation. For example, the effect of dimensional parameters such as flash thickness, t, and width, w, on the total forging load can be easily obtained (Fig. 15.3).

15.4.1 Simplification of Forging Geometry

Any arbitrary three-dimensional forging is transformed into a simplified forging model. This model, including the simplified plane-strain and axisymmetric metal flows, allows conducting slab analysis for load estimation. As shown in Fig. 15.5(b), it is assumed that a simplified forging model is divided into an axisymmetric component of radius L and a plane-strain component of length L_s. The plan area, A_s, of a simplified model is equal to the plan area of the actual forging and is represented by $A_s = \pi L^2 + 2L_sL$. Also, the perimeter, P_s, of a simplified model is equal to the perimeter of the plan area of the actual forging and is expressed by $P_s = 2\pi L + 2L_s$. A_s and P_s of an actual forging part can be determined from any solid modeling software commonly used in industry. Once A_s

Table 15.3 Estimated flow stresses and forging loads

	Axisymmetric portion		Plane-strain portion	
Component	**Cavity**	**Flash**	**Cavity**	**Flash**
Temperature, °F (°C)	2,055 (1124)	2,013 (1101)	2,063 (1128)	2,015 (1102)
Flow stress, psi (MPa)	15,298 (105)	18,467 (127)	14,958 (103)	18,409 (127)
Forging load, tons	34	18	241	47
	52 (cavity + flash)		288 (cavity + flash)	
Total forging load, tons	340 (axisymmetric + plane strain)			

and P_s are known, L and L_s are found. The cross section of the simplified model is simplified, as shown in Fig. 15.5(d). The cavity height is denoted by H and the radius (or half width of the cavity) by L, the flash thickness by t and the flash width by w. The cavity height, H, is the average height of the actual forging and is obtained by dividing the volume of the forging without flash by its plan area.

15.4.2 Metal Flow in the Cavity

The load estimation is made for the final stage of forging operation when the die is totally filled and the load has its maximum value. In the impression-die forging, an approximate metal flow in the die cavity at the final forging stage depends on the cavity height, the cavity width, and flash thickness.

For a large flash thickness (cavity width to cavity thickness ratio is greater than 2), material is assumed to flow into the flash by shearing along a fictitious disk having the same thickness as the flash, as shown in Fig. 15.6(a).

If the die cavity height is small in relation to its width, the material forms a dead metal zone at the die corners, and the material slides along AB and flows by internal shearing along the line BC (Fig. 15.6b). At the surface BC, the friction factor is then 1.

For a high cavity height to cavity width ratio, the material forms a shear surface ABC, and a sticking friction condition is assumed at this surface (friction factor, $m = 1$) (Fig. 15.6c). The shearing height, h_s, and shearing angle, α, is determined analytically. In the present method, the type of metal flow in the die cavity is determined based on the principle that material flows in a manner that consumes a minimum amount of plastic deformation energy. Table 15.1 shows the equations derived for each type of metal flow to calculate the load components.

Table 15.4 Comparison of forging load

	ForgePAL	Nagpal and Altan	Experimental results
Total forging load, tons	340	312	320

15.5 Example of Load Estimation

The introduced load estimation method has been programmed at the Engineering Research Center for Net Shape Manufacturing. ForgePAL is a computer program running on the programmable controller and calculates the forging load of impression-die forging. Examples of ForgePAL were shown in this section. A copy of ForgePAL is included in Appendix 15.A, given on the CD attached to this book.

15.5.1 Connecting Rod Forging

As discussed earlier in Appendix A of Chapter 14, Nagpal and Altan [Nagpal et al., 1975] calculated the forging load for the stainless steel connecting rod. The same dimensions are used in ForgePAL to estimate the forging load (Fig. 15.7). The other necessary forging conditions are approximated here. Based on the geometry transformation rule, the connecting-rod part was simplified. The perimeter, plan area, and volume of the forging were found by using a solid modeling software, and the cavity height was calculated (Table 15.2). However, this average height may not be realistic for certain forging parts. If the profile of the die is too complex (changing greatly in height), simply dividing volume by plan area may yield poor results. In such cases, the user should experiment with results and try to guess a more realistic height input.

Simplification of Forging Geometry. With the perimeter and projected area of the connecting rod, the half cavity length, L, and the depth of the plane-strain component, L_s, were found as follows:

$$L = 0.5482 \text{ in. and } L_s = 5.113 \text{ in.}$$

Estimation of Flow Stress. Using the material input data, such as density, specific heat, and heat-transfer coefficient of SS 304, ForgePAL first calculated the average forging temperature required for estimating the flow stress value at the end of the forging stroke. The results of the calculations are summarized in Table 15.3.

Table 15.5 Geometry input and estimated forging load for a lead disk forging

		...	ForgePAL	Measured
Flash thickness, t, in. (mm)	0.004 (0.10)	Flash load, ton	7	N/A
Flash width, w, in. (mm)	0.2 (5.1)	Cavity load, ton	35	N/A
Cavity height, h_f, in. (mm)	0.780 (19.8)	Total load, ton	42	41
Half die length, L, in. (mm)	0.877 (22.3)			

Estimation of Forging Load. The load calculation method for a complex part is based on the addition of the loads calculated for each component. The forging load of each component consists of the cavity load and the flash load. Finally, the total forging load for the connecting rod is equal to the combination of the load for the axisymmetric and plane-strain portions. As is shown in Table 15.4, a total forging load of 340 tons is predicted and its results are compared with experimental results.

15.5.2 Forging of a Lead Disk with Flash

[Schey et al.] measured separately the flash and the cavity loads in lead disk forging, as shown in Fig. 15.8. In this reference, the flow stress values in the cavity, $\bar{\sigma}_c$, and in the flash, $\bar{\sigma}_f$, are given as:

$$\bar{\sigma}_c = 2{,}300 \text{ psi and } \bar{\sigma}_f = 5{,}300 \text{ psi}$$

The part is symmetric and round about the center axis with flash at the periphery. Thus, the axisymmetric model is selected for load calculation in ForgePAL. The necessary axisymmetric geometry inputs obtained from Fig. 15.8 and estimated forging loads are summarized in Table 15.5.

REFERENCES

[Altan et al., 1983]: Altan, T., Oh, S.I., Gegel, H., *Metal Forming Fundamentals and Applications,* American Society for Metals, 1983.

[Mohammed et al., 1999]: Asaduzzaman, M., Demir, A., Vazquez, V., Altan, T., "Development of Computer Program for Estimation of Forging Load and Die Pressures in Hot Forging," Report No. F/ERC/NSM-99-R-18, Engineering Research Center for Net Shape Manufacturing, The Ohio State University, 1999.

[Nagpal et al., 1975]: Nagpal, V., and Altan, T., "Estimation of Forging Load in Closed-Die Forging," Battelle Columbus Laboratories, 1975.

[Schey et al.]: Schey, J.A., et al., "Metal Flow in Closed-Die Press Forming of Steel," Research Report to American Iron and Steel Institute, IIT Research Institute, Chicago, Illinois.

SELECTED REFERENCES

- **[Douglas et al., 1989]:** Douglas, J.R., and Altan, T., "Flow Stress Determination for Metals at Forging Rates and Temperatures," *Trans. ASME, J. Eng. Industry,* Feb. 1975, p 66, 1989.
- **[Lange, 1985]:** Lange, K., *Handbook of Metal Forming,* McGraw-Hill Book Company, 1985.
- **[Schultes et al., 1981]:** Schultes, T., Sevenler, K., Altan, T., "Prediction of Forging Load and Stresses Using a Programmable Calculator," Topical Report No. 4, Battelle Columbus Laboratories, 1981.
- **[Subramanian et al., 1980]:** Subramanian, T.L., and Altan, T., "Practical Method for Estimating Forging Loads with the Use of a Programmable Calculator," *Journal of Applied Metal Working,* No. 2, Jan. 1980, p 60.

CHAPTER 16

Process Modeling in Impression-Die Forging Using Finite-Element Analysis

Manas Shirgaokar
Gracious Ngaile
Gangshu Shen

16.1 Introduction

Development of finite-element (FE) process simulation in forging started in the late 1970s. At that time, automatic remeshing was not available, and therefore, a considerable amount of time was needed to complete a simple FE simulation [Ngaile et al., 2002]. However, the development of remeshing methods and the advances in computational technology have made the industrial application of FE simulation practical. Commercial FE simulation software is gaining wide acceptance in the forging industry and is fast becoming an integral part of the forging design and development process.

The main objectives of the numerical process design in forging are to [Vasquez et al., 1999]:

- Develop adequate die design and establish process parameters by:
  a. Process simulation to assure die fill
  b. Preventing flow-induced defects such as laps and cold shuts
  c. Predicting processing limits that should not be exceeded so that internal and surface defects are avoided
  d. Predicting temperatures so that part properties, friction conditions, and die wear can be controlled
- Improve part quality and complexity while reducing manufacturing costs by:
  a. Predicting and improving grain flow and microstructure
  b. Reducing die tryouts and lead times
  c. Reducing rejects and improving material yield
- Predict forging load and energy as well as tool stresses and temperatures so that:
  a. Premature tool failure can be avoided.
  b. The appropriate forging machines can be selected for a given application.

Process modeling of closed-die forging using finite-element modeling (FEM) has been applied in aerospace forging for a couple of decades [Howson et al., 1989, and Oh, 1982]. The goal of using computer modeling in closed-die forging is rapid development of right-the-first-time processes and to enhance the performance of components through better process understanding and control. In its earlier application, process modeling helped die design engineers to preview the metal flow and possible defect formation in a forging. After the forging simulation is done, the contours of state variables, such as effective strain, effective strain rate, and temperature at any instant of time during a forging, can be generated. The thermomechanical histories of selected individual locations within a forging can also be tracked [Shen et al., 1993]. These functions of process modeling provided an insight into the forging process that was not available in the old days. Integrated with the process modeling, microstructure modeling is a new area that has a bright future [Sellars, 1990, and Shen et al., 2000]. Microstructure modeling allows the right-the-first-time optimum metallurgical features of the forging to be previewed on the computer. Metallurgical aspects of forging, such as grain size and precipitation, can be predicted with reasonable accuracy using computational tools prior to committing the forging to shop trials. Some of the proven practical applications of process simulation in closed-die forging include:

- Design of forging sequences in cold, warm, and hot forging, including the prediction of forming forces, die stresses, and preform shapes
- Prediction and optimization of flash dimensions in hot forging from billet or powder metallurgy preforms
- Prediction of die stresses, fracture, and die wear; improvement in process variables and die design to reduce die failure
- Prediction and elimination of failures, surface folds, or fractures as well as internal fractures
- Investigation of the effect of friction on metal flow
- Prediction of microstructure and properties, elastic recovery, and residual stresses

16.2 Information Flow in Process Modeling

It is a well-known fact that product design activity represents only a small portion, 5 to 15%, of the total production costs of a part. However, decisions made at the design stage determine the overall manufacturing, maintenance, and support costs associated with the specific product. Once the part is designed for a specific process, the following steps lead to a rational process design:

1. Establish a preliminary die design and select process parameters by using experience-based knowledge.
2. Verify the initial design and process conditions using process modeling. For this purpose it is appropriate to use well-established commercially available computer codes.
3. Modify die design and initial selection of process variables, as needed, based on the results of process simulation.
4. Complete the die design phase and manufacture the dies.
5. Conduct die tryouts on production equipment.
6. Modify die design and process conditions, if necessary, to produce quality parts.

Hopefully, at this stage little or no modification will be necessary, since process modeling is expected to be accurate and sufficient to make all the necessary changes before manufacturing the dies.

Information flow in process modeling is shown schematically in Fig. 16.1 [Shen et al., 2001]. The input of the geometric parameters, process parameters, and material parameters sets up a unique case of a closed-die forging. The modeling is then performed to provide information on the metal flow and thermomechanical history of the forging, the distribution of the state variables at any stage of the forging, and the equipment response during forging. The histories of the state variables, such as strain, strain rate, temperature, etc., are then input to the microstructure model for microstructural feature prediction. All of the information generated is used for judging the closed-die forging case. The nonsatisfaction in any of these areas will require a new model with a set of modified process parameters until the satisfied results are obtained. Then, the optimum process is selected for shop practice.

16.3 Process Modeling Input

Preparing correct input for process modeling is very important. There is a saying in computer modeling: garbage in and garbage out. Sometimes, a time-consuming process modeling is useless because of a small error in input preparation. Process modeling input is discussed in terms of geometric parameters, process parameters, and material parameters [SFTC, 2002].

16.3.1 Geometric Parameters

The starting workpiece geometry and the die geometry need to be defined in a closed-die forging modeling. Depending on its geometrical complexity, a forging process can be simulated either as a two-dimensional, axisymmetric or plane-strain, or a three-dimensional problem. If the process involves multiple stations, the die geometry of each station needs to be provided. A typical starting workpiece geometry for a closed-die forging is a cylinder with or without chamfers. The diameter and the height of the cylinder are defined in the preprocessing stage. A lot of closed-die forgings are axisymmetric, which need a two-dimensional geometry handling. Boundary conditions on specific segments of the workpiece and dies that relate to deformation and heat transfer need to be defined. For example, for an axisymmetric cylinder to be forged in a pair of axisymmetric dies, the nodal velocity in the direction perpendicular to the centerline should be defined as zero, and the heat flux in that direction should also be defined as zero.

16.3.2 Process Parameters

The typical process parameters to be considered in a closed-die forging include [SFTC, 2002]:

- The environment temperature
- The workpiece temperature
- The die temperatures
- The coefficients of heat transfer between the dies and the billet and the billet and the atmosphere
- The time used to transfer the workpiece from the furnace to the dies
- The time needed to have the workpiece resting on the bottom die
- The workpiece and die interface heat-transfer coefficient during free resting
- The workpiece and die interface heat-transfer coefficient during deformation
- The workpiece and die interface friction, etc.

The die velocity is a very important parameter to be defined in the modeling of a closed-die forging. If a hydraulic press is used, depending on the actual die speed profiles, the die velocity can be defined as a constant or series of velocities that decrease during deformation. The actual die speed recorded from the forging can also be used to define the die velocity profile. If a mechanical press is used, the rpm of the flywheel, the press stroke, and the distance from the bottom dead center when the upper die touches the part need to be defined. If a screw press is used, the total energy, the efficiency, and the ram displacement need to be defined. If a hammer is used, the blow energy, the blow efficiency, the mass of the moving ram and die, the number of blows, and the time interval between blows must be defined. Forgings performed in different machines, with unique velocity versus stroke characteristics, have been simulated successfully using the commercial FE software DEFORM (Scientific Forming Technologies Corp.) [SFTC, 2002].

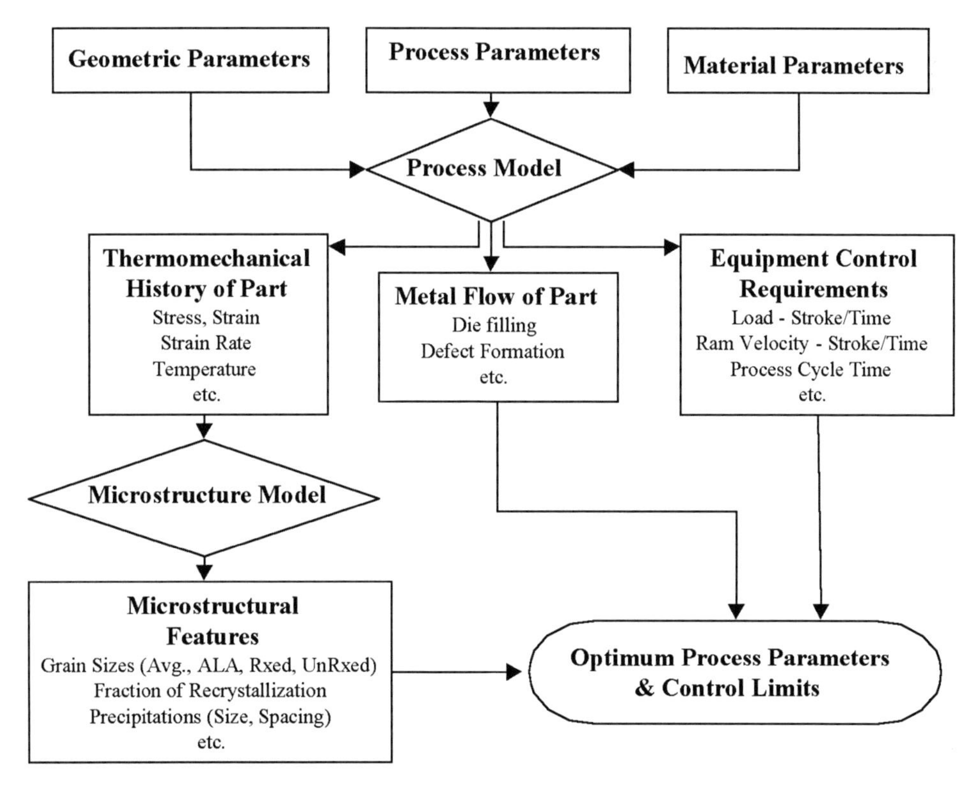

Fig. 16.1 Flow chart of modeling of closed-die forging [Shen et al., 2001]

16.3.3 Tool and Workpiece Material Properties

In order to accurately predict the metal flow and forming loads, it is necessary to use reliable input data. The stress-strain relation or flow curve is generally obtained from a compression test. However, the test is limited in achievable strains. In order to obtain the flow stress at large strains and strain rates, the torsion test can be used or, alternatively, the compression data is extrapolated with care.

In most simulations, the tools are considered rigid; thus, die deformation and stresses are neglected. However, in precision forging operations, the relatively small elastic deformations of the dies may influence the thermal and mechanical loading conditions and the contact stress distribution at the die/workpiece interface. Thus, die stress analysis is a crucial part of process simulation to verify the die design and the forging process parameters.

16.3.4 Interface Conditions (Friction and Heat Transfer)

The friction and heat-transfer conditions at the interface between the die and the billet have a significant effect on the metal flow and the loads required to produce the part. In forging simulations, due to the high contact stresses at the interface between the workpiece and the die, the constant shear friction factor gives better results than the coulomb friction coefficient.

The most common way to determine the shear friction factor in forging is to perform ring compression tests. From these tests, it is possible to estimate the heat-transfer coefficient, flow stress and friction as a function of temperature, strain rate, strain, and forming pressure, as discussed in Chapter 6, "Temperatures and Heat Transfer."

Friction factors measured with the ring compression test, however, are not valid for precision forging processes (hot, warm, and cold) where the interface pressure is very high and the surface generation is large. The friction conditions change during the process due to changes in the lubricant and the temperature at the die/workpiece interface. In such applications, the double cup extrusion test is recommended for estimation of the friction factor, as discussed in Chapter 7, "Friction and Lubrication."

16.3.5 Material Parameters

The closed-die hot forging modeling is a coupled heat-transfer and deformation simulation. Material parameters that relate to both heat transfer and deformation need to be defined. The material parameters commonly used for heat-transfer modeling are the thermal conductivity, heat capacity, and emissivity of the workpiece and die materials. These parameters are usually defined as a function of temperature, The flow stress of the workpiece material is very important for the correct prediction of metal flow behavior. It is usually defined as a function of strain, strain rate, temperature, and possible starting microstructures. The Young's modulus, the Poisson's ratio as a function of temperature, and the thermal expansion of the die materials are important parameters for die stress analysis.

16.4 Characteristics of the Simulation Code

16.4.1 Mesh Generation and Automatic Remeshing

In forging processes, the workpiece generally undergoes large plastic deformation, and the relative motion between the deforming material and the die surface is significant. In the simulation of such processes, the starting mesh is well defined and can have the desired mesh density distribution. As the simulation progresses, the mesh tends to get distorted significantly. Hence, it is necessary to generate a new mesh and interpolate the simulation data from the old mesh to the new one to obtain accurate results. Automated mesh generation (AMG) schemes have been incorporated in commercial FE codes for metal forming simulations. In DEFORM, there are two tasks in AMG: 1) determination of optimal mesh density distribution and 2) generation of the FE mesh based on the given density. The mesh density should conform to the geometrical features of the workpiece at each step of deformation [Wu et al., 1992]. In order to maximize the geometric conformity, it is necessary to consider mesh densities that take into account the boundary curvature and local thickness.

In DEFORM, two-dimensional (2-D) simulations use quadrilateral elements, whereas three-dimensional (3-D) simulations use tetrahedral elements for meshing and automatic remeshing [Wu et al., 1996]. With this automatic remeshing capability, it is possible to set up a simulation model and run it to the end with very little interaction with the user.

16.4.2 *Reliability and Computational Time*

Several FE simulation codes are commercially available for numerical simulation of forging processes, such as DEFORM (2-D and 3-D), FORGE (2-D and 3-D) (Ternion Corp.), Qform (2-D and 3-D), etc. In addition to a reliable FE solver, the accurate and efficient use of metal flow simulations require [Knoerr et al., 1992]:

- Interactive preprocessing to provide the user with control over the initial geometry, mesh generation, and input data; automatic remeshing to allow the simulation to continue when the distortion of the old mesh is excessive; interactive postprocessing that provides more advanced data analysis, such as point tracking and flow line calculation
- Appropriate input data describing the thermal and physical properties of die and billet material the heat transfer and friction at the die/workpiece interface under the processing conditions investigated, and the flow behavior of the deforming material at the relatively large strains that occur in practical forging operations
- Analysis capabilities that are able to perform the process simulation with rigid dies to reduce calculation time and to use contact stresses and temperature distribution estimated with the process simulation using rigid dies to perform elastic-plastic die stress analysis

The time required to run a simulation depends on the computer used and the amount of memory and workload the computer has. However, with today's computers, it is possible to run a 2-D simulation in a couple of hours, while a 3-D simulation can take anywhere between a day to a week, depending on the part complexity [Wu et al., 1996].

16.5 Process Modeling Output

The process modeling provides extensive information of the forging process. The output of process modeling can be discussed in terms of the metal flow, the distribution and history of state variables, the equipment response during forging, and the microstructure of the forging.

16.5.1 *Metal Flow*

The information on metal flow is very important for die design. Improper metal flow produces defects in the forging. In real closed-die forging, it is necessary to wait until the forging is finished to see the forged part and the defect, if there is one. The advantage of computer simulation of forging is that the entire forging process is stored in a database file in the computer and can be tracked. Whether there is a defect formed and how it is formed can be previewed before the actual forging. Figure 16.2 shows the lap formation for a rejected process in the design stage. The lap formation can be eliminated by changing the workpiece geometry (the billet or preform), or the die geometry, or both. The computer modeling can again indicate if the corrective measure works or not.

16.5.2 *Distribution and History of State Variables*

The distribution of the state variables, such as the strain, strain rate, and temperature, at any stage of a closed-die forging can be plotted from the database file saved for the forging simulation. The history of these state variables can also be tracked.

Figure 16.3(a) shows the effective strain distribution of a closed-die forging forged in an isothermal press. The effective strain has a value of 0.4 to 0.9 in the bore die lock region. The region that is in contact with the upper die has an effective strain value of 0.4 to 0.9, and the region that is in contact with the lower die, a value of 0.7 to 0.9. With an effective strain of 2.0 to 2.8, the bore rim transition region has the largest strain. The effective strain value is approximately 1.5 for both the rim and the midheight of the bore region. From the state variable distribution plot, the state variable at a specific stage of the forging is known. This specific stage,

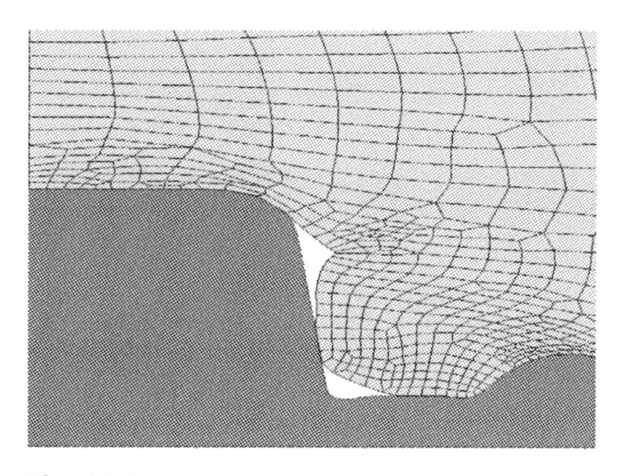

Fig. 16.2 Lap prediction using process modeling tool

shown in Fig. 16.3(a), is the end of the forging. The distribution of the state variables can be plotted for any other stages of forging as well.

Figure 16.3(b) shows the effective strain versus time of a material point located at midheight of the bore section of the forging, as shown in Fig. 16.3(a). In this isothermal forging case, a 20 min deformation time was used, as shown in the figure. The final strain value, 1.5, shown in Fig. 16.3(b) is in agreement with the value shown in the distribution plot in Fig. 16.3(a). The history plot of state variables (strain, strain rate, and temperature) provides valuable information on the thermomechanical history of the forging that determines its mechanical properties.

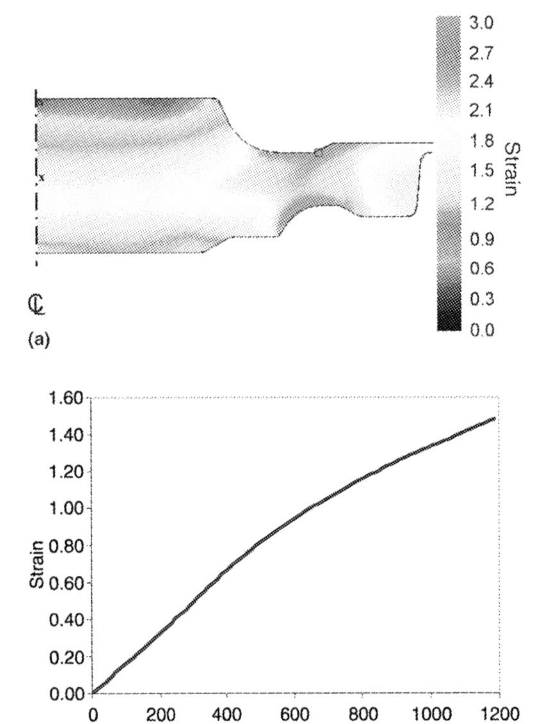

Fig. 16.3 (a) Effective strain distribution and (b) the effective strain history of the center location of a closed-die forging

16.5.3 *Equipment Response/Hammer Forging*

Process modeling also provides the information regarding the response of the equipment. Examples of equipment response discussed here are forging load and ram velocity of hammer forging. The information is usually not available in the hammer shop. However, it is useful for understanding the hammer response to a forging process.

Figure 16.4 shows the load versus stroke predicted for a hammer forging operation. The figure shows that there are eight blows in the hammer operation. Each ends with a zero load. The stroke in the figure is the stroke of the ram/die. The zero stroke refers to the position of the die, where the first die/workpiece contact occurs during forging. This zero position is the same for all of the eight hammer blows. With the increase in the number of blows, the load increases and the stroke per blow decreases. The last blow of the sequence has the shortest stroke. This behavior is very real for hammer forging operations. During a hammer forging operation, the workpiece increases its contact area with the dies, which increases the forging load. The total available blow energy is fixed for a hammer. With the increase in forging load, the length of

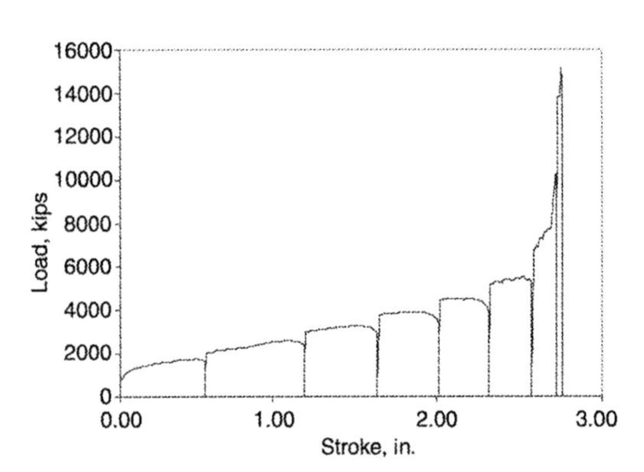

Fig. 16.4 Load versus stroke obtained from a hammer forging simulation

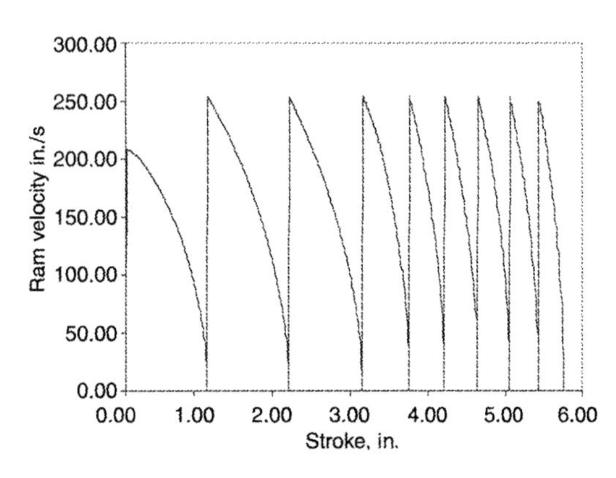

Fig. 16.5 Ram velocity versus stroke obtained from a hammer forging simulation

stroke is reduced. Moreover, the blow efficiency, which is the ratio between the energy used for deformation and the total blow energy, is also reduced with the increase in forging load. Thus, a smaller amount of energy is available toward the end of a blow sequence and with the decrease in the stroke per blow.

Figure 16.5 gives the ram velocity versus stroke obtained from a simulation of another hammer forging process. There are nine blows for this hammer operation. The velocity of the first blow was smaller than the other eight blows, because a soft blow was used initially to locate the workpiece. In a soft blow, there is only a portion of blow energy applied to the workpiece. Thus, the first blow has a smaller starting ram velocity. After the first blow, full energy was applied to the forging. Thus, the starting ram ve-

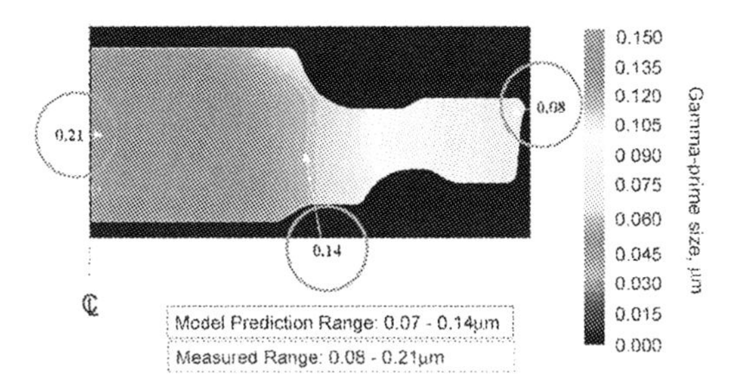

Fig. 16.6 Prediction of the distribution of the size (μm) of gamma prime for a Rene 88 experimental forging

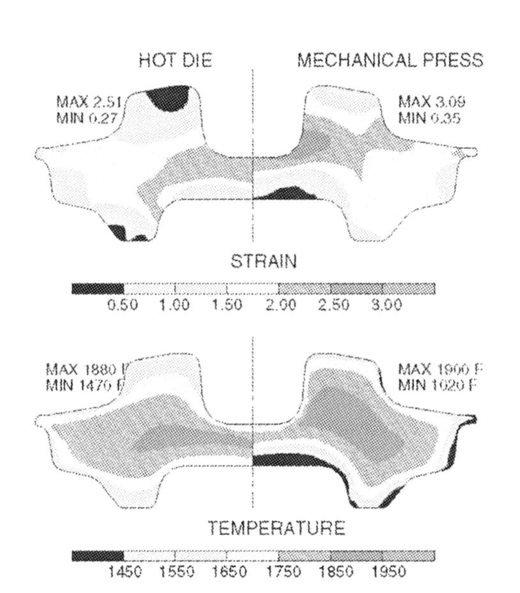

Fig. 16.7 Comparisons of hot-die forging and mechanical press forging of an experimental part using process modeling

Fig. 16.8 Rene 88 experimental part out of forging press [Hardwicke et al., 2000]

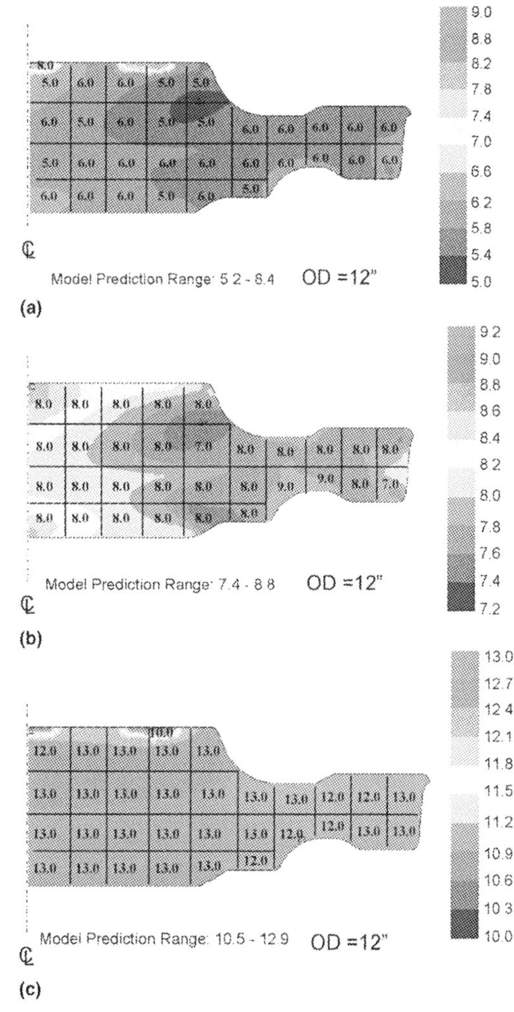

Fig. 16.9 Predicted model and optically measured grain sizes in the three developmental René 88DT disks with (a) coarse, (b) medium, and (c) fine grains [Hardwicke et al., 2000]

locity for the rest of the blows was the same. There is always an energy loss to surroundings in a hammer blow. Therefore, blow efficiency needs to be factored in for each hammer blow. However, the blow efficiency only has an effect after the ram/die workpiece are in contact. Hence, blow efficiency does not influence the starting velocity of the ram/die. It is factored in during the blow. The decay in ram velocity in each blow is a result of both the energy consumption in deforming the workpiece and the energy lost to the surroundings.

16.5.4 Microstructures in Superalloys

Microstructure and property modeling is now the major emphasis in advanced forging process design and improvement, especially in forging aerospace alloys such as nickel and titanium superalloys. The development and utilization of physical metallurgy-based microstructure models and the integration of the models with finite-element analysis has allowed for microstructure prediction by computer. Two important microstructural features of superalloy forgings are the grain size and the gamma-prime precipitation. The grain size modeling is discussed in detail in Chapter 19, "Microstructure Modeling in Superalloy Forging." The prediction of gamma-prime distribution is discussed here. Gamma prime is a very important precipitation phase in strengthening superalloys. The size and spacing are two features of interest in gamma-prime precipitation. Figure 16.6 shows the prediction of the distribution of the size of gamma prime of an experimental nickel-base superalloy forging, Rene 88, coupled with a few measurement points. The measurement made is in the range of 0.07 to 0.21 μm. The model predicts a range of 0.08 to 0.14 μm. The fine gamma prime was correctly predicted and the coarser gamma prime was underpredicted, which pointed out the need for further improvement of the gamma-prime model. The microstructure prediction feature is useful for the process development for closed-die forging.

16.6 Examples of Modeling Applications

One of the major concerns in the research of manufacturing processes is to find the optimum production conditions in order to reduce production costs and lead-time. In order to optimize a process, the effect of the most important process parameters has to be investigated. Conducting experiments can be a very time-consuming and expensive process. It is possible to reduce the number of necessary experiments by using FEM-based simulation of metal forming processes.

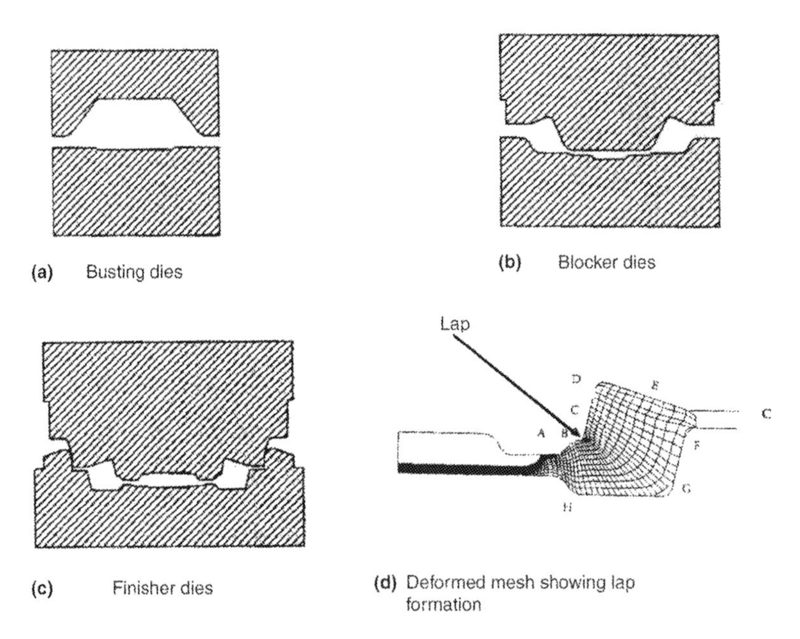

Fig. 16.10 Investigation of defects in ring gear forging using FEM [Jenkins et al., 1989]

16.6.1 Process Modeling for Equipment Selection

Figure 16.7 illustrates an example of the application of process modeling to select the most suitable equipment (a hot-die hydraulic press or a mechanical press) for forging a superalloy part in a hot-die hydraulic press or a mechanical press. Effective strain and temperature distributions are compared in this figure. There was a preferred strain and temperature window for the selection of the process to meet the customer's property requirements. In this example, the hot-die forging appears to generate better strain and temperature distributions than the mechanical press.

16.6.2 Optimization of Microstructure in Forging Jet Engine Disks

For manufacturing superalloy disks for jet engines, it is extremely important to meet specific grain size requirements. A disk of fine grain produces excellent tensile and fatigue properties but with reduction in creep property. A disk with a medium grain size yields balanced tensile and creep properties. A coarse grain disk provides excellent creep properties with a lower tensile strength. Disks used in different sections of a jet engine require different grain sizes and properties. A new disk product can be defined by meeting new requirements for distribution of grain size and related properties.

Process modeling coupled with microstructure modeling was used to develop processes to produce disks with different grain sizes for a potential new product. The specific goal was to produce Rene 88 (a nickel-base superalloy) disks with coarse (ASTM 6), medium (ASTM 8), and fine (ASTM 12) grain sizes. The microstructure modeling provided possible process windows for producing each disk with the targeted grain size. The process modeling, using FEM code DEFORM, provided the actual thermomechanical histories of each disk. The microstructure model developed for Rene 88 was integrated to DEFORM postprocessing module, where the user-defined subroutine can be linked with DEFORM. During the postprocessing in DEFORM, the entire thermomechanical history experienced in each local point and the starting grain size at that point were used to calculate the final grain size.

With the guidance from the process modeling integrated with microstructure modeling, after a couple of iterations, processing conditions were selected for producing disks with the three targeted grain sizes: ASTM 6 (coarse), ASTM 8 (medium), and ASTM 12 (fine). The actual forging process was performed for a reality check. One forged disk is shown in Fig. 16.8. The grain sizes obtained from three production processes of the disks were actually uniform ASTM 5–6, ASTM 7–9, and ASTM 12–13. Each disk hit its assigned grain size goal, respectively, and there were no abnormal grains observed on any of the

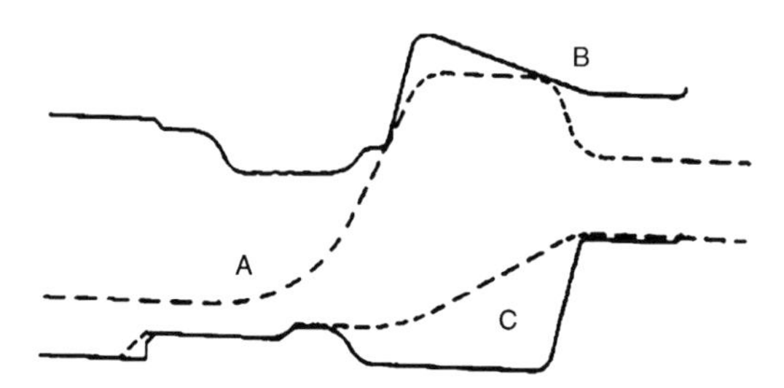

Fig. 16.11 Modified blocker design (broken lines) positioned in the open finisher dies (solid lines) [Jenkins et al., 1989]

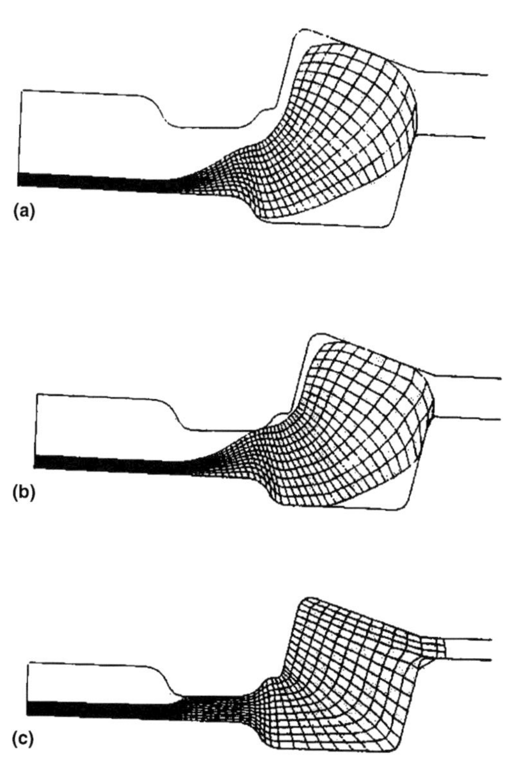

Fig. 16.12 Deformed mesh of the finishing simulation with the modified blocker design [Jenkins et al., 1989]

disks. Figure 16.9 compares the grain sizes obtained from the model prediction (color coded) and the optical rating (number in the block) for the three disks [Hardwicke et al., 2000]. The measurements agree well with the predictions in all of the three cases. In the coarse and fine-grained disks, the model predicted two small die lock regions (where metal flow is prevented), where the grain size is a little different than the bulk. The measured grain sizes in the disk proved this phenomenon. However, the complex friction phenomenon in real forgings makes it difficult to predict the exact die lock location. The success in producing disks with the required grain size was attributed to both the selection of proper process conditions aided by the process and grain size modeling as well as good production process control.

16.6.3 Investigation of Defect Formation in Ring Gear Forging

The process analyzed was the forging of an automotive ring gear blank [Jenkins et al., 1989]. In production, the part is hot forged from AISI 4320 steel in three sets of dies. The dies were of H11 steel, lubricated with a graphite and water mixture and maintained at approximately 300 °F (150 °C).

The first step in the manufacturing process involves cold shearing the billets from stock and induction heating them to 2200 °F (1200 °C). Next, a billet is placed in the busting dies and upset (Fig. 16.10a). It is then transferred to a blocker die and forged (Fig. 16.10b) and finally transferred to and forged in a finisher die (Fig. 16.10c). During initial forging trials, buckling flow in the blocker dies caused a lap to be formed intermittently around the circumference of the part (Fig. 16.10d). As the finish dies filled, the lap worsened. Because of this defect, the part was rejected, and hence, a new blocker die design was required.

The following observations were made during simulation of the process:

- The sharp corner radius and steep angle of the inside wall on the upper die resulted in the formation of a gap between the inside die wall and the workpiece.
- As the workpiece contacted the uppermost surface of the top die and began upsetting, the inside surface of the blocker began to buckle.
- The radial flow from the web region forced the buckle out toward the outer die walls, and as the upsetting and radial flow combined, the buckling became more severe.

To counter the above problem, the following modification was made to the original blocker design:

- The corner radius (region A of Fig. 16.11) was increased by a factor of 2 to aid the metal flow around the corner.
- The angle of the top surface of the upper die (region B of Fig. 16.11) was decreased until it was horizontal to increase the height of the blocker.
- The outer wall of the lower die (region C of Fig. 16.11) was modified so that upsetting flow from the top die would fill voids in the upper die cavity instead of voids in the lower die cavity.

Figure 16.12 shows the die fill in the simulation run with the new blocker design. At the start of the working stroke, the workpiece followed the walls of the upper and lower die. With

Fig. 16.13 Automotive component formed by forward/backward hot forging process [Brucelle et al., 1999]

Fig. 16.14 Cracks formed as a result of thermal cycling [Brucelle et al., 1999]

further deformation, the workpiece contacted the uppermost wall of the top die, and a gap formed between the inside wall of the top die and the workpiece. At the final stroke position, a small gap remained along the inside wall of the upper die, but no buckle was formed. Figure 16.12(a–c) shows the finish die operation with the modified blocker output. Upon deformation, the upper die pushes the workpiece down until contact is made with the outer wall of the lower die. With further reduction, the workpiece contacts the outer web region of the upper die. As the stroke continues, the inside corner fills up without any indication of defective flow patterns. With further upsetting of the workpiece, the uppermost fillet of the top die and the outside fillet of the bottom die continue to fill, and the die cavity fills up completely. Hence, the result from the finisher simulation indicates that the modified blocker workpiece fills the finisher die without defects.

16.6.4 Investigation of Tool Failure

Hot forging is a widely used manufacturing process in the automotive industry. High production rates result in severe thermomechanical stresses in the dies. Either thermal cracking or

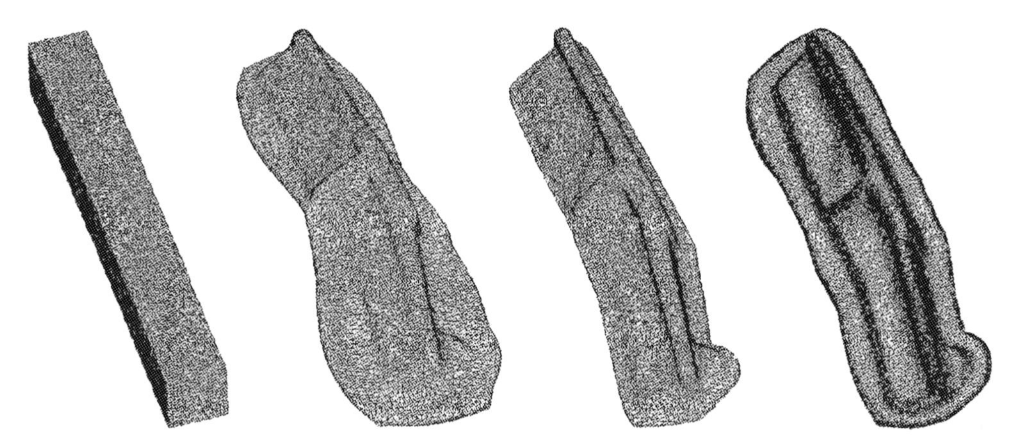

Fig. 16.15 Forging sequence of the titanium fitting. Part geometry courtesy of Weber Metals Inc.

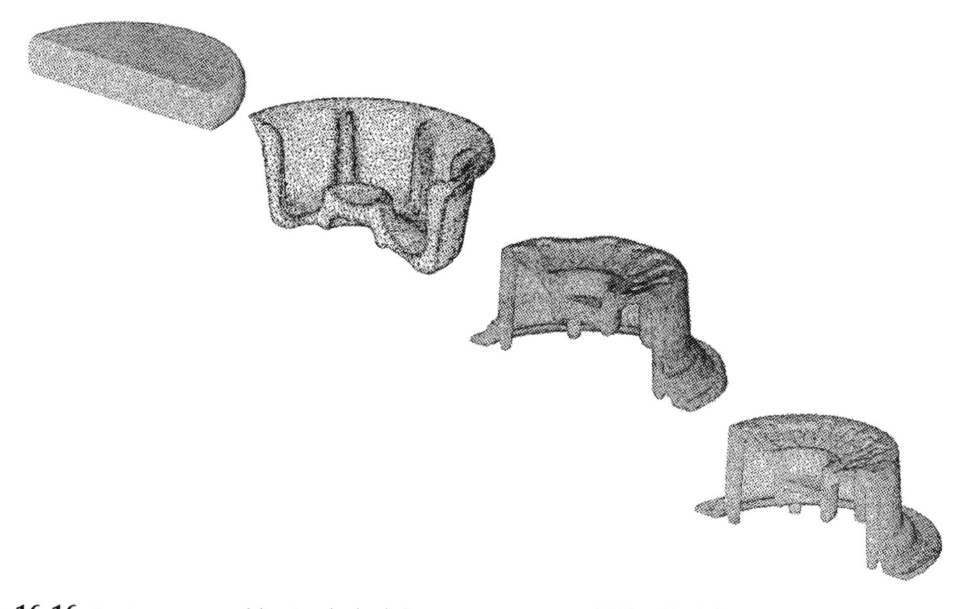

Fig. 16.16 Forging sequence of the aircraft wheel. Part geometry courtesy of Weber Metals Inc.

wear governs the life of the dies. In the forging industry, the tooling cost alone can constitute up to 10% of the total cost of the component.

This example deals with the investigation of the effect of thermomechanical stresses on the tool life in the hot extrusion of the automotive component shown in Fig. 16.13 [Brucelle et al., 1999]. The workpiece was from austenitic stainless steel AISI 316L. The punch was from tool steel (X85 WCrMoV6-5-4-2). The resulting stresses in this process are a combination of the purely mechanical stresses due to forging and the thermomechanical stresses as a result of thermal cycling of the punch surface due to the alternating hot forging and waiting periods. The stresses due to thermal cycling were found to comprise approximately 75% of the total stress field. This cycling causes tool damage, known as heat checking. Originally, the punch had to be changed approximately every 500 cycles due to cracking as a result of thermal cycling (Fig. 16.14). It is a commonly known fact that geometry changes are not the best way to reduce the stress level with regard to thermal stresses. From this study, it was determined that increased tool life could be achieved by modifying the hot forging process parameters such as billet temperature and the forging rate.

Finite-element modeling simulation and experimental work were used to conduct a parametric study to determine the optimum process parameters to achieve higher life expectancy of the tools. This combined numerical and experimental approach can be summarized as:

- A two-step numerical simulation:
  a. Process simulation to determine the purely mechanical stresses, forging loads, and thermal boundary conditions for the punch
  b. Thermoelastic simulation for thermal stress analysis of the punch
- A two-step experimental stage:
  a. Metallurgical validation of the constitutive laws of the workpiece material
  b. Industrial forging test validation of the thermal boundary conditions for the punch

The surface temperatures on the punch are a factor of the heat-transfer coefficient at the tool/workpiece interface. This coefficient is a function of various factors, such as surface topography, contact pressures, temperature difference, and duration of contact [Snaith et al., 1986]. Forging tests were conducted on an industrial press using a test punch with five thermocouples. Several numerical iterations (FEM simulations) were performed by using different heat-transfer coefficients until the calculated temperature distribution was in agreement with that from the experiments.

In order to reduce the thermal stresses, a reduction of the thermal gradient during forging must be obtained. There are two options: 1) modification of process parameters to decrease the temperature (reduction of the punch speed, thus reducing the flow stress, or decreasing workpiece temperature, resulting in an increase in flow stress) or 2) use of lubricating/insulating

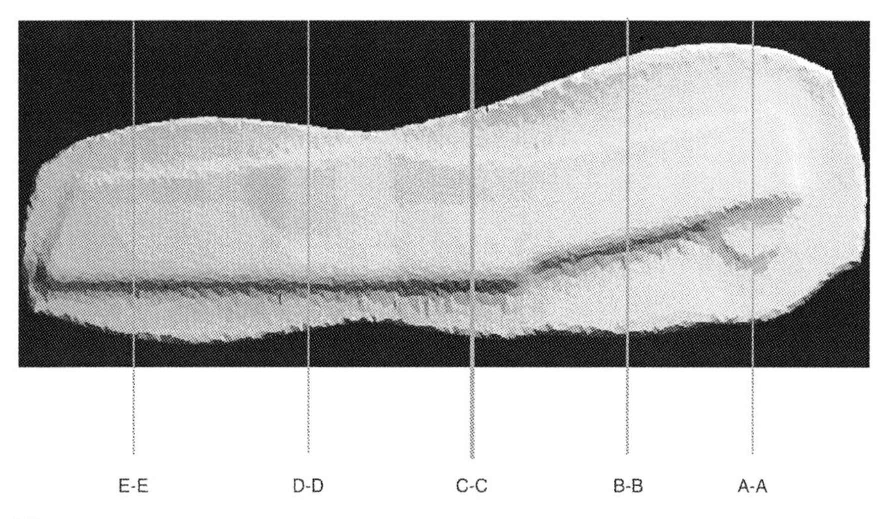

Fig. 16.17 Sections taken along the fitting to check for die filling at the blocker stage [Shirgaokar et al., 2002]

products during forging to reduce the heat transfer, which is an empirical approach. The first option was selected, since the available press could handle increased forging loads as a result of increased flow stress.

A parametric study was conducted to investigate the influence of forging speed and initial workpiece temperature on the final thermomechanical stresses. The optimum process parameters were thus determined, resulting in a 30% decrease in the stresses. Thus, a combination of process simulation and experimental verification resulted in an increase in the tool life for the punch in this hot forging process.

16.6.5 Multistage Forging Simulations of Aircraft Components

Multistage forging simulations of two aircraft components (a titanium fitting and an aluminum wheel) were run to study metal flow, temperature distribution, die filling, and die stresses [Shirgaokar et al., 2002]. The commercial FEM code DEFORM-3D was used for these simulations. The two components considered for this study are produced by closed-die forging with flash. Since the parts are forged at elevated temperatures, it was necessary to run nonisothermal simulations. Flash removal between the forging stages also had to be considered for the simulations in order to ensure appropriate material volume in the dies for the subsequent forging stage. Each of the components was forged in three stages, namely, two blocker stages followed by a finisher stage. Figures 16.15 and 16.16 show the forging sequence of the titanium fitting and the aluminum wheel, respectively. The results obtained at the end of the simulations were the effective stress distribution, die filling, metal flow during forging, temperature distribution, and strain distribution.

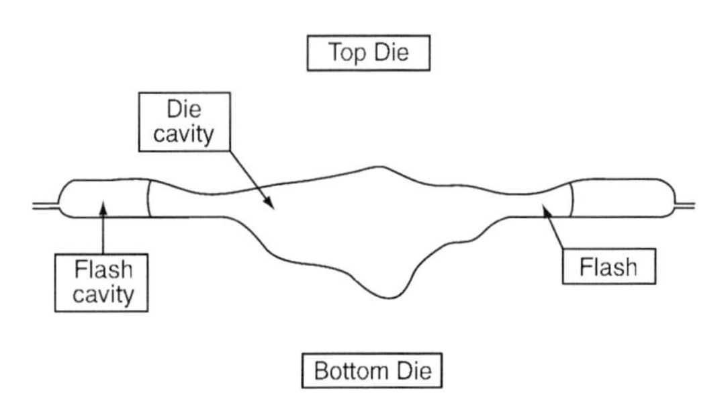

Fig. 16.18 Section A-A of the fitting after the first blocker operation [Shirgaokar et al., 2002]

The simulation strategy adopted for the two components was to remove the flash in-between stages. This was done by using the Boolean capability of DEFORM, i.e., volume manipulation. Die filling was checked by examining various cross sections along the length of the forging (Fig. 16.17 and 16.18).

The simulations were stopped when die filling was achieved, and it was this stage of the simulation that was used to determine the stresses in the dies. In order to reduce computational time, the dies were kept rigid throughout the simulation. At the last step, they were changed to elastic, and the stresses from the workpiece were

(a) Steel Forging

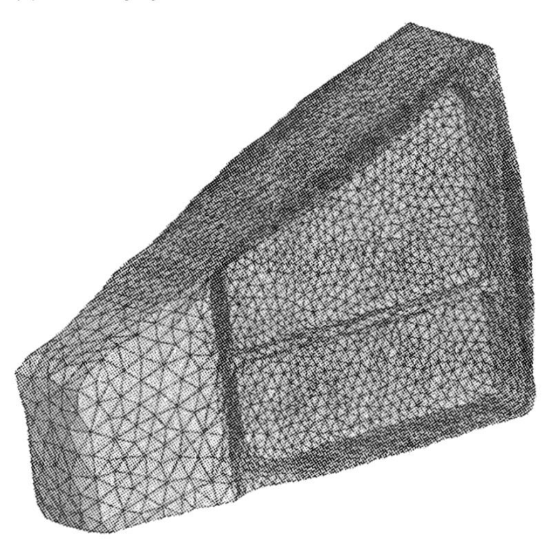

(b) Simulation Result

Fig. 16.19 Forging of AISI 4340 aerospace component

interpolated onto the dies. The results obtained from the die stress analysis simulations were the effective stress, the maximum principal stress, and the temperature distribution. Using these results, an effective die design was established.

16.6.6 Precision Forging of an Aerospace Component

The simulation of a precision-forged aerospace component was conducted as part of a study to develop guidelines for the design of prestressing containers for dies used in forging of complex parts. The simulated part is shown in Fig. 16.19(b). Experiments for this component were conducted in a previous study conducted at Batelle Columbus Laboratories [Becker et al., 1972]. The results from the simulations were compared to the experimentally determined loads and were found to be in good agreement. The simulation results are used to obtain the loading on the die and the punch. The load on the die is used to perform stress analysis and design a cost-effective prestressing container. The loading on the punch is used to perform a stress analysis and correlate the punch deflection to thickness measurements taken at various locations of the forgings.

16.6.7 Die Design for Flashless Forging of Connecting Rods

In conventional hot forging of connecting rods, the material wasted to the flash accounts for approximately 20 to 40% of the original workpiece. It is essential to accurately control the volume distribution of the preform to avoid overloading the dies and in order to fill the cavity. It is equally important that the preform be simple enough to be mass-produced. Thus, the design of a flashless forging operation is more complex than that of a conventional closed-die forging with flash. In order to accelerate the development process and reduce prototyping costs, it is essential to conduct a substantial amount of the design process on the computer.

The requirements for conducting a successful flashless forging process are [Vasquez et al., 2000]:

- The volume of the initial preform and the volume of the die cavity at the final forging stage must be approximately the same.
- There must be neither a local volume excess nor a shortage, which means that the mass distribution and positioning of the preform must be precise.
- If there is a compensation space provided in the die, the real cavity must be filled first.

The 2-D and 3-D FE simulations were used extensively to analyze and optimize the metal flow in flashless forging of the connecting rod shown in Fig. 16.20(a). In order to verify the applicability of the simulation results, physical modeling experiments were conducted using plasticine. Figure 16.20(b) shows the initial pre-

(a)

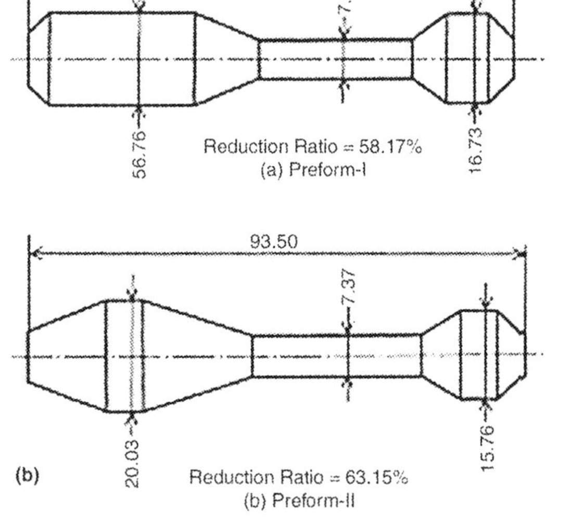

(b)

Fig. 16.20 Development of the preform shape for flashless forging of a connecting rod [Vasquez et al., 2000]

form design along with the final modified preform. Finite-element simulation predicted underfilling with the initial preform, which was modified accordingly to give a defect-free connecting rod (Fig. 16.21).

16.6.8 Integrated Heat Treatment Analysis

The forming of a medium-carbon manganese steel bevel gear was analyzed using DEFORM 3D. This gear was hot forged with flash. The simulation was conducted utilizing rotational symmetry. Thus, only ½0th of the total volume was simulated. Mesh density windows were used for local mesh refinement during simulation. Contours of effective strain (darker areas indicate higher strain) are shown in Fig. 16.22(a). After the forging simulation, the gear geometry was modified to account for the flash removal and drilling the inside diameter.

The modified gear geometry was used to simulate a heat treatment operation. In this simulation, the gear was austenized by heating to 1560 °F (850 °C) and cooled in 60 s with a heat-transfer coefficient representative of an oil quench. A time-temperature-transformation diagram for medium-carbon manganese steel determined the

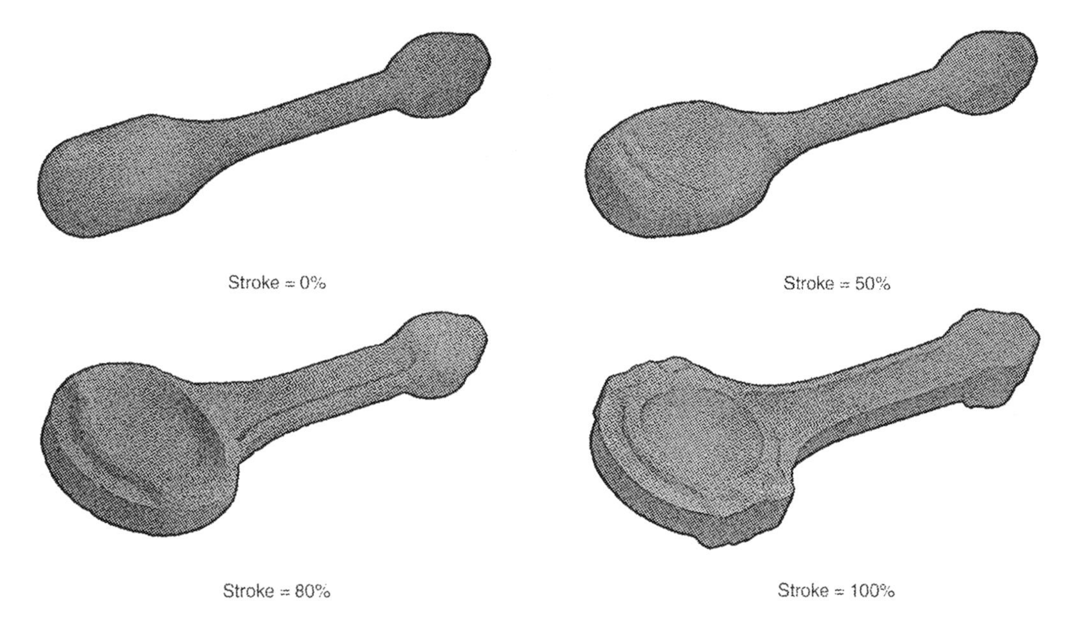

Fig. 16.21 Deformation sequence for flashless precision forging of a connecting rod [Takemasu et al., 1996]

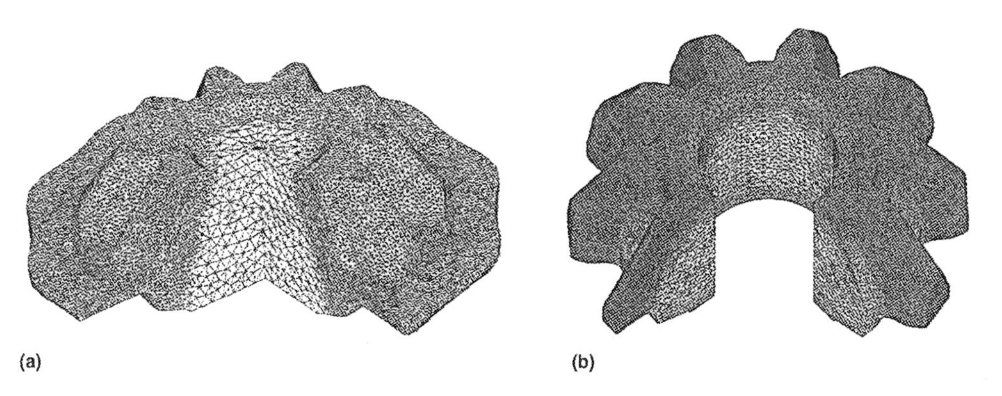

Fig. 16.22 (a) The deformation simulation of a hot forged gear with flash. (b) The volume fraction of martensite (dark is higher) in a steel gear after quenching [Wu et al., 2001]

diffusion behavior during the austenite-pearlite/bainite transformation, whereas the Magee equation was used to model the martensite response. In Fig. 16.22(b), the volume fraction of martensite is shown after quenching. The dark regions represent a more complete martensite transformation, and the light areas indicate a mixture of bainite and pearlite [Wu et al., 2001].

REFERENCES

[Becker et al., 1972]: Becker, J.R., Douglas, J.R., Simonen, F.A., "Effective Tooling Designs for Production of Precision Forgings," Technical Report AFML-TR-72-89, Battelle Columbus Laboratories, 1972.

[Brucelle et al., 1999]: Brucelle, O., and Bernhart, G.," Methodology for Service Life Increase of Hot Forging Tools," *Journal of Materials Processing Technology,* Vol. 87, 1999, p 237.

[Hardwicke et al., 2000]: Hardwicke, C., and Shen, G., "Modeling Grain Size Evolution of P/M Rene 88DT Forgings," *Advanced Technologies for Superalloy Affordability,* K.-M. Chang, S.K. Srivastava, D.U. Furrer, and K.R. Bain, Ed., The Minerals, Metals, and Materials Society, 2000.

[Howson et al., 1989]: Howson, T.E., and Delgado, H.E., "Computer Modeling Metal Flow in Forging," *JOM,* Feb. 1989, p 32–34.

[Jenkins et al., 1989]: Jenkins, B.L., Oh, S.I., Altan, T., "Investigation of Defect Formation in a 3-Station Closed Die Forging Operation," *CIRP Annals,* Vol. 381, p 243.

[Knoerr et al., 1992]: Knoerr, M., Lee, J., Altan, T., "Application of the 2D Finite Element Method to Simulation of Various Forming Processes," *Journal of Materials Processing Technology,* Vol. 33, 1992, p 31.

[Ngaile et al., 2002]: Ngaile, G., and Altan, T., "Simulations of Manufacturing Processes: Past, Present and Future," *Proceedings of the Seventh ICTP,* Oct 2002, Japan, p 271.

[Oh, 1982]: Oh, S.I., "Finite Element Analysis of Metal Forming Problem with Arbitrarily Shaped Dies," *Int. J. Mech. Science,* Vol. 24 (No. 4) 1982, p 479.

[Sellars, 1990]: Sellars, C.M., "Modeling Microstructural Development during Hot Rolling," *Materials Science and Technology,* Vol. 6, 1990, p 1072.

[SFTC, 2002]: *DEFORM 7.2 User Manuals,* Scientific Forming Technologies Corporation, Columbus, OH, 2002.

[Shen et al., 1993]: Shen, G., Shivpuri, R., Semiatin, S.L., Lee, J.Y., "Investigation of Microstructure and Thermomechanical History in the Hammer Forging of an Incoloy 901 Disk," *Annals of the CIRP,* Vol. 42/1, 1993, p 343–346.

[Shen et al., 2000]: Shen, G., "Microstructure Modeling of Forged Components of Ingot Metallurgy Nickel Based Superalloys," *Advanced Technologies for Superalloy Affordability,* K.M. Chang, S.K. Srivastava, D.U. Furrer, and K.R. Bain, Ed., TMS, 2000, p 223–231.

[Shen et al., 2001]: Shen, G., Denkenberger, R., Furrer, D., "Aerospace Forging—Process and Modeling," *Materials Design Approaches and Experiences,* J.C. Zhao, M. Fahrmann, and T.M. Pollock, Ed., TMS, 2001, p 347–357.

[Shirgaokar et al., 2002]: Shirgaokar, M., Ngaile, G., Altan, T., "Multi-Stage Forging Simulations of Aircraft Components," ERC/NSM–02-R-84, Engineering Research Center for Net Shape Manufacturing, 2002.

[Snaith et al., 1986]: Snaith, B., Probert, S.D., O'Callaghan, P.W., "Thermal Resistances of Pressed Contacts," *Appl. Energy,* Vol. 22 1986, p 31–84.

[Takemasu et al., 1996]: Takemasu, T., Vasquez, V., Painter, B., Altan, T., "Investigation of Metal Flow and Preform Optimization in Flashless Forging of a Connecting Rod," *Journal of Materials Processing Technology,* Vol. 59, 1996, p 95.

[Vasquez et al., 1999]: Vasquez, V., Walters, J., and Altan, T., "Forging Process Simulation—State of the Art in USA," *Proceedings of the Conference on New Developments in Forging Technology,* Stuttgart, Germany, May 19–20, 1999.

[Vasquez et al., 2000]: Vasquez, V., and Altan, T., "Die Design for Flashless Forging of Complex Parts," *Journal of Materials Processing Technology,* Vol. 98, 2000, p. 81.

[Wu et al., 1992]: Wu, W.T., Oh, S.I., Altan, T., Miller, R.A., "Optimal Mesh Density Determination for the FEM Simulation of Forming Processes," NUMIFORM '92, Sept 14–18, 1992, France.

[Wu et al., 1996]: Wu, W.T., Li, G.J., Arvind, A., Tang, G.P., "Development of a Three Dimensional Finite Element Based Process Simulation Tool for the Metal Forming Industry," *Proceedings of the Third Biennial Joint Conference on Engineering Systems*

Design and Analysis, Montpellier, France, 1996.
[Wu et al., 2001]: Wu, W.T., Oh, S.I., Arimoto, K., "Recent Developments in Process Simulation for Bulk Forming Processes," *Journal of Materials Processing Technology,* Vol. 111, 2001, p 2–9.

SELECTED REFERENCES

- **[Devadas et al., 1991]:** Devadas, C., Samarasekera, I.V., Hawbolt, E.B., "The Thermal and Metallurgical State of Strip during Hot Rolling: Part III, Microstructural Evolution," *Met. Trans. A,* Vol. 12, 1991, p 335–349.
- **[Shen et al., 1989]:** Shen, G., Im, Y.T., Altan, T., "Effect of Flash Dimensions and Billet Size in Closed-Die Forging of an Aluminum Alloy Part," *Transactions of the NAMRI of SME,* Society of Manufacturing Engineers, 1989, p 34–40.
- **[Shen et al., 1995]:** Shen, G., Semiatin, S.L., Shivpuri, R., "Modeling Microstructural Development during the Forging of Waspaloy," *Met. Trans. A,* 26, 1995, p 1795–1802.
- **[Shen et al., 2001]:** Shen, G., Kahlke, D., Denkenberger, R., Furrer, D., "Advances in the State-of-the-Art of Hammer Forged Alloy 718 Aerospace Components," *Superalloys 718, 625, 706 and Various Derivatives,* E.A. Loria, Ed., 2001, p 237–247.
- **[Vasquez et al., 1996]:** Vasquez, V., and Altan, T., "Investigation of Metal Flow and Preform Optimization in Flashless Forging of a Connecting Rod," *Journal of Materials Processing Technology,* Vol. 59, 1996, p 95.
- **[Walters et al., 1998]:** Walters, J., Wu, W.T., Arvind, A., Guoji, L., Lambert, D., Tang, J.P., "Recent Developments of Process Simulation for Industrial Applications," *Proceedings of the Fourth International Conference on Precision Forging—Cold, Warm and Hot Forging,* Oct 12–14, 1998, Columbus, Ohio.
- **[Wu et al., 1985]:** Wu, W.T., and Oh, S.I., "ALPID—A General Purpose FEM Code for Simulation of Non-isothermal Forging Processes," *Proceedings of North American Manufacturing Research Conference (NAMRC) XIII,* Berkeley, CA, 1985, p 449–460.

CHAPTER 17

Cold and Warm Forging

Prashant Mangukia

17.1 Introduction

Cold forging is defined as forming or forging of a bulk material at room temperature with no initial heating of the preform or intermediate stages. Cold extrusion is a special type of forging process wherein the cold metal flows plastically under compressive forces into a variety of shapes. These shapes are usually axisymmetric with relatively small nonsymmetrical features, and, unlike impression-die forging (see Chapter 14, "Process Design in Impression-Die Forging"), the process does not generate flash. The terms *cold forging* and *cold extrusion* are often used interchangeably and refer to well-known forming operations such as extrusion, upsetting or heading, coining, ironing, and swaging [Feldmann, 1961; Wick, 1961; Billigmann et al., 1973; Watkins, 1973; and Altan et al., 1983]. These operations are usually performed in mechanical or hydraulic presses, which are discussed in Chapter 10, "Principles of Forging Machines," and Chapter 11, "Presses and Hammers for Cold and Hot Forging." Several forming steps are used to produce a final part or relatively complex geometry, starting with a slug or billet of simple shape, as shown in Fig. 17.1. Some basic techniques of cold forging are illustrated in Fig. 17.2. Through a combination of these techniques, a very large number of parts can be produced, as illustrated schematically in Fig. 17.3.

In warm forging, the billet is heated to temperatures below the recrystallization temperature, for example, up to 1290 to 1470 °F (700 to 800 °C) for steels, in order to lower the flow stress and the forging pressures. In cold forging, the billet or the slug is at room temperature when deformation starts.

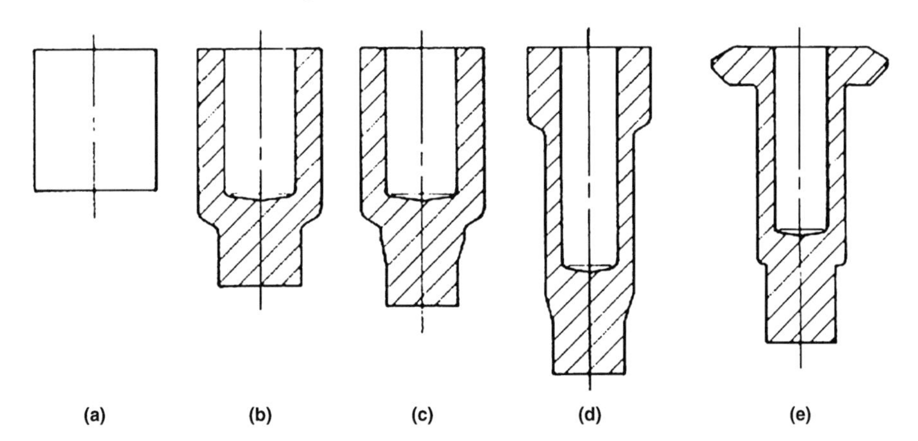

Fig. 17.1 Schematic illustration of forming sequences in cold forging of a gear blank. (a) Sheared blank. (b) Simultaneous forward rod and backward extrusion. (c) Forward extrusion. (d) Backward cup extrusion. (e) Simultaneous upsetting of flange and coining of shoulder [Sagemuller, 1968]

Cold and warm forging are extremely important and economical processes, especially for producing round or nearly round parts in large quantities. Some of the advantages provided by this process are:

- High production rates
- Excellent dimensional tolerances and surface finish for forged parts
- Significant savings in material and machining

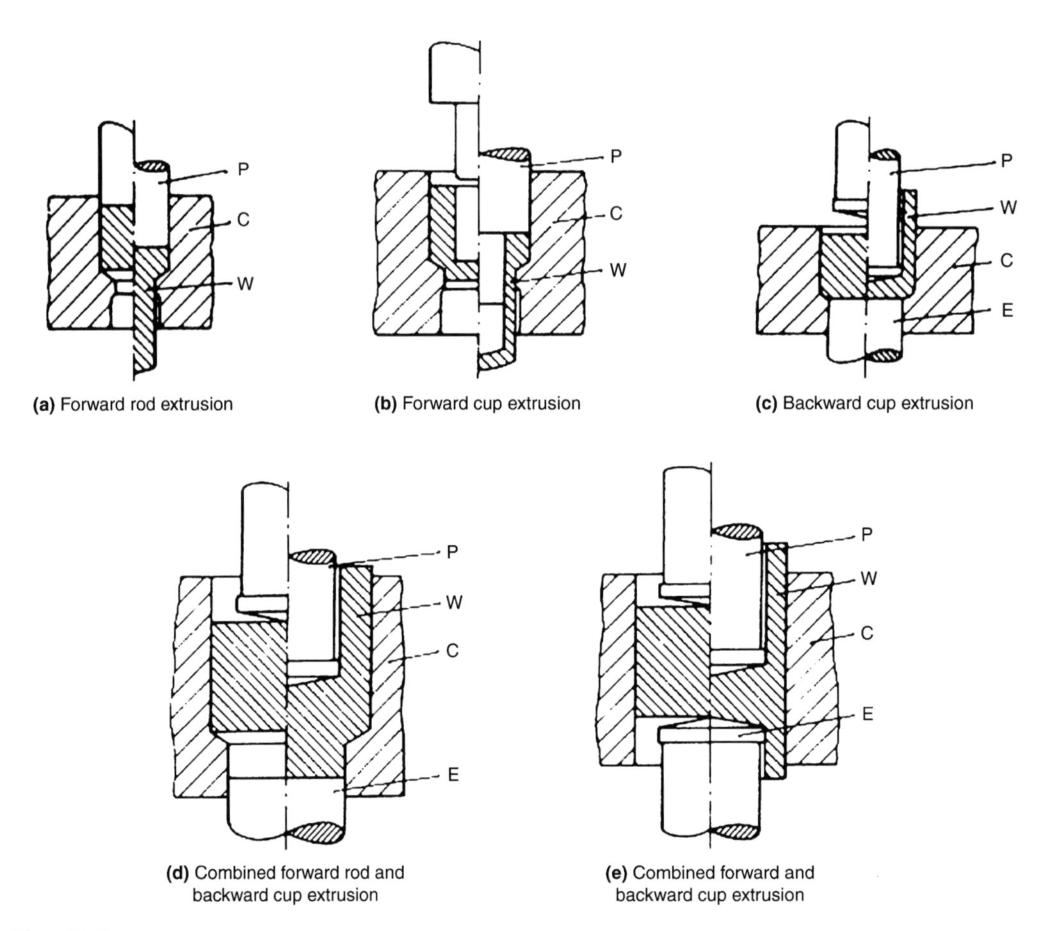

(a) Forward rod extrusion **(b)** Forward cup extrusion **(c)** Backward cup extrusion **(d)** Combined forward rod and backward cup extrusion **(e)** Combined forward and backward cup extrusion

Fig. 17.2 Various types of cold forging (extrusion) techniques (P, punch; C, container; W, workpiece; E, ejector) [Feldmann, 1977]

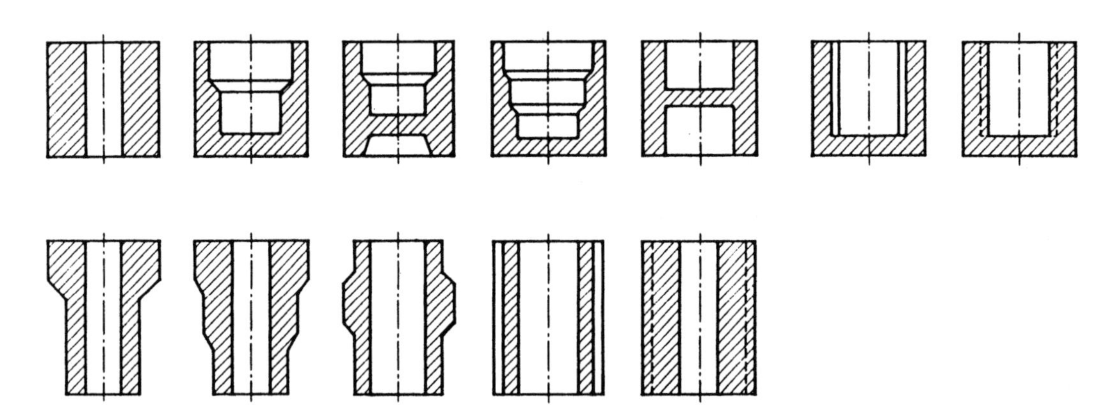

Fig. 17.3 Examples of cold forged tubular or cup-shaped parts [Feldmann, 1977]

- Higher tensile strengths in the forged part than in the original material, because of strain hardening
- Favorable grain flow to improve strength

By far the largest area of application of cold and warm forging is the automobile industry. However, cold forged parts are also used in manufacturing bicycles, motorcycles, farm machinery, off-highway equipment, and nuts and bolts.

17.2 Cold Forging as a System

A cold forging system, as shown in Fig. 17.4, comprises all the input variables such as billet or blank (geometry or material), the tooling (the geometry and material), the material at the tool/material interface, the mechanics of plastic deformation, the equipment used, the characteristics of the final product, and finally, the plant environment where the process is being conducted. The "systems approach" in forging allows study of the input/output relationships and the effect of process variables on product quality and process economics. The key to a successful metal forming operation, i.e., to obtaining the desired shape and properties, is the understanding and control of metal flow. The direction of metal flow, the magnitude of deformation, and the temperatures involved greatly influence the properties of the formed components. Metal flow determines both the mechanical properties related to local deformation and the formation of defects such as cracks or folds at or below the surface. The local metal flow is in turn influenced by the process variables.

17.3 Materials for Cold Forging

All metals that exhibit ductility at room temperature can be cold forged. This group consists primarily of steels and aluminum alloys. However, alloys of copper, zinc, tin, titanium, beryllium, and nickel are also cold forged for special applications [Gentzsch, 1967].

Examples of steels that are used extensively for producing cold extruded parts are:

- *Case hardening steels:* unalloyed: 1010, 1015; alloyed: 5115, 5120, 3115
- *Heat treatable steels:* unalloyed: 1020, 1035, 1045; alloyed: 5140, 4130, 4140, 8620
- *Stainless steels:* pearlitic: 410, 430, 431; austenitic: 302, 304, 316, 321

Stainless steels usually are not easily forged. Especially, cold forging of austentic or austentic-ferritic steels (which work harden very strongly) require high forces and tool pressures. Furthermore, these materials are difficult to lubricate. Cold forging of stainless steel is sometimes limited due to lack of information on the behavior of the material [ICFG, 2001a].

Examples of commonly cold forged aluminum alloys are:

- Pure or nearly pure aluminum alloys: 1285, 1070, 1050, and 1100
- Nonhardenable aluminum alloys: 3003, 5152, and 5052
- Hardenable aluminum alloys: 6063, 6053, 6066, 2017, 2024, and 7075

Aluminum may be used as an alternative to steel to save weight [ICFG, 2002]. Table 17.1 summarizes the properties of aluminum alloys suitable for cold forging.

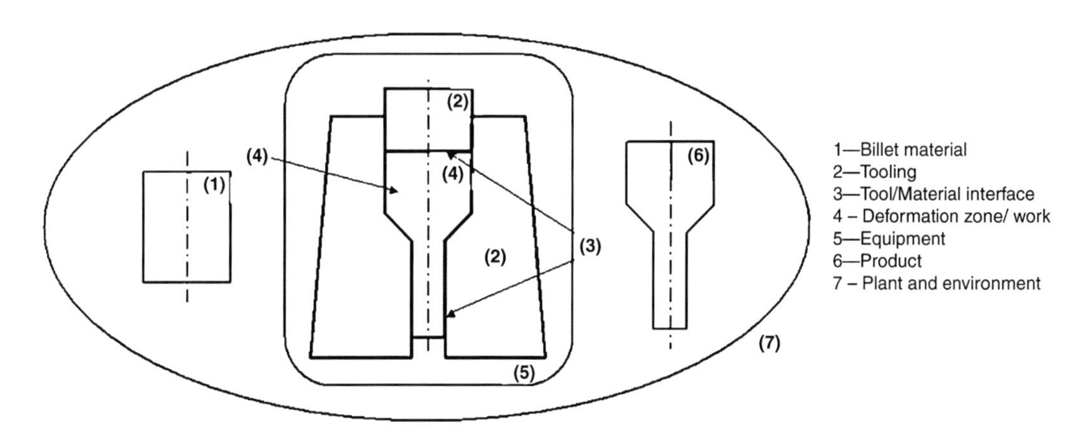

Fig. 17.4 Cold forging as a system

Table 17.1 Properties of aluminum alloys suitable for cold forging [ICFG, 2002]

Aluminum alloy	Heat treatable	Strength	Elongation	Corrosion resistance	Machinability	Weldability
1050	No	*	****	****	*	****
3103	No	**	*	****	*	****
5056	No	***	**	****	**	**
2014	Yes	***	**	*	***	*
6061	Yes	***	***	***	**	****
7075	Yes	****	**	**	***	*

Note: * = poor; **** = best

Materials for cold forging are supplied as rolled or drawn rod or wire as well as in the form of sheared or sawed-off billets. The dimensions, weight, and surface finish of the sheared (or sawed) billet or preform must be closely controlled in order to maintain dimensional tolerances in the cold forged part and to avoid excessive loading of the forging press and tooling.

17.4 Billet Preparation and Lubrication in Cold Forging of Steel and Aluminum

By far, the largest area of application for cold forging is the production of steel parts. Cold forging plants usually receive small-diameter material in coils and large-diameter stock in bars. In very large-volume production, horizontal mechanical presses, called headers or upsetters, are used. The coil, coated with lubricant, is fed into the machine, where it is sheared and forged in several steps. In forging of relatively small production lots, vertical presses are used, and individual billets (after being lubricated) are fed into the first die station. Billet volume or weight is closely controlled, and it is desirable to obtain square billet faces during shearing or sawing [Herbst, 1967].

In cold forging, the lubricant is required to withstand high pressures, on the order of 280 ksi (1930 Mpa) in extrusion of steel, so as to avoid metal-to-metal contact between the tool and the extruded material. In cold forging of low-carbon and low-alloy steels, it is accepted practice to coat the surface of the billet or coil with a lubricant carrier. This zinc phosphate coating provides a good substrate for lubricants that withstand high forming pressures. The phosphating and lubricating steps given in Table 17.2 are almost universally employed for cold extrusion of steels. The success of the zinc phosphatizing treatment is influenced by the composition of the steel, especially the chromium content. Consequently, special procedures and other conversion coatings, such as oxalates, are preferred for austenitic stainless steels. Stearate-type soaps, which adhere tenaciously to the phosphate coatings, are commonly used as lubricants for forging and extrusion of steel at room temperature. Solid lubricants such as MoS_2 and graphite have proved to be beneficial under severe forging conditions, where surface generation and forming pressures are large [Doehring, 1972].

Lubrication for cold forging of aluminum may be divided into two categories:

- Lubrication without conversion coatings
- Lubrication with a conversion coating

The lubricants applied without conversion coatings are oil, grease, or alkali stearates (especially zinc stearate). The most common conversion coatings are calcium aluminate and aluminum fluoride coatings. In both cases, the conversion coating is combined with a lubricant, normally zinc stearate. It is essential that the slug surfaces are completely clean when applying the conversion coating. This requires careful degreasing and pickling of the surface before coating [Bay, 1997, and Bay, 1994].

The types of alloys and the surface expansion have a major influence on the choice of the lu-

Table 17.2 Typical procedure for phosphating and lubricating billets of carbon and low-alloy steels for cold extrusion

1. Degrease and clean slugs in a hot alkaline solution for 1 to 5 min at 151 to 203 °F (66 to 95 °C).
2. Rinse in cold water.
3. Remove scale, usually by pickling.
4. Rinse in cold water.
5. Rinse in neutralizing solution if a pickling process was used.
6. Dip in a zinc phosphate solution (usually of a proprietary type) for approximately 5 min at 180 to 203 °F (82 to 95 °C) to develop a uniform coating of appropriate thickness.
7. Rinse in cold water; neutralize if necessary.
8. Lubricate the slugs, usually with stearate soap but sometimes with other types of lubricants.
9. Air dry the slugs to obtain a thin, adherent coating of lubricant adsorbed on the zinc phosphate coating.

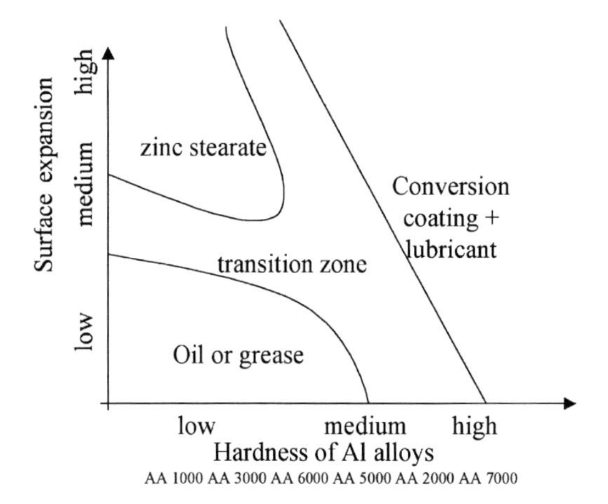

Fig. 17.5 Choice of lubricant system for different aluminum alloys for different processes [Bay, 1997]

bricant system. Figure 17.5 shows the appropriate lubricants for different aluminum alloys and surface expansions. The abscissa indicates the different series of aluminum alloys, arranged with increasing hardness corresponding to increasing difficulties in forging. The ordinate indicates the degree of surface expansion at low, medium, and high levels, depending on the forming process, as shown in the corresponding sketches. The lubricant system is divided into three groups:

- Oil or grease
- Zinc stearate
- Conversion coating + lubricant

Series AA 1000 and 3000 can be cold forged without conversion coatings, even in cases of large surface expansion (for example, can extrusion). For alloys in series AA 6000 and 5000, lubrication with oil, grease, or zinc stearate may be applied in cases of low to medium surface expansion. Large reductions can be achieved by using conversion coatings. For the series 2000, only light reductions are possible without conversion coatings, and for series 7000, conversion coatings are always necessary [Bay, 1997 and Bay, 1994].

17.5 Upsetting

Upsetting is defined as "free forming, by which a billet or a portion of a workpiece is reduced in height between usually plane, parallel platens." Upsetting is a basic deformation process that can be varied in many ways. A large segment of industry primarily depends on the upsetting process for producing parts such as screws, nuts, and rivets. A sketch of the upsetting process is shown in Fig. 17.6.

Successful upsetting mainly depends on two process limitations:

- Upset strain, $\bar{\varepsilon}$, which affects the forming limit or forgeability of the workpiece material:

$$\bar{\varepsilon} = \ln\left(\frac{h_0}{h_1}\right) \qquad \text{(Eq 17.1)}$$

- Upset ratio, which affects the buckling of the workpiece:

$$R_u = \left(\frac{h_0}{d_0}\right) \qquad \text{(Eq 17.2)}$$

In cold upsetting, a ratio of $R_u \leq 2.3$ can be achieved in one hit if the deformation occurs over a portion of the workpiece. Larger values of R_u require several deformation stages. Table 17.3 gives the recommended values for R_u.

In upsetting, the following parameters are significant: dimensions of the workpiece, its strength, its formability, the required upset ratio, the desired accuracy, and the surface quality. When forming in several stages, the design of the heading preforms affects the fiber structure

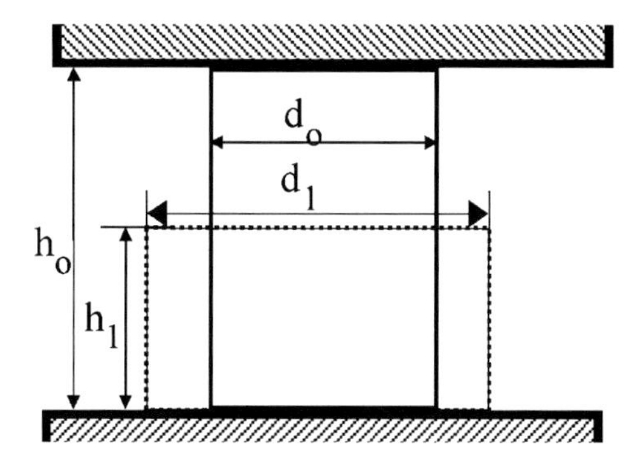

Fig. 17.6 Sketch of upsetting process [Lange et al., 1985]

Table 17.3 Recommended R_u values for cold upsetting [Lange et al., 1985]

Operation	R_u value
One operation (single-stroke process)	≤2.3
Two operations (two-stroke process)	≤4.5
Three operations (three-stroke)	≤8.0
Multistroke (more than three) with whole die (limited by difficulties arising during ejection)	≤10.0
Multistroke (more than three) with split die	≤20.0

of the final shape. Heading preforms are to be shaped such that the workpiece is guided correctly to avoid buckling and folding [Lange et al., 1985].

Figure 17.7 illustrates different techniques for upsetting. The limits on the length of the unsupported stock may vary, depending on the type of the heading die and the flatness and squareness of the end surface of the bar. Upsetting with tapered dies (Fig. 17.7c) is commonly used for the intermediate stages of a multistage upsetting process, because it allows greater upset ratios than cylindrical upsetting; thus, the number of stages required may be reduced. In order to apply the above rules to taper upsetting, an "equivalent diameter" must be calculated. The equivalent diameter (d_m) is calculated as:

$$d_m = \left(\frac{d_1^4 + d_2^4}{2}\right)^{1/4} \qquad \text{(Eq 17.3)}$$

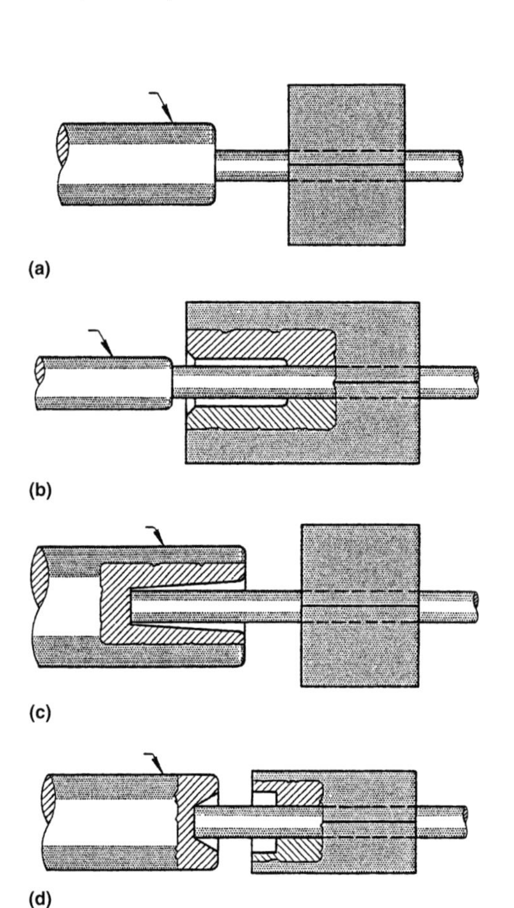

Fig. 17.7 Different techniques for upsetting. (a) Unsupported working stock. (b) Stock supported in die impression. (c) Stock supported in heading tool recess. (d) Stock supported in heading tool recess and die impression [ASM, 1970]

where d_1 and d_2 are diameters of the tapered die cavity. This design method allows greater material to be gathered in a single stage for taper upsetting, and it corresponds to the conditions found in practice.

17.6 Load Estimation for Flashless Closed-Die Upsetting

In cold forming operations, the two primary considerations in determining process feasibility are the load and energy required to form the part. These two factors determine the press size and the maximum production rate. It is important to correctly estimate the forming load for each die station. Oversizing a header will result in unnecessarily large machine cost, lower production rate, and higher part costs. Undersizing a header will lead to overloading the press, resulting in frequent downtime for maintenance and repairs [Altan et al., 1996]. The tooling used in this process is shown in Fig. 17.8. It consists of a flat-faced punch and a simple round die cavity. At the beginning of the stroke, the billet is cylindrical and undergoes a process similar to an open upset until it makes contact with the die casing (Fig. 17.8a). At this point, the forming load increases dramatically as the corner filling occurs, as seen in the load-stroke curve (Fig. 17.8b). The load required to fill the cavity is extremely high (approximately 3 to 10 times the load necessary for forging the same part in the cavity without corner filling) [Altan et al., 1982, and Lange et al., 1985].

The amount of corner filling is expressed as the ratio of the length of die wall in contact with the deformed billet to the length of the die wall and is termed in percentage of die wall contact (%DWC).

Slab method analysis is used to predict the forging load required, based on material properties and process geometries. The maximum tooling load, L, based on geometry and material properties is [Altan et al., 1996]:

$$L = \frac{\pi}{4} d_1^2 \sigma_f \left(1 + \frac{m d_1}{3\sqrt{3} h_1}\right) \qquad \text{(Eq 17.4a)}$$

where L = maximum load on tooling
d_1 = final upset head diameter
σ_f = material flow stress = $K\bar{\varepsilon}^n = \sigma_f = K(\ln h_0/h_1)^n$

m = shear friction factor
h_1 = final height of the upset head
h_0 = initial height of the billet
K = strength coefficient
n = material strain-hardening exponent

Equation 17.4(a) needs to be modified for closed-die upsetting. Since the material flow is restricted by the cavity in the closed-die upset, deformation is less than in the open upset. Hence, a modification of the traditional slab method is made to account for the reduced deformation zone to predict the forging load.

Equation 17.4(a) is modified and presented as [Altan et al., 1996]:

$$L = \frac{\pi}{4} d_1^2 \sigma_f \left(1 + \frac{md_1}{3\sqrt{32h_1'}}\right) \quad \text{(Eq 17.4b)}$$

where, in addition to the symbols given above:

σ_f = material flow stress = $K\bar{\varepsilon}^n = \sigma_f = K(\ln 2h_0'/2h_1')^n$
h_1' = final height of layer 1 (or 3), as shown in Fig. 17.9
h_0' = initial height corresponding to final height of layer 1 (or 3), as shown in Fig. 17.9

To use the modified slab analysis, the initial head diameter and the head height must be known. Also, the final head diameter, height, and amount of die wall contact must be known. The final slab height, h_1', is found by using:

$$h_1' = \left(\frac{100\% - \%DWC}{2}\right) h_1 \quad \text{(Eq 17.4c)}$$

where h_1 and h_1' are the same as in Eq 17.4(a) and (b), respectively.

To find the corresponding initial slab height, h_0', the volume constancy principle is used:

$$h_0' = \frac{4}{\pi d_0^4} V \quad \text{(Eq 17.4d)}$$

The validity of Eq 17.4 was verified by comparing the predictions with experiments [Altan et al., 1996].

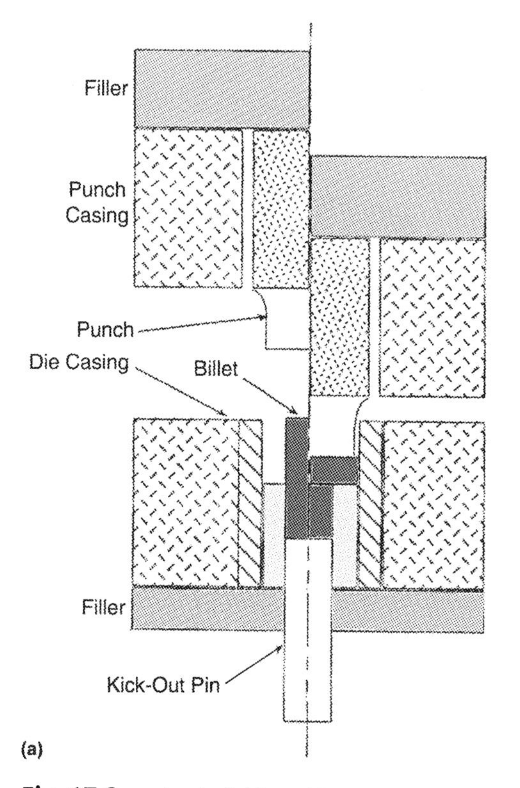

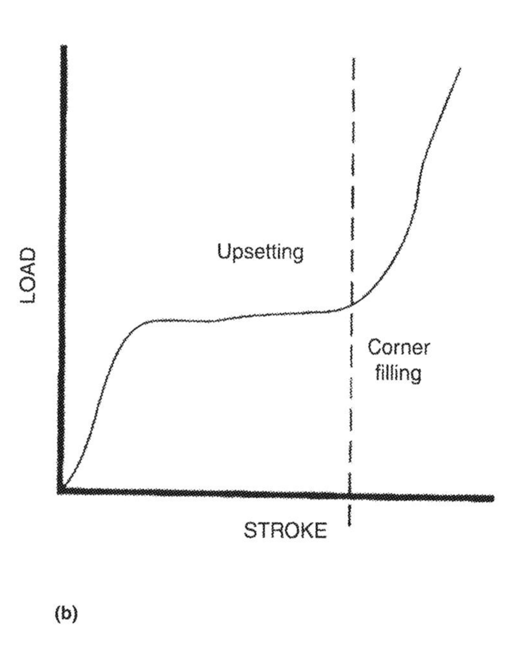

Fig. 17.8 Tooling for flashless cold upsetting process and load-stroke curve [Altan et al., 1996]

17.7 Extrusion

The two most commonly used extrusion processes, forward rod extrusion and backward cup extrusion, are illustrated in Fig. 17.10. The qualitative variations of the punch load versus punch displacement curves are shown in Fig. 17.11. The areas under these curves represent energy and can be easily calculated when the extrusion load in backward extrusion, which is constant, is estimated, or when, in forward rod extrusion, both the peak load at the beginning of the stroke and the end load at the end of the stroke are known.

17.7.1 *Variables Affecting Forging Load and Energy*

In cold extrusion, the material at various locations in the deformation zone is subject to different amounts of deformation. The values of strain, ε, and the corresponding flow stress, σ,

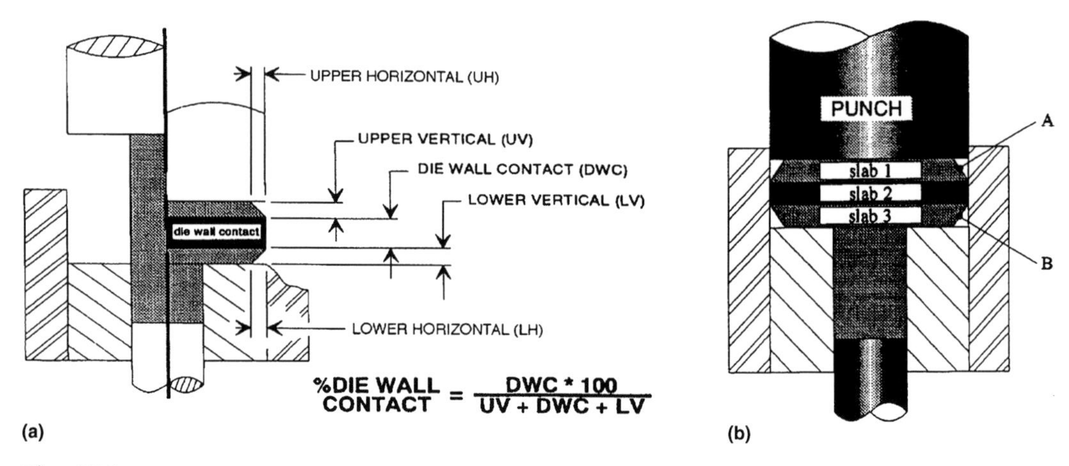

Fig. 17.9 Dividing layers for the modified slab analysis method [Altan et al., 1996]

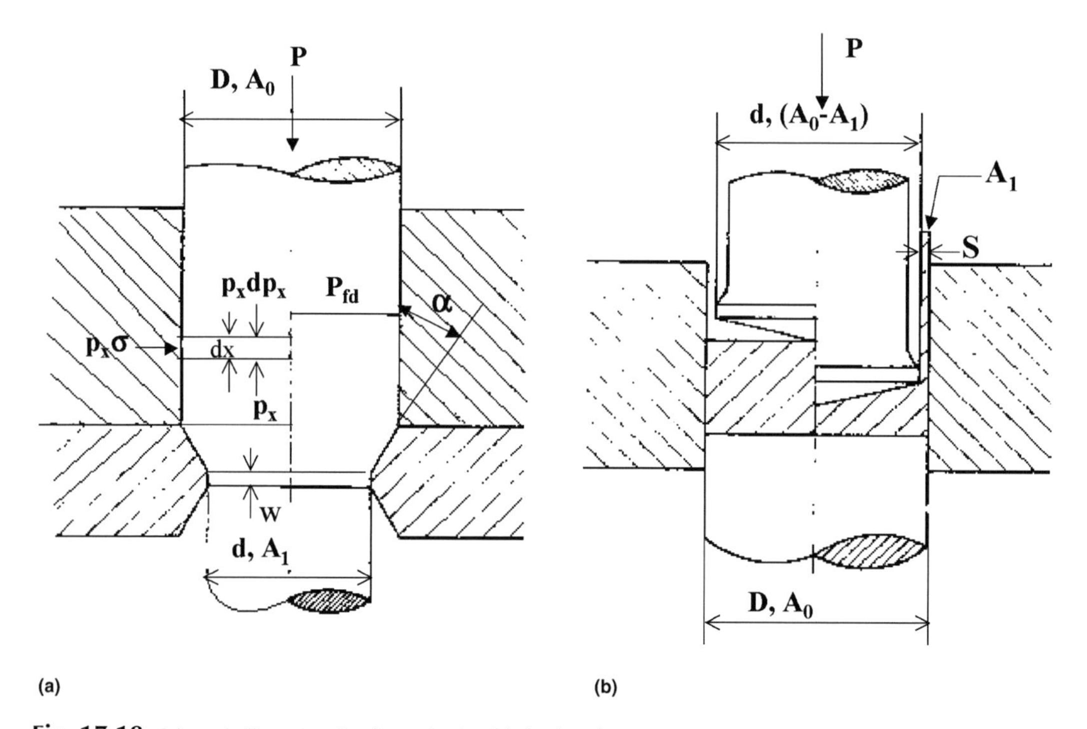

Fig. 17.10 Schematic illustration of (a) forward rod and (b) backward cup extrusion processes [Altan et al., 1983]

vary within the deformation zone. It is therefore necessary to use average values of flow stress, σ, and effective strain, ε, to characterize the total deformation of the material. The total forging load consists of the following components:

$$P = P_{fd} + P_{fc} + P_{dh} + P_{ds} \qquad \text{(Eq 17.5)}$$

where P_{fd} is the load necessary to overcome friction at the die surface (in forward extrusion) or at the die and punch surfaces (in backward extrusion), P_{fc} is the load necessary to overcome container friction in forward extrusion ($P_{fc} = 0$ in backward extrusion), P_{dh} is the load necessary for homogeneous deformation, and P_{ds} is the load necessary for internal shearing due to inhomogeneous deformation. The variations of the extrusion load for forward rod and backward cup extrusion are shown in Fig. 17.11. The loads are influenced by the following process variables:

- *Extrusion ratio, R:* The extrusion load increases with increasing reduction, because the amount of deformation, i.e., the average strain, increases with reduction.
- *Die geometry (angle, radii):* The die geometry directly influences material flow, and therefore, it affects the distribution of the effective strain and flow stress in the deformation zone. In forward extrusion, for a given reduction, a larger die angle increases the volume of metal undergoing shear deformation and results in an increase in shear deformation load, P_{ds}. On the other hand, the length of the die decreases, which results in a decrease in die friction load, P_{df}. Consequently, for a given reduction and given friction conditions, there is an optimum die angle that minimizes the extrusion load.
- *Extrusion velocity:* With increasing velocity, both the strain rate and the temperature generated in the deforming material increase. These effects counteract each other, and consequently, the extrusion velocity does not significantly affect the load in cold extrusion.
- *Lubrication:* Improved lubrication lowers the container friction force, P_{fc}, and the die friction force, P_{df}, of Eq 17.5, resulting in lower extrusion loads.
- *Workpiece material:* The flow stress of the billet material directly influences the loads P_{dh} and P_{ds} of Eq 17.5. The prior heat treatment and/or any prior work hardening also affect the flow stress of a material. Therefore, flow stress values depend not only on the chemical composition of the material but also on its prior processing history. The temperature of the workpiece material influences the flow stress, $\bar{\sigma}$.
- *Billet dimensions:* In forward extrusion, an increase in billet length results in an increase in container friction load, P_{fd}. In backward extrusion, the billet length has little effect on the extrusion load. This is illustrated in Fig. 17.11.

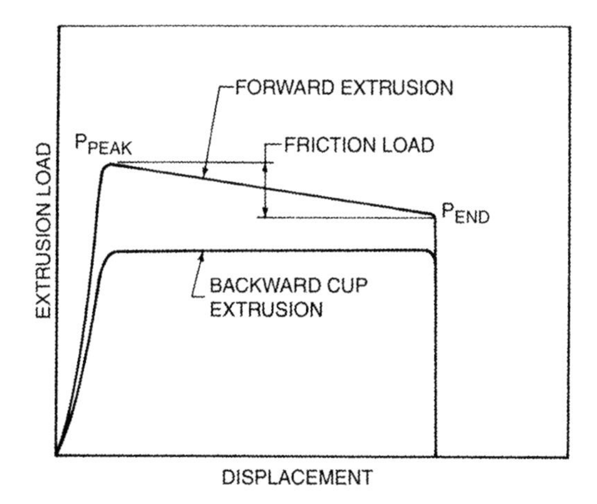

Fig. 17.11 Schematic illustration of punch load versus punch displacement curves in forward rod and backward cup extrusion processes

In forward extrusion, the most important rule is that the reduction in area cannot exceed some known limits. Higher reduction ratios can be obtained in trapped-die extrusion (Fig. 17.10a). In open-die extrusion (where the billet is not entirely guided in the container, Fig. 17.13a), the load required must be less than that causing buckling or upsetting of the unsupported stock. Typical limits of reduction in area for open-die extrusion are 35% for low-carbon steel, 25% for aluminum, and 40% for AISI 4140 steel. For trapped die, this limit is much higher, about 70 to 75% [Drozda, 1983]. The extrusion angle is also a function of the reduction in the area.

In backward extrusion, for low-to-medium carbon steel, maximum reduction in area of 70 to 75% is allowed. In backward extrusion, there is also a limit for the minimum reduction in area, which is 20 to 25% [Altan et al., 1987]. Another rule is that the maximum height of the cavity cannot exceed three times the punch diameter. Also, the bottom thickness cannot be less than 1 to 1.5 times the extruded wall thickness.

These rules are also affected by several factors, such as blank material, coating, lubrication, and final part shape.

17.7.2 Trapped-Die or Impact Extrusion

In this process (Fig. 17.12), higher strains are possible compared to open-die extrusion (Fig. 17.13). The billet is pressed through a die by a punch. Different types of these processes are described below and illustrated in Fig. 17.12.

- *Forward extrusion:* The metal flow is in the direction of action of machine motion.
- *Rod forward* (Fig. 17.12a): A solid component is reduced in cross section. The die determines the shape of the tool opening.
- *Tube forward extrusion* (Fig. 17.12b): A hollow cup or can of reduced wall thickness is produced from a hollow can or sleeve. Both the die and the punch determine the shape of the tool opening. The process is also known as hooker extrusion.
- *Can forward extrusion* (Fig. 17.12c): A hollow cup, can, or sleeve is produced from a solid component. The die and the counterpunch determine the shape of the tool opening.

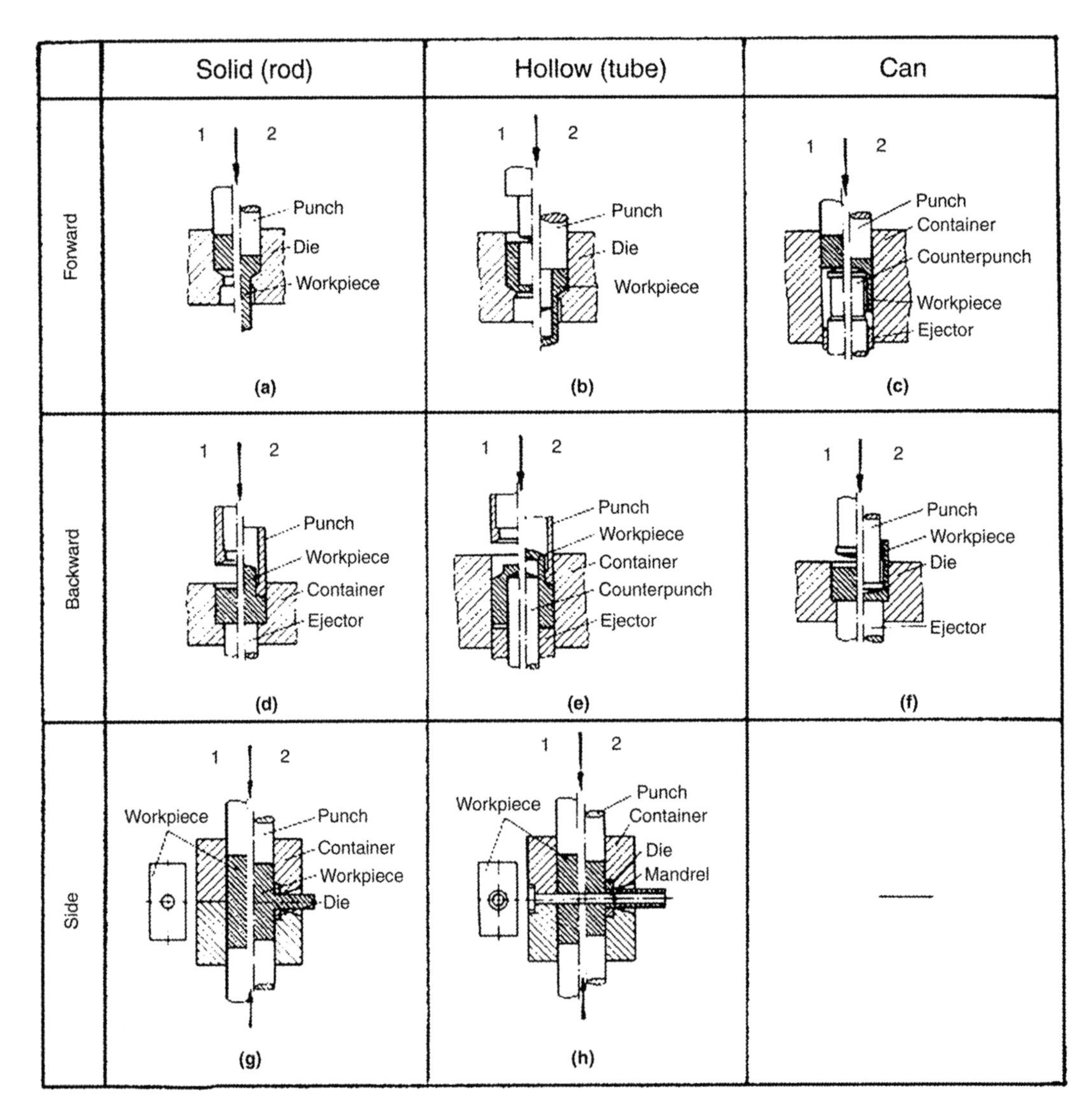

Fig. 17.12 Schematic representation of extrusion process for components. (a) Forward rod. (b) Forward tube. (c) Forward can. (d) Backward rod. (e) Backward tube. (f) Backward can. (g) Side rod. (h) Side tube. 1, initial shape of workpiece; 2, final shape of workpiece [Lange et al., 1985]

- *Backward extrusion:* The metal flow is opposite to the direction of the action of the machine.
- *Rod backward extrusion* (Fig. 17.12d): A solid component is reduced in cross section. The tool opening is determined by the punch only.
- *Tube backward extrusion* (Fig. 17.12e): A sleeve or a can with reduced wall thickness is produced from a sleeve or a can. Both the die and the punch determine the tool opening.
- *Can backward extrusion* (Fig. 17.12f): A thin-walled hollow body (can, sleeve, or cup) is extruded from a solid component. Both the die and the punch determine the tool opening.
- *Rod side extrusion* (Fig. 17.12g): A solid body with a solid protrusion of any profile is extruded. The split die determines the tool opening.
- *Tube side extrusion* (Fig. 17.12h): A workpiece with a hollow protrusion of any profile is extruded. The split die and the mandrel determine the tool opening.

17.7.3 Free or Open-Die Extrusion

The process of free or open-die extrusion is used for relatively small extrusion ratios. Different types of these processes are described below and illustrated in Fig. 17.13:

- *Free-die extrusions of solid bodies* (Fig. 17.13a): Reduction of the cross section of a solid body without supporting the undeformed portion of the component in the container. The unsupported portion should neither upset nor buckle during the process.
- *Free extrusion of hollow bodies, or nosing* (Fig. 17.13b and c): This process consists of extruding the hollow body (can, sleeve, or tube) at its end. The requirement of the container shape at the end of the hollow body depends on its wall thickness.

17.8 Estimation of Friction and Flow Stress

There are a number of formulas for predicting the pressures in forward and backward extrusion [Feldmann, 1961; Wick, 1961; and Gentzsch, 1967]. These formulas are derived either through approximate methods of plasticity theory or empirically from a series of experiments. In both cases, in addition to the approximation inherent in a given formula, estimation of the material flow stress and of the friction factor for a specific process also introduces inaccuracies into the predictions.

Friction is discussed in detail in Chapter 7, "Friction and Lubrication." The value of the friction factor, f, used in expressing the frictional shear stress:

$$\tau = f\bar{\sigma} = m\bar{\sigma}/\sqrt{3}$$

is in the order of 0.03 to 0.08 in cold extrusion. This value is approximately the same as the value of μ (used in expressing $\tau = \sigma_n\mu$, with σ_n normal stress) used in some literature references.

In the deformation zone, the strain and, consequently, the flow stress, $\bar{\sigma}$, vary with location. Due to inhomogeneous deformation and internal shearing, the volume of the material near the die

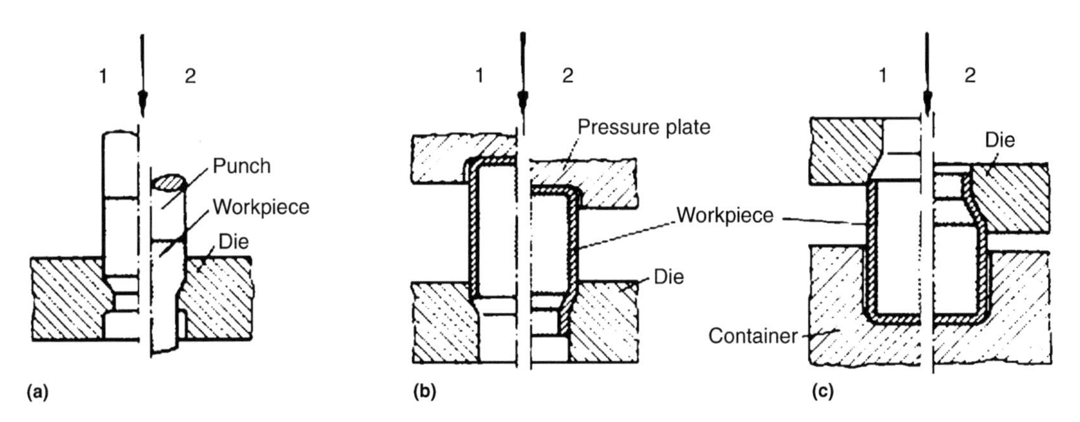

Fig. 17.13 Schematic representation of free-extrusion processes. (a) Solid bodies. (b) Hollow bodies (nosing). (c) Hollow bodies (sinking) with container. 1, initial shape of workpiece; 2, final shape of workpiece [Lange et al., 1985]

surface is subject to more severe deformation. Near the interface, therefore, the local strains and flow stresses are higher. Since the flow stress varies over the deformation zone, most formulas use a so-called "average" or "mean" flow stress, $\bar{\sigma}_a$, which is difficult to determine accurately. A reasonable approximation of the average flow stress, $\bar{\sigma}_a$, can be obtained from the curve for flow stress, $\bar{\sigma}$, versus effective strain, $\bar{\varepsilon}$, as follows:

$$\sigma_a = \frac{1}{\ln R}\int_0^{\ln R} \bar{\sigma} d\bar{\varepsilon} = \frac{a}{\ln R} \quad \text{(Eq 17.6)}$$

where R is the ratio of the initial cross-sectional area, A_0, to the final cross-sectional area, A_1, i.e., $R = A_0/A_1$:

$$a = \int_0^{\ln R} \bar{\sigma} d\bar{\varepsilon} \quad \text{(Eq 17.7)}$$

The value "a" of the integral is the surface area under the effective stress/effective strain curve and corresponds to the specific energy for homogeneous deformation up to the strain $\bar{\varepsilon}_i = \ln R$. If the flow stress can be expressed in the exponential form, then:

$$\bar{\sigma} = K\bar{\varepsilon}^n \quad \text{(Eq 17.8)}$$

and

$$\bar{\sigma}_a = \frac{1}{\ln R}\int_0^{\ln R} K\bar{\varepsilon}^n \sigma d\bar{\varepsilon} = \frac{K(\ln R)^n}{n+1} \quad \text{(Eq 17.9)}$$

where n is the strain-hardening exponent, and K is the flow stress at effective strain $\bar{\varepsilon} = 1$. Since the material flow is influenced by tool geometry and by interface lubrication conditions, the strain distribution and consequently the average strain, $\bar{\varepsilon}$, and the average flow stress, $\bar{\sigma}_a$, are also influenced by tool geometry and lubrication. This fact, however, is not reflected in the approximate estimation of the average flow stress, $\bar{\sigma}_a$, described above.

17.9 Prediction of Extrusion Loads from Selected Formulas

Various formulas for forward and backward extrusion were evaluated in predicting loads for 35 different material values (17 different steels with various heat treatments). Values of container friction were included in predictions of pressures in forward extrusion [Altan et al., 1972]. The formulas that gave the best results are summarized in Tables 17.4 and 17.5. The

Table 17.4 Formulas for calculation of load in forward rod extrusion

Source	Formula	Remarks
[Siebel, 1950] [Feldmann, 1961]	$P = A_0 \cdot \bar{\sigma}_a \ln R + \frac{2}{3}\alpha A_0\bar{\sigma}_a + \frac{A_0\bar{\sigma}_a \ln R\mu}{\cos\alpha \sin\alpha} + \pi D \cdot L\bar{\sigma}_0\mu$	P includes loads due to homogeneous deformation, shearing, die friction, container friction.
[P.E.R.A., 1973]	$P = A_0\bar{\sigma}_0(3.45 \ln R + 1.15)$ $P = \bar{\sigma}_a \cdot A_0(\ln R + 0.6) \cdot \left(1.25 + 2\mu\sqrt{\frac{\pi L}{A_0}}\right)$	For 0.1 to 0.3% C steels
[Pugh et al., 1966]	$P = 8.2A_0\sigma_u^{0.78}(\ln R)^{0.73} = 2.7A_0H^{0.78}(\ln R)^{0.73}$	Originally derived for steels with zinc phosphate + MoS_2. H = hardness of billet before extrusion, kg/mm^2. σ_u in tons/in.^{2}; 1 ton = 2240 lb (1016 kg); P in tons
[James and Kottcamp, 1965]	$P = 0.5A_0(\bar{\sigma}_0 + \sigma_u \cdot F_n)\bar{\varepsilon}_a \exp\frac{4\mu L}{D}$ $F_n = \left(\frac{e\bar{\varepsilon}_a}{n}\right)^n$ $\bar{\varepsilon}_a = 1.24 \ln R + 0.53$	Based on average strain, ε_a, determined in model test with lead and with $\alpha = 27°$ e = 2.71828

Table 17.5 Formulas for calculation of forming load in backward cup extrusion

Source	Formula	Remarks
[P.E.R.A., 1973]	$P = A_0\sigma_0(3.45 \ln (A_0/A_1) + 1.15)$	For 0.1 to 0.3% C steel
[Pugh et al., 1966]	$P = A_0 6.0\sigma_u^{0.8}(\ln R)^{0.72}$ $= A_0 2.8H^{0.72}(\ln R)^{0.72}$	Steels with zinc phosphate + Bonderlube 235 H, hardness in kg/mm_2 σ_0 in tons/in.^{2}
[James and Kottcamp, 1965]	$P = A_0\left(\frac{\bar{\sigma}_0 + \sigma_u F_n}{2.4}\right)\bar{\varepsilon}_a$ $\varepsilon_a = 2.36 \ln R + 0.28$ $F_n = (e\bar{\varepsilon}_a/n)^n$	Based on average strain, $\bar{\varepsilon}_a$, determined in model test with lead and with 5% cone-nosed punch. e = 2.71828
[Altan, 1970]	$P = A_0K_c\sigma_u\ln(A_0/A_1)$	K_c = 2.5 to 3 for low-carbon steel; used $K_c = 3$

flow stress data were obtained, in the form $\bar{\sigma} = K\bar{\varepsilon}^n$, from tensile tests. Whenever necessary, the average flow stress, $\bar{\sigma}_a$, used in the formulas was determined from Eq 17.9. In some formulas, values of tensile strength, $\bar{\sigma}_u$, yield stress, $\bar{\sigma}_0$, or hardness were used. The properties of the billet materials considered in this study are given in Table 17.6. A friction factor, f, or a coefficient of friction $\mu = 0.04$ was used in evaluating all formulas. The value $\mu = 0.04$ was selected on the basis of previous studies that indicated that, in cold forging (zinc phosphate coating + Bonderlube 235 lubricant, Henkel Surface Technologies), μ can be estimated to be between 0.03 and 0.08. As expected, the predicted extrusion pressures would vary considerably with the value of the friction factor.

The formulas that gave the best predictions for punch loads in forward extrusion are given in Table 17.4. The experimental results were obtained by extruding various steel billets 1 in. (2.5 cm) in diameter and 1.5 to 3.5 in. (3.8 to 8.9 cm) in length, with zinc-phosphate-stearate lubrication, at different reductions through a die with a 60° die half angle and a ⅛ in. (3.2 mm) die land.

The predicted extrusion pressures obtained by use of the formulas in Table 17.4 are compared with experimental data in Table 17.7. The predicted pressure values correspond to breakthrough pressures, i.e., they include the container friction. It can be seen in Table 17.4 that the simplest formulas, suggested by P.E.R.A. [P.E.R.A. 1973] and by Pugh [Pugh et al., 1966], give predictions approximately as good as those given by the other formulas in Table 17.4. The formulas that gave the best-predicted values of punch loads, or pressures, in backward cup extrusion are given in Table 17.5. The predicted and measured punch pressures are compared in Table 17.8 for five steels. The backward extruded billets were 1 in. (2.5 cm) in diameter and 1.0 in. (2.5 cm) long. The lubrication was the same as in forward extrusion, i.e., zinc-phosphate-stearate lubrication. The mechanical properties of backward extruded steels are given in Table 17.6.

None of the formulas given in Table 17.5 takes into account the tool geometry (punch angles, radii, die angles). The experimental results are obtained, in most backward extrusion trials, for two punch designs (with 0.09 in., or 2.3 mm, and 0.05 in., or 1.3 mm, punch edge radii). For comparing the predicted and measured punch pressures, an average of the pressures obtained

Table 17.6 Mechanical properties of forward and backward extruded steels

	K			σ_0		σ_u		
	10^3 psi	MPa	n	10^3 psi	MPa	10^3 psi	MPa	Hardness, HRB
1005, hot rolled	86	593	0.250	35	241	47	324	50
1018, hot rolled	117	807	0.224	45	310	68	469	70
12L14, annealed	115	793	0.312	43	296	59	407	60
1038, annealed	134	924	0.255	44	303	70	483	71
8620, subcritical annealed	120	827	0.173	55	379	74	510	82

Table 17.7 Comparison of measured and predicted breakthrough punch pressures in forward extrusion of various steels

		Measured pressure		Pressure predicted using formula from Table 17.4, ksi (MPa)				
Steel	Reduction, %	ksi	MPa	Siebel	P.E.R.A.	Billigmann	Pugh	James and Kottcamp
1005, hot rolled	20	68	469	84 (579)	70 (483)	64 (441)	60 (414)	84 (579)
	50	120	827	132 (910)	128 (883)	127 (876)	132 (910)	137 (945)
	60	144	993	153 (1055)	154 (1062)	155 (1069)	161 (1110)	153 (1055)
	70	161	1110	184 (1269)	189 (1303)	194 (1338)	195 (1344)	181 (1248)
1018, hot rolled	20	111	765	118 (814)	92 (634)	93 (641)	80 (552)	115 (793)
	50	186	1282	183 (1262)	166 (1145)	178 (1227)	176 (1213)	187 (1289)
	60	205	1413	212 (1462)	201 (1386)	216 (1489)	214 (1475)	208 (1434)
12L14, annealed	20	96	662	99 (683)	87 (600)	75 (517)	72 (496)	111 (765)
	50	172	1186	165 (1138)	158 (1089)	159 (1096)	158 (1089)	185 (1276)
	60	187	1289	194 (1338)	191 (1317)	197 (1358)	192 (1324)	209 (1441)
1038, annealed	20	103	710	124 (855)	87 (600)	99 (682)	82 (565)	121 (834)
	50	190	1310	200 (1379)	161 (1110)	197 (1358)	181 (1248)	198 (1365)
	60	210	1448	234 (1613)	195 (1344)	241 (1662)	220 (1517)	221 (1524)
8620, subcritical annealed	20	98	676	137 (945)	112 (772)	106 (731)	86 (593)	124 (855)
	50	178	1227	202 (1393)	202 (1393)	194 (1338)	188 (1296)	195 (1344)
	60	205	1413	230 (1586)	245 (1689)	232 (1600)	230 (1586)	216 (1489)

Table 17.8 Comparison of measured and predicted punch pressures in backward cup extrusion of various steels

Steel	Reduction, %	Measured pressure, ksi	Measured pressure, MPa	Pressure predicted using formula from Table 17.5, ksi (MPa): P.E.R.A.	Pugh	James and Kottcamp	Schoffmann
1005, hot rolled	50	223	1538	245 (1689)	236 (1627)	214 (1475)	194 (1338)
	60	228	1572	248 (1710)	240 (1655)	237 (1634)	213 (1469)
	70	249	1717	260 (1793)	249 (1717)	270 (1862)	240 (1655)
8620, subcritical annealed	50	305	2103	389 (2682)	340 (2344)	300 (2068)	307 (2117)
	60	309	2130	393 (2710)	345 (2379)	326 (2248)	337 (2324)
	70	341	2351	412 (2841)	359 (2475)	366 (2523)	377 (2599)
1038, annealed	50	304	2096	309 (2130)	325 (2241)	309 (2130)	290 (1999)
	60	313	2158	312 (2151)	331 (2282)	343 (2365)	319 (2199)
	70	327	2255	327 (2255)	344 (2372)	392 (2703)	357 (2461)
1018, hot rolled	50	286	1972	320 (2206)	317 (2186)	290 (1999)	281 (1937)
	60	293	2020	323 (2227)	322 (2220)	319 (2199)	308 (2124)
	70	313	2158	338 (2330)	335 (2310)	362 (2496)	345 (2379)
12L14, annealed	50	274	1889	303 (2089)	284 (1958)	293 (2020)	245 (1689)
	60	282	1944	306 (2110)	288 (1986)	329 (2268)	269 (1855)
	70	316	2179	321 (2213)	300 (2068)	380 (2620)	301 (2075)

with both punches was used whenever experimental data were available for both cases.

17.10 Prediction of Extrusion Loads from Model Test

Except where exaggerated inhomogeneities are present in the deforming material, metal flow in extrusion is influenced mainly by tool geometry and lubrication conditions. The effects of material properties on metal flow are relatively insignificant [Altan, 1970, and Sashar, 1967].

If a strain-hardening material is considered and the friction at the tool/material interfaces of the deformation zone is neglected, the external mechanical energy is equal to the internal deformation energy [Altan, 1970], i.e.:

$$p_a A_0 v \Delta t = \int_V \bar{\sigma} d\bar{\varepsilon} dV \qquad \text{(Eq 17.10)}$$

or

$$p_a V = \int_0^{\bar{\varepsilon}_h} \bar{\sigma} d\bar{\varepsilon} \qquad \text{(Eq 17.11)}$$

where, in addition to the symbols previously defined, p_a is average punch pressure, V is volume of deforming material, v is punch velocity, and Δt is time increment. The left side of Eq 17.10 represents the amount of mechanical energy necessary for deformation. This energy is introduced by the punch, which moves at a velocity (v) during the time (Δt) necessary to extrude the volume of material equal to the volume of the deformation zone ($V = A_0 v \Delta t$). The right side of Eq 17.10 represents the total deformation energy obtained by adding the deformation energies consumed within each volume element in the deformation zone. This total deformation energy can be calculated only if the flow stress, $\bar{\sigma}$, and effective strain, $\bar{\varepsilon}$, at each volume element are known. Since this information is not usually available, the following averaging method can be used.

Assuming that every volume element in the deformation zone has the same average strain, $\bar{\varepsilon}$, and the same average flow stress, $\bar{\sigma}_a$, then Eq 17.11 can be written as:

$$p_a V = \bar{\sigma}_a \bar{\varepsilon}_a V \qquad \text{(Eq 17.12)}$$

and $\bar{\sigma} = K\bar{\varepsilon}^n$ with Eq 17.12 gives:

$$p_a = K\bar{\varepsilon}_a^{\,n+1} \qquad \text{(Eq 17.13)}$$

or

$$\bar{\varepsilon}_a = \left(\frac{p_a}{K}\right)^{1/n+1} \qquad \text{(Eq 17.14)}$$

Equations 17.13 and 17.14 can be used for predicting extrusion pressures from a model test. First, a model material (plasticine, aluminum, or mild steel) is extruded using certain tool geometry, and the extrusion pressure, p_a, is measured. By use of Eq. 17.14 and the known K and n values of the model material, the average strain, $\bar{\varepsilon}_a$, is calculated. Then, the calculated values of $\bar{\varepsilon}_a$ and the K and n values of the real material are used with Eq 17.13 to estimate the real extrusion pressure.

Using 1005 hot rolled steel ($\bar{\sigma} = 86{,}000\ \bar{\varepsilon}^{0.25}$ psi) as the model material, the model test method described above was applied in order to predict punch pressures in forward extrusion for a series of the other steels. A comparison of the experimental and predicted values is given in Table 17.9. It can be seen that the largest difference between the two values does not exceed 20%, while predictions are within 10% for most extrusions. It should be noted that the extrusion pressure, p_a, in Eq 17.13 is the same as the ejector pressure. The punch pressure is obtained from the extrusion ratio. For instance, for R = 80%, punch pressure = $p_a/0.80$.

In forward extrusion, only the deformation pressure, i.e., the end pressure, can be predicted from a model test. In order to predict the maximum load, the friction in the extrusion container must be considered. With the symbols listed in Fig. 17.10(a), the equilibrium of forces in the axial direction gives:

$$dp_x\left(\frac{\pi D^2}{4}\right) = dx\tau_f\pi D \qquad \text{(Eq 17.15)}$$

Assuming a constant frictional shear stress, τ_f, at the tool/material interface, $\tau_f = f\bar{\sigma}_0$, Eq 17.15 transforms into:

$$\frac{dp_x}{dx} = \frac{4f\bar{\sigma}_0}{D} \qquad \text{(Eq 17.16)}$$

by integrating, Eq 17.16 gives:

$$p_x = \frac{4f\bar{\sigma}_0}{D}x + C \qquad \text{(Eq 17.17)}$$

The integration constant C is determined from the condition for x = 0, $p_x = p_e$ in Eq 17.13, or, in forward extrusion:

$$p_x = p_e + \frac{4f\bar{\sigma}_0}{D}L \qquad \text{(Eq 17.18)}$$

Equation 17.18 illustrates that the peak pressure, p_p, in forward rod extrusion is equal to the sum of the end pressure, p_e, necessary for deformation and the additional pressure, $p_{fc} = 4f\bar{\sigma}_0\ L/D$, necessary to overcome the container friction.

The shear friction factor, f, is to be determined from Eq 17.18 for a given billet length, L, and for the known yield stress, $\bar{\sigma}_0$, of the model material, in this case, 1005HR steel. It is reasonable to assume that the friction factor, f, would not significantly change from one steel to another. The average strain, $\bar{\varepsilon}$, is determined from Eq 17.14, using the data for 1005 HR steel. The end pressure, p_e, is then determined from Eq 17.13. Finally, the peak pressures, p_p, are calculated for different steels using Eq 17.18. The results are compared with experimental values in Table 17.10. The agreements between prediction and experiment are, in most cases, within 10 to 15% and can be considered acceptable for practical purposes.

17.11 Tooling for Cold Forging [Lange et al., 1985, and ICFG, 1992]

A tooling setup for cold forging is shown in Fig. 17.14.

17.11.1 Punch

The punch is the portion of the tool that forms the internal surface of the workpiece in a can extrusion, or that pushes the workpiece through a die in a rod, tube, or open-die extrusion.

Table 17.9 Comparison of measured and predicted (model test) punch pressures in backward extrusion of various steels (model material, hot rolled 1005 steel; $\bar{\sigma} = 86{,}000\bar{\varepsilon}^{0.25}$ psi)

	K				Measured pressure		Predicted pressure	
Steel	ksi	MPa	n	Reduction, %	ksi	MPa	ksi	MPa
8620, subcritical annealed	120	827	0.173	50	305	2103	304	2096
				60	309	2130	308	2124
				70	341	2351	334	2303
1038, annealed	134	924	0.255	50	304	2096	347	2392
				60	313	2158	357	2461
				70	327	2255	387	2668
1018, hot rolled	117	807	0.244	50	286	1972	300	2068
				60	293	2020	305	2103
				70	313	2158	335	2310
12L14, annealed	115	793	0.312	50	274	1889	302	2082
				60	282	1944	315	2172
				70	316	2179	343	2365

Table 17.10 Comparison of measured and predicted (model test) peak punch pressures in forward extrusion of various steels

	K				Measured pressure		Predicted pressure	
Steel	ksi	MPa	n	Reduction, %	ksi	MPa	ksi	MPa
1018, hot rolled	117	807	0.224	20	111	765	95	655
				50	186	1282	165	1138
				60	205	1413	190	1310
12L14, annealed	115	793	0.312	20	86	593	90	621
				50	172	1186	158	1089
				60	187	1289	192	1324
8620, subcritical annealed	120	827	0.173	20	98	676	100	689
				50	178	1227	182	1255
				60	205	1413	220	1517
1038, annealed	134	924	0.255	20	103	710	100	689
				50	190	1310	180	1241
				60	210	1448	220	1517
4340, annealed	167	1151	0.193	20	122	841	132	910
				50	228	1572	232	1600
				60	251	1731	275	1896

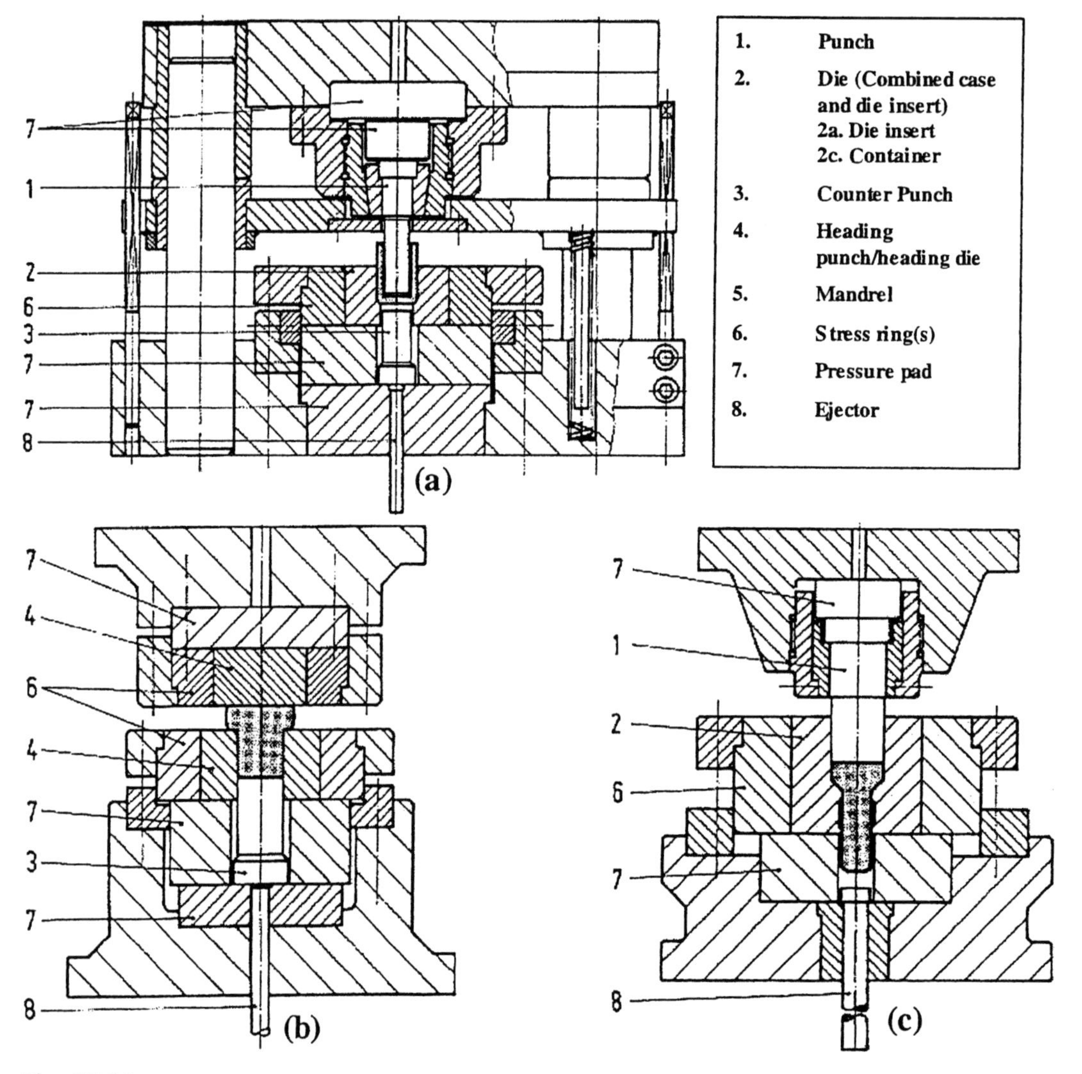

Fig. 17.14 Tooling setup for cold forging. (a) Can extrusion. (b) Upsetting. (c) Forward extrusion [ICFG, 1992]

In forward rod extrusion, wear of the punch is normally not a problem. Compressive stresses range up to 350 ksi (2413 MPa) and sometimes slightly more. In can extrusion, the punch is subjected to compressive and bending load as well as heavy wear, which increases the temperature at the punch nose, especially in high-speed presses. During stripping, tensile stresses occur. In ironing, the punch pushes the workpiece through the die and forms its internal surface. Depending on size, shape, loading, and wear, both tool steels and cemented carbides may be selected as tool materials.

17.11.2 Die, Die Insert, Container, Case (Combined Case and Die Insert)

The die is the portion of the tool assembly that contains the workpiece and forms the external surfaces. The choice of the material depends on the maximum internal pressure and fatigue as well as the toughness and wear-resistance requirements.

Normally, dies are reinforced by one or more stress rings. Depending on the stress state, principal designs used in the industry are as follows:

- Die
- Die with one-piece insert
- Die with axially split insert
- Die with transversely split insert

Depending on size and shape, both tool steels and cemented carbides may be selected for the die insert.

17.11.3 Counter Punch

The counter punch forms the base shape and is usually used to eject the workpiece from the die. In both can extrusion and coining or sizing, the counter punch is subjected to compressive stresses similar to those occurring in punches. Both tool steels and cemented carbides may be selected.

17.11.4 Heading Punch/Heading Die

The heading punch upsets the head of the workpiece. The heading die contains the shank and forms the underhead surface of the workpiece. Choice of the tool material and its heat treatment depends on the compressive stresses (up to 300 ksi, or 2068 MPa, and more), section size, and the demand for wear resistance and toughness. Besides tool steels, for a range of small and medium section sizes, cemented carbides may be used when long runs demand high wear resistance.

17.11.5 Mandrel (Pilot)

The mandrel or pilot is that part of the punch assembly that enters a hollow billet and forms the inside wall. The mandrel is subjected to wear and high tensile stresses, and therefore, the choice of material is aimed at properties such as wear resistance and/or high yield strength. Due to high tensile stresses, tool steels are normally used.

17.11.6 Stress Rings

The stress rings form the intermediate and outer portions of the die assembly to prevent the die from bursting. The tensile stresses will amount to 200 ksi (1379 MPa) and more. Normally, tool steels are used.

17.11.7 Pressure Pads

The pressure pad is the block of material that supports and spreads the load behind the pad or die. Sufficient thickness is important in order to spread the load properly and to avoid bending. An extreme degree of parallelism and squareness is essential. The tool material should provide high compressive strength up to 220 ksi (1517 MPa). Normally, tool steels are used.

17.11.8 Ejector

The ejector is the part of the tooling that ejects the workpiece from the die, usually without taking part in the forming operation. The choice of the material depends on the section used and the ejection pressure. Normally, tool steels are used.

17.12 Punch Design for Cold Forging

Figure 17.15 shows some punches and mandrels used for solid and hollow forward extrusion.

The design process for punches and mandrels can be divided into the following stages:

- Determining the forming load
- Deciding the overall shape and proportions of the tool. For can extrusion, the detailed shape of the punch nose is to be determined.
- Choosing an appropriate tool material, considering the stresses, manufacturability, tool life, availability, and costs

- Verifying the design with model tests, prototyping, or finite-element simulations

For punches, an approximate design guideline is to determine the maximum compressive stress expected in the process and comparing this with the compressive properties of the tool material. The punch face shape has a significant effect on the friction force and the material flow. Figure 17.16 shows different types of punch face shapes. The advantage of using a flat/conical punch geometry (Fig. 17.16c) is that the lubricant carrier and the lubricant cannot be separated easily or overextended locally, as is the case with conical faces.

Some recommendations for designing punches are as follows:

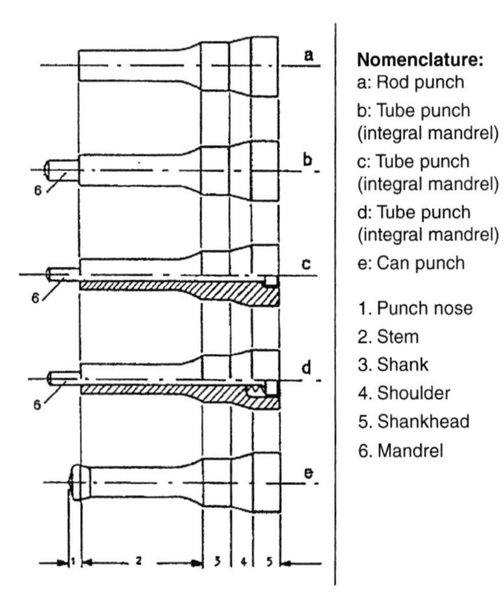

Fig. 17.15 General designs for punches used in cold extrusion [ICFG, 1992]

- Punches should be made as short as possible. The punch will buckle if the stress exceeds a certain level for a given length-to-diameter ratio. The design should also avoid excessive bending stress that may occur due to buckling of the punch. Punches for can extrusion are more susceptible to buckling failure than those for rod or tube extrusion.
- Large and abrupt changes in cross sections should be avoided over the whole punch length in order to minimize undesirable stress concentration. The changes in the cross sections should be done using small cone angles and large transition radii, nicely rounded and polished.
- In forward extrusion, the diameter of the punch at the deformation zone should be chosen such that:
  a. There is sufficient entry clearance for the punch.
  b. The clearance between the punch and the die should not be very large, to avoid burr formation.
  c. The clearance should not be too small, or else the punch will wear out due to elastic deflection of the tooling.

Figure 17.17 shows the parameters involved in design for backward can extrusion punches. The details of punch design and parameters calculation can be found in the literature [ICFG, 1992, and Lange et al., 1985].

17.13 Die Design and Shrink Fit [Lange et al., 1985, and ICFG, 1992]

Dies for cold extrusion can be of different configurations, depending on the design preferences or operational requirements. In general,

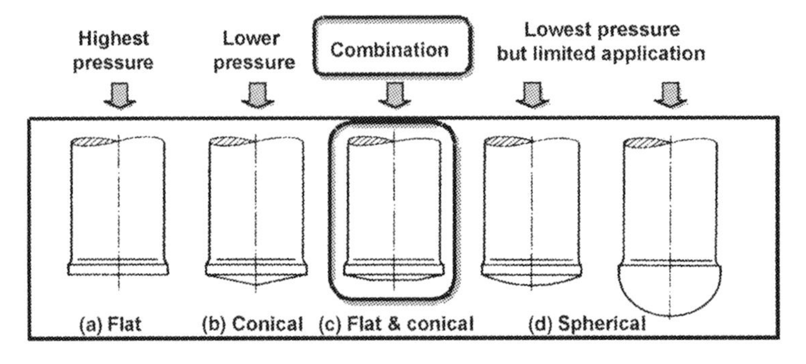

Fig. 17.16 Punch face shapes for backward can extrusion [Lange et al., 1985]

the design of dies and die assemblies requires consideration of the following:

- The flow stress of the workpiece material, which is affected by strain, temperature, and strain rate
- The type of the process
- The geometry of the die and the slug or preform
- Friction and lubrication

Figures 17.8(a) and (b) show the general nomenclature for extrusion dies. One half of each figure illustrates the use of split dies. The recommended design guidelines for the parameters in Fig. 17.18 are as follows:

- The die entry radius should be as large as possible so as to minimize high stress concentrations.
- The included die angle for minimum extrusion pressure should lie between 50 and 70° for extrusion ratios of approximately 2 and 4, respectively.
- The die land radius should be fairly small, i.e., between 0.008 and 0.06 in. (0.2 and 1.5 mm), depending on the diameter of the die throat.
- The length of the die land should be between 0.08 and 0.16 in. (2 and 4 mm), depending on the throat diameter.
- The die relief should be kept small to provide guidance for the extruded rod and should be higher than the permitted wear allowance on the die throat.

Extrusion dies are generally very highly stressed. One or more rings are used to assemble the container with interference fits in order to apply compressive stresses to the die ring or liner. The number and diameter of the rings depend on the magnitude of stresses needed and the overall die space available in the press. Figure 17.19 shows the dies with two stress rings. Heat treatment of stress rings is very important to develop the desired mechanical properties. Table 17.11 gives guidelines for the diameters of the stress rings.

In applications where the extrusion pressure is very high, in excess of 300 ksi (2068 MPa), it is recommended to use stress rings that are strip wound. Plastic deformation in conventional stress rings makes it difficult to control the compressive prestress and results in die failures. Strip-wound containers have strength that is twice or three times higher than the conventional stress rings. This high strength of strip-wound containers makes it possible to provide optimum prestress of the die inserts, leading to an improvement of two to ten times in die life, depending on the cold forging operation [Groenbaek, 1997].

17.14 Process Sequence Design

In practice, cold forming requires several stages to transform the initial simple billet ge-

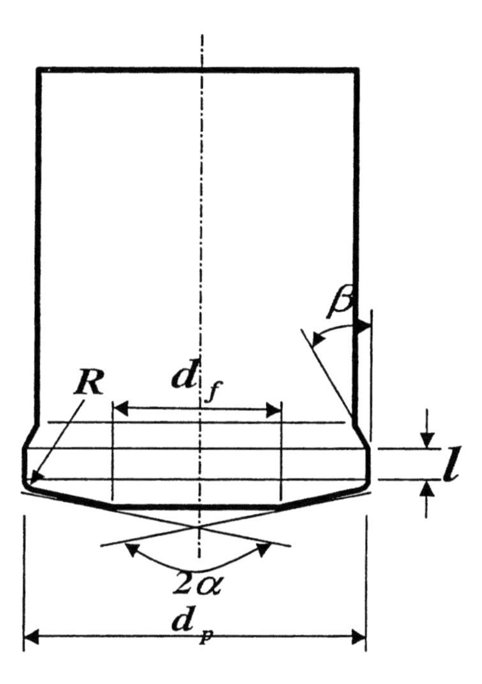

Based on Punch Diameter, d_P

Flat Diameter,

$$d_f = d_P - [2R + (0.2 \sim 0.3)d_P]$$

Included Face Angle,

$$2\alpha = 160^o \sim 170^o$$

Punch Land,

$$l = 0.3 \sim 0.7 \times \sqrt{d_P}$$

Punch Radius,

$$R = 0.05 \sim 0.1 \times d_P$$

Relief Angle,

$$\beta = 4^o \sim 5^o$$

Fig. 17.17 Can extrusion punch design guidelines [ICGF, 1992]

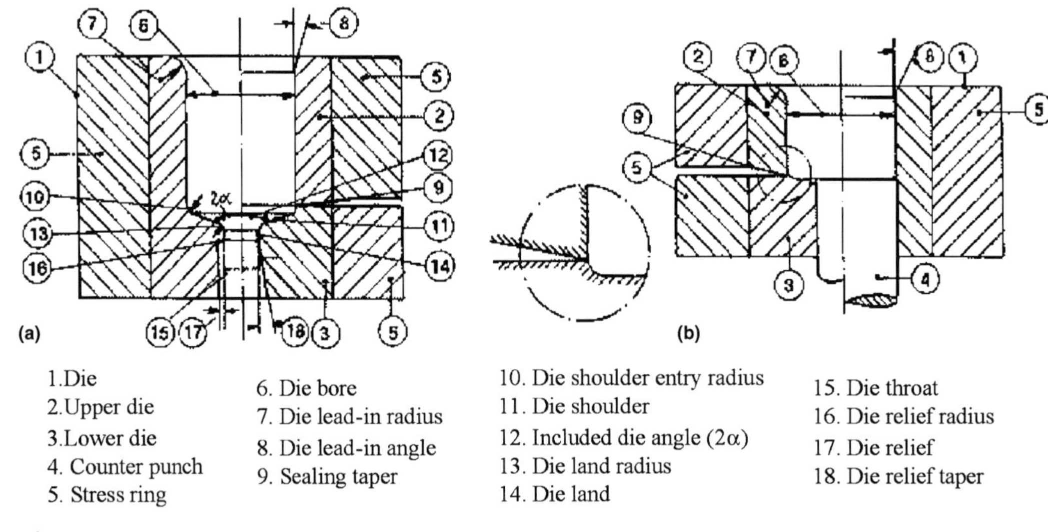

Fig. 17.18 General nomenclature for extrusion dies. (a) Rod/tube extrusion. (b) Can extrusion [ICFG, 1992]

ometry into a more complex product. One of the major efforts required in applying the use of the cold forging process successfully is the design of the forming sequences for multistage forging operations. Highly experienced die designers do this work, and it requires both experienced judgment and the application of established design rules [Sevenler et al., 1987]. Computer-aided tools, such as finite-element modeling (FEM) simulations, are used to assist the die designers while they establish the forming sequences in cold forging [Altan et al., 1992].

In the early stages of forming sequence design, the designer draws rough sketches of the preforms for each station. The designer then calculates the volumes and tries to establish the accurate workpiece dimensions. The designer may then want to change some part or preform dimensions to examine various design alternatives while keeping the workpiece volume constant. The FEM simulation of metal flow and tool deflection may be required at this stage of design. The FEM can be used as a validation tool before tryouts. Figures 17.20 through 17.22 show some examples of forging sequences. More examples can be found on the CD available with this book.

17.15 Parameters Affecting Tool Life

Tool life in cold forging is important, as it affects the tool cost and thereby increases the process and parts costs. Figure 17.23 shows the various parameters that affect tool life [Yamanaka et al., 2002].

Tool life depends on the production conditions, work material, human skills, and tool quality. Tool quality is affected by how well the tool is designed, the manufacturing of the tool, the material selected, and the heat and surface treatment conditions. Figure 17.24 shows the typical tool defects encountered in cold forging tooling.

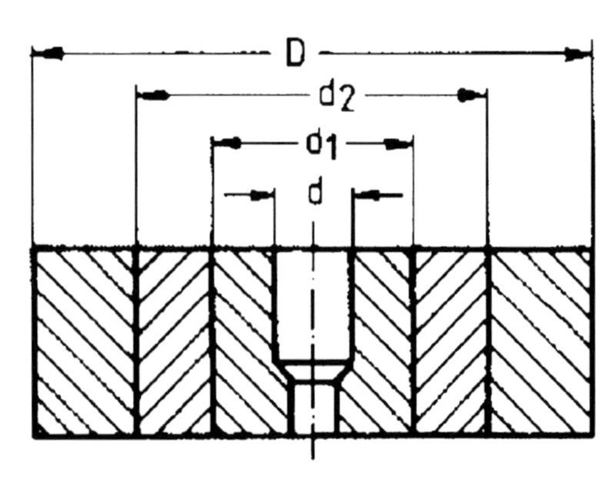

D = overall diameter

= second stress ring outer diameter

d = die bore diameter

d_1 = die diameter

d_2 = first stress ring outer diameter

Fig. 17.19 Die with two stress rings [ICFG, 1992]

17.15.1 *Die Failure*

In general, die failure may occur in one or more of the following forms.

Pickup and Wear [ICFG, 1992]. If extruded components show bright areas, scratches, and score lines in the direction of extrusion, then unfavorable frictional conditions between the die and workpiece result in pickup on the die. These conditions may be caused by insufficient lubrication, a rough slug surface, too rough die surface (owing to incorrect manufacture or heavy wear), too low die insert hardness, or unsuitable tool geometry.

The die should be taken out for service as soon as the effects of pickups are observed, so that regrinding and finishing can be carried out. Tool steel dies are more prone to pickup and heavy wear than cemented carbide dies.

Fracture of Inserts [ICFG, 1992]. Die inserts may fail in different ways.

Axial Cracks. Fractures of this type are usually the result of overstressing, which may arise from insufficiently high pressures.

Table 17.11 Guidelines for stress rings diameter calculations [ICFG, 1992]

Internal pressure, ksi	Internal pressure, MPa	Number of stress rings required	Required ratio of overall diameter to die bore diameter (D/d) (approx.)	Intermediate diameter (approx.)
Up to 150	Up to 1034	None	4 to 5	. . .
150 to 250	1034 to 1724	One	4 to 6	$d_1 \sim 0.9\ \nu$ Dd
250 to 300	1724 to 2068	Two	4 to 6	d: d_1: d_2: D ≈ 1: 1.6 to 1.8: 2.5 to 3.2: 4 to 6

Note: ν is the Poisson's ratio, and D, d, d_1, and d_2 are the same as in Fig. 17.19.

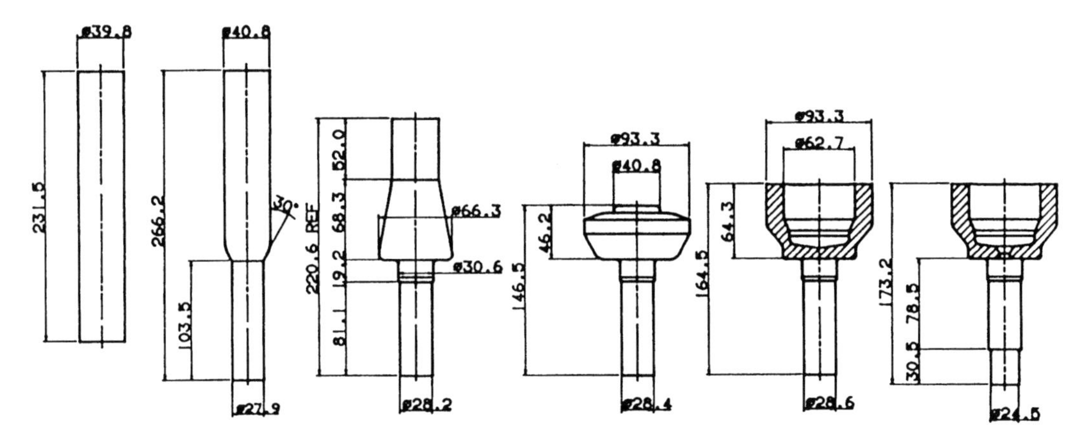

Fig. 17.20 Forging sequence (example 1)

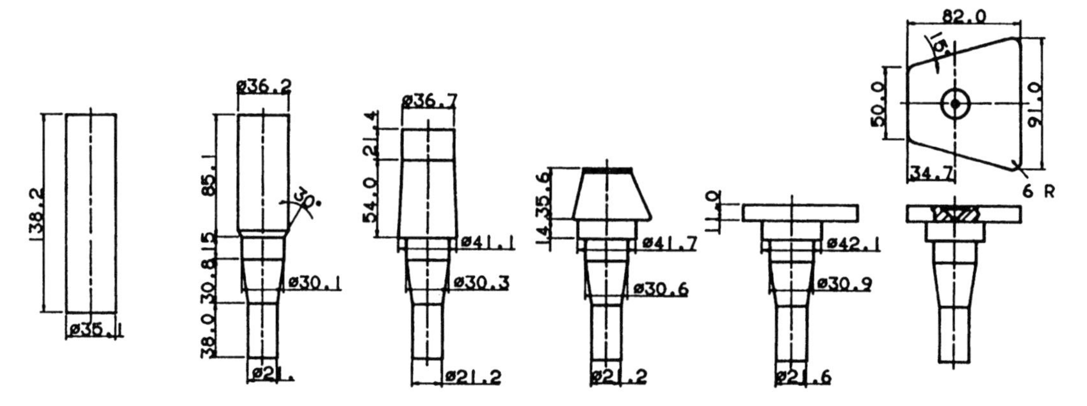

Fig. 17.21 Forging sequence (example 2)

If insufficient prestressing is suspected, then it could be increased by using additional stress rings, increasing the overall diameter of the assembly, or using stress rings of greater strength. In all the situations, the interference must be adjusted appropriately.

If the die design is already considered as the best possible, then fracture is due to excessive extrusion pressure. The most frequent reason for this could be high flow stress of the slug material as a consequence of inadequate annealing or departure from correct material composition.

Transverse Cracks. A transverse crack, which is most common, is usually in one-piece rod extrusion die inserts and occasionally in tube extrusion dies in the vicinity of the die shoulder entry. A similar type may also occur in the die bore at locations corresponding to the position of the rear face of the slug at the beginning and end of the extrusion stroke. These are due to stress concentration or triaxial stresses involving a high component of axial tensile stress. In most cases, fatigue failure (after the production of a fairly large number of parts) is responsible.

To prevent transverse cracks, use of a tougher die material, decreasing the hardness of a tool steel insert by 2 to 5 points in the Rockwell C scale, splitting the die or decreasing the die angle, and/or increasing the die entry radius could be helpful [ICFG, 1992].

17.15.2 Procedure to Improve Tool Life

Figure 17.25 describes a typical procedure to improve tool life in the cold forging process. Observation and proper inspection of the failed tool must be done. Based on the causes of the failures discussed above, knowledge, and experience, a tool failure mechanism can be assumed. Using experience and finite-element simulations, a new

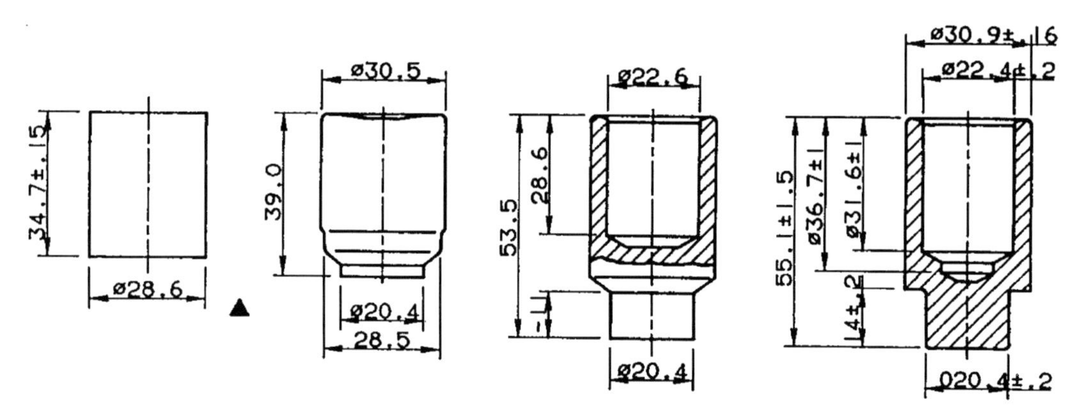

Fig. 17.22 Forging sequence (example 3)

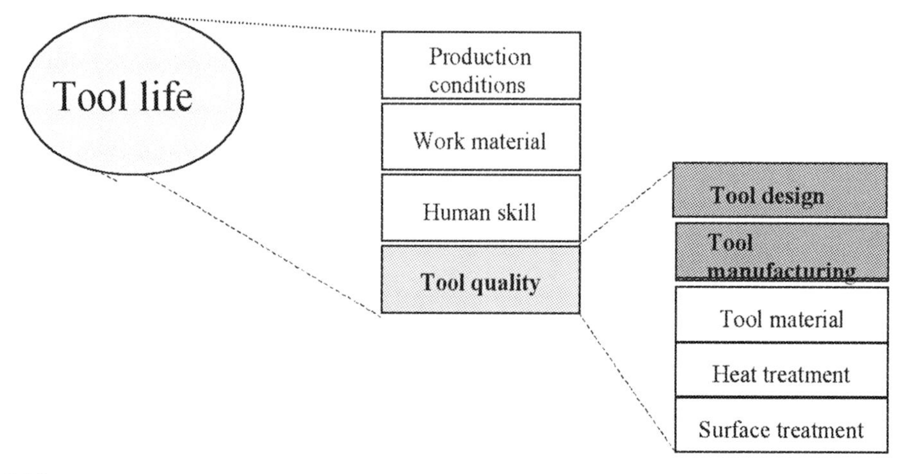

Fig. 17.23 Parameters affecting tool life in cold forging [Yamanaka, 2002]

design can be suggested to overcome the tool failure. If the new design gives a satisfactory tool life, then it could be finalized for the process.

17.16 Warm Forging

In cold forging of parts with relatively complex geometries from high-carbon and alloy steels, forging pressures are extremely high and the ductility of the material is low. As a result, short tool life and defects formed during forging limit the economic use of the cold forging processes. Consequently, in many cases, warm forging, i.e., forging at temperatures below recrystallization temperature, is commonly used [Altan et al., 1983]. For warm forging, steels are usually heated between room temperature and usual hot forging temperature. The normal temperature range is considered to be 1110 to 1650 °F (600 to 900 °C). An exception is the warm forging of austentic stainless steels, which usually are forged between 390 and 570 °F (200 and 300 °C) [ICFG, 2001b]. The process may be interpreted broadly as thermomechanical processing at elevated temperature to achieve the following advantages:

- A reduction in flow stress. This is applicable, in particular, to high-alloy steels. As a result, tool stresses and forging loads are reduced (Fig. 17.26).
- Greater ductility of the forged part. This allows more complex shapes to be forged.
- A reduction in strain hardening. This may reduce the number of forming and annealing operations.
- Greater toughness of the forged part
- Improved accuracy as compared to hot forging
- Enhanced product properties through grain refinement and controlled phase transformations in heat treatable steels

As an example, variations of tensile stress and ductility (as indicated by reduction of area) with temperature are shown in Fig. 17.26 for 1045 steel. It can be seen that the tensile stress does not decrease continuously with temperature. There is a temperature range, in this case, at approximately 400 to 800 °F (205 to 425 °C), wherein forging would not be recommended. Warm forging requires determination of the optimum forging temperature and the suitable lubricant. Selection of warm forging lubricants has proved to be especially difficult. The factor that limits the use of warm forging is that the tech-

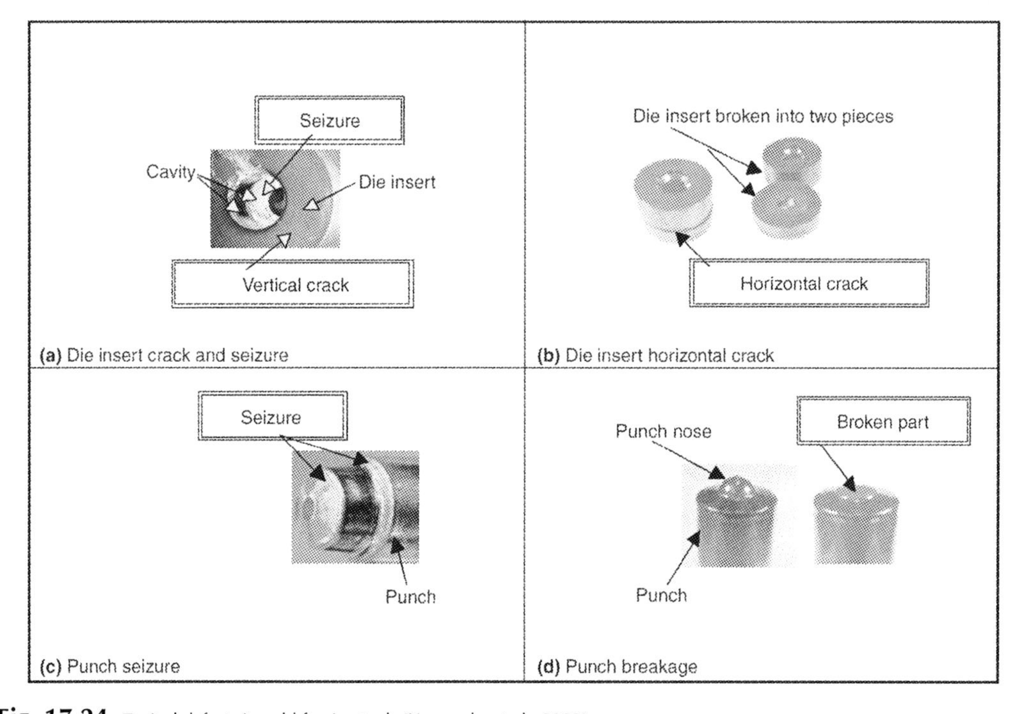

Fig. 17.24 Typical defects in cold forging tools [Yamanaka et al., 2002]

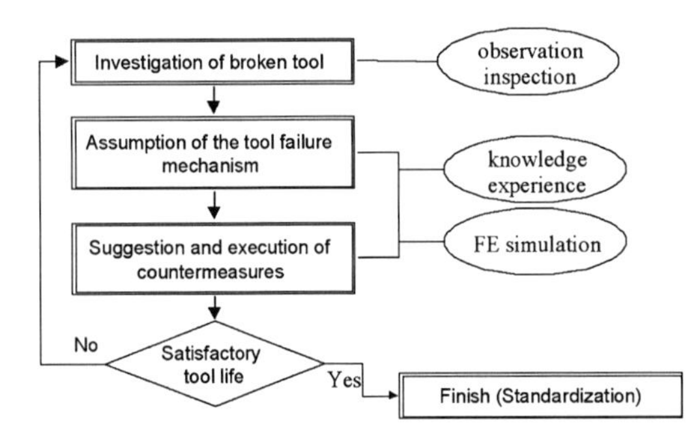

Fig. 17.25 Procedure for improving tool life [Yamanaka et al., 2002]

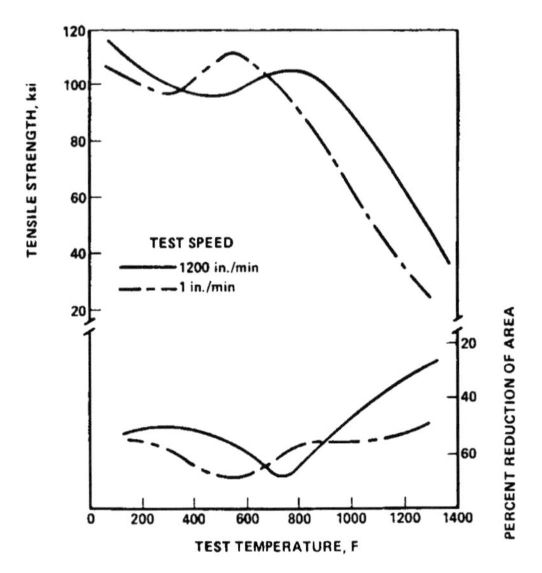

Fig. 17.26 Effects of test temperature and test speed (strain rate) on tensile strength and reduction of areas of hot rolled type 1045 [Altan et al., 1983]

nology is still undergoing development, particularly in aspects of surface treatment, lubrication, and tooling.

Tooling for warm forging is similar to that for cold forging, with some modifications made in the die to allow increased temperatures, internal die cooling, and venting of coolants.

REFERENCES

[Altan, 1970]: Altan, T., "The Use of Model Materials in Predicting Forming Loads in Metal Working," *Transactions ASME,* May 1970, p 444.

[Altan et al., 1972]: Altan, T., and Beckner, J.R., "Prediction of Punch Loads and Pressures in Cold Extrusion of Steel," SME Technical Paper MF72-142, Society of Manufacturing Engineers, 1972.

[Altan et al., 1982]: Raghupathi, P.S., Oh, S.I., Altan, T., "Methods of Load Estimation in Flashless Forging Processes," Topical Report No. 10, Battelle Columbus Laboratories, 1982, p 94.

[Altan et al., 1983]: Altan, T., Oh, S.-I., Gegel, H.L., "*Metal Forming Fundamentals and Applications,*" American Society for Metals, 1983.

[Altan et al., 1987]: Sevenler, K., Raghupathi, P.S., Altan, T., "Forming Sequence Design for Multistage Cold Forging," *Journal of Mechanical Working Technology,* Vol. 14, 1987, p 121–135.

[Altan et al., 1992]: Kim, H., Sevenler, K., Altan, T., "Computer Aided Part and Processing Sequence Design in Cold Forging," *J. of Materials Processing Technology,* Vol 33 (No. 1–2), 1992, p 57–76.

[Altan et al., 1996]: Altan, T., O'Connell, M., Painter, B., Maul, G., "Flashless Closed-Die Upset Forging-Load Estimation for Optimal Cold Header Selection," *Journal of Materials Processing Technology,* Vol 59, No. 1–2, May 1996, p 81–94.

[ASM, 1970]: *Forging and Casting,* Vol. 5, *Metals Handbook,* 8th ed., American Society for Metals, 1970, p 72.

[Bay, 1994]: Bay, N., "The State of the Art in Cold Forging Lubrication," *Journal of Materials Processing Technology,* Vol 46, 1994, p 19.

[Bay, 1997]: Bay, N., "Cold forming of Aluminum—State of the Art," *Journal of Materials Processing Technology,* Vol 71, 1997, p 76–90.

[Billigmann et al., 1973]: Billigmann, J., and Feldmann, H.D., "Upsetting and Pressing," Carl Hansen Verlag, Munich, 1973 (in German).

[Doehring, 1972]: Doehring, R.C., "New Developments in Cold and Warm Forging," SME Technical Paper MF72-526, Society of Manufacturing Engineers, 1972.

[Drozda, 1983]: Drozda, T.J., "Money Saving Innovations in Automatic Forming," *Manufacturing Engineering,* Vol 90, (No. 2), 1983, p. 32–39.

[Feldmann, 1961]: Feldmann, H.D., "*Cold Forging of Steel,*" Hutchinson and Company Ltd., London, 1961.

[Feldmann, 1977]: Feldmann, H.D., "Cold Extrusion of Steel," *Merkblatt 201,* prepared for

Beratungsstelle fuer Stahlverwendung, Dusseldorf, 1977 (in German).

[Gentzsch, 1967]: Gentzsch, G., "Cold Upsetting, Cold Extrusion and Coining," Vol. 1, *Literature Review;* Vol. 2, *References,* VDI Verlag, Dusseldorf, 1967 (in German).

[Groenbaek, 1997]: Groenbaek, J., and Nielsen, E.B., "Stripwound Containers for Combined Radial and Axial Prestressing," *Journal of Material Processing Technology,* Vol 71 (No. 1), 1997, p 30.

[Herbst, 1967]: Herbst, V., "Accurate Shearing of Workpiece Materials for Cold and Warm Forging," Doctoral dissertation, Technical University, Hanover, Germany, 1967 (in German).

[ICFG, 1992]: "Objectives, History and Published Documents," ISBN 3-87525-058-3, International Cold Forging Group, 1992.

[ICFG, 2001a]: "Steels for Cold Forging: Their Behavior and Selection," Document No. 11/01, International Cold Forging Group, 2001.

[ICFG, 2001b]: "Warm Forging of Steels," Document No. 12/01, International Cold Forging Group, 2001.

[ICFG, 2002]: "Cold Forging of Aluminum," Document No. 13/02, International Cold Forging Group, 2001.

[James and Kottcamp, 1965]: James, C.T., and Kottcamp, E.H., "Selection of Steel and Heat Treatment for Ease of Cold Extrusion," paper presented at Chicago Technical Meeting of the American Meeting of the American Iron and Steel Institute, Oct. 20, 1965.

[Lange et al., 1985]: Lange, K., et al., "*Handbook of Metal Forming,*" McGraw-Hill, 1985, p 2.3, 9.19.

[P.E.R.A., 1973]: "Cold Extrusion of Carbon Steels," P.E.R.A. Report No. 69, 1973.

[Pugh et al., 1966]: Pugh, H., et al., "Some Aspects of Cold Extrusion of Steel," *Sheet Metal Ind.,* Vol 43, 1966, p 268–305.

[Sagemuller, 1968]: Sagemuller, F., "Cold Impact Extrusion of Large Formed Parts," *Wire,* No. 95, June 1968, p 2.

[Sahar, 1967]: "Prediction of Extrusion Pressures in the Cold Forging of Steel," *Trans. Metall. Society AIME,* Vol 239, 1967, p 1461–1468.

[Sevenler et al., 1987]: Sevenler, K., Raghupathi, P.S., Altan, T., "Forming Sequence Design for Multi-Stage Cold Forging," *J. of Mechanical Working Technology,* Vol. 14, 1987, p 121–135.

[Siebel, 1950]: Siebel, E., "Fundamentals and Concepts of Forming," *Werkstattstechnik and Maschienbau,* Vol 40, 1950, p 373 (in German).

[Watkins, 1973]: "Cold Forging and Extrusion of Steel," Review 176, *International Metallurgical Review,* Vol 18, 1973 (Part I: Basic Principles, p 123; Part II: Properties and Tooling, p 147; Part III: Economics and Future Applications, p 162).

[Wick, 1961]: Wick, C.H., "*Chipless Machining,*" Industrial Press, New York, 1961.

[Yamanaka et al., 2002]: Yamanaka, M., and Sunami, F., "Tool Design for Precision Forging," presented at Cold and Warm Precision Forging Workshop (Canton, MI), Yamanaka Engineering Co., Ltd., Nov 14, 2002.

SELECTED REFERENCES

- **[Wick, 1984]:** Wick, C.H., "*Tools and Manufacturing Engineers Handbook: Volume I, Forming,*" Society of Manufacturing Engineers, 1984.
- **[Witte, 1967]:** "Investigation on the Variations of Loads and Energies in Cold Extrusion under Production Conditions, Influence of Phosphate Coating Thickness and Lubricants," Report No. 6, Institute for Forming, Technical University, Stuttgart, 1967, Verlag Girardet, Essen. (in German).

CHAPTER 18

Process Modeling in Cold Forging Using Finite-Element Analysis

Prashant Mangukia

18.1 Introduction

The finite-element method (FEM) is one of the numerical techniques used for solving differential equations governing engineering problems. This method has been applied to various engineering processes, including metal forming. The effectiveness of this method is now widely recognized.

Finite-element methods for simulations of metal forming processes are classified into those for the elastic-plastic and rigid-plastic analyses. In the elastic-plastic simulation, material is modeled as deforming elastic-plastically. Thus, the results are given for not only the plastic deformation but also for elastic deformation such as residual stresses and springback. Rigid-plastic simulation assumes material to deform only plastically, and, in comparison to elastic-plastic simulations, it results in shorter computing time.

18.2 Process Modeling Input

As discussed in Chapter 16, "Process Modeling in Impression-Die Forging Using Finite-Element Analysis," the accuracy of finite-element (FE) process simulation depends heavily on the accuracy of the input data, namely, flow stress as a function of temperature, strain, strain rate, and microstructure, as well as friction characteristics at the interface. Other inputs required are the geometric parameters of the objects and the process parameters. Several issues, such as material properties, geometry representation, computation time, and remeshing capability, must be considered in cost-effective and reliable application of numerical process modeling. The inputs to FEM are discussed in detail in section 16.3, "Process Modeling Input," in Chapter 16 of this book. A brief summary is presented below in terms of important inputs for modeling the cold forging process.

18.2.1 Geometric Parameters

Depending on its geometrical complexity, a forging process can be simulated either as a two-dimensional, axisymmetric or plane-strain, or a three-dimensional problem. In general, in order to have an efficient simulation it is necessary to remove all minor geometrical features, such as small radii in the dies that do not have a significant effect on the metal flow [Altan et al., 1990]. However, for some specific applications, such as microforming processes, the size effects should be taken into account in the simulation [Messner et al., 1994].

18.2.2 Tool and Workpiece Material Properties

In most cold forming simulations, tools are considered as rigid. Temperature increases in cold forging have little influence on the process. Therefore, the simulations are usually carried out under isothermal conditions. The stress-strain relation of materials, commonly referred to as the flow stress curve, is usually obtained by the cylinder compression test. To be applicable without errors or corrections, the cylindrical sample must be upset without any barreling (Fig. 18.1), i.e., a state of uniform stress must be maintained in the sample at all times during the test. Use of adequate lubrication prevents barreling to some extent. However, in practice, barreling is inevitable, and procedures for correcting flow stress errors due to barreling may be necessary (see Chapter 4, "Flow Stress and Forgeability," in this book).

Figure 18.1(a) shows a sample that was upset to 62% reduction (true strain of 0.96). This sample shows significant barreling, implying that the lubricant applied was ineffective at this high level of strain. The failure of the lubricant is evident by the two different regions (shiny and nonshiny areas) depicted on the surface of the sample (Fig. 18.1a). In general, for process modeling of precision forging, the experimental flow stress curves were found to be reliable only up to a strain level of 0.5 [Altan et al., 2001].

One way of correcting the flow stress, or determining approximately the error magnitude at high strain levels caused by inadequate lubrication (barreling), is to simulate the compression process by FEM using various friction factors (Fig. 18.1c) [Altan et al., 2001]. By measuring the amount of barreling (Fig. 18.1a) and comparing the experimental data with FEM predictions, it is possible to obtain more reliable flow stress data at higher strain levels. This method, called the inverse analysis technique, has been used to simultaneously determine friction and stress/strain data using a ring test [Altan et al., 2003].

Attention should also be given to other sources of flow stress error determined through compression tests, such as variation in coil material properties, specimen preparation, surface defects, parallelism of platens, and specimen size.

Though in most FE simulations the tools are considered rigid, this assumption may not hold for complicated forgings with tight tolerances, particularly in microforming applications. Hence, depending on the part complexity, elastic-plastic properties of the tools/dies become an important component in the simulations. The details of flow stress determination and inverse analysis can be found in Chapter 8, "Inverse Analysis of Simultaneous Determination of Flow Stress and Friction," of this book.

18.2.3 Interface Conditions (Friction and Heat Transfer)

The friction and heat-transfer coefficient are not readily available in literature. The most common way to determine the shear friction factor in forging is to perform ring compression tests. Friction factors measured with the ring compression test, however, are not valid for precision cold forging processes (hot, warm, and cold) where the interface pressure is very high and the surface generation is large. The friction conditions change during the process due to changes in the lubricant and the temperature at the die/workpiece interface. In such applications, the double-cup extrusion test is recommended for estimation of the friction factor (see Chapter 7, "Friction and Lubrication," in this book).

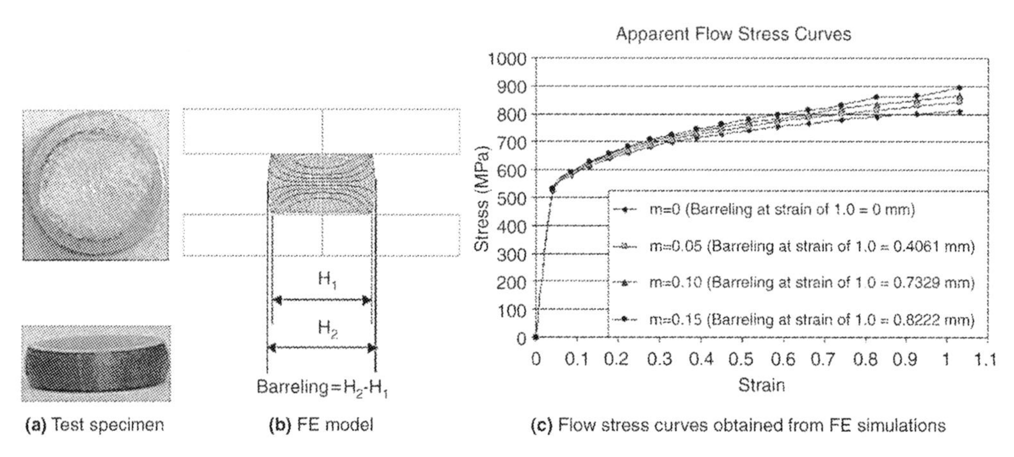

Fig. 18.1 Correction of flow stress data obtained from a compression test [Altan et al., 2001]

18.3 Process Modeling Output

The process modeling provides extensive information on the forging process. The output of process modeling can be discussed in terms of the metal flow, the distribution and history of state variables, the equipment response during forging, and the microstructure of the forging, as discussed in Chapter 16, "Process Modeling in Impression-Die Forging Using Finite-Element Analysis," in this book.

18.4 Process Modeling Examples

A few case studies from the literature are discussed below to examine the applications of FEM in cold forging processes.

18.4.1 Prediction and Elimination of Defects in Cold Extrusion [Hannan et al., 2000]

In the automotive industry, many shaft and shaftlike components, including fasteners, are produced by forward extrusion. Some of these components are critical for vehicle safety and must be free of defects. These defects could be visible external ones, such as laps or cracks, or nonvisible internal defects, such as chevron cracking (Fig. 18.2) [Hannan et al., 2000].

Spherical dies are sometimes used in extrusion of safety parts, because they provide more uniform grain flow around corners and improved dimensional controls on net shape surfaces. Therefore, it is beneficial to investigate the occurrences of chevron cracks in extrusion with spherical dies.

The objectives of this study were to measure the critical damage values for typical materials used in cold extrusion and to develop criteria or guidelines for designing forward extrusion dies for producing chevron-free extrusions. Ductile fracture can be defined as a fracture that occurs after a component experiences a significant amount of plastic deformation and is influenced by numerous parameters, including the deformation history of the workpiece material and the process conditions (i.e., rate of deformation, lubrication, and friction). Other factors that influence fracture include chemical composition, microstructure, surface conditions, and homogeneity.

At the Engineering Research Center for Net Shape Manufacturing (ERC/NSM), a methodology for predicting ductile fracture via FE simulations was developed, as shown in Fig. 18.3. It was based on the modified Cockroft and Latham criterion. Assuming the critical damage value is a material constant, several tests should be performed to obtain, for a given material, the flow stress and the critical damage value (CDV). Once the CDV for a specific material is obtained, it is possible to predict through FEM simulations the formation of cracks in forming operations. Two different materials were selected, details of which could be found in [Hannan et al., 2000]. Figure 18.4 shows that FEM can successfully predict the formation of chevron cracks in forward extrusion using this methodology. The commercial FE code DEFORM (Scientific Forming Technologies Corp.) was used for this study.

In summary, the CDV concept can be used to predict the possibility of fracture formation in cold extrusion. Thus, using this concept and FEM simulation, die design and process sequence can be modified to avoid fracture defects in cold extrusion.

18.4.2 Modeling of Golf Ball Mold Cavities [Hannan et el., 1999]

Golf ball mold cavities are manufactured by the hobbing process. This process allows the

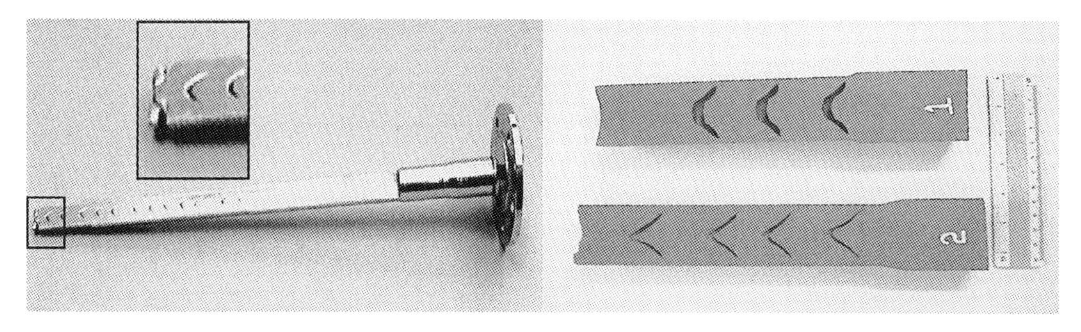

Fig. 18.2 Automotive axle shaft with chevrons [Hannan et al., 2000]

molds to be manufactured more efficiently and reduces the cost. Finite-element modeling was used in this study to design the tooling. A parametric analysis was conducted to evaluate the effect of various process conditions on the material flow and forming load for the hobbing process. For the parametric study, the tool design shown in Fig. 18.5 was used.

The results of FEM simulations showed that a design can be formulated that resulted in a forming load that was realistic and did not surpass the yield strength of the tool material. However, parametric analysis also showed problems with material flow of the preform. In each simulation, dimples did not fill completely, as shown in Fig. 18.6.

Using the information from this study, a design that resulted in good material flow (all dimples filled) and low forming loads was developed. Two-dimensional (2-D) and three-di-

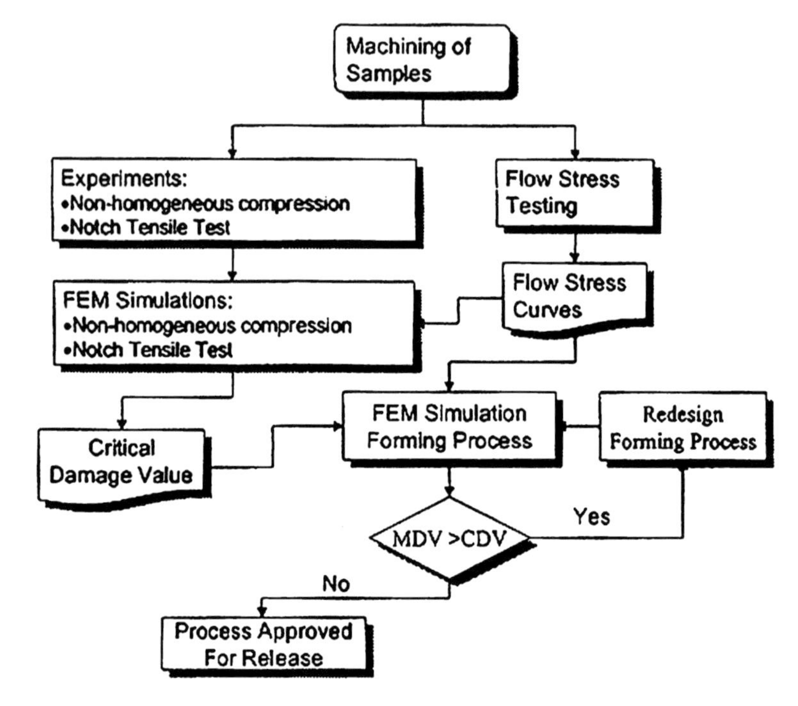

Fig. 18.3 Methodology to predict and prevent the formation of cracks in metal forming operations [Hannan et al., 2000]

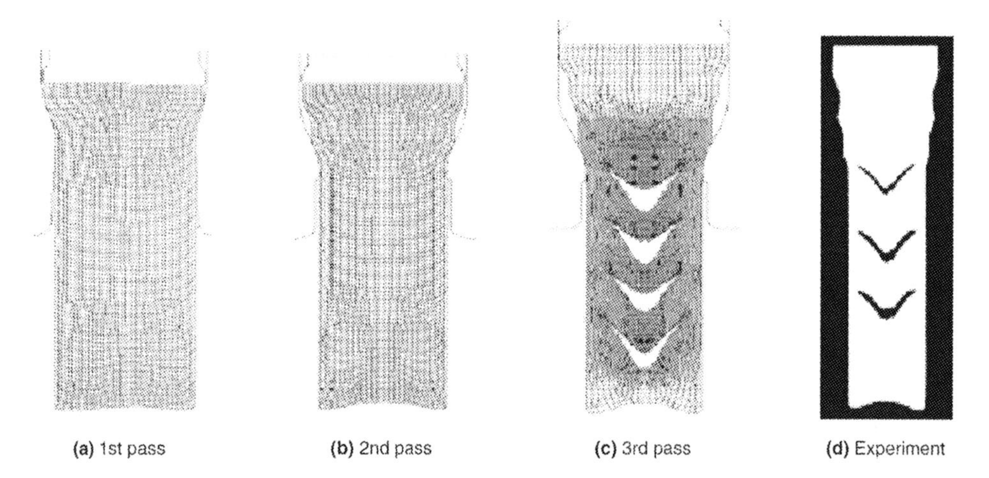

Fig. 18.4 Simulation of chevron cracks and experimental validation [Hannan et al., 2000]

mensional (3-D) FEM simulations were used to analyze the modified design that fills the dimples of the lower punch completely and at realistic loads. The details of the study can be found in [Hannan et al., 1999].

18.4.3 Design of Automotive Parts

Net shape forging of power steering pinions with helical teeth, helical gears, and inner races presents a great challenge. However, the use of commercial 3-D FE software in process development has drastically reduced the production costs by eliminating multiple trials and, at the same time, has improved part quality, tool life, and geometric complexity of forgings. Figure 18.7 shows the application of FE simulations in the design and development of the helical extrusion process for manufacturing the pinion shown. The advantages of using FEM to simulate and hence design the optimum process were [Yamanaka et al., 2002]:

- Obtain a better drive feeling
- Generate the tooth profile freely
- Establish an iterative technique for die and process design as well as heat treatment

Figure 18.8 shows the simulation of cold forging of an automotive part. The actual part is shown, which has underfill and fracture. The one-die two-blow process used to manufacture this part was successfully simulated, and the results show similar underfill and explain the fracture in the part.

18.4.4 Microforming of Surgical Blades

The trend in miniaturization allows the production of cold forged parts with dimensions

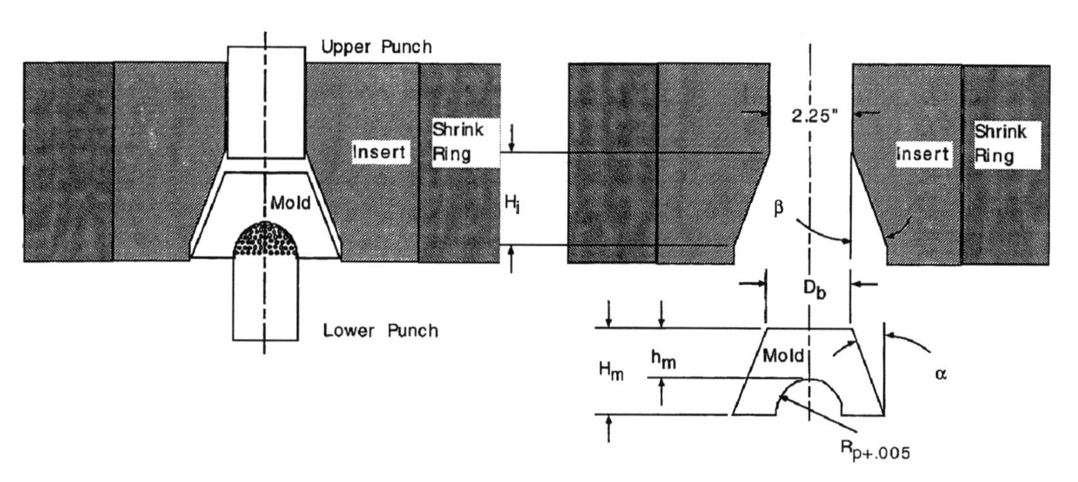

Fig. 18.5 Tool setup and studied parameters in FEM [Hannan et al., 1999]

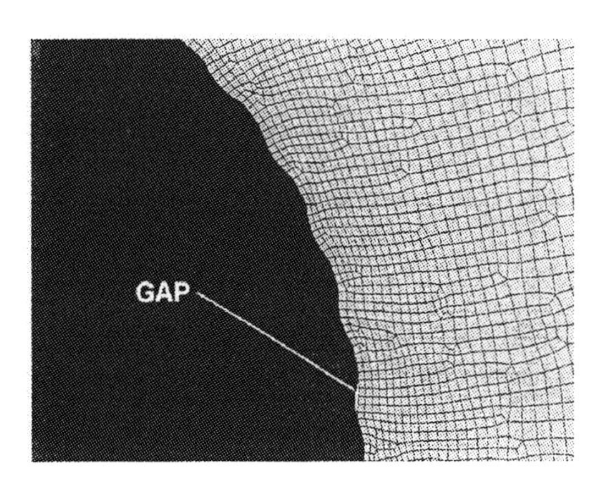

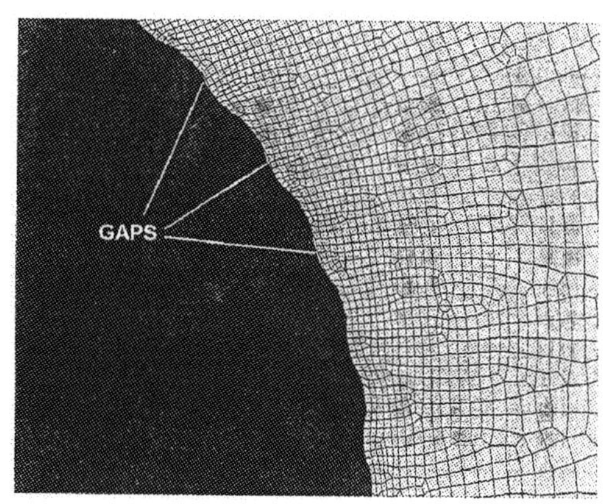

Fig. 18.6 Incomplete filling of the dimples, as predicted by two-dimensional FEM simulations [Hannan et al., 1999]

less than 0.04 in. (1 mm) range for electronics and biomedical applications. These parts are currently produced by 3-D etching and other metal-removal processes. Microforming is a potential process for mass production of net shape/near-net shape microcomponents. However, for microforming to be cost-effective and competitive, a comprehensive knowledge pertaining to the following factors is needed: scale effects/microplasticity, effect of microstructure on the process, relative stiffness of the tooling, and process control and capability.

With the aid of FEM, the interrelationships between these variables can be studied so as to

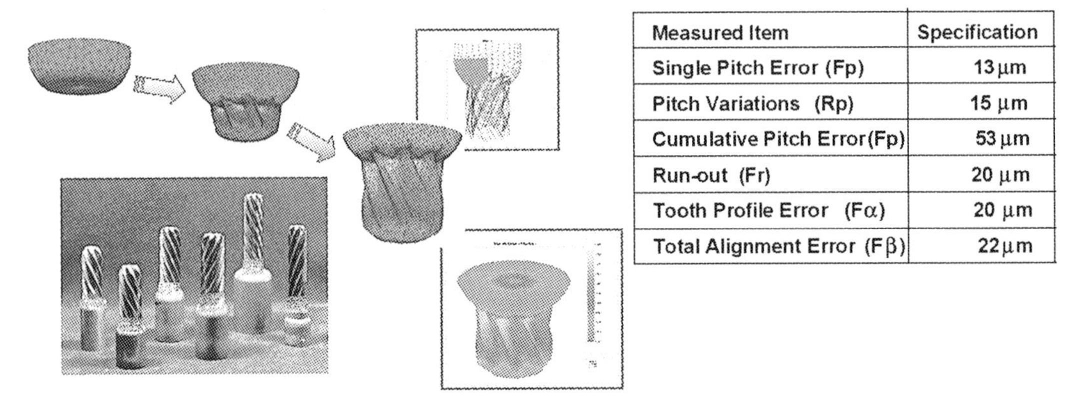

Measured Item	Specification
Single Pitch Error (Fp)	13 μm
Pitch Variations (Rp)	15 μm
Cumulative Pitch Error(Fp)	53 μm
Run-out (Fr)	20 μm
Tooth Profile Error (Fα)	20 μm
Total Alignment Error (Fβ)	22 μm

Fig. 18.7 Development of steering pinion with the aid of FE simulation [Yamanaka et al., 2002]

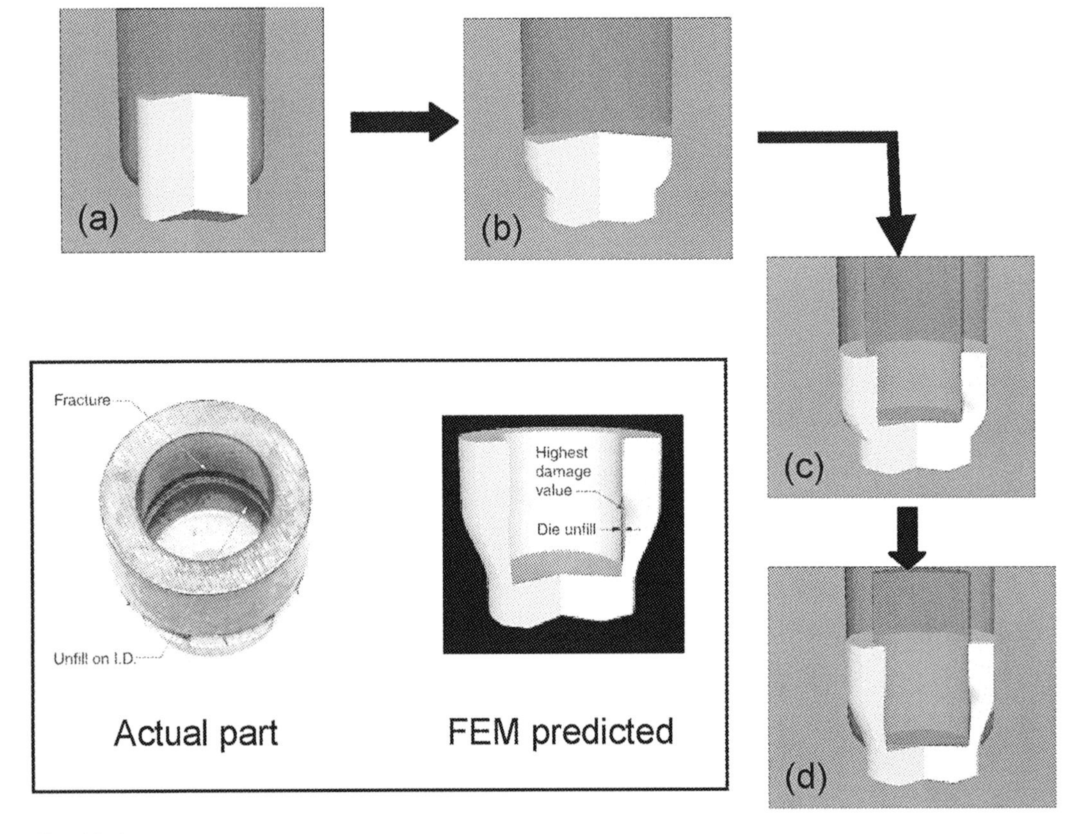

Fig. 18.8 Cold forging of an automotive part (a, b, c, and d are the stages of forming). Courtesy of DEFORM

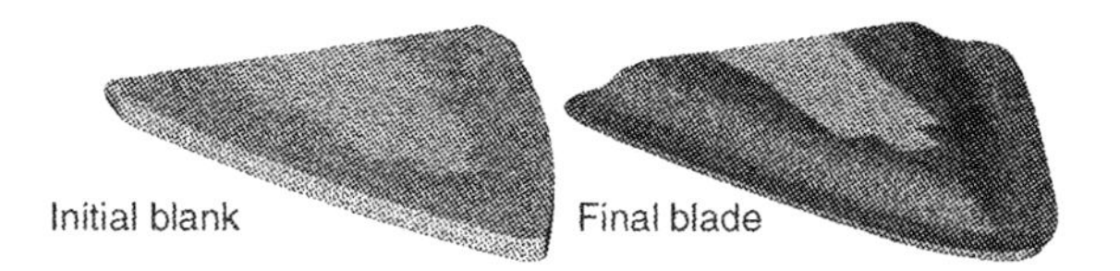

Fig. 18.9 Microforming of surgical blades. Blank thickness = 0.004 in. (0.1 mm); final blade thickness = 0.0004 in. (0.01 mm) [Palaniswamy et al., 2001]

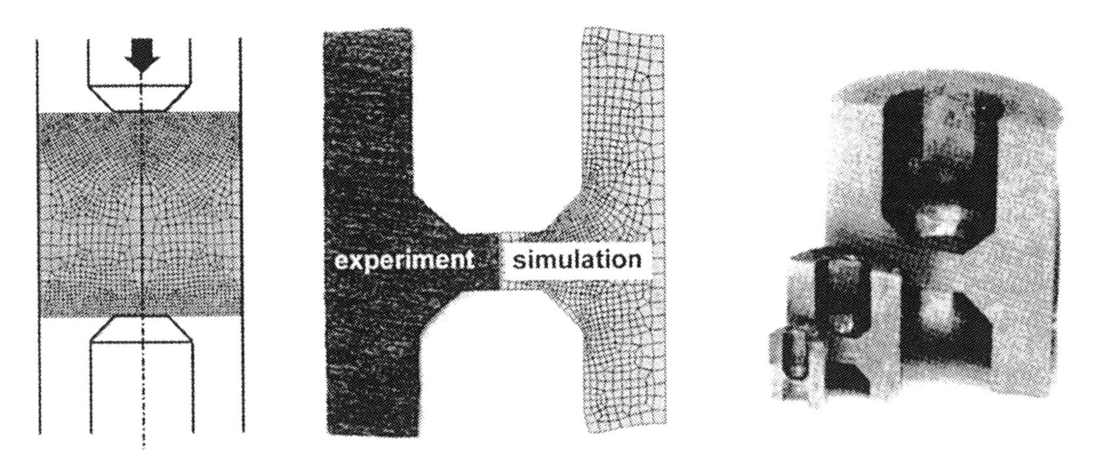

Fig. 18.10 Double-cup extrusion test for determination of friction conditions in microforming [Tiesler et al., 1999]

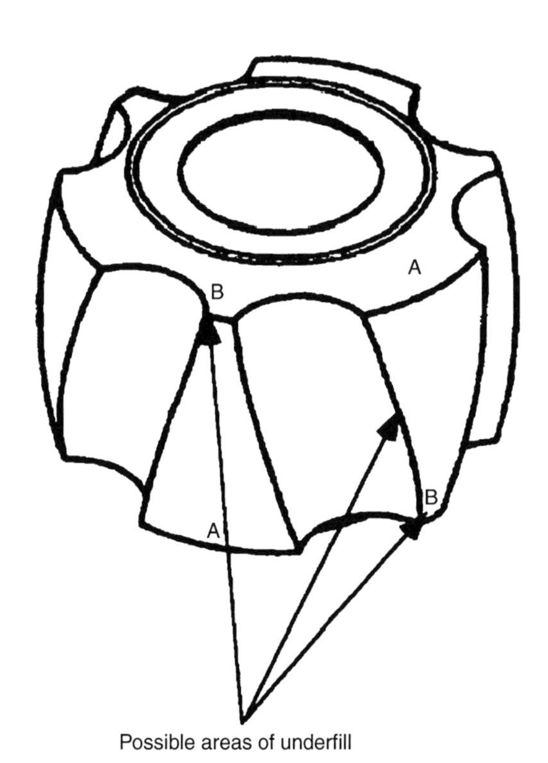

Fig. 18.11 Possible areas of underfill in cold forging of a cross-groove inner race [Vazquez et al., 1996]

provide guidelines for developing microforming processes. In cooperation with industry, the ERC/NSM has developed a microforming process for making surgical blades. Figure 18.9 shows an example of 3-D FE simulations for a surgical blade with initial blank thickness of 0.004 in. (0.1 mm) and final edge thickness of 0.0004 in. (0.01 mm) [Palaniswamy et al., 2001].

Due to the high surface-to-volume ratio in the microforming process, friction becomes even more important than in conventional forging. Effects of miniaturization on friction have been investigated by using the double-cup backward extrusion test, with oil as a lubricant (Fig. 18.10). Comparisons between experiment and simulation show that friction effects increase with a decrease in size of the specimen. The friction observed for a 0.04 in. (1 mm) diameter billet was four times that for a 0.16 in. (4 mm) diameter billet.

18.4.5 FEM Analysis of Process Design to Cold Forge a Cross-Groove Inner Race [Vazquez et al., 1996]

Figure 18.11 shows the cross-groove inner race. Cold forging of this part is very difficult, as the grooves of the inner race act as undercuts. Thus, the objective of this study was to develop a new design concept for tooling to improve the tolerance that is achievable in the cold forged groove. Figure 18.11 shows the possible areas (marked as B) of underfill. Two plausible designs were chosen for investigation from several other designs. Physical modeling and FE analysis were used to analyze the designs selected. Two-dimensional FEM simulations were performed for preliminary investigation of the metal flow. DEFORM was used for both 2-D and 3-D FEM simulations. The results of 2-D simulations give a window of operation for further investigation, though they may not accurately predict all the parameters. In order to obtain a more accurate prediction of the metal flow, strains, and the contact pressure distribution between the billet and the die, 3-D FEM simulations (one-sixth of the billet, due to symmetry) were conducted. The calculations showed that, at the end of the stroke, the die cavity pressure increases considerably. Since the maximum pressure should not exceed a certain limit, it is necessary to control the motion and pressures of the inner punches. This study clearly illustrated that in the multiaction forming process of the cross-groove inner race, a compromise between cavity filling and pressure must be achieved. Three-dimensional FEM simulations of different concepts for the punch motion help to select the best tool design before the actual tooling is built. The details of this study are given in [Vazquez et al., 1996].

18.4.6 Computer-Aided Design System and FEM Application for Process Design in Cold Forging [Kim et al., 1996]

Instead of trying to develop computer systems to automate the forming sequence design, the goal should be to develop computer-aided tools to assist die designers, while they "do" the design. Such a computer-aided design (CAD) system was developed at ERC/NSM. It consists of three parts: a computer program called FORMEX, a commercial CAD program, and a commercial FE program, DEFORM. In order to increase the capabilities of FORMEX, forging sequences collected from references and industry were entered in the "sequence library." Finite-element simulations were conducted to evaluate the sequence generated. Figure 18.12 shows an example of FE simulation of such a sequence. This shows that FE simulations of

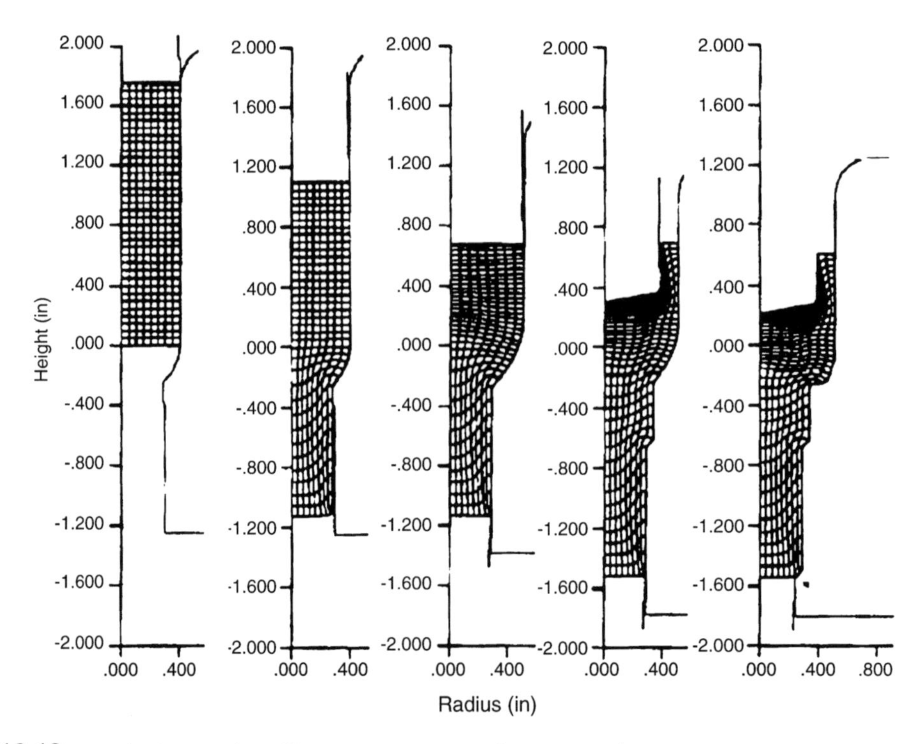

Fig. 18.12 Example of FEM analysis of forging process sequence design [Kim et al., 1996]

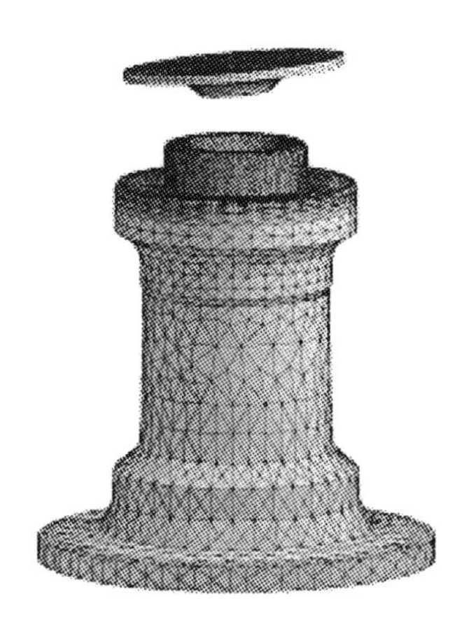

Fig. 18.13 Three-dimensional finite-element model for orbital forming simulation [Cho et al., 2003]

metal forming processes can assist the designers in establishing and optimizing process variables and die design. As a result, process development effort and cost can be reduced.

18.4.7 Advanced Orbital Forming Simulations [Cho et al., 2003]

Orbital forging is an incremental forging process with a complicated die movement that can be used to reduce axial load requirements for axisymmetric or near-axisymmetric forging operations. At the ERC/NSM, orbital forging simulations were conducted to study and develop a robust assembly process of an automobile spindle. Orbital forming requires a number of revolutions to form a part. Therefore, in FEM simulation, a relatively large number of simulation steps are required to minimize the errors in solution accuracy. Also, a full 3-D modeling is required, due to an asymmetric rotational tool movement. Figure 18.13 shows the FE model. Figure 18.14 shows the predicted metal flow at intermediate stages of forming. Solution accuracy was verified by comparing the predicted forces and tab outer diameters with experimental measurements, and good agreement was obtained.

REFERENCES

[Altan et al., 1990]: Altan, T., Oh, S.I., "Application of FEM to 2-D Metal Flow Simulation: Practical Examples," *Advanced Technology of Plasticity,* Vol 4, p 1779, 1990.

[Altan et al., 2001]: Dixit, R., Ngaile, G., Altan, T., "Measurement of Flow Stress for Cold Forging," ERC/NSM-01-R-05, Engineering Research Center for Net Shape Manufacturing.

[Altan et al., 2003]: Cho, H., Ngaile, G., Altan, T., "Simultaneous Determination of Flow Stress and Interface Friction by Finite Element Based Inverse Analysis Technique," Engineering Research Center for Net Shape Manufacturing, submitted for publication to *CIRP.*

[Cho et al., 2003]: Cho, H., Kim, N., Altan, T., "Simulation of Orbital Forming Process Using 3-D FEM and Inverse Analysis for Determination of Reliable Flow Stress," submitted

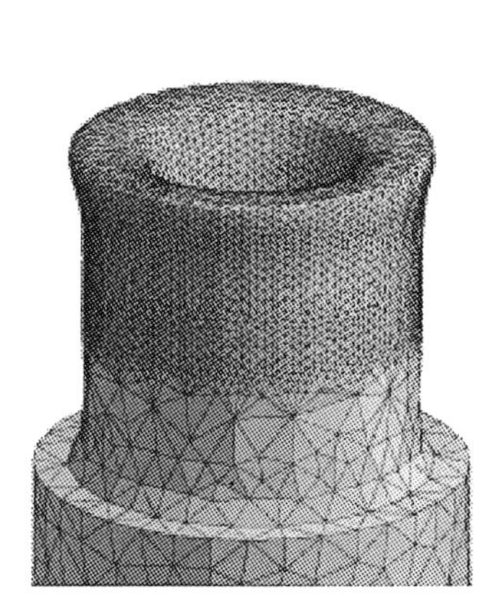

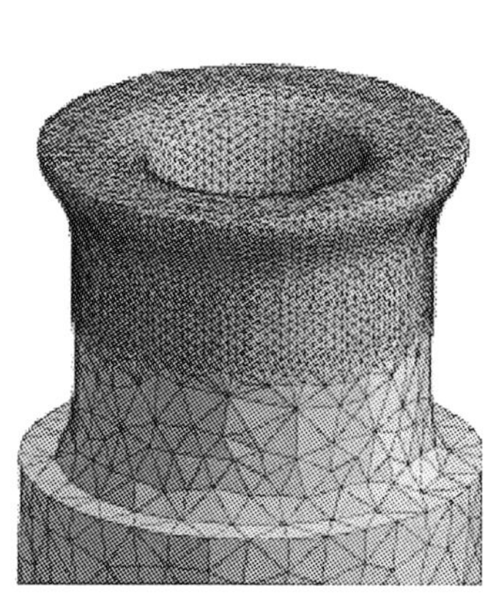

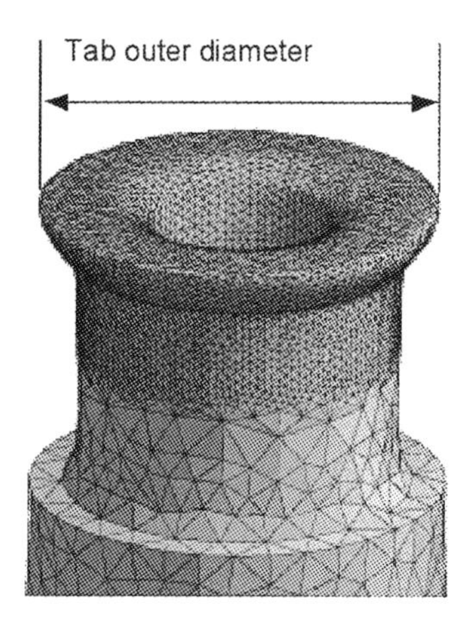

(a) After 40% punch stroke (b) After 80% punch stroke (c) At the end of stroke

Fig. 18.14 Predicted metal flow at intermediate stages of simulation [Cho et al., 2003]

to Third International Seminar for Precision Forging, Japan Society for Technology of Plasticity (Nagoya, Japan), 2003.

[Hannan et al., 1999]: Hannan, D., Vazquez, V., Altan, T., "Precision Forging of Golf Ball Mold Cavities," Report No. PF/ERC/NSM-99-R-12A, Engineering Reseach Center for Net Shape Manufacturing.

[Hannan et al., 2000]: Hannan, D., and Altan, T., "Prediction and Elimination of Defects in Cold Forging Using Process Simulation," Tenth International Cold Forging Congress, International Cold Forging Group, Sept 13–15, 2000 (Stuttgart).

[Kim et al., 1996]: Kim, H., and Altan, T., "Cold Forging of Steel—Practical Examples of Computerized Part and Process Design," *Journal of Materials Processing Technology,* Vol 59, 1996, p 122.

[Messner et al., 1994]: Messner, A., Engel, U., Kals, R., Vollertsen, F., "Size effect in the FE simulation of microforming processes," *JMPT,* Vol 45, p 374, 1994

[Palaniswamy et al., 2001]: Palaniswamy, H., Ngaile, G., Altan, T., "Coining of Surgical Slit Knife," F/ERC/NSM-01-R-26, Engineering Research Center for Net Shape Manufacturing.

[Tiesler et al., 1999]: Tiesler, N., Engel, U., Geiger, M., "Forming of Microparts—Effects of Miniaturization on Friction," *Advanced Technology of Plasticity,* Vol. 2, Proceedings of the Sixth ICTP, Sept. 19–24, 1999.

[Vazquez et al., 1996]: Vazquez, V., Sweeney, K., Wallace, D., Wolff, C., Ober, M., Altan, T., "Tooling and Process Design to Cold Forge a Cross Groove Inner Race for a Constant Velocity Joint—Physical Modeling and FEM Process Simulation," *JMPT,* Vol 59, p 144–157.

[Yamanaka et al., 2002]: Yamanaka, M., and Sunami, F., "Tool Design for Precision Forging," Cold and Warm Precision Forging Workshop, Schuler Inc., Yamanaka Engineering, Scientific Forming Technologies Corp., and ERC/NSM, Nov 14, 2002 (Canton, MI).

SELECTED REFERENCE

- **[Hayama, 1969]:** Hayama, M., *Plasticity and Plastic Working,* Ohmusha, 1969, p 125.

CHAPTER 19

Microstructure Modeling in Superalloy Forging

Gangshu Shen

19.1 Introduction

In aerospace forging, in addition to meeting the geometry requirements, meeting microstructure and mechanical property requirements of components plays a major role in the development of the forging processes and sequences. The understanding and prediction of microstructures that develop during the hot deformation processing has long been the "art" of forging metallurgists. Many tools, such as wedge tests, were used in the past to understand the impact of temperature, strain, and strain rate on microstructures and properties and to guide metallurgists in process development [Deridder and Koch, 1979]. These methods were good, but quantitative extraction of various processing parameters on the resulting microstructures and properties was not effectively accomplished. The application of finite-element modeling (FEM) to forging made it possible to obtain the detailed thermomechanical histories at each individual location of the forged components. Microstructure modeling procedures were developed to make use of this information for grain size and property predictions for steels [Sellars, 1979] and superalloys [Shen et al., 1995, and Shen and Hardwicke, 2000].

In this chapter, the experimental procedure, microstructure model formulation, and the application of the microstructure model in superalloy forgings are discussed.

19.2 Experiments for Microstructure Model Development

Hot forging involves preheating, deformation, transfer, and dwell (resting) between operations and final postforging cooldown. Small-scale experiments were run to simulate these operations and to establish the relationship between variables and microstructures and to generate data for the development of a microstructure model. Real-scale tests were run to compare the model prediction with reality. Three sets of experiments were used in the model development.

19.2.1 Preheating Tests

Heat treatment studies were conducted with different temperatures and hold times to produce the as-preheated grain size for a forging operation. From these tests, the grain growth model for preheating of a particular billet pedigree was developed. This model was then used to establish the preheated grain size (d_0) just prior to subsequent forging operations.

19.2.2 Compression Tests

Laboratory upset tests were conducted with different temperatures, strains, strain rates, as-preheated grain sizes, and postdeformation hold times to characterize dynamic recrystallization during forging and meta-dynamic recrystallization and static grain growth during postforging cooldown. Both the MTS Systems Corporation compression stand and Gleeble test unit were used to perform these compression tests. The advantage of the MTS test stand is that it can provide more uniform temperature of the workpiece. The advantage of the Gleeble test is that it can perform a fast postforging cooling with a controlled manner. From these compression tests with rapid postdeformation cooling, information related to dynamic recrystallization kinetics was obtained. The kinetics information included: peak strain for dynamic recrystallization ($\bar{\varepsilon}_p$), which is related to the critical strain for dynamic recrystallization ($\bar{\varepsilon}_c$); the strain that corresponds to 50% (0.5 fraction) dynamic recrystallization ($\bar{\varepsilon}_{0.5}$); the fraction of dynamic recrystallization (X_{dyn}); and the size of dynamically recrystallized grains (d_{dyn}). Information regarding meta-dynamic recrystallization and postforging grain growth was also obtained from compression tests with controlled postforge hold times. This information included: time for 50% meta-dynamic recrystallization ($t_{0.5}$), fraction of meta-dynamic recrystallization ($X_{m\text{-}dyn}$), meta-dynamically recrystallized grain size ($d_{m\text{-}dyn}$), and grain growth at a given temperature and time after the completion of meta-dynamic recrystallization.

19.2.3 Pancake and Generic Forgings

In addition to laboratory tests, large pancakes and generic component configurations were produced on production equipment under various forging conditions and methods to assess the microstructure model.

Finite-element analysis was used for each experiment to provide detailed information for each test. Thus, accurate thermal-mechanical histories of local points of forged samples were used to develop the models for the microstructure evolution of Waspaloy during the forging process.

19.3 Microstructure Model Formulation

The processes that control grain structure evolution during hot working of superalloys were found to be dynamic recrystallization, meta-dynamic recrystallization, and static grain growth. Microstructure model formulation is discussed in these three categories. Waspaloy formulas are used as examples here [Shen et al, 1995].

19.3.1 Dynamic Recrystallization

Dynamic recrystallization happens instantaneously during high-temperature deformation. The fraction of dynamic recrystallization can be obtained by examining micrographs obtained from samples quenched after the deformation. Under production conditions, pure dynamic recrystallization is difficult to achieve. This is because meta-dynamic recrystallization often follows immediately. The amount of dynamic recrystallization is related to the as-preheated grain size (d_0), effective strain ($\bar{\varepsilon}$), temperature (T), and effective strain rate ($\dot{\bar{\varepsilon}}$) in a hot deformation process. There are four important parameters related to dynamic recrystallization: the peak strain ($\bar{\varepsilon}_p$), the strain for 50% dynamic recrystallization ($\bar{\varepsilon}_{0.5}$), the fraction of dynamic recrystallization (X_{dyn}), and the size of dynamically recrystallized grains (d_{dyn}).

Peak Strain. The strain corresponding to the peak stress ($\bar{\varepsilon}_p$) in the flow stress curve is an important measure for the onset of dynamic recrystallization. The occurrence of dynamic recrystallization modifies the appearance of flow curves. At the strain rates typical for forging of Waspaloy, single-peak stress-strain curves are most common. As a result of dynamic recrystallization, the stress diminishes to a value intermediate between the yield stress and the peak stress once past the peak strain. The reason for this curve following a single peak is that under the condition of high Z (Zener-Hollomon parameter, $Z = \dot{\bar{\varepsilon}} \exp[468000/RT]$ for Waspaloy), the dislocation density can be built up very fast. Before recrystallization is complete, the dislocation densities at the center of recrystallized grains have increased sufficiently that another cycle of nucleation occurs, and new grains begin to grow again. Thus, average flow stress intermediate between the yield stress and the peak stress is maintained. The equations developed for the peak strain for Waspaloy are:

$$\bar{\varepsilon}_p = 5.375 \times 10^{-4} d_0^{0.54} Z^{0.106} \quad \text{(sub- and in-}\gamma'\text{ solvus)} \quad \text{(Eq 19.1)}$$

$$\bar{\varepsilon}_p = 1.685 \times 10^{-4} d_0^{0.54} Z^{0.106} \quad \text{(super-}\gamma'\text{ solvus)} \quad \text{(Eq 19.2)}$$

Strain for 50% Dynamic Recrystallization. Micrographs taken from quenched compression samples show that dynamic recrystallization progresses in a sigmoidal manner with respect to strain. The Avrami equation can be used to describe a sigmoidal curve for the fraction of dynamic recrystallization versus strain:

$$X_{dyn} = 1 - \exp\{-\ln 2[\bar{\varepsilon}/\bar{\varepsilon}_{0.5}]^{n}\} \quad \text{(Eq 19.3)}$$

When the constants $\bar{\varepsilon}_{0.5}$ and n are determined, the relation for the fraction of dynamic recrystallization is determined. The strain for 50% recrystallization, $\bar{\varepsilon}_{0.5}$, can be obtained from compression tests with different magnitudes of strain for a given condition of temperature, strain rate, and as-preheated grain size, as shown in Fig. 19.1. The exponent can be obtained by taking the logarithm of Eq 19.3. $\bar{\varepsilon}_{0.5}$ is related to as-preheated grain size, d_0, and Z by:

$$\bar{\varepsilon}_{0.5} = 0.145\, d_0^{0.32} Z^{0.03} \quad \text{(sub-}\gamma'\text{ solvus)} \quad \text{(Eq 19.4)}$$

$$\bar{\varepsilon}_{0.5} = 0.056\, d_0^{0.32} Z^{0.03} \quad \text{(in-}\gamma'\text{ solvus)} \quad \text{(Eq 19.5)}$$

$$\bar{\varepsilon}_{0.5} = 0.035\, d_0^{0.29} Z^{0.04} \quad \text{(super-}\gamma'\text{ solvus)} \quad \text{(Eq 19.6)}$$

Fraction of Dynamic Recrystallization. After the strain for 50% dynamic recrystallization and the exponent n for Eq 19.3 are determined, equations for the fraction of dynamically recrystallized grains can be formulated for Waspaloy as below:

$$X_{dyn} = 1 - \exp\{-\ln 2[\bar{\varepsilon}/\bar{\varepsilon}_{0.5}]^{3.0}\} \quad \text{(sub-}\gamma'\text{ solvus)} \quad \text{(Eq 19.7)}$$

$$X_{dyn} = 1 - \exp\{-\ln 2[\bar{\varepsilon}/\bar{\varepsilon}_{0.5}]^{2.0}\} \quad \text{(in-}\gamma'\text{ solvus)} \quad \text{(Eq 19.8)}$$

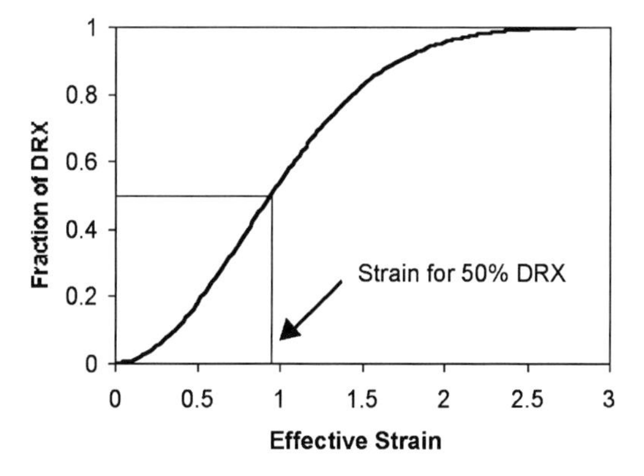

Fig. 19.1 Schematic of strain corresponding to 50% (0.5 fraction) dynamic recrystallization (DRX) for a given condition of temperature, strain rate, and as-preheated grain size

$$X_{dyn} = 1 - \exp\{-\ln 2[\bar{\varepsilon}/\bar{\varepsilon}_{0.5}]^{1.8}\} \quad \text{(super-}\gamma'\text{ solvus)} \quad \text{(Eq 19.9)}$$

Figures 19.2(a) and (b) summarize the experimental data and the fitted model for Waspaloy dynamic recrystallization at 1850 and 1951 °F (1010 and 1066 °C), respectively.

The critical strain for the start of dynamic recrystallization usually follows the relationship [Sellars, 1979]:

$$\bar{\varepsilon}_c = 0.8\, \bar{\varepsilon}_p \quad \text{(Eq 19.10)}$$

The Size of Dynamically Recrystallized Grain. The dynamically recrystallized grain size is the function of the Zener-Hollomon parameter, Z, only. This is because Z defines the density of the subgrains and the nuclei. Though the dynamically recrystallized grain size is not related to strain, the strain has to reach the value of steady-state strain to result in full dynamic recrystallization. The relationship between dynamically recrystallized grain, d_{dyn}, and Z is shown as follows:

$$d_{dyn} = 8103\, Z^{-0.16} \quad \text{(sub- and in-}\gamma'\text{ solvus)} \quad \text{(Eq 19.11)}$$

$$d_{dyn} = 108.85\, Z^{-0.0456} \quad \text{(super-}\gamma'\text{ solvus)} \quad \text{(Eq 19.12)}$$

Figure 19.3 shows the correlation between the experimental data and Eq 19.11 and 19.12. It is seen that there is a difference between subsolvus forging and supersolvus forging in terms of the sizes of the dynamically recrystallized grains. The subsolvus forging results in finer grain sizes, while supersolvus forging results in coarse grain sizes. However, subsolvus forging needs large strains to finish dynamic recrystallization, as shown in Eq 19.4 and 19.7.

19.3.2 Meta-Dynamic Recrystallization

Meta-dynamic recrystallization is important in the determination of the grain size obtained under practical forging conditions. Meta-dynamic recrystallization occurs when a deformation stops at a strain that passes the critical strain for dynamic recrystallization but does not reach the steady-state strain for dynamic recrystallization [McQueen and Jonas, 1975], which is the case for most regions in a forged part. Under meta-dynamic recrystallization conditions, the

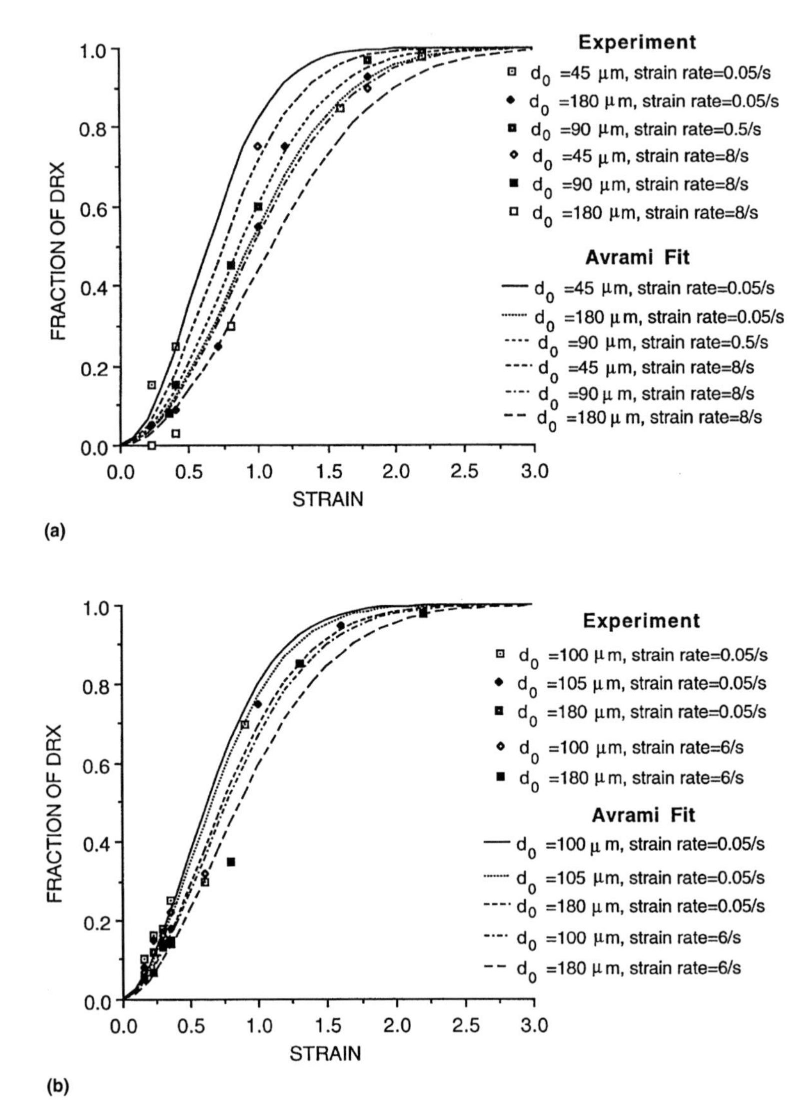

Fig. 19.2 Measured (data points) dynamic recrystallization (DRX) kinetics for hot deformation of Waspaloy at (a) 1850 °F (1010 °C) and (b) 1951 °F (1066 °C) and fitted curves

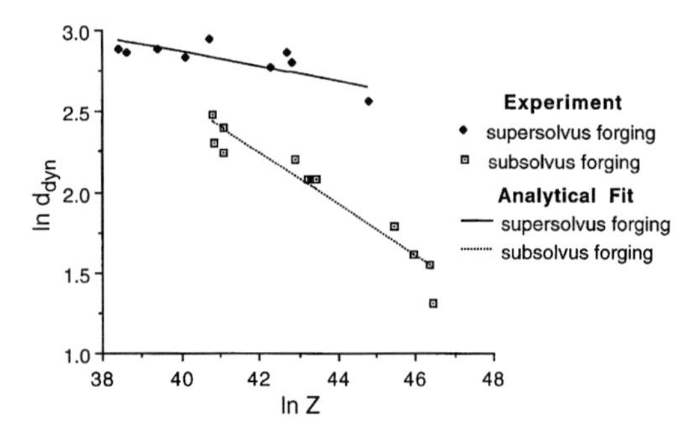

Fig. 19.3 Logarithm of dynamically recrystallized grain size (in µm) versus ln Z obtained from compression tests

partially recrystallized grain structure that is observed right after deformation (Fig. 19.4a) changes to a fully recrystallized grain structure (Fig. 19.4b) by continuous growth of the dynamically recrystallized nuclei at a high temperature. The meta-dynamically recrystallized grains are coarser than the dynamically recrystallized grains. However, they can often provide the uniformity of the grains under a production condition. The amount of meta-dynamic recrystallization is related to the as-preheated grain size, the strain, the temperature, the strain rate, and the holding time in a hot deformation process.

Important parameters related to meta-dynamic recrystallization are the time for 50% meta-dynamic recrystallization ($t_{0.5}$), the fraction of meta-dynamic recrystallization ($X_{m\text{-}dyn}$), and the size of the meta-dynamically recrystallized grain ($d_{m\text{-}dyn}$).

Time for 50% Meta-Dynamic Recrystallization. Meta-dynamic recrystallization is time dependent. For a given strain, strain rate, and as-preheated grain size, meta-dynamic recrystallization progresses in the following manner with respect to time:

$$X_{m\text{-}dyn} = 1 - \exp\{-\ln 2[t/t_{0.5}]^n\} \quad \text{(Eq 19.13)}$$

The $t_{0.5}$ can be obtained from compression tests with different holding times for a given temperature, strain, strain rate, and as-heated grain size. The empirical $t_{0.5}$ for meta-dynamic recrystallization follows:

$$t_{0.5} = 4.54 \times 10^{-5}\, d_0^{0.51} \bar{\varepsilon}^{-1.28} \dot{\bar{\varepsilon}}^{-0.073} \exp(9705/T) \quad \text{(Eq 19.14)}$$

The exponent, n, for meta-dynamic recrystallization is found to be 1 for Waspaloy. This number is typical for meta-dynamic recrystallization [Jonas, 1976, and Devadas et al., 1991].

Fraction of Metadynamic Recrystallization. The fraction of meta-dynamic recrystallization progresses according to:

$$X_{m\text{-}dyn} = 1 - \exp\{-\ln 2[t/t_{0.5}]^{1.0}\} \quad \text{(Eq 19.15)}$$

Figures 19.5(a) and (b) show the fraction of meta-dynamic recrystallization versus time obtained from the experiments and the predictions at 1951 °F (1066 °C) with different as-preheated grain size, strain, and strain-rate conditions. The meta-dynamic recrystallization finishes sooner for cases of larger strains, finer as-preheated grain sizes, higher strain rates, and higher temperatures.

Meta-Dynamic Recrystallized Grain Size. The grain size obtained at the end of meta-dynamic recrystallization is found to have the following relationship with the strain, the as-preheated grain size, and the Zener-Hollomon parameter, Z:

$$d_{m\text{-}dyn} = 14.56 d_0^{0.33} \bar{\varepsilon}^{-0.44} Z^{-0.026} \quad \text{(Eq 19.16)}$$

The meta-dynamic recrystallized grain size in ASTM number versus strain under conditions with different as-preheated grain sizes, temperatures, and strain rates is shown in Fig. 19.6.

19.3.3 Grain Growth

Under high-temperature deformation conditions, grain growth happens rapidly after the completion of meta-dynamic recrystallization. Grain-boundary energy is the driving force causing grain-boundary motion at high temperature. Grain-boundary energy is comparable to the surface energy; i.e., it tends to minimize itself whenever possible by decreasing the grain-boundary area. In general, grain growth will continue to occur at elevated temperatures until

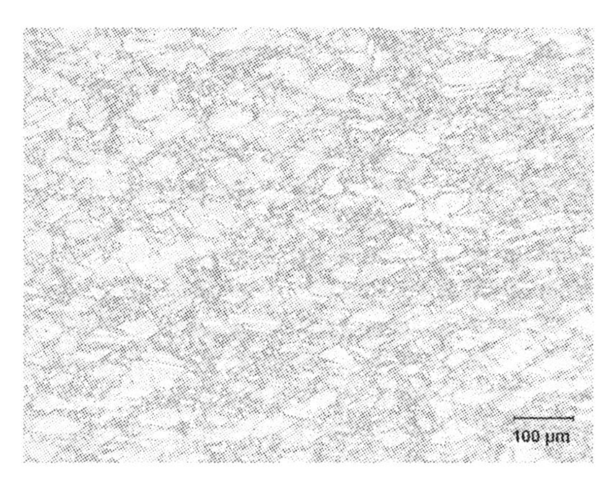

(a)

(b)

Fig. 19.4 Micrographs obtained from Waspaloy samples with different cooling histories after forging. (a) Rapidly cooled immediately after deformation. (b) Rapidly cooled after a 5 s hold at deformation temperature (1951 °F, or 1066 °C)

the balance between the grain-boundary energy and the pinning effects of precipitates (precipitate size and spacing) is reached.

Grain growth is characterized by compression tests with different postforging hold times. From the micrographs obtained from these tests, the microstructural evolution from partial dynamic recrystallization to full meta-dynamic recrystallization and to grain growth was observed. The change in grain size versus time after the completion of meta-dynamic recrystallization is found to follow:

$$d^3 - d_{m\text{-}dyn}^{\ 3} = 2 \times 10^{26}\ t\ \exp(-595000/[RT]) \quad \text{(Eq 19.17)}$$

The form of Eq 19.17 is well known for the characterization of grain growth. The reason for emphasizing that the $d_{m\text{-}dyn}$ is the grain size after complete meta-dynamic recrystallization is that after the completion of meta-dynamic recrystallization, the dislocations have essentially disappeared, and the driving force for grain size changes is the grain-boundary energy only.

The experimentally obtained data and the model prediction for the short-time grain growth are shown in Fig. 19.7. It is seen from this figure that the grain growth at a temperature of approximately 2050 °F (1121 °C) is very fast. There was not much difference in grain size after the completion of meta-dynamic recrystallization between the two samples obtained from compression tests at 1951 and 2050 °F (1066 and 1121 °C) (Fig. 19.3 and 19.7). However, the grain growth results in a large difference in the final grain size between the two sets of tests.

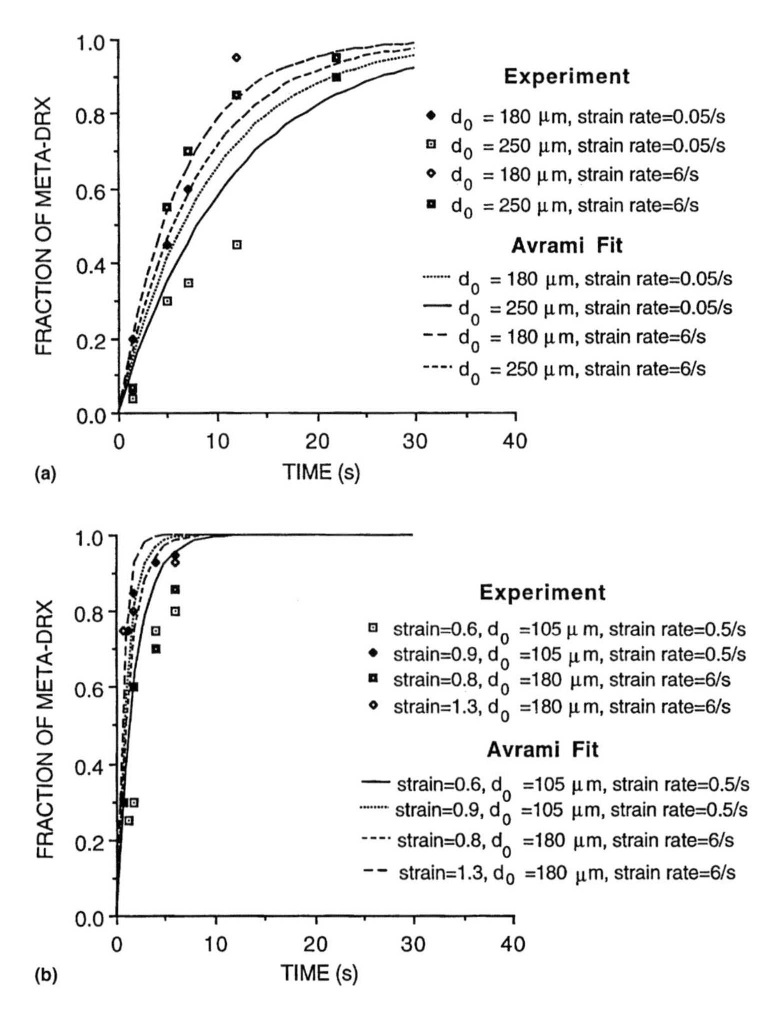

Fig. 19.5 Fraction of meta-dynamic recrystallization (DRX) versus time at 1951 °F (1066 °C) with a strain of (a) 0.22 and (b) 0.6 to 1.3

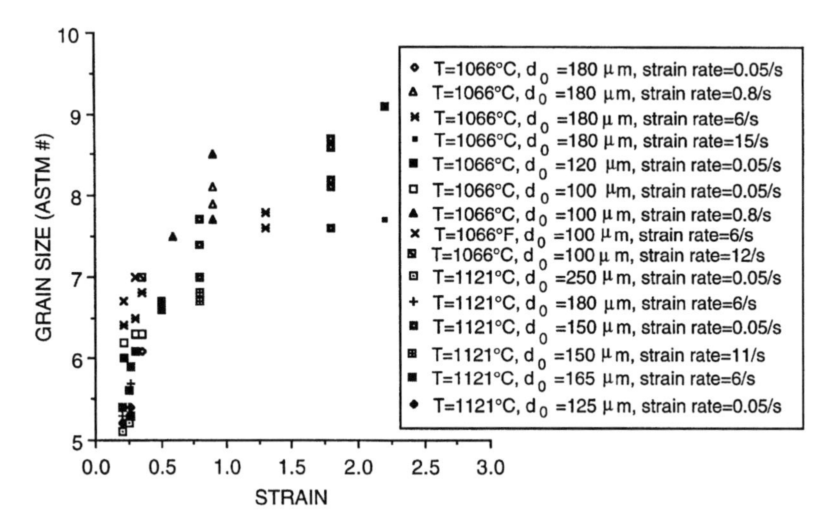

Fig. 19.6 Meta-dynamic recrystallized grain size versus strain for various process conditions

19.3.4 Model Summary

It is seen from these equations that the major factors in the control of the grain size in the forging of Waspaloy are strain, temperature-compensated strain rate, and the as-preheated grain size.

Strains create localized high densities of dislocations. To reduce their energy, dislocations rearrange into subgrains. When the subgrains reach a certain size, the nuclei of new grains form. The higher the strain, the greater the amount of dislocations and the greater the number of cycles of recrystallization. Hence, larger strains result in a higher percentage of recrystallization.

The reason that deformation under high-Z conditions gives finer grain size is that an increase in Z results in the increase in the subgrain density, which gives a higher density of nuclei. There are also more cycles of recrystallization present under high values of Z. Thus, the size of the recrystallized grains decreases [McQueen and Jonas, 1975].

At a given strain, the as-preheated grain size plays an important role in the determination of the fraction of recrystallization and recrystallized grain size, because when polycrystalline metal is deformed, the grain boundaries interrupt the slip processes. Thus, the lattice adjacent to grain boundaries distorts more than the center of the grain. The smaller the as-preheated grain, the larger the grain-boundary area and the volume of distorted metal. As a consequence, the number of possible sites of nucleation increases, the rate of nucleation increases, and the size of the recrystallized grains decreases. Moreover, the uniformity of distortion increases with the decrease in as-preheated grain size. Therefore, having a fine as-preheated initial grain size is very important for obtaining a fine recrystallized final grain size.

The equations developed for the quantitative prediction of these phenomena for Waspaloy are summarized in Table 19.1.

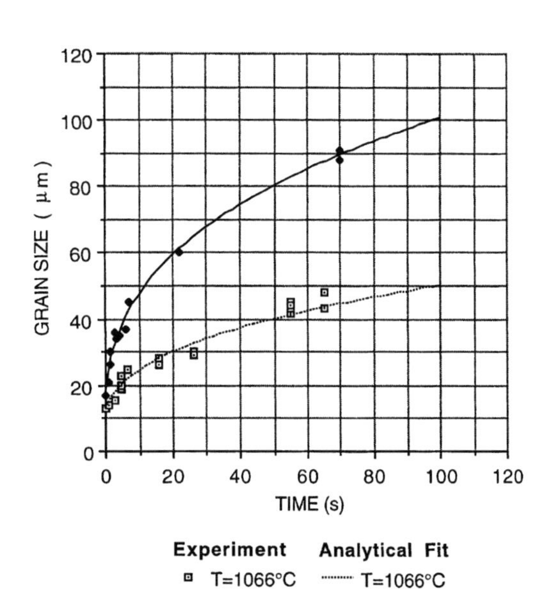

Fig. 19.7 Grain growth versus time after the completion of meta-dynamic recrystallization in Waspaloy forging

19.4 Prediction of Microstructure in Superalloy Forging

The methodology used in Waspaloy microstructure model development has also been used for superalloy 718 [Shen, 2000]. The models are integrated into finite-element software DEFORM (Scientific Forming Technologies Corp.) to predict microstructures developed for different forging processes [Wu and Oh, 1985, and Scientific Forming Technologies Corp., 2002].

Table 19.1 Mathematical model for microstructure development in Waspaloy forging

Dynamic recrystallization

$$Z = \dot{\bar{\varepsilon}} \exp(468000/RT)$$
$$\bar{\varepsilon}_c = 0.8\bar{\varepsilon}_p$$

Subsolvus forging

$$\bar{\varepsilon}_p = 5.375 \times 10^{-4}\, d_0^{0.54}\, Z^{0.106}$$
$$\bar{\varepsilon}_{0.5} = 0.145\, d_0^{0.32}\, Z^{0.03}$$
$$X_{dyn} = 1 - \exp\{-\ln 2[\bar{\varepsilon}/\bar{\varepsilon}_{0.5}]^{3.0}\}$$
$$d_{dyn} = 8103\, Z^{-0.16}$$

In-solvus forging

$$\bar{\varepsilon}_p = 5.375 \times 10^{-4}\, d_0^{0.54}\, Z^{0.106}$$
$$\bar{\varepsilon}_{0.5} = 0.056\, d_0^{0.32}\, Z^{0.03}$$
$$X_{dyn} = 1 - \exp\{-\ln 2[\bar{\varepsilon}/\bar{\varepsilon}_{0.5}]^{2.0}\}$$
$$d_{dyn} = 8103\, Z^{-0.16}$$

Supersolvus forging

$$\bar{\varepsilon}_p = 1.685 \times 10^{-4}\, d^{0.54}\, Z^{0.106}$$
$$\bar{\varepsilon}_{0.5} = 0.035\, d_0^{0.29}\, Z^{0.04}$$
$$X_{dyn} = 1 - \exp\{-\ln 2[\bar{\varepsilon}/\bar{\varepsilon}_{0.5}]^{1.8}\}$$
$$d_{dyn} = 108.85\, Z^{-0.0456}$$

Meta-dynamic recrystallization

$$t_{0.5} = 4.54 \times 10^{-5}\, d_0^{0.51}\, \bar{\varepsilon}^{-1.28} \dot{\bar{\varepsilon}}^{-0.073} \exp(9705/T)$$
$$X_{m\text{-}dyn} = 1 - \exp\{-\ln 2[t/t_{0.5}]^{1.0}\}$$
$$d_{m\text{-}dyn} = 14.56 d_0^{0.33} \bar{\varepsilon}^{-0.44} Z^{-0.026}$$

Grain Growth

$$d^3 - d_{m\text{-}dyn}^3 = 2 \times 10^{26}\, t \exp(-595000/[RT])$$

Figures 19.8 and 19.9 are the comparison of model prediction and experimentally obtained values (numbers shown on the contours) of recrystallization and average ASTM grain size of a hammer-forged 718 disk [Shen et al., 2001]. Multiple dies were used in this hammer forging. The blow-by-blow simulation was run for the hammer forging in the FEM code DEFORM that allows the entire thermal-mechanical history of the hammer forging to be stored in the computer. The microstructure model uses the thermal-mechanical history of the hammer forging to predict the recrystallization and grain size of the hammer-forged 718 disk. Figures 19.8 and 19.9 show that the fraction of recrystallization and the ASTM average grain size predicted by the model agree well with the experimentally measured values.

Figure 19.10 shows the model-predicted ASTM grain sizes and the experimentally measured ASTM grain sizes for an experimental Waspaloy disk. The forging process involved a hydraulic press isothermal forging followed by a hammer forging [Stewart, 1988]. Again, the model predicted well the average grain size of the forged Waspaloy disk. These examples show that the microstructure model is capable of predicting the recrystallization and grain size for quite complex processes, such as hammer forging with multiple blows and die sets and the combination of press and hammer forgings.

19.5 Nomenclature of Microstructure Model

- d final grain size, μm
- d_0 as-preheated grain size, μm

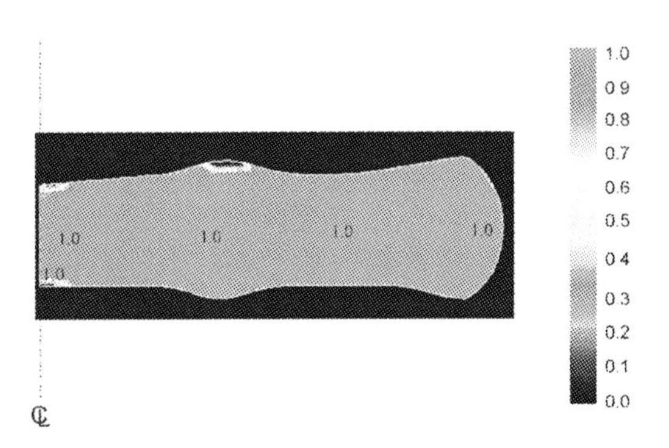

Fig. 19.8 Comparison of model prediction and experimental results. Fraction of recrystallization of a 718 developmental forging

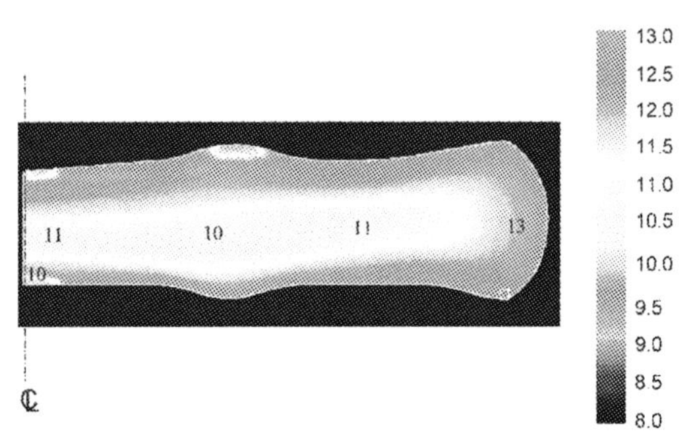

Fig. 19.9 Comparison of model prediction and experimental results. Average ASTM grain size of a 718 developmental forging

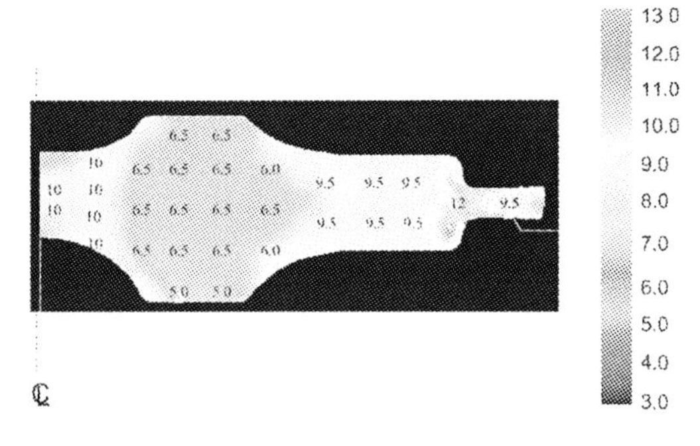

Fig. 19.10 Comparison of model predicted and experimentally obtained. Average ASTM grain size of a Waspaloy developmental forging

d_{dyn}	dynamically recrystallized grain, μm
$d_{m\text{-}dyn}$	meta-dynamically recrystallized grain, μm
n	exponent for Avrami equation
Q	activation energy, J/mol
R	gas constant, 8.314 J/(mol·K)
T	temperature, K
t	time, s
$t_{0.5}$	time for 50% meta-dynamic recrystallization
X_{dyn}	fraction of dynamic recrystallization
$X_{m\text{-}dyn}$	fraction of meta-dynamic recrystallization
Z	Zener-Hollomon parameter, 1/s
$\bar{\varepsilon}$	effective strain
$\bar{\varepsilon}_{0.5}$	effective strain for 50% dynamic recrystallization
$\bar{\varepsilon}_c$	critical strain for dynamic recrystallization
$\bar{\varepsilon}_p$	peak strain for dynamic recrystallization
$\dot{\bar{\varepsilon}}$	effective strain rate

REFERENCES

[Deridder and Koch, 1979]: Deridder, A.J., and Koch, R., "Forging and Processing of High Temperature Alloys," *MiCon 78: Optimization of Processing, Properties, and Service Performance through Microstructural Control,* STP 672, H. Abrams et. al, Ed., American Society for Testing and Materials, p 547–563.

[Devadas et al., 1991]: Devadas, C., "The Thermal and Metallurgical State of Strip during Hot Rolling: Part III, Microstructural Evolution," *Met. Trans. A.,* p 335–349.

[Jonas, 1976]: Jonas, J.J., "Recovery, Recrystallization and Precipitation under Hot Working Conditions," *Proceedings of the Fourth International Conference on Strength of Metals and Alloys* (Nancy, France), p 976–1002.

[McQueen and Jonas, 1975]: McQueen, H.J., and Jonas, J.J., "Recovery and Recrystallization during High Temperature Deformation," *Treatise on Materials Science and Technology,* Vol. 6, Academic Press, p 393–493.

[Scientific Forming Technologies Corp., 2002]: *DEFORM 7.2 User Manual,* Scientific Forming Technologies Corp., Columbus, OH.

[Sellars, 1979]: Sellars, C.M., *Hot Working and Forming Processes,* C.M. Sellars and G.J. Davies, Ed., TMS, London, p 3–15.

[Shen, 2000]: Shen, G., "Microstructure Modeling of Forged Components of Ingot Metallurgy Nickel Based Superalloys," *Advanced Technologies for Superalloy Affordability,* K.M. Chang et al., Ed., The Minerals, Metals, and Materials Society.

[Shen et al., 2001]: Shen, G., et al., "Advances in the State-of-the-Art of Hammer Forged Alloy 718 Aerospace Components," *Superalloys 718, 625, 706 and Various Derivatives,* E.A. Loria, Ed., TMA.

[Shen et al., 1995]: Shen, G., et al., "Modeling Microstructural Development during the Forging of Waspaloy," *Met. Trans. A.,* p 1795–1802.

[Shen and Hardwicke, 2000]: Shen, G., and Hardwicke, C., "Modeling Grain Size Evolution of P/M Rene 88DT Forgings," *Advanced Technology for Superalloy Affordability,* K.M. Chang et al., Ed., The Minerals Metals and Materials Society.

[Stewart, 1988]: Stewart, D., "ISOCON Manufacturing of Waspaloy Turbine Discs," *Superalloys,* S. Reichman et al., Ed., TMS, p 545–551.

[Wu and Oh, 1985]: Wu, W.T., and Oh, S.I., "ALPIDT—A General Purpose FEM Code for Simulation of Nonisothermal Forging Processes," *Proceedings of North American Manufacturing Research Conference (NAMRC) XIII* (Berkley, CA), p 449–460.

SELECTED REFERENCE

- **[Shen et al., 1996]:** Shen, G., et al., "Microstructure Development in a Titanium Alloy," *The Proceedings of the Symposium on Advances in Science and Technology of Titanium* (Anaheim, CA), TMS.

CHAPTER 20

Isothermal and Hot-Die Forging

Manas Shirgaokar
Gangshu Shen

20.1 Introduction

The manufacturing processes used to produce aerospace forgings depend on several factors, including component geometry, property requirements, and economics [Williams, 1996, and Noel et al., 1997]. With regard to economics, all forging operations must be optimized to reduce the amount of metal required to make the final component shape. Titanium and nickel-base superalloys, two widely used materials for aerospace components, are relatively expensive and hard to machine. Adding extra metal to the forging configuration results in added input metal and machining costs to produce the final components. Isothermal forging was developed to provide a near-net shape component geometry and well-controlled microstructures and properties with accurate control of the working temperature and strain rate. Isothermal forging, however, requires a large initial capital investment for equipment. Hot-die forging was developed to make some sacrifice in die temperature and net shape capability while lowering the initial investment. All of the isothermal and hot-die forged aerospace parts require a stringent postforging heat treatment to produce the optimum microstructure and properties of the components.

20.2 Isothermal Forging

Isothermal hydraulic press forging has been developed for near-net shape forging of materials that are difficult to process. Isothermal forging was first practiced in production using IN-100 material. Alloys such as IN-100, Rene 95, Rene 88, N 18, U-720, Waspaloy, Gamma-TiAl, Ti-6-2-4-6, IMI 834, Ti-6Al-4V, and others can be forged economically only by using the isothermal forging technique.

In an isothermal forging process, the dies are heated to the same temperature as the workpiece, allowing near-net shape configurations to be formed, which results in the use of less raw material and minimum postforging machining. In isothermal forging of titanium and nickel-base superalloys, since the forging temperatures are very high (1700 to 2200 °F, or 925 to 1205 °C), superalloys and molybdenum alloys are often used as tooling materials. Since molybdenum alloys are prone to rapid oxidation at high temperatures, a vacuum or inert atmosphere is used, which also increases the capital investment in isothermal forging.

Nickel-base superalloys were developed for superior elevated-temperature strength and creep resistance. Therefore, they are difficult to process in the wrought form. Highly alloyed cast and wrought superalloys have a very narrow processing window. The new, even higher-alloy-content powder metallurgy (P/M) materials are even more temperature and strain-rate sensitive and require a low strain-rate superplastic forging condition to avoid abnormal structure being developed during postforging heat treatments [Soucail et al., 1996]. The strain rate used in isothermal forging is usually low to reduce the adiabatic heating, maintain superplastic material behavior, and meet the processing requirements of P/M superalloys. Though the strain rate is low and the forging time is long in isothermal forging, due to the fact that the dies are heated to the same temperature as the workpiece, there is no die chilling, as in conventional forging. Successful application of these processes demands accurate temperature and strain-rate control, advances in titanium and nickel-base superalloy materials, and stringent requirements on component microstructure and properties.

20.2.1 Gatorizing

Alloys destined for use in rotating machinery such as gas turbines are designed to have high strength at elevated temperatures. Their performance characteristics make them a challenge to forge or machine using conventional techniques. Thus, an improved method for forging high-temperature alloys, known as Gatorizing, was patented by Moore et al. in 1970. This method consists of [Altan et al., 1973]:

- Preconditioning the stock under controlled conditions to secure a temporary condition of low strength and high ductility
- Hot working to the desired shape while maintaining those attributes
- Restoring normal properties to the workpiece through heat treatment

With Gatorizing, hard-to-work nickel alloys were forged for the first time, and the higher strength that could be obtained from these alloys resulted in higher strength-to-weight ratios in aircraft components such as jet engine disks. This process also led to development of techniques for isothermal forging of integrally bladed engine rotors (disks of superalloys forged integrally with ceramic blades) [Walker et al., 1976, and McLeod et al., 1980].

20.2.2 IsoCon

A process that combines isothermal forging and hammer forging has been developed for the processing of Waspaloy, a nickel-base superalloy material. This technique is referred to as IsoCon and is made up of a controlled isothermal forging operation followed by a controlled conventional hammer forging operation [Stewart, 1988]. The isothermal step results in the required microstructure and general component configuration, while the final hammer forging step results in controlled cold strain, which significantly increases the properties of the final part and imparts the final geometry refinement. This metallurgical-based process requires a high degree of process control for both the isothermal and hammer forging operations. Utilization of deformation and metallurgical models for IsoCon process design optimization has been beneficial in obtaining the final component results.

20.3 Hot-Die Forging

In addition to isothermal forging, hot-die hydraulic press forging is also widely used for producing aerospace components. The die temperatures used in hot-die forging are usually a few hundred degrees lower than the workpiece temperature. However, they are much higher than the die temperatures used in conventional forging, ranging from 400 to 800 °F (205 to 425 °C). Higher die temperatures require stronger materials for the dies. Superalloys are often used as die materials for hot-die forging.

To keep a constant high temperature of the dies, consistent heating of the die is required. Induction, resistance, and radiant systems are usually used for die heating in hot-die forging. The strain rate used is usually an order of magnitude higher than that used in isothermal forging, to reduce die chilling. Due to the higher strain rates used in hot-die forging, P/M superalloys are usually not forged by this process. On the other hand, titanium alloys and cast and wrought superalloys are often forged by hot-die forging processes.

Since hot-die forging is not performed in a vacuum/inert environment, fast post-forging cooling processes, such as water quench and oil quench, can be applied immediately after forging, which is not usually possible for isothermal forging. Whether an isothermal forging or a hot-die forging should be selected for a component depends on the material, the microstructure and property requirements, and the economics. In the forging of high-alloy-content P/M superalloys, the isothermal forging process is the only proven method for success.

20.4 Benefits of Isothermal and Hot-Die Forging

Forging basically involves the deformation of a metal billet or a preform between two or more dies in order to obtain the final part. The deformation may be carried out by means of various machines, such as hammers or mechanical, hydraulic, or screw presses. Parts with complex geometries are forged in impression dies that have the shape of the desired part. This may be done in one step or a sequence of steps (i.e., preforming and blocking), depending on the part complexity. This process results in very high die stresses and hence calls for the material to be forged at elevated temperatures in order to reduce the flow stress, increase formability, and thus reduce die stresses. During hot forging, the heat transfer from the hot material to the colder dies can result in die chilling. Dies are thus normally heated to approximately 600 to 800 °F (315 to 425 °C) to reduce the heat transfer. Conventional die materials do not allow the use of temperatures higher than 800 °F (425 °C), since they lose their strength and hardness above this temperature range [Semiatin et al., 1983]. Aluminum alloys are generally isothermally forged, since the dies are readily heated to the same temperature range as the material (approximately 800 °F, or 425 °C). Since isothermal forging of aluminum alloys is a well-established state-of-the-art process, it is more appropriate to concentrate on high-temperature alloys such as titanium and nickel.

Forging of steels and high-temperature alloys can be an expensive process due to the preforming and blocking operations needed to achieve the desired part geometry, especially in aerospace forging. This industry produces a relatively small number of parts, which does not justify the high die costs. Thus, it is considered economical to forge a part with high machining allowances to reduce die costs while simultaneously increasing machining costs and material losses. This approach is, however, not suitable for titanium and nickel alloys that are more expensive in comparison to steel and aluminum. This cost ratio has been a driving factor for the development of isothermal (dies and workpiece at the same temperature) and hot-die (die temperature close to that of the workpiece) forging methods.

Isothermal and hot-die forging offer the following advantages [Semiatin et al., 1983]:

- Closer tolerances than conventional forging due to elimination of die chilling. Reduction of machining costs and material losses. These forging processes allow for smaller corner and fillet radii, smaller draft angles, and smaller forging envelope, consequently leading to materials savings and a reduced forging weight (Fig. 20.1 and 20.2).
- Elimination of die chilling also results in reduction of preforming and blocking steps, thus reducing die costs.
- Since die chilling is not a problem, slow deformation speeds can be used (e.g., a hydraulic press). This reduces the strain rate and flow stress of the forged material. Due to reduction of forging pressures, larger parts can be forged using existing equipment.

Figure 20.3 shows a comparison between isothermal forging and other conventional methods of producing a typical rib-web-type aircraft structural component.

20.5 High-Temperature Materials for Isothermal and Hot-Die Forging

An in-depth understanding of the plastic properties of the forging material is extremely crucial for the design of an isothermal forging process, since the selection of the other elements, such as equipment, die material, and lubricants, is based on this knowledge.

At hot forging temperatures ($T \geq 0.6 T_m$, where T_m is the melting point or solidus temperature of the material), the flow stress is usually a function of strain rate, except at very low strains. Consequently, the deformation behavior of a material is best understood by examining the flow stress dependence on strain rate [Chen et al., 1980, and Greenwood et al., 1978]. The dependence of the flow stress on temperature is minimal at very low (creep) as well as high (conventional hot forging) strain rates. At intermediate strain rates, however, the flow stress dependence on $\bar{\varepsilon}$ is often large, and it is in this region that superplastic behavior is predominant (Fig. 20.4). The high strain-rate sensitivity of superplastic materials promotes die filling and resists localized plastic deformation, making these materials extremely suitable for isothermal and hot-die forging. Also, the reduction of flow stress with decreasing strain rate lowers the forging loads, thus making it possible to form a part isothermally with a smaller press than in conventional hot forging. Isothermal forging also reduces the number of forging steps and ancillary operations that are necessary in conventional forging.

20.5.1 Titanium Alloys

Titanium alloys are among the most commonly used materials for isothermal and hot-die forging. Unalloyed titanium occurs in two forms, namely, the alpha phase (hexagonal close-packed structure), which is stable up to approximately 1615 °F (880 °C), and the beta phase (body-centered cubic structure), which is stable from 1615 °F (880 °C) to the melting point. The temperature at which this transformation occurs is known as the beta transus temperature. It is possible to increase or lower this temperature by the use of alloying elements known as stabilizers (Table 20.1).

Titanium alloys are classified as either alpha (α) alloys, beta (α) alloys, or alpha + beta (α + α) alloys. Table 20.2 lists some of the commercially used titanium alloys with their classification. The flow stress of these alloys is very sensitive to temperature and strain rate, especially for the alpha + beta alloys and near-alpha alloys below the beta transus temperature. The beta transus temperatures for common titanium alloys are given in Table 20.3.

The data given in Fig. 20.5 are indicative of the advantages of isothermal forging over conventional nonisothermal methods for alpha, near-alpha, and alpha + beta alloys. If Ti-6Al-4V were forged conventionally at a temperature of 1725 °F (940 °C), chilling of the workpiece surface to a temperature of 1600 °F (870 °C) could lead to severe deformation inhomogeneities. This is because the flow stress of this alloy at 1600 °F (870 °C) is approximately three times that at the desired forging temperature of 1725 °F (940 °C). Also, a substantial reduction in the forging pressures is possible by reducing the strain rate by a factor of a hundred or a thousand. Thus, for a given press capacity, parts can be forged isothermally at lower strain rates and lower preheat temperatures than in conventional forging processes with higher strain rates, even if die chilling could be avoided. This results in lower energy expenditure and less contamination through alpha case formation at lower preheat temperatures [Semiatin et al., 1983]. Figures 20.6 and 20.7 illustrate these points. The forging pressures near the transus temperature are similar to those encountered at a temperature several hundred degrees lower when the strain rate is slowed down to those employed in creep forging. The practical use of these forging rates, however, is restricted by economic considerations.

Compared to the alpha, near-alpha, and alpha + beta alloys, the beta and near-beta titanium alloys were developed and put into commercial

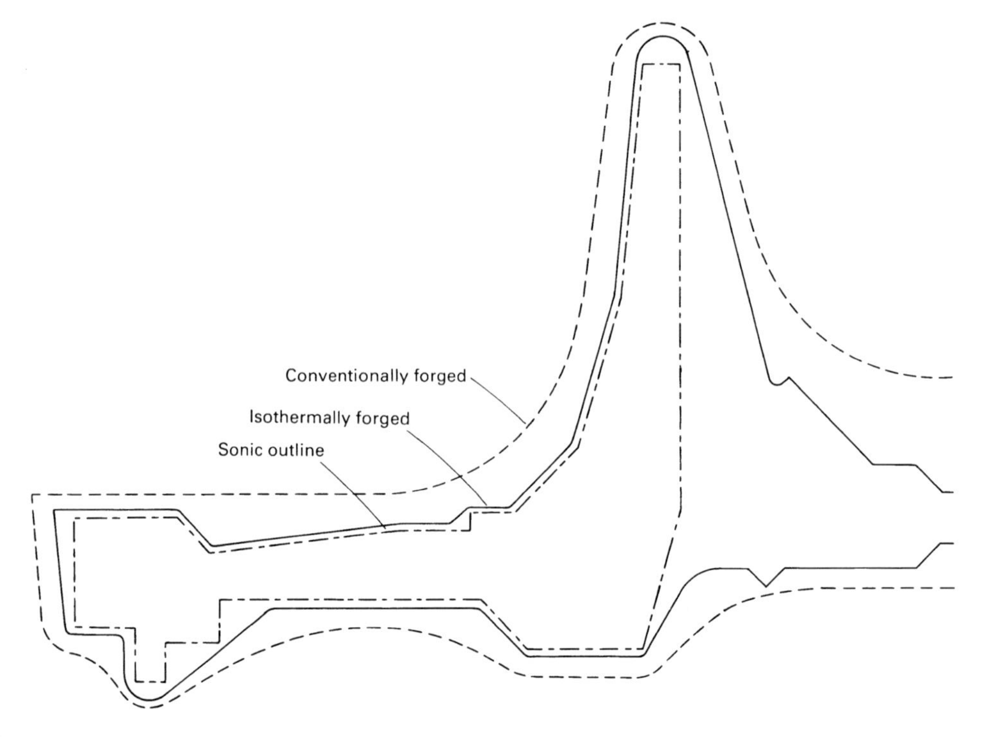

Fig. 20.1 Weight reduction obtained by forging a disk by isothermal methods rather than conventional forging. A 60 lb (27 kg) weight reduction was obtained [Shah, 1988].

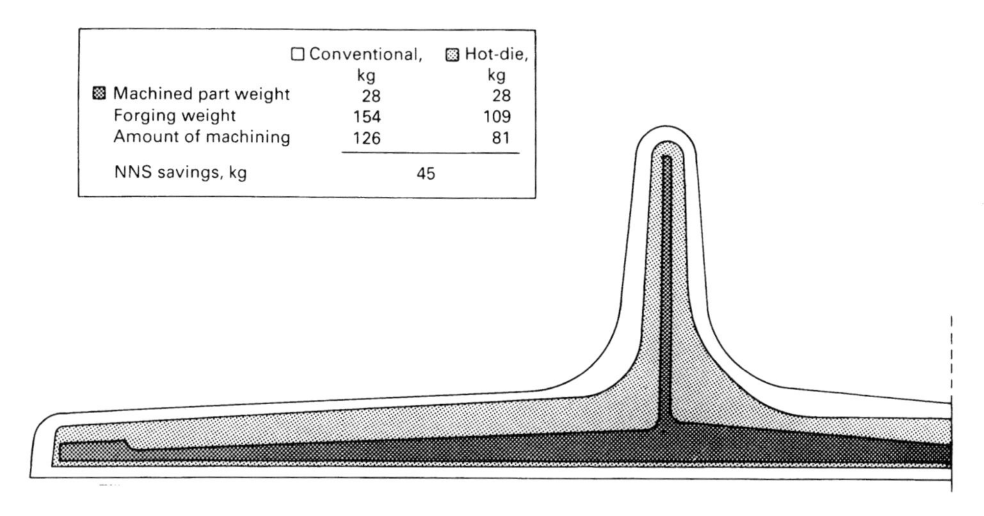

Fig. 20.2 Comparison between conventional and hot-die forging of a Ti-6Al-4V structural part on the basis of raw material saved [Shah, 1988]

application much later. These alloys contain molybdenum, vanadium, or iron as beta stabilizers, i.e., to stabilize the high-temperature body-centered cubic phase at room temperature. However, it should be noted that the beta phase in these alloys is only metastable. As shown in Table 20.3, the beta transus temperatures of these alloys are relatively low, with a lower percentage of the alpha (hexagonally close-packed) phase.

The lower transus temperatures of the beta alloys may lead one to conclude that these alloys would have a lower flow stress compared to the alpha and alpha + beta alloys [Semiatin et al., 1983]. However, the data provided in Fig. 20.5 does not support this conclusion. It is found that the beta alloy Ti-13V-11Cr-3Al does not show this behavior, which could be due to high levels of vanadium and chromium [Altan et al., 1973]. The data for Ti-10V-2Fe-Al is, however, significantly lower in magnitude compared to that of Ti-6Al-4V (Fig. 20.8). The figure shows that the flow stress data for the beta alloy at 1500 °F (815 °C) are comparable to that of Ti-6Al-4V at the

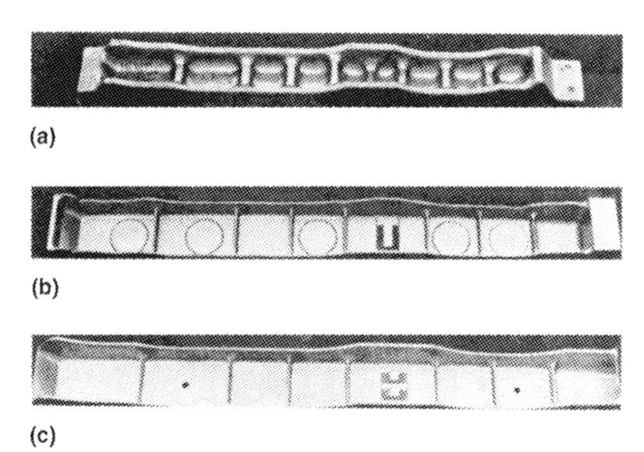

Fig. 20.3 Comparison of various methods of producing torque ribs [DuMond, 1975 and 1976]

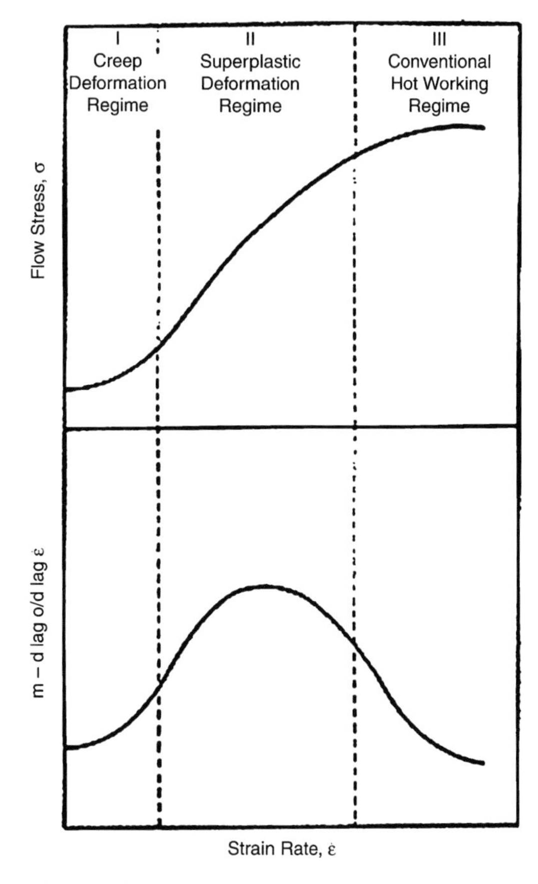

Fig. 20.4 Deformation regimes for superplastic materials [Chen et al., 1980]

higher temperature of 1700 °F (925 °C) for a given strain rate. This observation is of significance to the isothermal and hot-die forging technology, since it implies cost savings by the use of less expensive die materials, less die wear, and fewer problems with lubrication as a result of lower processing temperatures.

20.5.2 Nickel Alloys

Nickel-base superalloys have a degree of alloying in order to increase strength as well as creep and fatigue resistance at elevated temperatures, which are often a few hundred degrees below the working temperature. This is done through a combination of solid-solution strengthening, precipitation of gamma prime [Ni_3(Ti, Al)] and various carbides, and formation of phases containing boron and zirconium [Altan et al., 1973]. As a result of this alloying, nickel-base alloys are among the most difficult-to-work alloys in commercial use today. This also presents problems when it comes to producing ingots or bar stock suitable for forging. This situation arises due to macro- and microsegregation during solidification of the superalloy compositions [Daykin et al., 1972]. During solidification, regions low in alloy content, particularly the precipitation-hardening elements, freeze first, resulting in regions with higher alloy content as well. The working temperature regime of the former is lower and wider than that of the highly alloyed regions.

In forging nickel alloys, problems arising from high alloy content and segregation become worse when conventional hot forging is employed [Cremisio et al., 1972]. Die chilling can lower the workpiece temperature to below the solutioning temperature, leading to precipitation and the subsequent drop in workability and the possibility of fracture. Also, temperature rise during deformation, a result of the higher strain rates of conventional forging, could lead to melting, particularly at the grain boundaries where lower-melting-point phases are found. Thus, isothermal and hot-die forging have clear advantages over conventional hot forging of superalloys.

Figures 20.9 and 20.10 show the flow stress data for nickel-base superalloys such as Waspaloy and Inconel 718. These materials show a strong dependence of the flow stress on temperature, strain rate, and microstructure. It is observed that the curves show a maximum, followed by a drop and eventually a steady-state regime. This behavior is typical of materials that undergo dynamic recrystallization, and it promotes ductility and workability. The decrease in the flow stress following the maxima can be attributed to deformation heating. It was also found experimentally that the flow stress data could be considerably different for materials with different grain sizes or grain structure.

Flow stress data for fine-grained superalloys produced by P/M show characteristics that make them desirable for isothermal forging. In one study, P/M materials have been found to show lower flow stress than cast products, with high m values of 0.5 at strain rates of 10^{-3} and lower [Moskowitz et al., 1972].

Table 20.1 Alpha and beta stabilizers for titanium alloys

Alpha stabilizers (increase the beta transus temperature)	Aluminum, tin
Beta stabilizers (decrease the beta transus temperature)	Vanadium, molybdenum, chromium, copper

Table 20.2 Common titanium alloys used in isothermal and hot-die forging

Classification	Titanium alloy
Alpha and near alpha	Commercial-purity titanium and Ti-5Al-2.5Sn, Ti-8Al-1Mo-1V, Ti-2.5Cu, Ti-6242, Ti-6Al-2Nb-1Ta-0.8 Mo, Ti-5Al-5Sn-2Zr-2Mo
Alpha/beta	Ti-6Al-4V, Ti-6Al-4V-2Sn, Ti-6Al-2Sn-4Zr-2Mo, Ti-3Al-2.5V
Beta	Ti-13V-11Cr-3Al, Ti-8Mo-8V-2Fe-3Al, Ti-3Al-8V-6Cr-4Mo-4Zr, Ti-11.5Mo-6Zr-4.5Sn, Ti-10V-2Fe-3Al

Table 20.3 Beta transus temperatures and forging temperatures for titanium alloys [Altan et al., 1973]

	Beta transus temperature		Forging temperature	
Alloy	°F	°C	°F	°C
Commercially pure	1760	960	1600/1700	870/927
Alpha alloys				
Ti-5Al-2.5Sn	1900	1040	1725/1850	940/1010
Ti-8Al-1Mo-1V	1860	1015	1725/1850	940/1010
Alpha + beta alloys				
Ti-6Al-4V	1820	993	1550/1800	843/982
Ti-6Al-4V-2Sn	1735	945	1550/1675	843/915
Ti-6Al-2Sn-4Zr-2Mo	1825	995	1700/1800	927/982
Ti-6Al-2Sn-4Zr-6Mo	1750	955	1625/1700	885/927
Beta alloys				
Beta III	1400	760	1550/1650	843/900
Ti-13V-11Cr-3Al	1325	718	1600/1800	870/982
Ti-3Al-8V-6Cr-4Mo-4Zr	1475	800	1500/1600	815/870
Ti-10V-2Fe-3Al	1475	800	1400/1600	760/870

The main characteristics of the materials discussed above can be summarized as [Semiatin et al., 1983]:

- Strong dependence of the flow stress on temperature
- High rate sensitivity at low strain rates
- Fine grain size

These properties make the titanium and nickel-base superalloys more suitable for isothermal forging than conventional hot forging. Isothermal forging avoids the problems in metal flow and microstructure arising out of die chilling. It is possible to forge parts at lower strain rates, leading to reduced forging loads and better die filling. However, in order to maintain a high rate sensitivity during forging, it is necessary that the preform or billet have a fine grain size, which should be retained during forging. Thus, alloys with a homogeneous two-phase structure are most likely to meet this requirement. Wrought titanium alloys and wrought and P/M nickel-base superalloys are therefore among the most appropriate high-temperature alloys for isothermal and hot-die forging.

20.6 Equipment and Tooling

Of all the equipment available to carry out a forging operation, the slow speeds of hydraulic presses make them more suitable for isothermal and hot-die forging operations. These machines make it possible to use the high strain-rate sensitivity of the forging materials at low strain rates as an advantage. Isothermal forging processes are carried out under constant load at approximately 0.1 to 1 in./min (0.25 to 2.5 cm/min), with hot-die forging being slightly faster to avoid die chilling [Semiatin et al., 1983].

The flow stress is dependent on the strain rate (press speed), which in turn determines the forging pressure. Hence, the maximum allowable die pressures dictate the forging speed. The flow stress of the material and hence the die pressures can be maintained within allowable limits by lowering the strain rate and press speed. Practically, it is possible to start the forging stroke with a higher ram speed, since the part is initially thick with a relatively small surface area and requires little pressure for deformation. Thus, it is necessary to control the press ram throughout the forging stroke so that the desired strain rates can be applied at different stages. This is an essential requirement, since the part geometry, metal flow, and forging pressure vary throughout the stroke.

The size of the part and the allowable forging pressure on the dies can be used as the criteria for selecting the hydraulic press for the desired

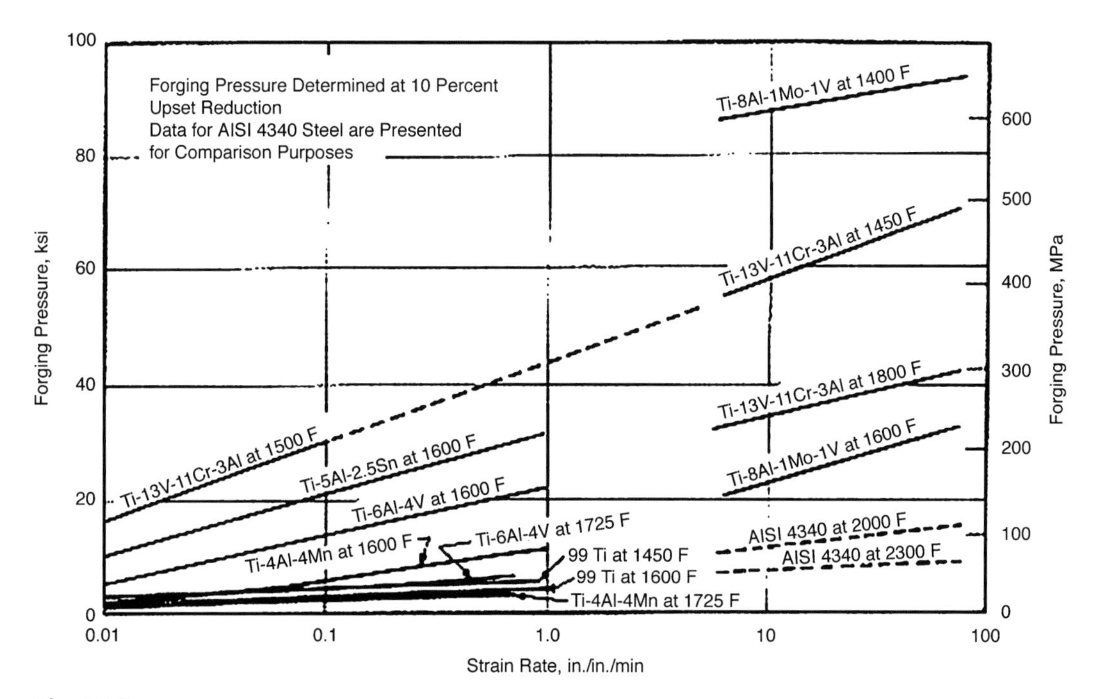

Fig. 20.5 Effect of strain rate on forging pressures for titanium alloys at different temperatures [Altan et al., 1973]

isothermal forging process. The size of the dies is governed by the limitations of the technique used to heat them as well as the cost of the die materials, which are usually very expensive.

Sensors are used in conjunction with press equipment to form a closed-loop control system in the isothermal forging press. The high levels of process control that are achieved through isothermal forging have allowed near-net forging of complex components from alloys with very narrow processing windows. Isothermal forging can be a very cost-effective manufacturing process for the manufacture of critical components that are required to be produced from very expensive and difficult-to-process materials. Vast quantities of quality jet engine parts have been produced by isothermal forging. Figure 20.11 shows a jet engine disk being forged in an isothermal press.

Ladish Company designed and constructed two isothermal presses (5,000 and 10,000 ton). Between the two presses, the 10,000 ton press is the largest isothermal press in the world (Fig. 20.12). Pratt & Whitney-Georgia has one 3,000 ton and two 8,000 ton isothermal presses. Wyman-Gordon has one 8,000 ton isothermal press. Isothermal presses have been utilized for the most demanding aerospace component applications.

20.6.1 Die Materials

Due to the very high temperatures of the dies in isothermal and hot-die forging, selection of the die materials is an extremely critical part of the forging process design. The various factors to be taken into consideration during die material selection are wear and creep resistance, hot

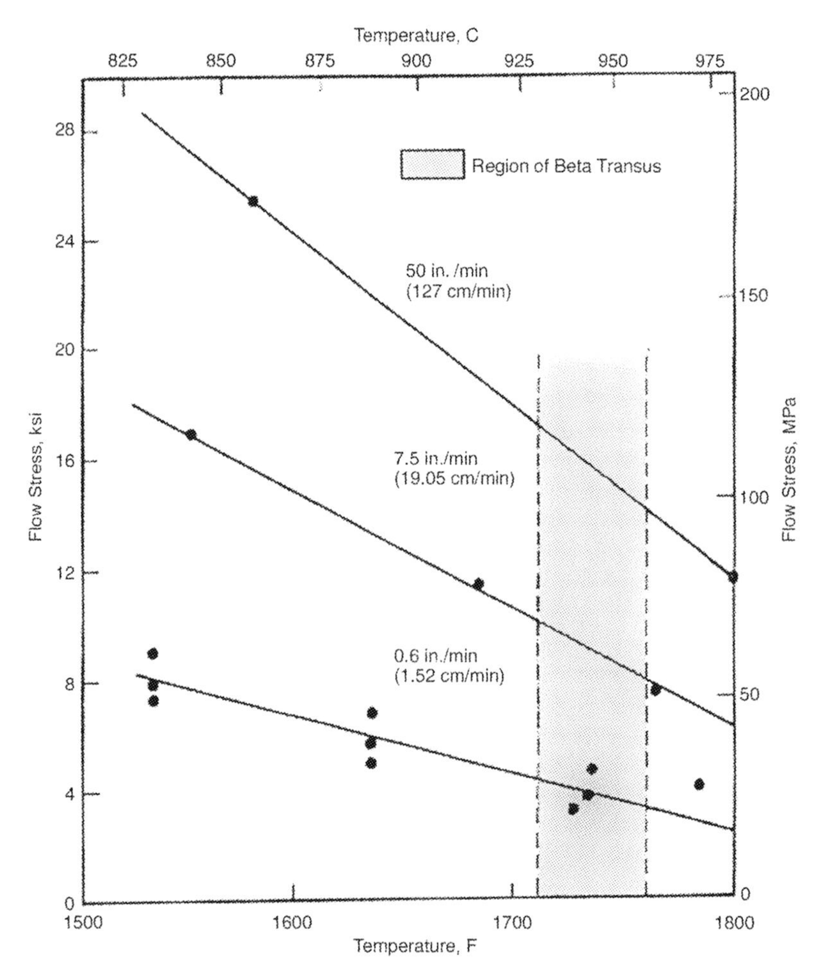

Fig. 20.6 Effect of deformation rate and temperature on flow stress of Ti-6Al-6V-2Sn alloy under isothermal forging conditions [Fix, 1972]

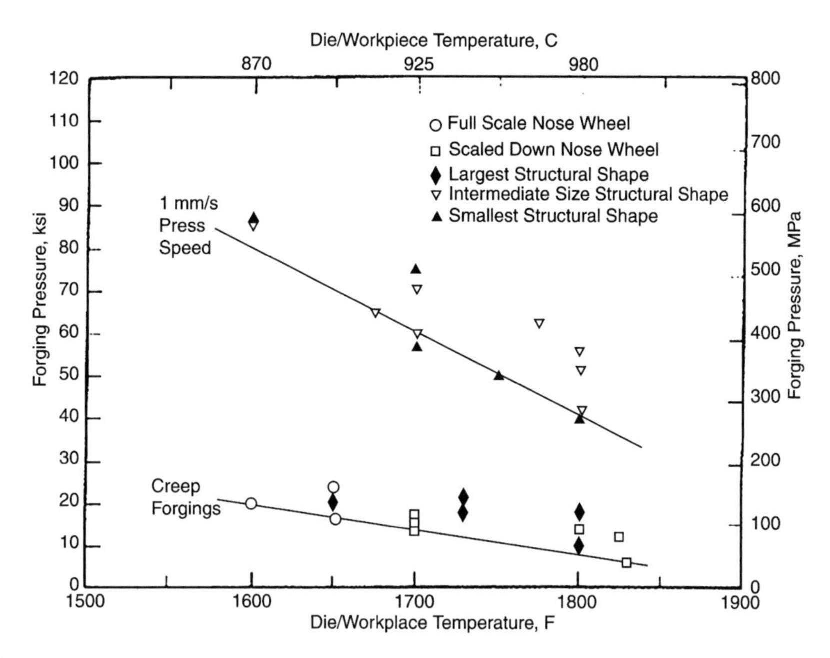

Fig. 20.7 Effect of temperature on forging pressure in isothermal forging of Ti-6Al-6V-2Sn [Kulkarni et al., 1972]

hardness, toughness, and overall structural integrity. When die temperatures are in the range of those encountered in conventional hot forging, it is possible to use a variety of low-alloy tool steels or hot working die steels. However, when forging titanium or nickel-base alloys, where temperatures are in the range of 1500 to 2000 °F (815 to 1095 °C), one is restricted to a limited range of die materials, ranging from superalloys and refractory metal alloys to ceramics [Semiatin et al., 1983].

Due to their low strength and creep resistance above 1500 °F (815 °C), iron- and cobalt-base superalloys have been found, in general, to be unsuitable as die materials for isothermal forging. Wrought nickel-base superalloys such as Waspaloy, Udimet 700, Astroloy, and Unitemp AF2-1DA are well suited for isothermal forging of titanium alloys at 1700 °F (925 °C) and lower. What makes these wrought nickel-base superalloys attractive as die materials is their structural integrity and resistance to defects such as porosity in large forged die blocks. They also retain their strength in the temperature range of 1500 to 1700 °F (815 to 925 °C), unlike the iron-base materials (Fig. 20.13). However, in practice, the temperatures for nickel-base superalloy dies are maintained at approximately 1650 °F (900 °C) or lower to extend die life in case of large production runs. This makes it a case of near-isothermal forging instead of isothermal forging. Figure 20.13 shows a drastic drop in the strength of these die materials after a temperature of 1700 °F (925 °C) because of reversion of the gamma-prime precipitates. Cast nickel-base superalloys such as IN-100, MAR-M-200, Inconel 713C, and TRW-NASA VIA [Simmons, 1971] have also found acceptance as die materials. Figure 20.14 shows that these alloys are comparable and sometimes superior in strength to the wrought alloys. Isothermal forging of wrought nickel-base superalloys requires special die materials, such TZM molybdenum, TZC, and Mo-Hf-C [Clare et al., 1977–78]. TZM mo-

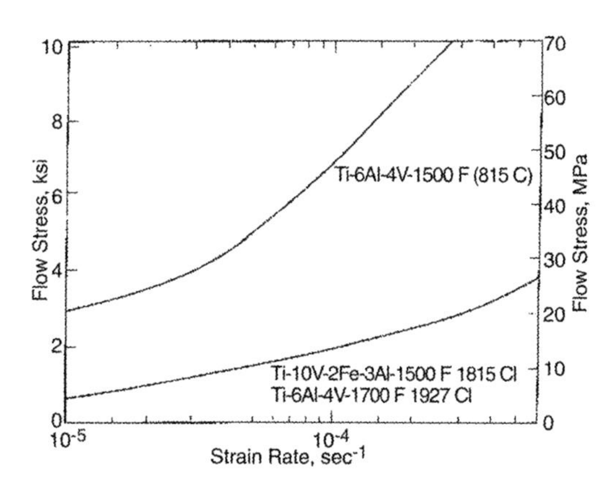

Fig. 20.8 Flow stress data for Ti-10V-2Fe-Al compared with that of Ti-6Al-4V [Rosenberg, 1978]

lybdenum has good strength and stability up to temperatures of 2190 °F (1200 °C) as well as good resistance to fatigue crack initiation and crack propagation [Hoffelner et al., 1982]. The major drawback of this alloy is that it readily reacts with oxygen, making it necessary to have a surrounding inert atmosphere or a vacuum. This increases the cost of the forging process, making the isothermal forging of wrought nickel-base superalloys very expensive. Both TZC and Mo-Hf-C have significantly higher strength at isothermal forging temperatures compared to TZM. These alloys have higher resistance to plastic deformation and better wear resistance.

20.6.2 Die Heating Techniques

One of the greatest challenges to be overcome in isothermal and hot-die forging is the development of methods for heating the dies to a uniform elevated temperature. There are several options available for heating the die holders and inserts, such as gas-fired burners, resistance heater bands, and induction heating. Of these, gas-fired burners tend to be bulky, requiring the

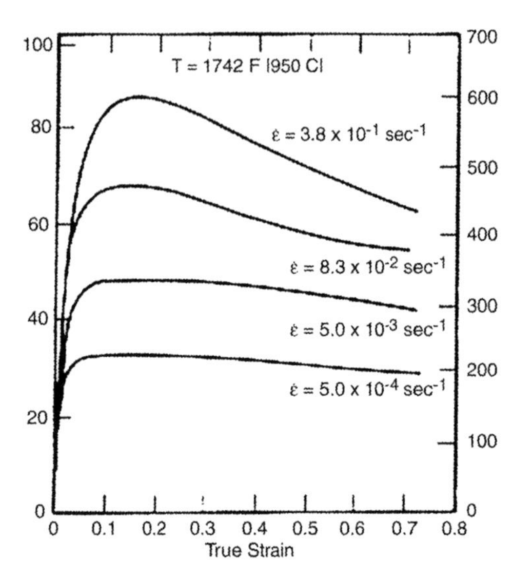

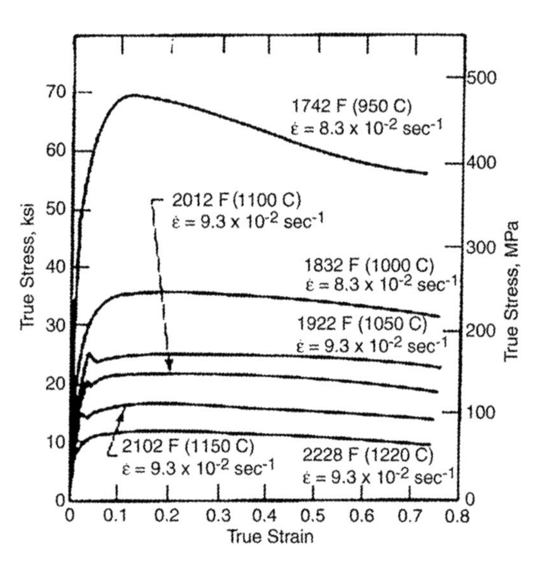

Fig. 20.9 Flow stress data for Waspaloy [Guimaraes et al., 1981]

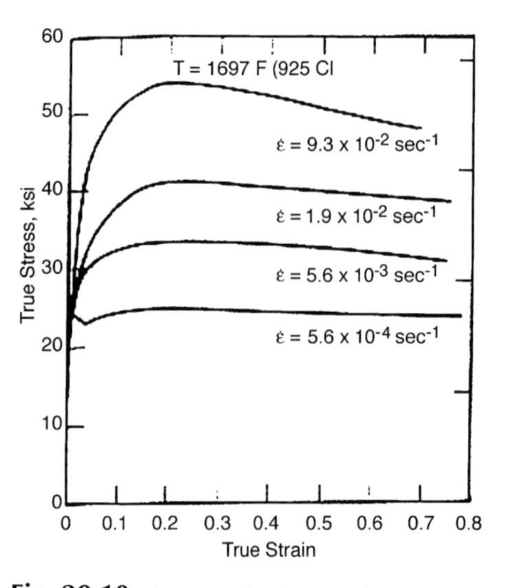

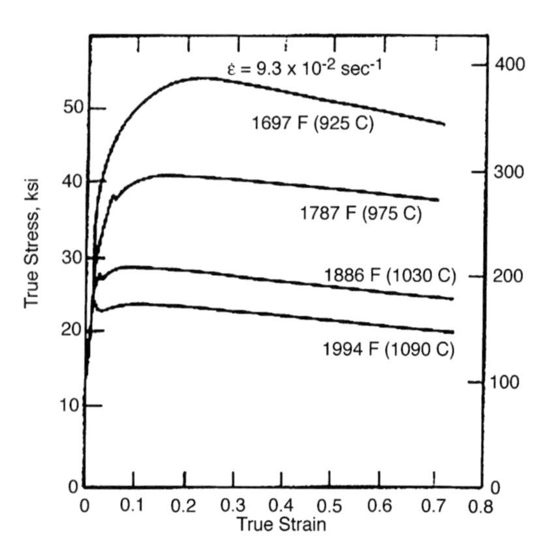

Fig. 20.10 Flow stress data for Inconel 718 [Guimaraes et al., 1981]

use of nozzles and additional equipment, which can hamper easy access to the actual forging area by occupying a large amount of space.

Figure 20.15 shows the heating setup for isothermal forging of titanium alloys using resistance heaters, which tend to be one of the simplest to implement [Antes, 1980]. In this particular case, die inserts of IN-100 are held in place by die holders (IN-100) and backed by bolsters of IN-100 and H21 hot work die steel.

Fig. 20.11 A jet engine part being forged in an isothermal press at Ladish. Courtesy of Ladish Co.

Fig. 20.12 10,000 ton isothermal forging press, designed, constructed, and operated by Ladish. Courtesy of Ladish Co.

This combination of back-up materials is necessary because of the conduction of heat into the die stack. Depending on the die geometry and operating temperature, it is possible to use nickel-base superalloys as bolster materials, whereas the remainder of the die assembly can be made from tool steels and stainless steels. The use of insulators, such as metallic sheets coated with ceramics such as zirconium oxide, and water cooling reduces heat losses and prevents damage to the press. Alumina blocks and spacer plates are also effective means of reducing heat losses.

Induction coils made of copper tubing are another simple means of die heating (Fig. 20.16) [Prasad et al., 1969]. These coils are usually water cooled and can be bent to various shapes depending on the die geometry. In induction heating, the rate of heat generation is higher at die circumference than the cavity. In order to get uniform heating, low frequencies of alternating current of 40 to 100 Hz are used. This makes use of the property that difference in local heating rates reduces when the frequency is reduced. The temperature gradients arising out of nonuniform heating may induce large thermal stresses, which may cause cracking. Susceptors, such as graphite used in the Gatorizing process, can also be used in conjunction with the induction coils to promote uniform heating through radiation.

It has been found that the use of a single method of die heating often leads to nonuniform temperature distribution in the die. Hence, more sophisticated heating systems utilizing two or more heating methods can be implemented (Fig. 20.17). In this setup, a portable dummy resistance furnace is used to heat the outer portion of the dies, thus making it possible to remove it during the forging operation. The second part of the heating system consists of a set of gas burners that are aimed toward the center of the die. Temperature variations were limited to within ± 50 °F (28 °C) for dies with an outer diameter of 42 in. (107 cm). A similar two-part system was implemented at Wyman-Gordon, consisting of devices surrounding the horizontal and vertical portions of the die assembly [Chen et al., 1977]. The vertical part contains gas-fired, infrared heating elements, whereas the horizontal part contains resistance heater rods set into Waspaloy heater plates directly above the top die insert and die holder and below the bottom die insert and holder (Fig. 20.18). This system enabled precise temperature control over even the largest die areas.

20.6.3 Lubrication Part Separation Systems in Isothermal and Hot-Die Forging

Lubrication systems in isothermal forging are required to provide low friction for good metal flow, ease of release of the forging from the dies, and to obtain a good surface finish. These systems should not lead to any sort of a buildup of the lubricant, which might cause problems in achieving the desired finished part tolerances. Billet lubricants must provide a protective coating to prevent the surface from oxidation during heating or forging. At the same time, these lubricants should not react with either the workpiece or the dies [Semiatin et al., 1983]. Lubricants in isothermal forging are not required to have the same characteristics as those used in conventional forging, since they do not operate over a range of temperatures; thus, these lubricants can be developed to deliver an optimum performance at the operating temperature of the isothermal forging process or a narrow temperature range. Petroleum- and graphite-based lubricants as well as molybdenum disulfide are unsuitable for isothermal and hot-die forging, since they rapidly decompose at elevated temperatures. Various glass mixtures, in the form of frits

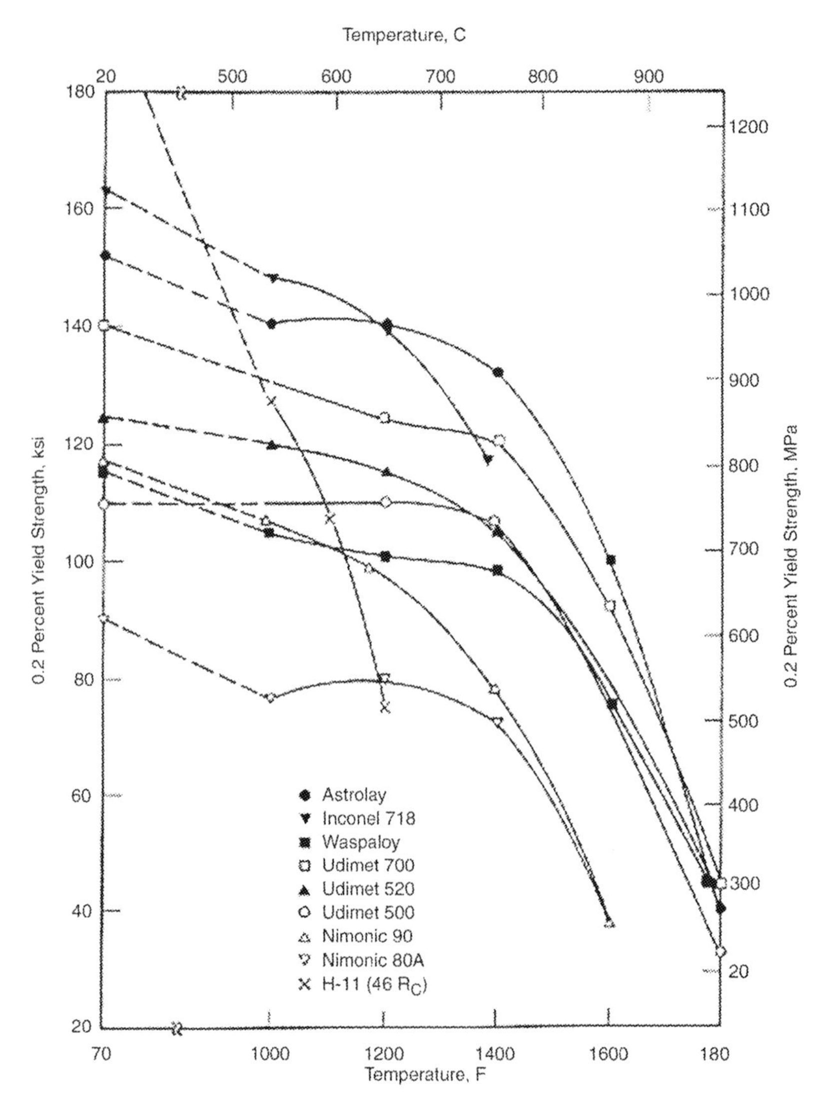

Fig. 20.13 Yield strengths of some wrought nickel-base superalloys, with H-11 being included for comparison [International Nickel Co., 1977]

(composed of a variety of glass-forming oxides) or premixed compounds with their own aqueous or organic solvent carriers, provide the best lubrication in isothermal forging. Frits are normally ground into a fine powder, which is then converted into a slurry (alcohol bath) used for dipping the forging billets. The alcohol evaporates, leaving behind a powderlike coating on the billets. This coating, under forging temperatures, becomes a viscous, glassy layer providing lubrication and oxidation protection. Figure 20.19 shows frit selection on the basis of forging temperature. Glass of 200 to 1000 P viscosity has been found to give good lubricity and a good continuous film characteristic required to prevent galling under high pressures [Semiatin et al., 1983].

20.7 Postforging Heat Treatment

Postforging heat treatments are always used for isothermal and hot-die forged aerospace components. The heat treatment of titanium and nickel-base superalloys, like most engineering materials, is critical to overall component performance. The microstructures produced from hot working operations alone are not optimized for most applications, so the microstructure of forging must be tailored by heat treatment to re-

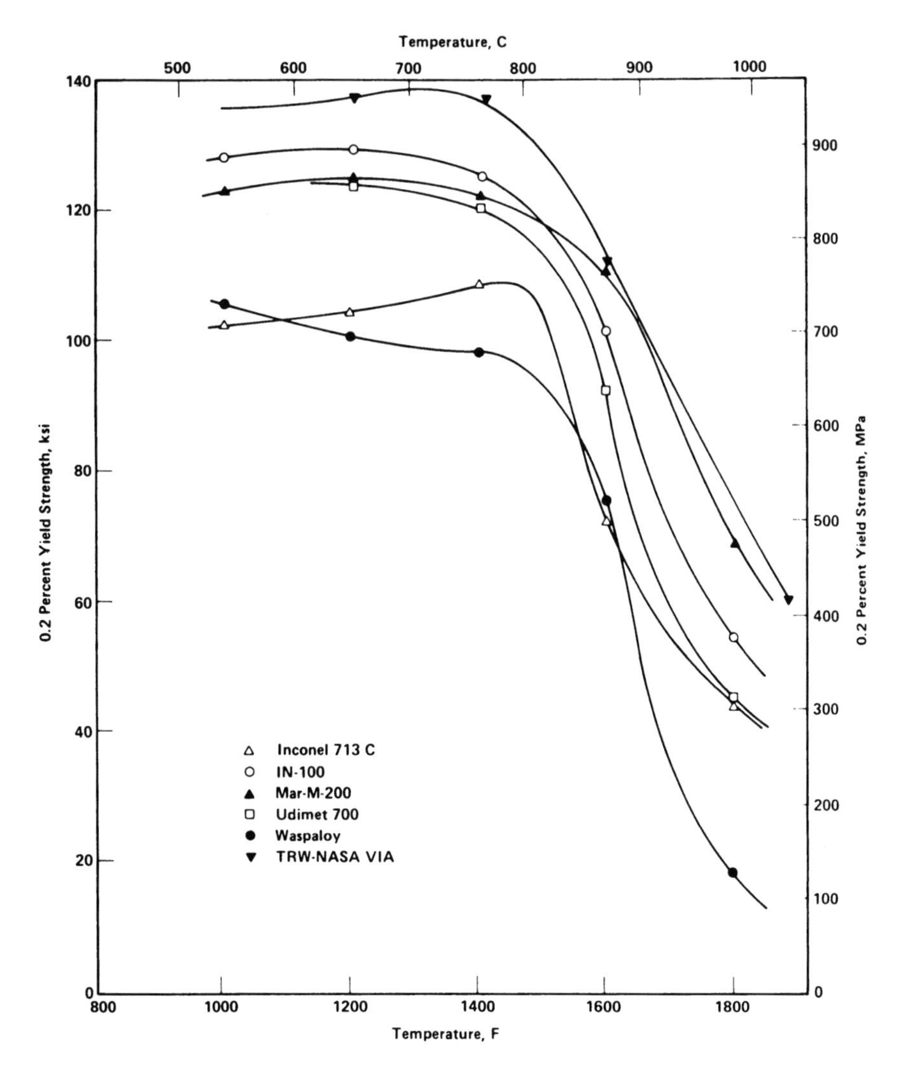

Fig. 20.14 Yield strengths of several cast and wrought nickel-base superalloys [International Nickel Co., 1977, and Simmons, 1971]

sult in the optimum grain size, grain-boundary morphology, phases, and phase distributions.

One of the most important parameters in heat treatment is the cooling method and part cooling rate profile. The imposed cooling rate, along with the material transformation and phase growth kinetics, controls the development of the required mechanical properties. Often, high cooling rates are required to produce high strengths [Chang, 1997]. Other times, controlled, slower cooling rates are desired to develop serrated grain boundaries and optimum microstructures in nickel-base materials [Myagawa, 1976, and Koul et al., 1985].

Although mechanical properties are often the most significant driving force for selecting various cooling methods and rates, other significant factors deserve consideration. Large variations in cooling rate within a given part will create large thermal stresses and hence high residual stresses [Chang, 1997 and 1996]. Thermal and residual stresses can result in problems with quench cracking, machining distortion, or unexpected mechanical properties and performance in the final component. This issue has led to engineering of specialty heat treat cooling methods for forgings. Various quench media (air, oil, polymer, water), flow patterns imparted on the component, sequences of quench media and method used, and the component heat treat geometry are all utilized to tailor the cooling rate and subsequently, the microstructure, mechanical properties, and residual stresses of a forging. The goal of modern heat treatment practices for aerospace forgings is to achieve the best balance

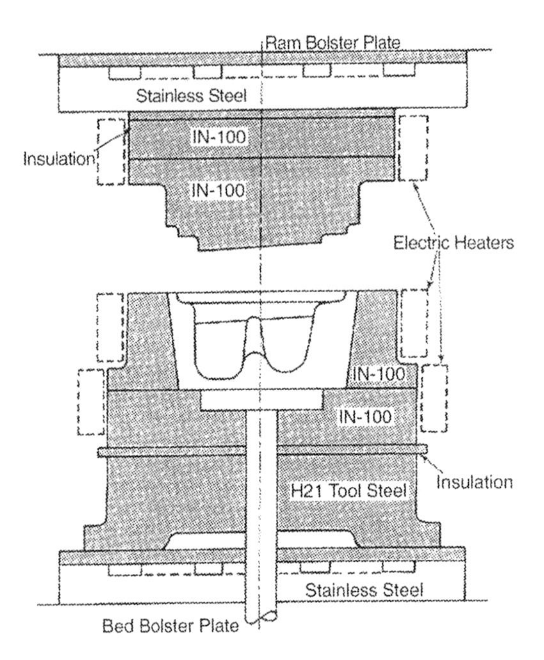

Fig. 20.15 Die setup for isothermal/hot-die forging with resistance heaters [Antes, 1980]

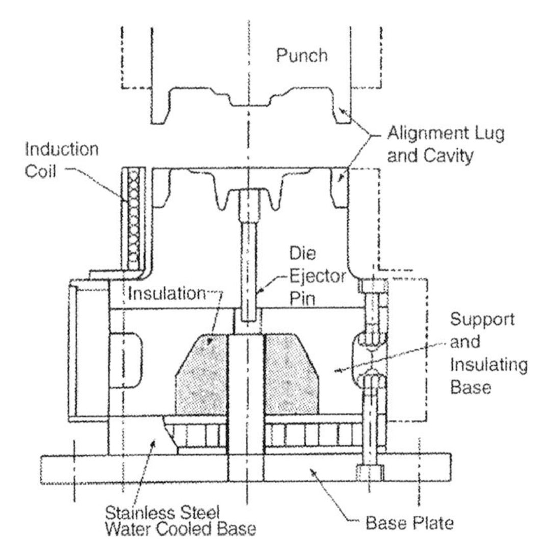

Fig. 20.16 Die setup for isothermal/hot-die forging with induction coils [Prasad et al., 1969]

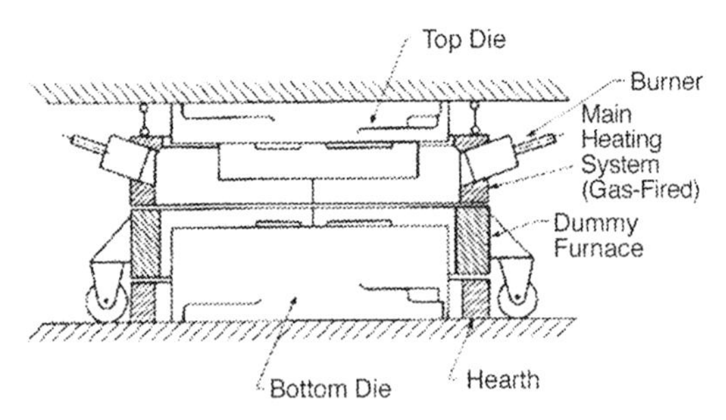

Fig. 20.17 Die heating system utilizing resistance heaters and gas burners [Kulkarni, 1978]

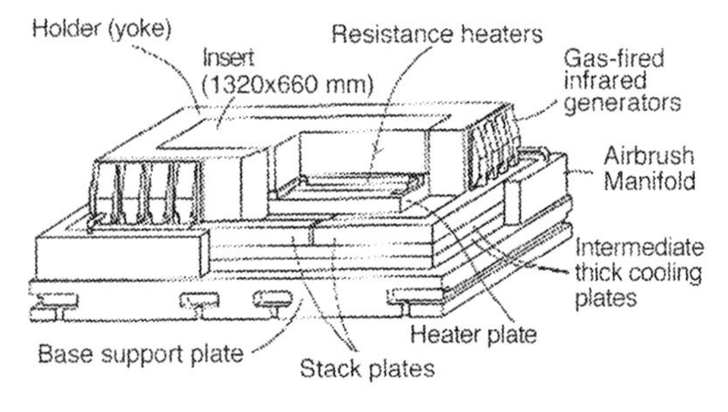

Fig. 20.18 A two-part die heating system comprising gas-fired infrared heater and resistance heaters [Shah, 1988]

of mechanical properties with the minimum residual stress. Process modeling is vital to design of the optimum combination of forge geometry, heat treatment geometry, and heat treat cooling method [Shen et al., 2000 and 2001].

20.8 Production of Isothermal/Hot-Die Forgings

The development of the isothermal and hot-die forging technology started in the early 1970s and was put into production in approximately the late 1970s. Some examples are shown in Fig. 20.20.

Figure 20.20(a) shows Ti-6Al-4V hot-die forged F-15 bearing supports weighing 46.5 lb (21.1 kg) forged in Astroloy dies at 1700 °F (925 °C). This part was forged in three stages:

- Preblock and block, with conventional forging processes
- Finish forging as doubles in Astroloy hot dies

Figure 20.20(b) shows an engine mount that was hot-die forged to net dimensions on the surface shown. The backside, which is flat, was machined during the final machining operations. The forged IN-100 disk shown in Fig. 20.20(c) was forged in one step from a billet using TZM

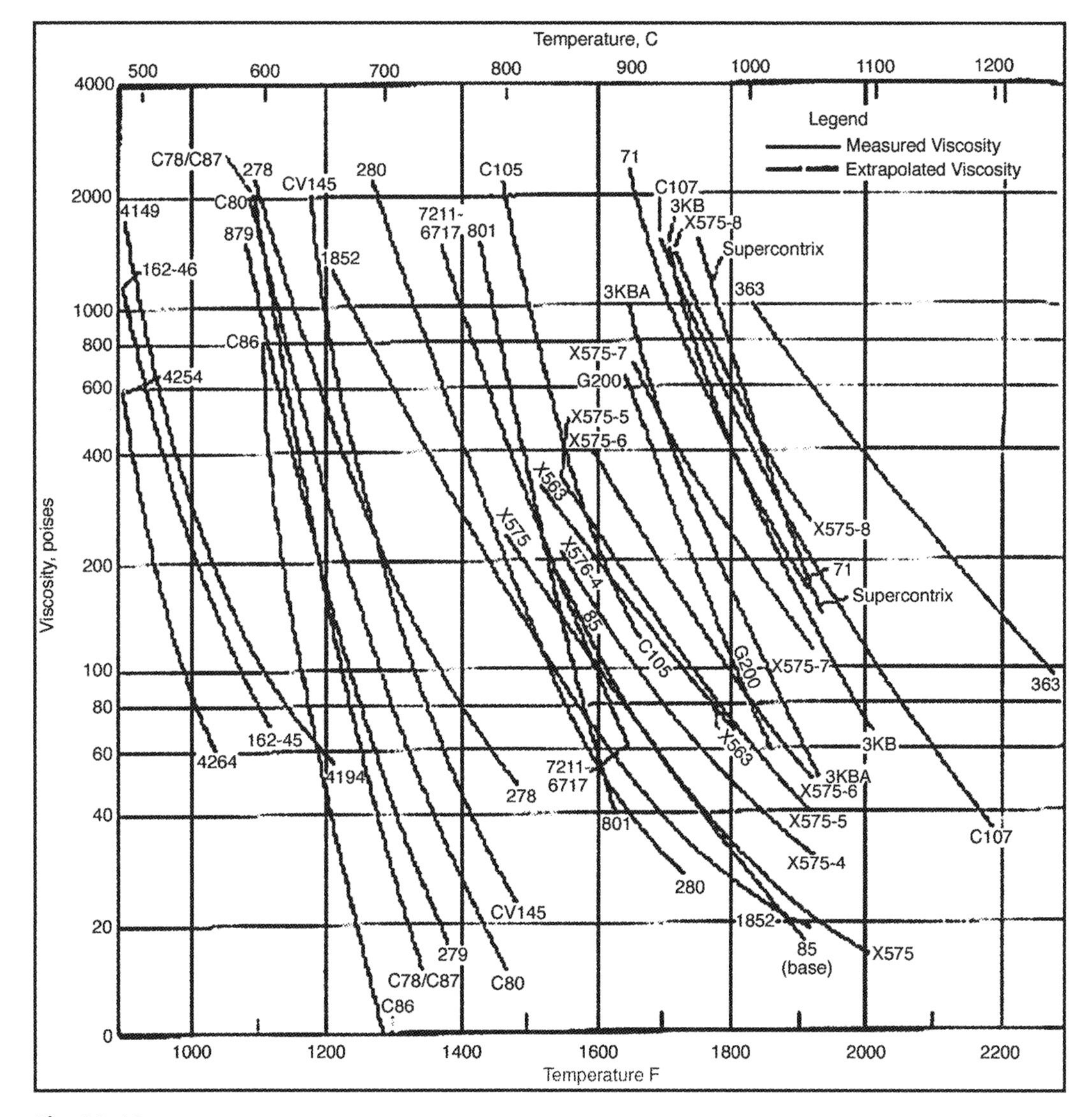

Fig. 20.19 Viscosity temperature curves for various metalworking glasses [Semiatin et al., 1983]

dies. This part was machined all over to yield the final shape, i.e., it had no net surfaces. Forgeability and savings in material costs were the two driving forces behind selecting isothermal forging for this component.

20.9 Economic Benefits of Isothermal and Hot-Die Forging

Isothermally forged parts are similar in properties to conventionally forged parts. Thus, the main advantage of using isothermal forging stems from the differences in the manufacturing costs. The cost of making a forging can be split up into the following categories [Semiatin et al., 1983]:

- Die material
- Die machining and die installation (labor costs)
- Workpiece materials
- Lubricants and lubrication systems
- Preheating and forging (labor and energy costs)
- Rough and finish machining after forging (labor costs)

Die materials for isothermal forging are much more expensive and difficult to machine than conventional die materials. However, isothermal forging requires fewer number of forging stages and therefore fewer die sets. This reduces the overall tooling costs when compared to conventional forging. Another factor to be taken into consideration is the production run and the performance of the dies in runs of varying dura-

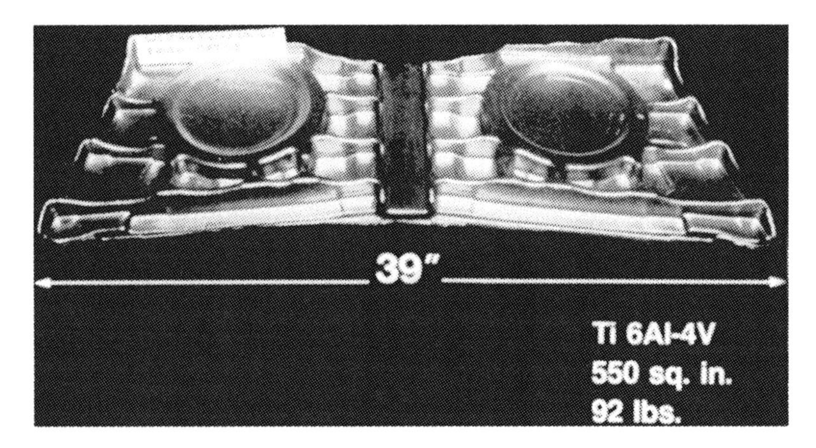

(a) F-15 bearing support of Ti-6Al-4V finish forged with a hot-die near-net forging process.

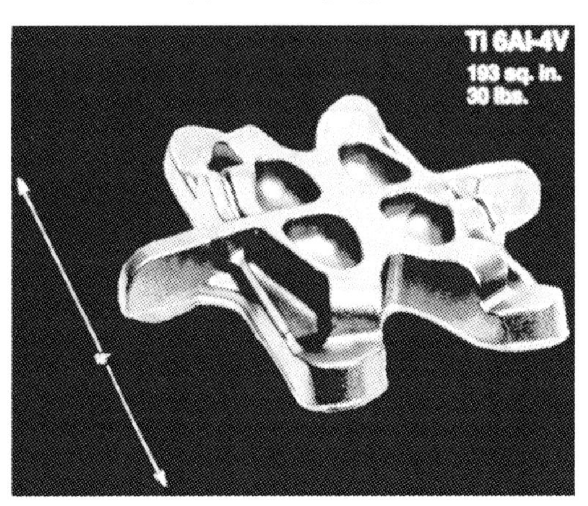

(b) Hot-die forged Ti-6Al-4V engine mount with net surfaces.

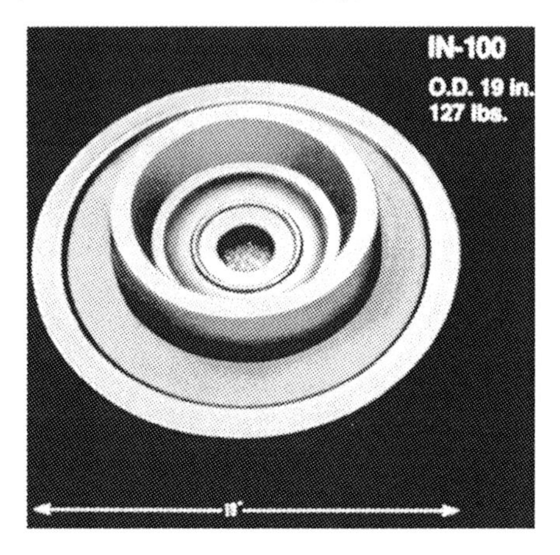

(c) Isothermically forged IN 100 disk.

Fig. 20.20 Production examples of isothermal/hot-die forged aircraft components [Shah, 1988]

tions. Though the die costs are high in isothermal forging, the material input and costs are lower. The same is true for the machining costs required to produce the finished forging.

Figures 20.21(a) and (b) show the cost comparison for two aircraft parts, namely, a connecting link and a bearing support [Shah, 1988]. It was seen that the forging for the connecting link (Fig. 20.21a) using the conventional method weighed 38.3 lb (17.4 kg), whereas a hot-die forged part weighed in at 29 lb (13 kg). The hot-die forging process used a die stack with Astroloy dies at approximately 1700 °F (925 °C), with some net surfaces. Figure 20.21(a) shows significant savings in initial tooling costs and that it took over 500 forgings for the higher tooling costs of hot-die forging to be justified. Thus, hot-die forging was not economically viable for production runs of less than 500 parts in this case.

Figure 20.21(b) shows a similar cost comparison for the bearing support shown in Fig. 20.20(a), which was also forged in Astroloy dies at 1700 °F (925 °C). However, due to its larger size compared to the link, the difference in die costs between conventional and hot-die forging was greater. Due to significant reduction in material and machining costs, the break-even point for this part was less than 200 [Shah, 1988].

20.10 Summary

Isothermal and hot-die forging technologies were developed in the 1970s and are used in current applications in production of precision forgings, mostly in the aircraft industry. These processes have proven to be economically feasible for forging expensive high-temperature titanium and nickel-base alloys contingent on:

- Proper understanding of the workpiece material properties and deformation characteristics
- Proper tool material selection and die design as well as process conditions such as temperatures and die-workpiece lubrication
- Costs, as shown in the examples of the comparative studies between isothermal/hot-die forging and conventional methods

Titanium and nickel-base superalloys are excellent candidates for isothermal forging due to the strong dependence of their flow stress on temperature and strain rate. The high die costs associated with isothermal and hot-die forging are justified by the subsequent savings in material and machining.

Due to the advantages discussed earlier, isothermal and hot-die forging have found wide ac-

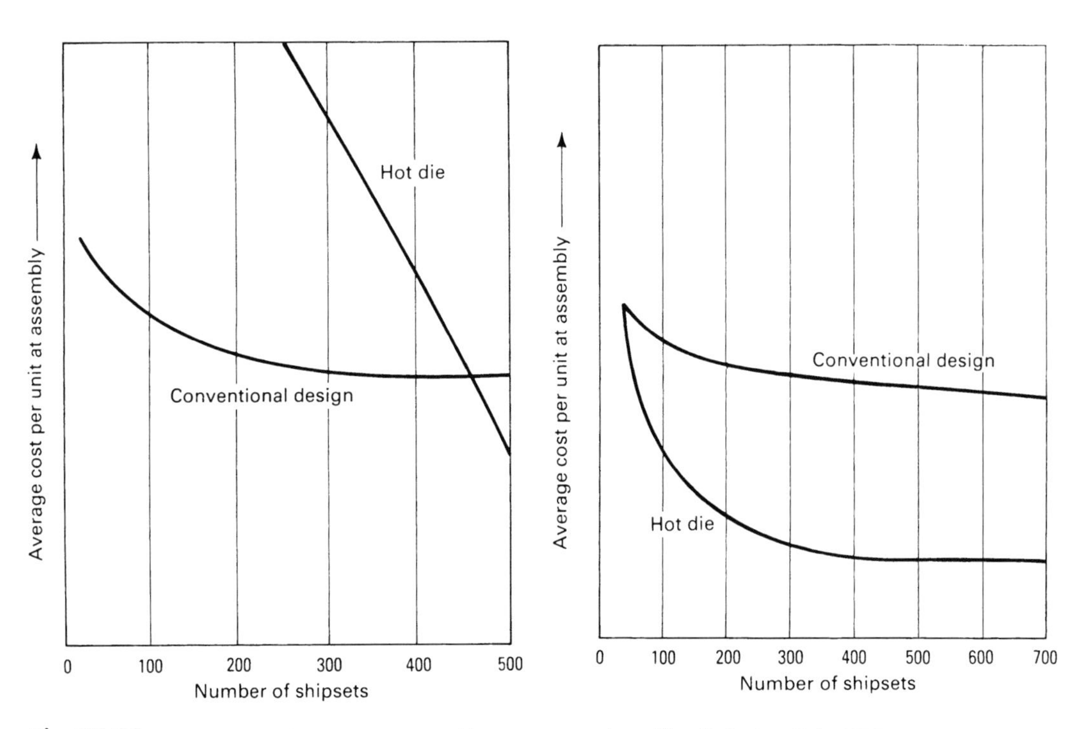

Fig. 20.21 Cost comparison between conventional forging versus isothermal/hot-die forging [Shah, 1988]

ceptance in the production of aircraft components such as titanium structural members and turbine engine components made of titanium and nickel-base superalloys.

REFERENCES

[Altan et al., 1973]: Altan, T., Boulger, F.W., Becker, J.R., Akgermann, N., Henning, H.J., "Forging Equipment, Materials and Practices," MCIC Report HB-03, Metals and Ceramics Information Center, Battelle's Columbus Laboratories, Columbus, OH, Oct 1973.

[Antes, 1980]: Antes, H.W., "Complex Parts of Titanium Alloy by Isothermal Forging," Industrial Research/Development, Vol. 22 (No. 11), Nov 1980, p 116–119.

[Chang, 1996]: Chang, K.-M., "Critical Issues of Powder Metallurgy Turbine Disks," *Acta Metallurgica Sinica,* Vol. 9, 1996, p 467–471.

[Chang, 1997]: Chang, K.-M., "Physical Simulation of Quench Cracking for P/M Superalloy Turbine Disks," *Proceedings of Seventh International Symposium on Physical Simulation of Casting, Hot Rolling and Welding,* National Research Institute for Metals (Japan), Jan 21–23, 1997, p 509–512.

[Chen et al., 1977]: Chen, C.C., Couts, W.H., Gure, C.P., Jain, S.C., unpublished research, Wyman-Gordon Co., Worcester, MA, Oct 1977.

[Chen et al., 1980]: Chen, C.C., and Coyne, J.E., "Recent Developments in Hot Die Forging of Titanium Alloys," *Titanium'80,* Vol. 4, The Metallurgical Society of AIME, 1980, p 2513–2522.

[Clare et al., 1977]: Clare, L.P., and Rhodes, R.H., "Superior Powder Metallurgy Molybdenum Die Alloys for Isothermal Forging," *High Temperatures—High Pressures,* Vol. 10, 1978, p 347–348.

[Clare et al., 1978]: Clare, L.P., and Rhodes, R.H., "Superior P/M Mo Die Alloys for Isothermal Forging," *Proceedings of the Ninth Plansee Seminar (III),* Metallwerk Plansee (Reutte, Austria), 1977.

[Cremisio et al., 1972]: Cremisio, R.S., and McQueen, H.J., "Some Observations of Hot Working Behavior of Superalloys According to Various Types of Hot Workability Tests," *Superalloys Processing, Proceedings of the Second International Conference,* MCIC Report 72-10, Section F, Metals and Ceramics Information Center, Battelle's Columbus Laboratories, Columbus, OH, Sept 1972.

[Daykin et al., 1972]: Daykin, R.P., and DeRidder, A.J., "Primary Working of Superalloys," *Superalloys Processing, Proceedings of the Second International Conference,* MCIC Report 72-10, Section F, Metals and Ceramics Information Center, Battelle's Columbus Laboratories, Columbus, OH, Sept 1972.

[DuMond, 1975]: DuMond, T.C., "Putting the Squeeze on Titanium Forging Costs," *Iron Age Metalworking International,* Vol. 216 (No. 22), Dec 1975, p 46–48.

[DuMond, 1976]: DuMond, T.C., "Isothermal Forging—New Ti Forming Process," *Iron Age Metalworking International,* Vol. 15 (No. 7), July 1976, p 31–32.

[Fix, 1972]: Fix, D.K., "Titanium Precision Forgings," *Titanium Science and Technology,* Vol. 1, Plenum Press, 1972, p 441–451.

[Greenwood et al., 1978]: Greenwood, G.W., Seeds, W.E., Yue, S., "Isothermal Forging of Titanium Alloys for Aerospace Applications," *Proceedings of the Metals Society Conference on Forging and Properties of Aerospace Materials,* The Metals Society, London, 1978, p 255–265, 341–349.

[Guimaraes et al., 1981]: Guimaraes, A.A., and Jonas, J.J., "Recrystallization and Aging Effects Associated with High Temperature Deformation of Waspaloy and Inconel 718," *Metallurgical Transactions A,* Vol. 12, 1981, p 1655–1666.

[Hoffelner et al., 1982]: Hoffelner, W., Wuthrich, C., Schroder, G., Gessinger, G.H., "TZM Molybdenum as a Die Material for Isothermal Forging of Titanium Alloys," *High Temperatures—High Pressures,* Vol. 14, 1982, p 33–40.

[International Nickel Co., 1977]: "High Temperature, High Strength, Nickel-Base Alloys," The International Nickel Co., Inc., New York, 1977.

[Koul et al., 1985]: Koul, A.K., and Thamburaj, R., "Serrated Grain Boundary Formation Potential of Ni-Based Superalloys and Its Implications," *Met. Trans. A,* Vol. 16, Jan 1985, p 20–26.

[Kulkarni, 1978]: Kulkarni, K.M., "Isothermal Forging—From Research to a Promising New Manufacturing Technology," *Proceedings of the Sixth North American Metalworking Research Conference,* Society of Manufacturing Engineers, 1978, p 24–32.

[Kulkarni et al., 1972]: Kulkarni, K.M., Parikh, N.M., and Watmough, T., "Isothermal Hot Die Forging of Complex Parts in a Titanium Alloy," *Journal of Institute Metals,* Vol. 100, May 1972, p 146–151.

[McLeod et al., 1980]: McLeod, S.A., Walker, B.H., Mendelson, M.I., "Development of an Integral Ceramic Blade Metal Disk with Circumferential Blade Attachment," *Ceramics for Turbine Blade Applications,* AGARD Conference Proceedings No. 276, AGARD, France, 1980.

[Moore et al., 1970]: Moore, J.R., and Athey, R.L., U.S. Patent No. 3,519,503.

[Moskowitz et al., 1972]: Moskowitz, L.N., Pelloux, R.M., Grant, N., "Properties of IN-100 Processed by Powder Metallurgy," *Superalloys—Processing, Proceedings of the Second International Conference,* MCIC Report 72-10, Section Z, Metals and Ceramics Information Center, Battelle's Columbus Laboratories, Columbus, OH, Sept 1972.

[Myagawa et al., 1976]: Myagawa, O., et al., "Zigzag Grain Boundaries and Strength of Heat Resisting Alloys," *Proceedings of the Third International Symposium on Metallurgy and Manufacturing of Superalloys,* TMS, 1976, p 245–254.

[Noel et al., 1997]: Noel, R., Furrer, D., Lemsky, J., Shen, G., Hoffman, R., "Forging—Business & Technology Perspective," *Proceedings of the Third ASM International Paris Conference* (Paris, France), 25–27 June, 1997.

[Prasad et al., 1969]: Prasad, J.S., and Watmough, T., "Precision Cast Superalloy Dies for Isothermal Forging of Titanium Alloys," *Transactions of American Foundrymen's Society,* Vol. 77, May 5–9, 1969, p 289–296.

[Rosenberg, 1978]: Rosenberg, H.W., "Ti-10V-2Fe-3Al—A Forging Alloy Development," *Proceedings of the Metal Society Conference on Forging and Properties of Aerospace Materials,* The Metals Society, London, 1978, p 279–299.

[Semiatin et al., 1983]: Semiatin, L., and Altan, T., "Isothermal and Hot-Die Forging of High Temperature Alloys," MCIC Report 83-47, Metals and Ceramics Information Center, Battelle's Columbus Laboratories, Columbus, OH, Oct 1983.

[Shah, 1988]: Shah, S., "Isothermal and Hot-Die Forging," *Forming and Forging,* Vol. 14, *ASM Handbook,* ASM International, 1988.

[Shen et al., 2000]: Shen, G., and Furrer, D., "Manufacturing of Aerospace Forgings," *Journal of Materials Processing Technology,* Vol. 98, 2000, p 189–195.

[Shen et al., 2001]: Shen, G., Denkenberger, R., Furrer, D., "Aerospace Forging—Process and Modeling," *Materials Design Approaches and Experiences,* J.-C. Zhao, M. Fahrmann, and T.M. Pollock, Ed., TMS, 2001.

[Simmons, 1971]: Simmons, W.F., "Description and Engineering Characteristics of Eleven New High Temperature Alloys," DMIC Memorandum 255, Defense Metal Information Center, Battelle's Columbus Laboratories, Columbus, OH, June 1971.

[Soucail et al., 1996]: Soucail, M., Marty M., Octor, H., "The Effect of High Temperature Deformation on Grain Growth in a PM Nickel Base Superalloy," *Superalloys 1996,* R.D. Kissinger, D.J. Deye, D.L. Anton, A.D. Cetel, M.V. Nathal, T.M. Pollock, and D.A. Woodford, Ed., TMS, 1996, p 663–666.

[Stewart, 1988]: Stewart, D., "ISOCON Manufacturing of Waspaloy Turbine Discs," *Superalloys 1988,* S. Reichman, D.N. Duhl, G. Maurer, S. Antolovich, and C. Lund, Ed., TMS, 1988, p 545–551.

[Walker et al., 1976]: Walker, B.H., and Carruthers, W.D., "Development of a Ceramic Blade Superalloys Disk Attachment for Gas Turbine Rotors," Paper No. 760240, Society of Automotive Engineers, 1976.

[Williams, 1996]: Williams, J.C., "Business Directions and Materials Challenges for the Aircraft Engine Industry," *Acta Metallurgica Sinica,* Vol. 9 (No. 6), Dec 1996, p 407.

SELECTED REFERENCES

- **[DeRidder et al., 1979]:** Deridder, A.J., and Koch, R., "Forging and Processing of High Temperature Alloys," *MiCon 78: Optimization of Processing, Properties, and Service Performance Through Microstructural Control,* STP 672, H. Abrams, G.N. Maniar, D.A. Nail, and H.D. Soloman, Ed., American Society for Testing and Materials, 1979, p 547–563.
- **[Kuhlmann, 1988]:** Kuhlmann, G.W., "Forging of Titanium Alloys," *Forming and Forging,* Vol. 14, *ASM Handbook,* ASM International, 1988.
- **[Lawley, 1986]:** Lawley, A., "Trends in Atomization and Consolidation of Powders for High-Temperature Aerospace Materials," *MICON 86,* ASTM, 1986, p 183–201.
- **[Spiegelberg, 1977]:** Spiegelberg, W.D., "Interface Separation—Lubrication Substances for Isothermal Forging at 1300 °F to 1500 °F," Technical Report AFML-TR-77-87, Westinghouse Electric Corporation, Pittsburgh, PA, April 1977.

CHAPTER 21

Die Materials and Die Manufacturing

Prashant Mangukia

21.1 Introduction

The design and manufacture of dies and the selection of the die materials are very important in the production of discrete parts by forging. The dies must be made by modern methods from appropriate die materials in order to provide acceptable die life at a reasonable cost. Often, the economic success of a forging process depends on die life and die costs per piece produced. For a given application, selection of the appropriate die material depends on three variables [Altan et al., 1983]:

- Variables related to the process itself, including factors such as size of the die cavity, type of machine used, deformation speed, initial stock size and temperature, die temperature to be used, lubrication, production rate, and number of parts to be produced
- Variables related to the type of die loading, including speed of loading, i.e., impact or gradual contact time between dies and deforming metal (this contact time is especially important in hot forging), maximum load and pressure on the dies, maximum and minimum die temperatures, and number of loading cycles to which dies will be subjected
- Mechanical properties of the die material, including hardenability, impact strength, hot strength (if hot forming is considered), and resistance to thermal and mechanical fatigue

These factors are summarized in Fig. 21.1, primarily for hot forging applications. Proper selection of the die materials and the die manufacturing technique determines, to a large extent, the useful life of forming dies. Dies may have to be replaced for a number of reasons, such as changes in dimensions due to wear or plastic deformation, determination of the surface finish, breakdown of lubrication, and cracking or breakage. Classification of tool steels by the American Iron and Steel Institute (AISI) is presented in Table 21.1.

21.2 Die and Tool Materials for Hot Forging

Table 21.2 provides a good summary of certain material properties of dies for hot forging and the corresponding failure mechanisms that they affect.

Die materials commonly used for hot forging can be grouped in terms of alloy content; these materials are listed in Tables 21.3 through 21.5. Low-alloy steels are listed in Table 21.3. Steels with ASM designations 6G, 6F2, and 6F3 possess good toughness and shock-resistance qualities, with reasonable resistance to abrasion and heat checking. However, these steels are tempered at relatively low temperatures, for example, die holders for hot forging or hammer die blocks. Low-alloy steels with higher (2 to 4%) nickel contents, with ASM designations 6F5 and 6F7, have higher hardenability and toughness and can be used in more severe applications than steels 6G, 6F2, and 6F3. The precipitation-hardening steel 6F4 can be hardened by a simple aging operation (950 to 1050 °F, or 510 to 565 °C) without any cracking or distortion. In hot forging in presses, heat transfer from the hot stock to the dies causes this steel to harden and become more abrasion resistant.

Hot work die steels are used at temperatures of 600 to 1200 °F (315 to 650 °C) and contain

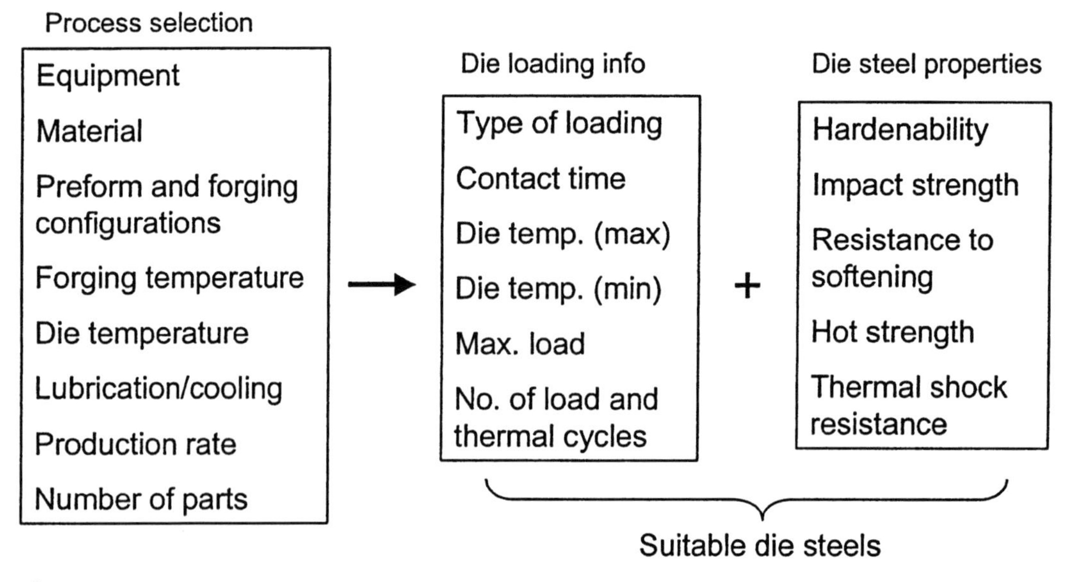

Fig. 21.1 Factors affecting die steel selection [Nagpal et al., 1980]

chromium, tungsten, and, in some cases, vanadium or molybdenum or both. These elements induce deep hardening characteristics and resistance to abrasion and softening. Generally, these steels are hardened by quenching in air or molten salt baths. The chromium-base steels contain 5% Cr. (Table 21.4). High molybdenum content gives these steels high resistance to softening; vanadium content increases resistance to heat checking and abrasion. Tungsten improves toughness and hot hardness; however, steels containing tungsten are not resistant to thermal shock and cannot be cooled intermittently with water. The tungsten-base hot work steels contain 9 to 18% W (Table 21.5); they also contain 2 to

Table 21.1 AISI classification and composition of tool steels [Krauss et al., 1998]

Group	Identifying symbol
Water-hardening tool steels	W
Shock-resisting tool steels	S
Oil-hardening cool work tool steels	O
Air-hardening, medium-alloy cold work tool steel	A
High-carbon, high-chromium cold work tool steels	D
Mold steels	P
Hot work tool steels, chromium, tungsten, and molybdenum	H
Tungsten high-speed tool steels	T
Molybdenum high-speed tool steels	M

Table 21.2 Critical die-related factors and corresponding properties in the die material used in hot forging [Krishnadev et al., 1997]

Critical die-related factor	Critical properties in die material
Heat checking	Hot strength (hot hardness)
	Tempering resistance
	Toughness/ductility
	Thermal expansion
	Thermal conductivity
Gross cracking	Toughness/ductility
Erosion	Hot strength (hot hardness)
	Chemical resistance
Plastic deformation	Hot strength (hot hardness)
	Tempering resistance

Table 21.3 Low-alloy steels for hot forging [Altan et al., 1983]

Designation (ASM)	Nominal composition, %								
	C	Mn	Si	Co	Cr	Mo	Ni	V	W
6G	0.55	0.80	0.25	...	1.00	0.45	...	0.10	...
6F2	0.55	0.75	0.25	...	1.00	0.30	1.00	0.10	...
6F3	0.55	0.60	0.85	...	1.00	0.75	1.80	0.10	...
6F4	0.20	0.70	0.25	...	...	3.35	3.00	...	...
6F5	0.55	1.00	1.00	...	0.50	0.50	2.70	0.10	...
6F7	0.40	0.35	...	...	1.5	0.75	4.25	...	...

Table 21.4 Chromium-base hot work die steels [Altan et al., 1983]

Designation (ASM)	Nominal composition, %								
	C	Mn	Si	Co	Cr	Mo	Ni	V	W
H10	0.40	0.40	1.00	...	3.30	2.50	...	0.50	...
H11	0.35	0.30	1.00	...	5.00	1.50	...	0.40	...
H12	0.35	0.40	1.00	...	5.00	1.50	...	0.50	1.50
H13	0.38	0.30	1.00	...	5.25	1.50	...	1.00	...
H14	0.40	0.35	1.00	...	5.00	...	...	...	5.00
H19	0.40	0.30	0.30	4.25	4.25	0.40	...	2.10	4.10

Table 21.5 Tungsten-base hot work die steels [Altan et al., 1983]

Designation (ASM)	Nominal composition, %								
	C	Mn	Si	Co	Cr	Mo	Ni	V	W
H21	0.30	0.30	0.30	...	3.50	...	...	0.45	9.25
H22	0.35	0.30	0.30	...	2.00	...	...	0.40	11.00
H23	0.30	0.30	0.30	...	12.00	...	...	1.00	12.00
H24	0.45	0.30	0.30	...	3.00	...	...	0.50	15.00
H25	0.25	0.30	0.30	...	4.0	...	...	0.50	15.00
H26	0.50	0.30	0.30	...	4.0	...	...	1.00	18.00

12% Cr and may have small amounts of vanadium. The high tungsten content provides resistance to softening at high temperatures while maintaining adequate toughness; however, it also makes it possible to water cool these die steels. High-speed steels, originally developed for metal cutting, can also be used in warm or hot forging applications. There are two types of high-speed steels: molybdenum-type high-speed steels, designated by the letter M, and tungsten-type high-speed steels, designated by the letter T (Table 21.6). These steels offer good combinations of hardness, strength, and toughness at elevated temperatures.

21.2.1 Comparisons of Die Steels for Hot Forging

Properties of materials that determine their selection as die materials for hot forging are:

- Ability to harden uniformly
- Wear resistance (this is the ability of a die steel to resist the abrasive action of hot metal during forging)
- Resistance to plastic deformation (this is the ability of a die steel to withstand pressure and resist deformation under load)
- Toughness
- Ability to resist thermal fatigue and heat checking
- Ability to resist mechanical fatigue

Ability to Harden Uniformly. The higher the hardenability, the greater the depth to which a material can be hardened. Hardenability depends on the composition of the tool steel. In general, the higher the alloy contents of a steel, the higher its hardenability, as measured by the D_I factor (in inches). The D_I of steel is the diameter of an infinitely long cylinder that would just transform to a specific microstructure at the center, if heat transfer during cooling were ideal, i.e., if the surface attained the temperature of the quenching medium instantly. A larger hardenability factor D_I means that the steel will harden to a greater depth on quenching, not that it will have a higher hardness. For example, the approximate nominal hardenability factors D_I (inches) for a few die steels are as follows: ASM-6G, 15; ASM-6F2, 15; ASM-6F3, 36; AISI-H10, 128; and AISI-H12, 88 [Altan et al., 1983].

Wear Resistance. Wear is a gradual change in dimensions or shape of a component caused by corrosion, dissolution, abrasion, and removal or transportation of the wear products. Abrasion resulting from friction is the most important of these mechanisms in terms of die wear. The higher the strength and hardening of the steel near the surface of the die, the greater its resistance to abrasion. Thus, in hot forging, the die steel should have a high hot hardness and should retain this hardness over extended periods of exposure to elevated temperatures.

Figure 21.2 shows the hot hardness of six hot work die steels at various temperatures. All of these steels were heat treated to about the same initial hardness. Hardness measurements were made after holding the specimens at testing temperatures for 30 min. Except for H12, all the die steels considered have about the same hot hardness at temperatures less than about 600 °F (315 °C). The differences in hot hardness show up only at temperatures above 900 °F (482 °C).

Figure 21.3 shows the resistance of some hot work die steels to softening at elevated temperatures for 10 h of exposure. All of these steels have about the same initial hardness after heat treatment. For the die steels shown, there is not much variation in resistance to softening at tem-

Table 21.6 Approximate relative rankings of 15 selected high-speed steels for three properties

Material evaluation	Wear resistance	Hot hardness	Toughness
Highest	T15; M15; M4; M42	T15; M15; T6; T5; M42	M2; T1; M1
Medium	M3; T6; M2 (H.C.); M10 (H.C.); M7; M2; T5	M4; T4; M3	M3; M4; M10; M10 (H.C.); M7; M2 (H.C.)
Lowest	T1; M10; T4; M1	T1; M2; M10; M1; M7; M10 (H.C.); M2 (H.C.)	T4; T5; M42; T15; M15; T6

T, tungsten-type high-speed steels; M, molybdenum-type high-speed steels; (H.C.), high carbon

peratures below 1000 °F (538 °C). However, for longer periods of exposure at higher temperatures, high-alloy hot work steels, such as H19, H21, and H10 modified, retain hardness better than the medium-alloy steels such as H11.

Resistance to Plastic Deformation. As can be seen in Fig. 21.4, the yield strengths of steel decrease at higher temperatures. However, yield strength is also dependent on prior heat treatment, composition, and hardness. The higher the

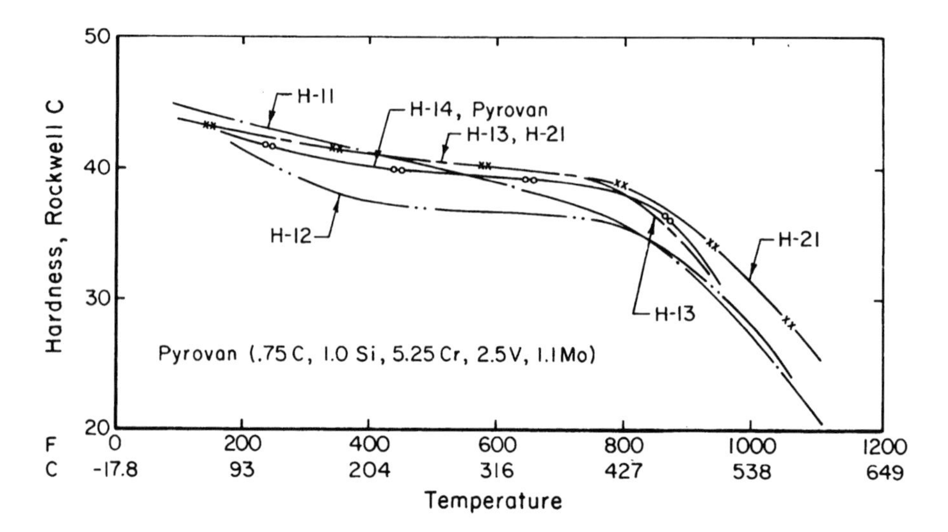

Fig. 21.2 Hot hardness of hot work die steels (measurements made after holding at testing temperature for 30 min.). Courtesy of Latrobe Steel Co.

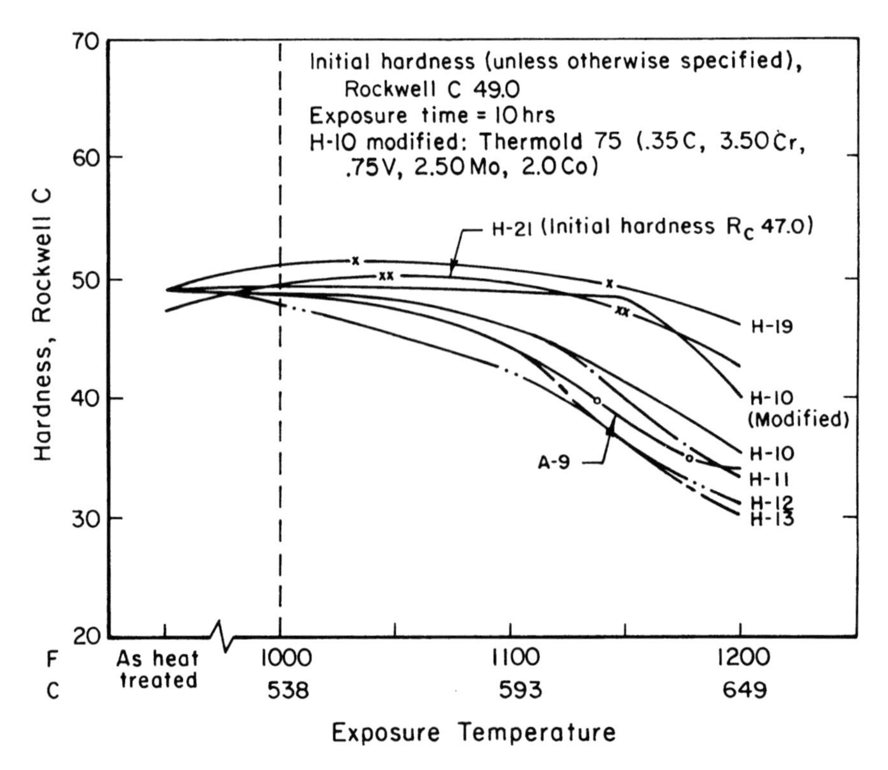

Fig. 21.3 Resistance of hot work die steels to softening during elevated-temperature exposure as measured by room-temperature hardness. Courtesy of Universal Cyclops Steel Corp.

initial hardness, the greater the yield strengths at various temperatures. In normal practice, the level to which die steel is hardened is determined by the toughness of steel. Thus, in forging applications, the die block is hardened to a level at which it should have enough toughness to avoid cracking. Figure 21.5 shows that, for the same initial hardness, 5% Cr-Mo steels (H11, etc.) have better hot strengths than 6F2 and 6F3 at temperatures greater than 700 °F (371 °C). Tensile strength, creep properties, and toughness of various die steels are given in Fig. 21.6 and 21.7.

Toughness is the ability of the material to absorb energy without fracture. It is a combination of strength and ductility. It increases with increasing temperature and is important in avoiding brittle fracture (Fig. 21.7 and 21.8). Most materials exhibit a ductile-to-brittle transition temperature. To avoid immediate and catastrophic failure, dies must be put into service above this temperature. Ductility, as measured by reduction in area measured in a tensile test, also increases with temperature (Fig. 21.8).

Figure 21.8 shows the ductility of various hot work steels at elevated temperatures, as measured by percent reduction in area of a specimen before fracture in a standard tensile test. As the curves show, high-alloy hot work steels, such as H19 and H21, have less ductility than medium-alloy hot work steels such as H11. This explains why H19 and H21 have lower toughness than that of H11.

Fracture toughness and resistance to shock loading are often measured by a notched-bar Charpy test. This test measures the amount of energy absorbed in introducing and propagating fracture, or toughness, of a material at high rates of deformation (impact loading). Figure 21.5 shows the results of V-notch Charpy tests on some die steels. The data show that toughness decreases as the alloy content increases. Medium-alloy steels such as H11, H12, and H13 have better resistance to brittle fracture in comparison with H14, H19, and H21, which have higher alloy contents. Increasing the hardness of the steel lowers its impact strength, as shown by data on 6F7 steel hardened to two different levels. On the other hand, wear resistance and hot strength decrease with decreasing hardness. Thus, a compromise is made in actual practice, and dies are tempered to near-maximum hard-

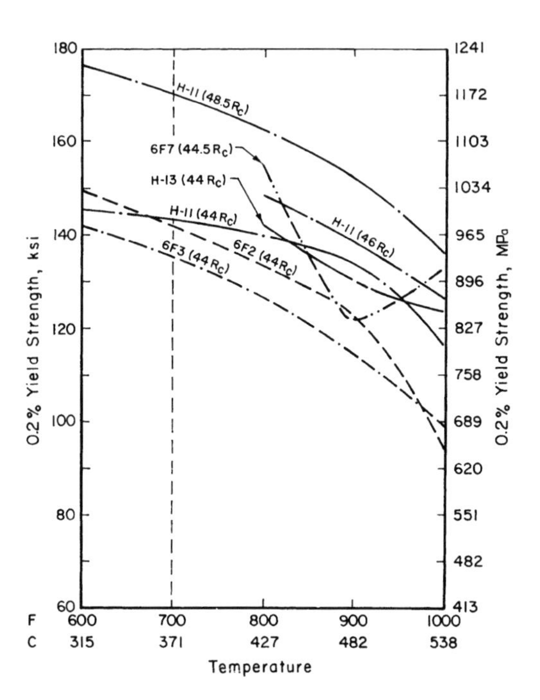

Fig. 21.4 Resistance of die steels to plastic deformation at elevated temperatures (values in parentheses indicate hardness at room temperature). Courtesy of Universal Cyclops Steel Corp. and A. Finkl and Sons Co.

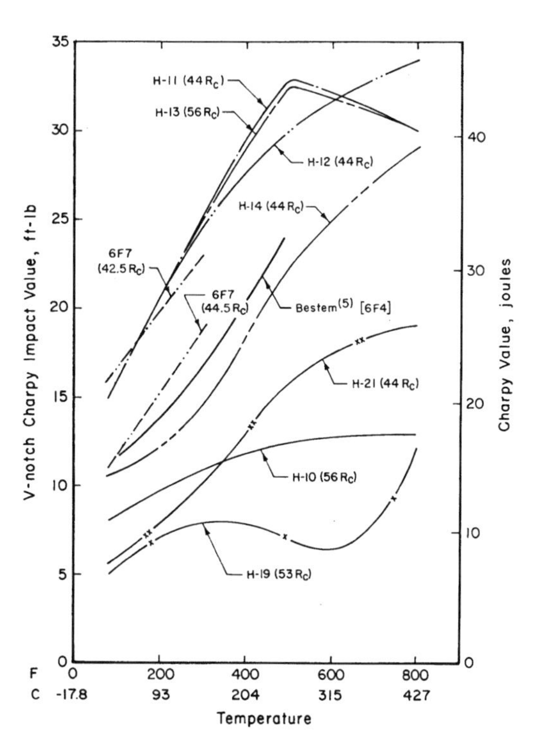

Fig. 21.5 Variation of Charpy toughness with different hardness levels and testing temperatures on hot work die steels (values in parentheses indicate hardness at room temperature) [Nagpal, 1976b]

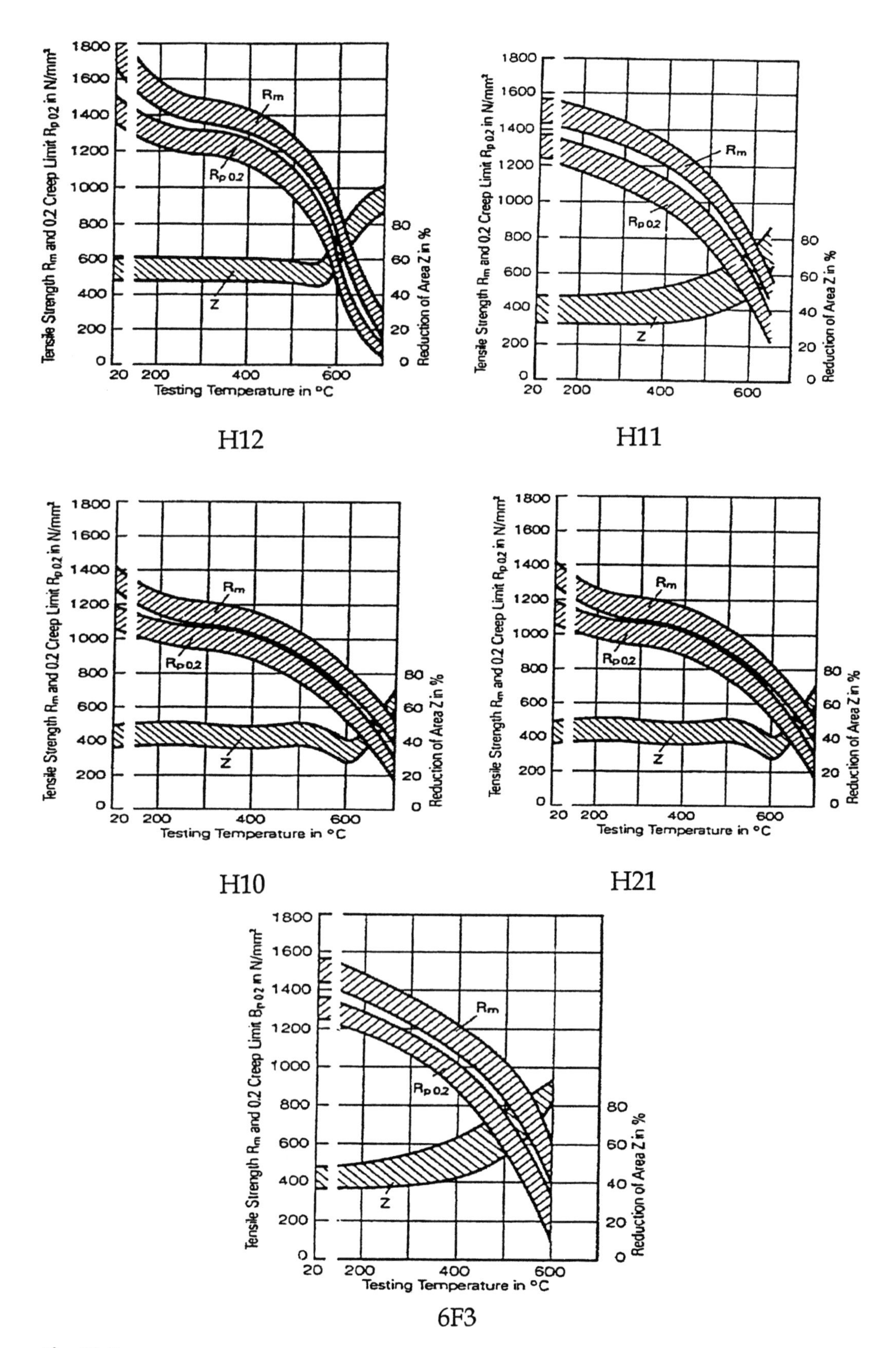

Fig. 21.6 Tensile strength and ductility versus test temperatures for selected die materials [Thyssen]

ness levels, at which they have sufficient toughness to withstand loading.

The data shown in Fig. 21.5 also point to the importance of preheating the dies prior to hot forging. Steels such as H10 and H21 attain reasonable toughness only at higher temperatures and require preheating. For general-purpose steels, such as 6F2 and 6G, preheating to a minimum temperature of 300 °F (150 °C) is recommended; for high-alloy steels, such as H14, H19, and 6F4, a higher preheating temperature of 480 °F (250 °C) has been recommended [Altan et al., 1983].

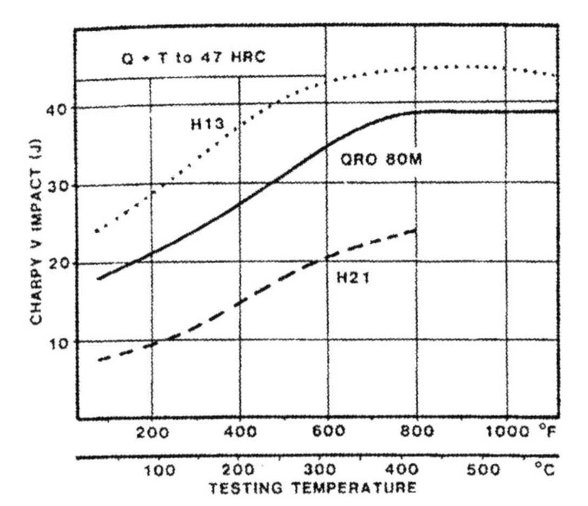

Fig. 21.7 Comparison of toughness properties for H13, H21, and a new hot work tool steel, QRO80M, in function of test temperature [Johansson et al., 1985]

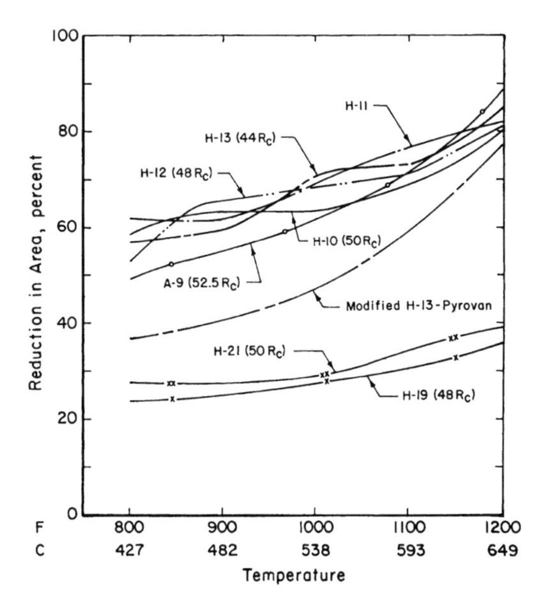

Fig. 21.8 Ductility of various die steels at high temperatures [Nagpal, 1976a]

Resistance to Heat Checking. Nonuniform expansion, caused by thermal gradients from the surface to the center of a die, is the chief factor contributing to heat checking. Therefore, a material with high thermal conductivity will make dies less prone to heat checking by conducting heat rapidly away from the heat surface, reducing surface-to-center temperature gradients. The magnitudes of thermal stresses caused by nonuniform expansion or temperature gradients also depend on the coefficient of thermal expansion of the steel: the higher the coefficient of thermal expansion, the greater the stresses. Thermal conductivities for some hot working steels are given in Table 21.7. From tests in which the temperature of the specimen fluctuated between 1200 °F (650 °C) and the water-quench bath temperature, it was determined that H10 was slightly more resistant to heat checking or cracking, after 1740 cycles, compared with H11, H12, and H13. After 3488 cycles, H10 exhibited significantly more resistance to cracking than did H11, H12, and H13.

In Table 21.8, die materials are rated relative to each other on resistance to wear, resistance to thermal shrinking, impact toughness, and hardenability. Comparison of die materials based on material property and their response to surface treatments is listed in Table 21.9.

21.2.2 Maraging Steels

Maraging steels are steels that have high nickel, cobalt, and molybdenum content but very little carbon. These steels are mostly used in die casting and very rarely in forging. Some of the compositions, of common maraging steels are given in Table 21.10.

21.2.3 Superalloys

Superalloys are based on nickel, cobalt, and iron. They are resistant to thermal softening.

Table 21.7 Thermal conductivities of different hot work die steels [Kesavapandian et al., 2001]

	Composition, wt%						Thermal conductivity,
Alloy No.	C	Cr	Mo	V	W	Co	W/cm · K
2581	0.30	2.65	...	0.35	8.50	...	0.23
2662	0.30	2.30	...	0.25	8.50	2.0	0.26
2567	0.30	2.35	...	0.60	4.25	...	0.27
2365	0.32	3.00	2.80	0.55	...	...	0.32
2344	0.40	5.30	1.30	1.00	...	...	0.33
2343	0.38	5.30	1.10	0.40	...	...	0.34

Their strength is due to the precipitation strengthening of intermetallic compounds. Thus, these materials are used in selected forging applications even though they are expensive.

Superalloys are of four types, namely:

- Iron-base alloys comprise die materials such as H46 and Inconel 706 and contain over 12% Cr. They also contain molybdenum and tungsten to provide the matrix with high-temperature strength. Iron-base superalloys also include austenitic steels with high chromium and nickel content. These dies can be used up to a temperature of 1200 °F (650 °C).
- Nickel-iron-base alloys contain 25 to 27% Ni, 10 to 15% Cr, and 50 to 60% Fe, with small quantities of molybdenum, titanium, and vanadium. The carbon content in these alloys is very low.
- Nickel-base alloys contain 50 to 80% Ni, 20% Cr, and a combination of molybdenum, aluminum, tungsten, cobalt, and niobium (Waspalloy, Udimet 500, and Inconel 718). The iron content is very low in these types of alloys. These alloys can be put in service up to a temperature of 2200 °F (1205 °C).
- Cobalt-base alloys are primarily made up of nickel, iron, chromium, tungsten, and cobalt. These alloys are more ductile and can be used up to a temperature of 1900 °F (1040 °C). Selected properties of various superal-

Table 21.8 Comparative properties for the selection and use of die steels [Nagpal et al., 1980]

Material	Relative wear resistance	Relative resistance to thermal shrinking	Relative impact strength (toughness)	Relative hardenability
6C				
6F2				
6F3				
6F4				
6F5				
6F7				
H10				
H11				
H12				
H13				
A8				
A9				
6H2				
6H1				
H14				
H19				
H21				
H26				

Table 21.9 Relative ranking of die materials and their response to surface engineering [Krishnadev et al., 1997]

Material	Impact toughness	Hot hardness	Resistance to die softening	Thermal checking resistance	Wear resistance	Resistance to surface engineering(a)
H13	Medium	Medium	Medium	High	Medium	Ion nitriding, laser, PFS, PVD, TD-VC
ORVAR Supreme	High	Medium	Medium	High	Medium	Ion nitriding, laser, PFS, PVD, TD-VC
QRO Supreme	High	High	High	Very high	Medium	Ion nitriding, laser, PFS, PVD, TD-VC
AerMet 100	Very high	Low	Low	NA	Medium	Laser, PFS, PVD, TD-VC
Matrix 11	Medium	Very high	Very high	Very high	Medium	Ion nitriding, laser, PFS, PVD, TD-VC
D2	Low	NA	NA	NA	High	Ion nitriding, PFS, PVD, TD-VC

(a) PFS, powder flame spray; PVD, physical vapor deposition; TD-VC, thermal diffusion-vanadium carbide

Table 21.10 Composition of common maraging steels [Kesavapandian et al., 2001]

Type	Composition, wt%									
	Ni	Co	Mo	Ti	Al	C	Si	Mn	S	P
I-VascoMax C-200	18.5	8.5	3.25	0.2	0.1	0.03	0.1	0.1	0.1	0.1
II-VascoMax C-250	18.5	7.5	4.8	0.4	0.1	0.03	0.1	0.1	0.1	0.1
III-VascoMax C-300	18.5	9.0	4.8	0.6	0.1	0.03	0.1	0.1	0.1	0.1
IV-VascoMax C-350	18.0	11.8	4.6	1.35	0.1	0.03	0.1	0.1	0.1	0.1
HWM	2.0	11.0	7.5	...	...	0.05	0.1	0.1	0.1	0.1
X2NiCoMoTi 12 8 8 Thyrotherm 2799	12.0	8.0	8.0	0.5	0.5	0.03	0.1	0.1	0.1	0.1
Marlock (Cr 0.2)	18.0	11.0	5.0	0.3	...	0.01	...	0.1	0.01	0.01

loys are given in the literature [Semiatin et al., 1981].

21.2.4 Ceramic Dies

Some of the newer nonferrous alloys have marked improvements over the traditional die materials (Cr-Mo-V-base steels) used in hot forging. For example, Nissan Motor Co. is using cermet dies in extrusion [Kesavapandian et al., 2001]. The material is composed of MoB (ceramic) and nickel (metal.) Boron improves resistance to oxidation at high temperatures. The material is powder formed and sintered. Die life improvements over traditional materials is 2 to 1.

Hot forging wear tests [Bramley et al., 1989] compare silicon-base ceramics to conventional die steels. Silicon-base ceramics exhibit far less wear than the conventional die steels. These materials are very pure, with fine grains and uniform microstructures. They have high hardness, strength, and resistance to mechanical and thermal shock. However, these materials are brittle and must not be subjected to tensile stresses; in addition, they are expensive. Therefore, they are used only in selected applications.

21.2.5 Nickel Aluminides

Nickel aluminides are attractive because they derive their strength from their ordered microstructure instead of heat treatments. Their yield strength increases with temperature up to about 1560 °F (850 °C), and they have excellent resistance to oxidation up to 1830 °F (1000 °C). Information on the composition and properties of various nickel aluminides is available in the literature.

[Maddox et al., 1997] documented the results of a series of performance tests between the nickel aluminide alloys 221M-T and more traditional die steels. Whereas H11 and H13 lose strength and hardness above 1000 °F (540 °C), nickel aluminide alloys have improved high-temperature strength, oxidation resistance, and thermal stability at temperatures ranging from 1500 to 2300 °F (815 to 1260 °C). The composition of this alloy is Ni_3Al, with additions of chromium for ductility and resistance to oxidation; molybdenum for strength; zirconium for strength, resistance to oxidation, weldability, and castability; and boron for ductility. The alloy is stronger than Inconel 718 in both tension and compression. The results of the performance test are impressive for the nickel aluminide die material. For the particular part being forged (at 2300 °F, or 1260 °C), conventional dies typically failed after 5000 parts due to erosion, whereas the nickel aluminide dies lasted for 35,000 parts (a sevenfold increase). After resinking the cavity, the nickel aluminide dies lasted for 50,000 parts (a tenfold increase). Hardness in the flash actually also went up due to work hardening. This alloy keeps its hardness at high temperatures from the cast microstructure and not through a thermal treatment. Also important is that, while heat checking was detected in the dies, it was not important, because of the high ductility of the material.

21.3 Heat Treatment

Heat treatment of die steels involves the following steps:

1. Austenitizing temperature of a hot work tool steel ranges from 1830 to 2730 °F (1000 to 1500 °C). During this phase, the structure of steel transforms from a ferrite-pearlite structure to austenite. The dies are held at these temperatures for a long time, which is called a "soak" or "hold" time, to convert the entire structure uniformly to austenite. Carbides or alloying elements go into solution [Krauss et al., 1998 and Kesavapandian et al., 2001].
2. After soaking, the dies are quenched in a quench medium to a temperature below the transformation temperature. Based on the cooling rate, the die transforms into different phases. Martensite is the ideal final structure, but lower bainite, upper bainite, pearlite, or retained austenite can be present in the structure.
3. During tempering, the martensite is tempered to a tougher structure. This can be done in several stages to maximize toughness without reducing hardness.

Hardening and tempering temperatures for various tool steels are available in literature and from die steel suppliers [Roberts et al., 1980].

21.4 Die and Tool Materials for Cold Forging

The tooling for cold forging is described in Chapter 17, "Cold and Warm Forging," in this book. The back or pressure plates, which must have high compressive strength, are made from steels hardened up to 62 HRC, as given in Table 21.11. Selection of the punch material depends on the type of deformation. For example, in forward extrusion, the punch material must have high compressive strength, whereas in backward extrusion, the punch must also have very good wear resistance, as a considerable amount of metal flow occurs along the punch surface. Cold work steels are a class of tool steels in which the surface temperature does not exceed 390 °F (200 °C) during use. These steels offer high hardness, good toughness, and good resistance to shock, pressure, or wear. The subdivisions of this class are water-hardened steels, medium-and high-alloyed cold-worked steels, and high-carbon chromium steels. Tool steels commonly used for cold forging punches are given in Table 21.12. The dies are subjected to high cycling pressure as well as abrasion. Therefore, die materials must have high fatigue strength and good resistance to wear (Table 21.13). In cold extrusion, the die inserts are prestressed with one or two shrink rings so that they can withstand the high stresses present in the die cavity. Materials suitable for shrink rings are given in Table 21.14. Tool steels used for ejectors and counterpunches are given in Table 21.15 [Lange, 1976]. Table 21.16 gives the relative performance of different cold forging tools.

The special-purpose steels are the materials that cannot be classified as either cold-worked or high-speed steels. They are required when certain properties are particularly important, such as corrosion resistance. Hardenable hard materials are a class between high-speed steels and carbides. They have good machinability, as long as they have not been hardened, and a very high resistance to wear in the hardened state. They are used to form very large cold forgings [ICFG, 1992].

Table 21.11 Tool steels for pressure plates [Lange, 1976]

Required strength		Designation	Required hardness	
ksi	MPa	(AISI)	HRC	HB
240	1700	A2	58 to 62	...
		D2	58 to 62	...
		D3	58 to 62	...
		O1	58 to 61	...
200	1400	H13	50 to 54	...
		O1	50 to 54	...
155	1100	A8	40 to 44	...
		O1	40 to 44	...
100	700	4340	...	270 to 330
		4140	...	270 to 330

Table 21.12 Tool steels for cold extrusion punches [Lange, 1976]

Required strength		Tool steel	
ksi	MPa	Designation (AISI)	Required hardness, HRC
Forward extrusion			
300	2100	M2	62 to 64
285	2000	D2	60 to 62
		O1	60 to 62
230	1600	S1	56 to 58
Backward extrusion			
315	2200	M4	63 to 65
300	2100	M2	62 to 64
285	2000	D2	60 to 62

Table 21.13 Tool steels and tungsten carbides for die inserts [Lange, 1976]

Tool steels		Tungsten carbides		
Designation (AISI)	Required hardness, HRC	Co, wt%	Density, g/cm^3	Hardness, DPH
D2	60 to 62	25 to 30	13.1 to 12.5	950 to 750
M2	60 to 64	18 to 42	13.6 to 13.2	1050 to 950
		15 to 18	14.0 to 13.7	1200 to 1100

Table 21.14 Steels for shrink rings [Lange, 1976]

Designation (AISI)	Required hardness, HB
H13	470 to 530
	440 to 510
	330 to 390
4340	330 to 390
	270 to 330
4140	270 to 330

21.4.1 Cemented Carbide

Cemented carbides are composite materials consisting of hard, wear-resistant material in a more ductile metallic matrix. The two elements are formed into one through a sintering process. Tungsten carbide, which consists of tungsten in a cobalt matrix, is typical in tool and die applications.

As manufacturing processes have improved, the use of tungsten and other cemented carbides

Table 21.15 Tool steels for counterpunches and ejectors [Lange, 1976]

Designation (AISI)	Required hardness, HB
M2	62 to 64
D2	60 to 62
A2	60 to 62
O1	60 to 62
S1	56 to 58

has grown drastically. Today, the material is used for more complex shapes and many forms of extrusion tooling. Carbides are characterized by high hardness, high modulus of elasticity, and great compressive strength, while the tensile strength is limited. The hardness is typically measured using the Vickers or Rockwell A scale. It is important to note that Rockwell C is completely inadequate for measuring carbide hardness, and an attempt to do so could likely damage the diamond tip on the hardness machine.

The International Cold Forging Group has broken cemented carbides into six groups: A, B, C, D, E, and F. These classifications are based on composition as relevant to cold forging applications [ICFG, 1992]. Table 21.17 displays typical ranges for the mechanical properties of each carbide class.

21.4.2 Material Selection for Cold Forging Tools

The material used to produce each tool component can be selected once the dimensions of the tool have been selected. Important factors to consider are the forming process, press parameters, batch size, and forming load. Table 21.18 displays common tool steels for each component in the two most common processes. Table 21.19 indicates which carbide groups are appropriate for the individual components.

When choosing a die or insert material, the first decision is whether to use tool steel or carbide. When deciding between the two, it is especially important to consider the size of the die insert, the geometry of the die profile, the tolerances of the component to be manufactured, the batch size, and the operating conditions. Carbide inserts must be under compressive stresses, since carbide cannot sustain tensile stresses.

In general, if the die insert is 1.2 in. (30 mm) in diameter or less, carbide can be used. Processes such as free extrusion permit die of greater size to be made from cemented carbide. Also, carbide is used almost exclusively for dies in high production.

With respect to geometry, all corners of a carbide die must have generous radii. In general, the shapes must be simple in order to use carbide. The ratio D/d_1 that compares the overall diameter of the die, D, and bore diameter of the die insert, d_1, must be greater than 4.5 for carbide to be applicable. In the case of free extrusion, the ratio can be reduced to 3. If the ratio is less than the critical value or the geometry is too complex for carbide, tool steel should be used.

Considerations for selecting the punch material are very similar to those accounted for in die material selection. Since cemented carbide has a very high compressive strength, it should be used if the maximum compressive stress is greater than 365 ksi (2500 MPa). If the punch will be used for can extrusion, cemented carbide may be used if the H/D ratio is greater than 2.5, where H is the depth of the can bore hole, and D is the diameter of the punch. When carbide is used for backward extrusion, the length of the punch should be kept as short as possible [ICFG, 1992].

When selecting a material for warm forging dies, the preliminary concerns must be the ther-

Table 21.16 Relative performance of tool steels [ICFG, 1992]

Tool steels	Designation (AISI)	Hardness, HRC	Wear-resistance index	Toughness index	Machinability index	Grindability index
Cold worked (CW)						
Plain carbon	W1	60	7	5	9	10
	W2	60	7	...	9	10
Medium/high-alloy CW	A2	60	7	6	8	7
	O1	60	7	7	8	9
	S1	55	5	8	7	8
	6F7	55	4	10	6	8
High-carbon chromium	D2	60	8	4	3	4
	D3	60	8	3	6	3
Hot worked (HW)						
Medium-alloy HW	6F2	45	2.5	10	6	8
High-alloyed HW	H13	45	4	9	8	8
	H10	45	4	8	8	8
	H19	45	4	8	5	7
High-speed steel	M2	60	9	4	4	5
	M3	64	10	2	3	2
	T15	64	10	2	1	1
	T42	64	10	2	1	1

Note: 10 = best; 1 = poor

Table 21.17 Cemented carbide properties [ICFG, 1992]

	Composition, wt%			WC grain size, µm	Density, g/cm^3	Hardness				Transverse rupture strength				Weibull exponent	Compressive strength				0.2% proof stress	Young's modulus		Poisson's ratio	Coefficient of thermal expansion
						Room temp.		570 °F (300 °C)		Room temp.		570 °F (300 °C)			Room temp.		570 °F (300 °C)						
Group	Co	C	TiC			HV	HRA	HV	TRE	ksi	MPa	ksi	MPa		ksi	MPa	ksi	MPa		ksi	MPa		
A	5–7	0–2.5	1	1–5	14.7–15.1	1450–1550	90.5–91	1200	>0.5	>260	>1800	>230	>1600	>9	>595	>4100	>450	>3100	>70	91,000	630,000	0.22	4.9×10^{-6}
B	8–10	0–2.5	1	1–5	14.4–14.7	1300–1400	89–90	1050	>0.7	>290	>2000	>260	>1800	>9	>550	>3800	>420	>2900	>65	88,000	610,000	0.22	5.0×10^{-6}
C	11–13	0–2.5	1	1–5	14.1–14.4	1200–1250	88–88.5	1000	>1.0	>320	>2200	>290	>2000	>10	>520	>3600	>405	>2800	>60	83,000	570,000	0.22	5.4×10^{-6}
D	14–17	0–2.5	1	1–5	13.8–14.2	1100–1150	87–87.5	900	>1.2	>360	>2500	>335	>2300	>10	>480	>3300	>390	>2700	>55	78,000	540,000	0.22	5.6×10^{-6}
E	18–22	0–2.5	1	1–5	13.3–13.8	950–1000	85.5–86	750	>1.5	>360	>2500	>320	>2200	>11	>450	>3100	>360	>2500	>45	73,000	500,000	0.23	6.3×10^{-6}
F	23–30	0–2.5	1	1–5	12.6–13	800–850	83.5–84	600	>1.9	>335	>2300	>275	>1900	>11	>390	>2700	>335	>2300	>35	65,000	450,000	0.24	7.0×10^{-6}

mal conditions. When considering the material properties, the same approach can be applied as in cold forging. Due to the importance of the thermal properties, minimal applications exist for carbide dies in warm forging; thus, tool steels are the most applicable material.

21.5 Die Manufacture

Forging dies or inserts are machined from solid blocks or forged die steels. By using standard support components such as die holders and guide pins, which assure the overall functionality of tooling assembly, the time necessary for manufacturing a die set is reduced, and machining is mainly devoted to producing the cavities or punches.

The information flow and processing steps used in die manufacturing may be divided into die design (including geometry transfer and modification), heat treatment, tool path generation, rough machining (of die block and/or electrical discharge machining, or EDM, electrode), finish machining (including semifinishing where necessary), manual finishing, or benching (including manual or automated polishing), and tryout. An information flow model (including the use of a coordinate measuring machine but neglecting heat treatment and coating) is presented in Fig. 21.9. As seen in this figure, forging dies are primarily manufactured by three- or four-axis computer numerical controlled (CNC) milling, EDM, or a combination of both.

Table 21.18 Tool component materials [ICFG, 1992]

Forward rod extrusion			Backward extrusion		
Punch	Container	Die insert	Punch	Container or die insert	Counterpunch
6F7	O1	S1	A2	O1	O1
D2	6F7	6F7	D3	S1	D2
D3	D2	D2	H13	6F7	D3
M2	D3	6F2	M2	D3	H13
M3	M2	H13	M3	H13	M2
T42	M3	M2	T42	M2	M3
		M3		M3	

Table 21.19 Recommended carbide groups [ICFG, 1992]

Tool part	Carbide group A	B	C	D	E	F
Extrusion punch	X	X	X	X	...	...
Die insert	...	...	...	X	X	X
Heading punch	...	...	X	X	X	X
Heading die insert	...	...	X	X	X	X
Reducing die insert	...	...	X	X	X	X
Ironing die insert	X	X	X	X	...	...
Ironing punch	X	X	X	...	...	...

21.5.1 High-Speed and Hard Machining

In a typical conventional die-making operation, the die cavity is usually rough machined to about 0.01 in. (0.3 mm) oversize dimensions. The die is then hardened, which may cause some distortion, and then EDMed to final dimensions. The trend in die manufacturing today is toward hard machining, both in roughing and finishing, and in replacing EDM whenever possible. The term *hard* refers to the hardness of the die material, which is usually in the range of 45 to 62 HRC. By using hard machining, the number of necessary machine setups is reduced and throughput is increased. Recent technological advances in high-speed machining of hardened die steels make this trend feasible and economical [Altan et al., 2001].

The main object of high-speed machining of hardened dies is to reduce benching by improving the surface finish and distortion; thus, quality is increased and costs are reduced. High-speed machining of hardened dies (40 to 62 HRC) has, within an approximate range, the following requirements and characteristics [Altan et al., 2001]:

- Feed rates: 50 ft/min (15 m/min) or higher when appropriate-pressured air or coolant mist is provided, usually through the spindle
- Spindle rpm: 10,000 to 50,000, depending on tool diameter
- Surface cutting speeds: 985 to 3280 ft/min (300 to 1000 m/min), depending on the hardness of the die/mold steel and the chip load
- High-speed control with high-speed data and look-forward capability to avoid data starvation. The look-forward capability tracks surface geometry, allowing the machine to accelerate and decelerate effectively for maintaining the prescribed surface contour.
- High acceleration and deceleration capabilities of the machine tool in the range of 2.6 to 3.9 ft/s^2 (0.8 to 1.2 m/s^2)

Hard machining requires cutting tools that can withstand very high temperatures and provide long tool life. Most commonly used tools have indexable cutting inserts from cemented carbide coated with TiN, TiCN, and TiAlN; polycrystalline cubic boron nitride (PCBN); or cubic boron nitride.

Ball-nosed end mills, from solid-coated carbide or with inserts, are commonly used in the production of die cavities. The ball nose allows the machining of complex curves and surfaces in the die cavity. Using an appropriate step-over distance and computer-aided tool-path generation, most cavities can be machined with acceptable surface finish.

In manufacturing cold forging dies from tool steels, hard-turning PCBN tools are used. How-

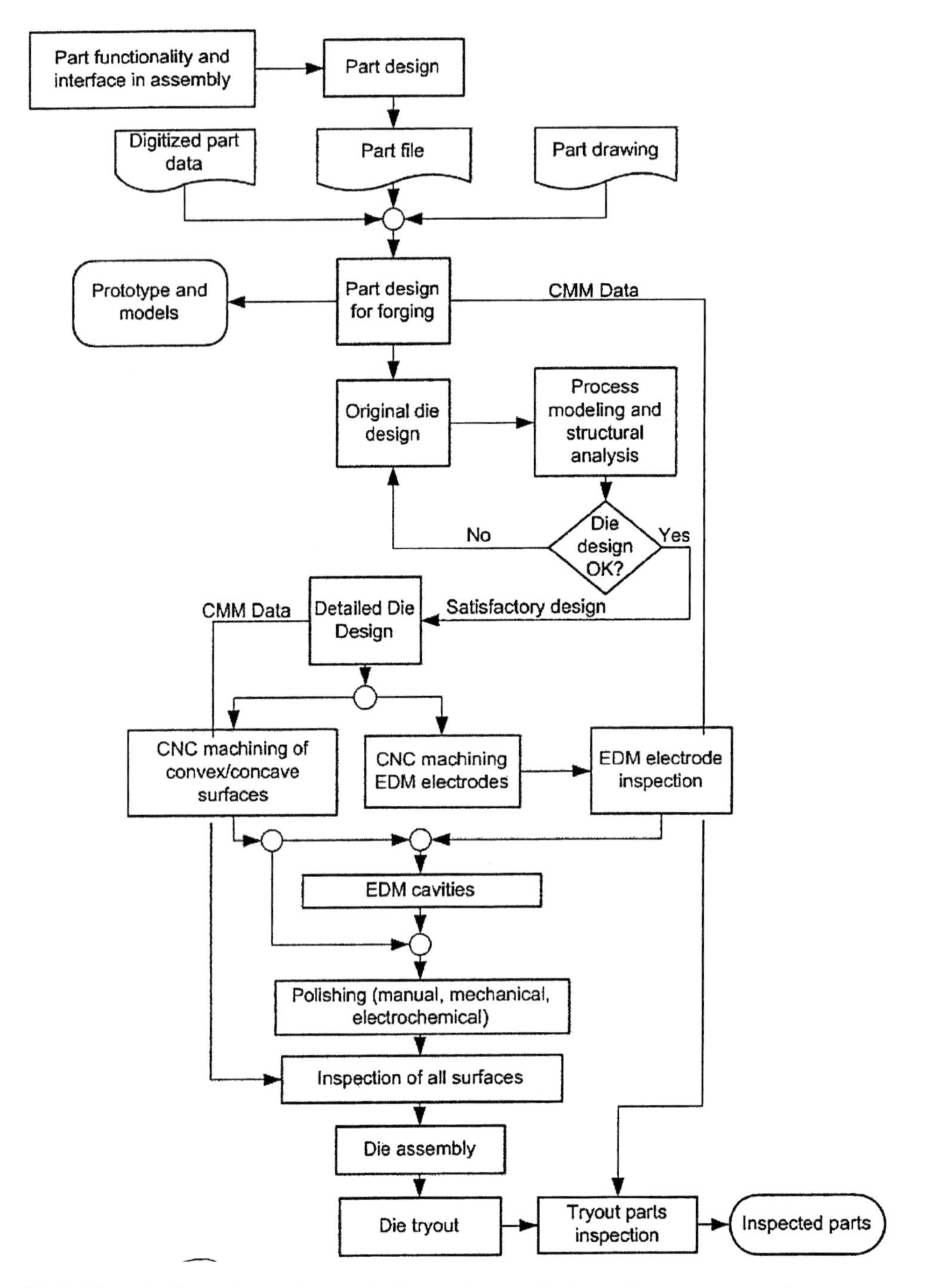

Fig. 21.9 Information flow and processing steps in die manufacturing. CMM, coordinate measuring machine; CNC, computer numerical control; EDM, electrical discharge machining

ever, in manufacturing carbide inserts, EDM must be utilized.

21.5.2 Electrodischarge Machining (EDM)

The EDM process is a versatile process for die manufacturing. The process consists of a power supply that passes a current through an electrode, creating a voltage potential between it and a conductive workpiece. As the electrode moves closer to the workpiece, the gap between the two will become sufficiently small, and a spark will pass from one material to the next. The spark vaporizes portions of both the electrode and the workpiece, and the removed material is washed away by a dielectric flushing fluid. There are two types of electrodischarge machines: sink and wire.

The sink EDM (SEDM), as shown in Fig. 21.10, generates internal cavities by lowering a graphite or copper electrode into the die block. As the spark removes material, the workpiece begins to take the form of the electrode. The electrodes are most frequently created by CNC milling. Advantages of the SEDM are that it creates a good surface finish; accuracy and repeatability are high; and the hardness of the material does not influence the efficiency of the process. Therefore, the majority of die makers use EDM mainly for finishing of dies from already rough-machined and hardened die steels and for manufacturing of carbide die inserts.

The disadvantages of the SEDM process begin with the lead time required to design and manufacture the electrodes. Process concerns include electrode wear, material removal rate, and particle flushing. Electrode wear is caused, because the process consumes it in addition to removing the workpiece material. This limits the repeatability of the process. The EDM technology functions on a much smaller scale than conventional machining, which produces inherently lower material removal rates. This reduces the efficiency of the process. Typically, the SEDM is used because the desired geometry requires the technology. As the workpiece material is removed, a flushing fluid is passed through the gap to remove the particles. The fluid also cools the workpiece, the electrode, and the removed particles. Particles resolidify to the workpiece surface during cooling and form a martensitic layer referred to as the white layer. This surface is composed of high tensile residual stresses from the thermal cycle and may have numerous surface cracks, which severely reduce the fatigue life of the die material. The dielectric fluid used in the EDM process is filtered to collect the particles flushed during the process. The cost of discarding this residue is high, because it is environmentally hazardous.

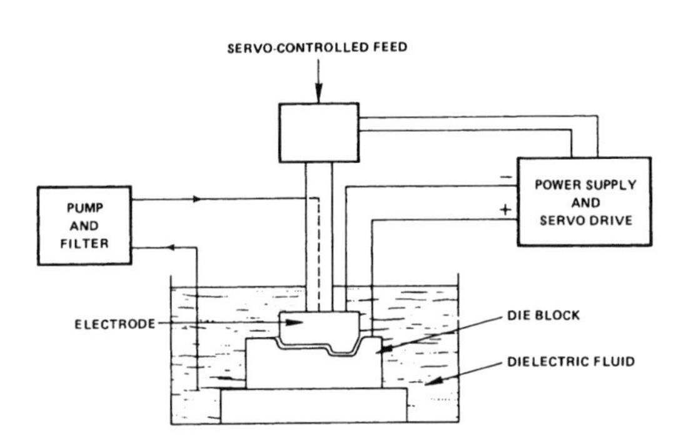

Fig. 21.10 Schematic of electrodischarge machining [Altan et al., 1983]

In the United States, the large majority of the electrodes are made from graphite. Graphite is used because it is a soft material that can easily be polished with sandpaper. When a better surface finish is required, copper is selected for the electrode material. Copper is the second most common material used in the United States, and more common in Europe and Japan. In addition to an improved surface finish, copper electrodes can produce much tighter tolerances.

The wire EDM (WEDM) functions in much the same way as the SEDM. The primary difference is that the electrode is a wire ranging in diameter from 0.002 to 0.012 in. (0.05 to 0.30 mm). Today, almost all machines are four-axis centers in which the upper and lower wire guides can move independently of one another in both the x- and y-directions. In addition, the table that holds the workpiece can move in the same directions. Movement of the guides and worktable is controlled by numerical-controlled programming. In forging, WEDM is mainly used for manufacturing trimming dies.

21.5.3 Hobbing

In hobbing, a hardened (58 to 62 HRC) punch is pressed into an annealed, soft die steel block, using a hydraulic press. The process may be cold or at elevated temperatures (warm or hot hobbing). A single hobbing punch can be used to manufacture a large number of cavities. This process is particularly attractive for making dies with shallow cavities or dies that can be hardened after the cavity has been produced. Major

Table 21.20 Relative ranking of various surface treatments and coatings [Krishnadev et al., 1997]

Surface treatment/coating	Impact toughness	Maximum exposure temperature	Thermal checking resistance	Wear resistance
Ion nitriding	Low	Same as base	Medium	Medium
CrN	High	1380 °F (750 °C)	Very high	High
TiN	High	840 °F (450 °C)	Very high	High
TiAlN	High	1470 °F (800 °C)	Very high	High
TD-VC	High	1470 °F (800 °C)	Very high	Very high
Laser	High	Same as base	High	High
PFS	High	Tailorable	Not applicable	High to very high

examples are coining dies and dies for hot and cold forging of knives, spoons, forks, handtools, etc.

21.5.4 Other Die-Making Methods [Altan et al., 1983]

In addition to the methods discussed above, there are a few other methods commonly used for die making. Cast dies, although not extensively used in practice, have been used successfully in some applications. This alternative may be attractive where many dies of the same geometry are to be made. Special cases in which cast dies are made cost-effective is isothermal or hot-die forging. In this application, the dies are made from nickel- and cobalt-base high-temperature alloys. Because these alloys cannot be machined easily, it is best to cast these dies and obtain the finished die cavity geometry by EDM.

Electrochemical machining is similar to EDM but does not use sparks for material removal. Only direct current between the metal electrode and the die steel is used for material removal. This method is more efficient than EDM in terms of metal removal rate; however, electrode wear is also quite large and, more importantly, difficult to predict. As a result, this method is used for die making only in selected applications.

21.6 Surface Treatments

Sometimes, after the die cavities of hot forging dies are formed using EDM or high-speed machining (HSM), the surface of the die is treated with special methods to increase hardness, die life, or surface finish. Some common methods of surface treatments include physical vapor deposition (PVD) (CrN, TiAlN, FUTURA multilayer) and duplex treatments, which include PVD and plasma nitride welding of the surface [Navinsek et al., 2001]. Physical vapor deposition is a process that coats the surface of the die with a hard, abrasive-resistant coating processed at a relatively low temperature (around 930 °F, or 500 °C). Plasma nitride surface welding is a process that adds nitrogen to the die cavity surface to increase its wear capabilities.

Because these surface treatments can significantly increase (50 to 100%) the tool life of hot forging dies, there exist certain drawbacks as well. One drawback is the added operation to the die-making process, which can be costly in terms of both money and time. Another drawback is that having a surface treatment on the die decreases its potential to be remachined. This is especially a drawback when it comes to HSM, because the surface treatments can be detrimental to tool inserts. Electrodischarge machining has a significant advantage to HSM in that even after coatings have been applied, it has been shown that the die cavity can be resunk using EDM.

The simultaneous need for high hardness to resist wear and high ductility to prevent fracture can be achieved using surface treatments. The common coatings used in forging are reactive coatings (nitriding or nitrocarburizing, boriding, vanadizing, carburizing, and ion implantation) and deposited coatings (hard chromium coating; nickel plating; hard facing; chemical vapor deposition coating, i.e., VC, WC, TiC, TiN, TiCN, TiC/TiN, Cr_xC_y; PVD coating, i.e., TiN, TiCN, TiAlN, CrN) [Krisnadev et al., 1997]. The qualities of various surface treatment techniques are given in Table 21.20

REFERENCES

[Altan et al., 1983]: Altan, T., Oh, S., and Geghel, H., *"Metal Forming: Fundamentals and Applications,"* American Society for Metals, 1983.

[Altan et al., 2001]: Altan, T., Lilly B., Yen, Y.C., "Manufacturing of Dies and Molds," Keynote Paper, *Annals of the CIRP,* Vol 50, Feb 2001.

[Blau, 1992]: Blau, P., *Friction and Wear of Ordered Intermetalic Alloys of Ni3Al,* ASM International, 1992.

[Bramley et al., 1989]: Bramley, A., Lord, J., and Davies, B., "Determination of Wear Resistance of Hot Work Die Materials," *Annals of the CIRP,* Vol 38/1, 1989, p 231–234.

[ICFG, 1992]: "Objectives, History and Published Documents," ISBN 3-87525-058-3, International Cold Forging Group, 1992.

[Johansonn et al., 1985]: Johansson, B., Jonsson, L., et al., "Some Aspect in the Properties of QRO 80M and Die Casting Die Performance," presented at the 13th International Die Casting Congress and Exposition (Milwaukee, WI), North American Die Casting Association, 1985.

[Kesavapandian et al., 2001]: Kesavapandian, G., Ngaile, G., Altan, T., "Evaluation of Die Materials and Surface Treatments for Hot Precision Forging," ERC/NSM Report No. B/ERC/NSM-01-R-42A, The Ohio State University, 2001.

[Krauss et al., 1998]: Roberts, G., Krauss, G., et al., *Tool Steels,* ASM International, 1998.

[Krishnadev et al., 1997]: Krishnadev, and Jain, C., "Enhancing Hot Forging Die Life," *Forging,* winter 1997, p 67–72.

[Lange, 1976]: Lange, K., "Massive Forming," *Textbook of Metal Forming,* Vol. 2, Springer-Verlag, New York, 1976 (in German).

[Maddox et al., 1997]: Maddox, G., and Orth, J., "Performance of Nickel Aluminide Forging Dies," *Forging,* winter 1997, p 75–78.

[Nagpal, 1976a]: Nagpal, V., "Selection of Steels for Forging Dies," Battelle, Columbus, OH, 1976.

[Nagpal, 1976b]: Nagpal, V., Selection of Tool Steels for Forging Dies, Battelle, Columbus, OH, 1976.

[Nagpal et al., 1980]: Nagpal, V., and Lahoti, G.D., "Application of the Radial Forging Process to Cold and Warm Forging of Common Tubes, Vol 1, Selection of Die and Mandrel Materials," Final report, Contract No. DAAA22-78-C-0109, prepared for Watervliet Arsenal by Battelle-Columbus Laboratories, Columbus, OH, May 1980.

[Navinsek et al., 2001]: Navinsek, B., et al., "Improvement of Hot Forging Manufacturing with PVD and DUPLEX Coatings," *Surface and Coatings Technology,* Vol 137, 2001, p 255.

[Roberts et al., 1980]: Roberts, G.A., and Robert, A.C., *Tool Steels,* American Society for Metals, 1980.

[Semiatin et al., 1981]: Semiatin, S.L., and Lahoti, G.D., "Forging Die Materials, Coatings, and Surface Treatments," Battelle, Columbus, OH, 1981.

[Thyssen]: Thyssen, B., series of brochures from Thyssen Company.

SELECTED REFERENCES

- **[Aston et al., 1969]:** Aston, J., and Barry, E., "A Further Consideration of Factors Affecting the Life of Drop Forging Dies," *J. Iron Steel Inst.,* Vol 210 (No. 7), July 1969, p 520–526.
- **[Blackwood, 1973]:** Blackwood, T.B., "The Role of Tool Steels in Extrusion Tooling," SME Technical Paper MF73-565, Society of Manufacturing Engineers, 1973.
- **[Bramley et al., 1975]:** Bramley, A., Lord, J., Beeley, P., *"The Friction and Lubrication of Solids,"* Vol I, Oxford University Press, London, 1975, p 38.
- **[Cerwin, 1995]:** Cerwin, N., *"Finkl Die Handbook,"* Vol 1–3 A. Finkl and Sons Co., 1995.
- **[Cser et al., 1993]:** Cser, L., Geiger, M., Lange, K., Hansel, M., "Tool Life and Tool Quality in Bulk Metal Forming," *Proc. Instn. Mech. Engr.,* Vol 207, 1993.
- **[Guobin et al., 1998]:** Guobin, L., Jianjun, W., Xiangzhi, L., Guiyun, L., "The Properties and Application of Bi-Metal Hot-Forging Die," *J. Mat. Proc. Tech.,* Vol 75, 1998, p 152–156.
- **[Horton et al.]:** Horton, J., Liu, C., et al., "Microstructures and Mechanical Properties of Ni3Al Alloyed with Iron Additions," Oak Ridge National Laboratory, Oak Ridge, TN.
- **[Klocke, 2000]:** Klocke, F., "Fine Machining of Cold Forging Tools," *Proceedings, Tenth ICFG Congress* (Tagung Fellbach, Germany), Sept 3–15, 2000, International Cold Forging Group, p 175.

CHAPTER 22

Die Failures in Cold and Hot Forging

Mark Gariety

22.1 Introduction

In forging, die failure can be a significant portion of the overall production cost. The dies that fail must be either repaired or replaced. The cost to replace a die includes the basic cost of the die, which encompasses the costs of material, machining, coating, and surface treatment as well as the cost of labor. In any case, the most important cost of die failure is related to the downtime of the manufacturing system, which reduces the overall productivity [Liou et al., 1988].

Precision forging processes are utilized to produce complex-shaped parts that require little or no finishing, thereby reducing manufacturing costs. In order to achieve both accuracy and complexity in a forged part, the dies must be produced with higher accuracy and tighter tolerances than the part to be forged. As a consequence, the manufacturing time for the dies increases considerably, making the dies more expensive than those needed to produce parts with conventional tolerances [Dahl et al., 1998, and Dahl et al., 1999].

Complex precision-forged parts are almost exclusively produced in closed dies. Therefore, the dies have to withstand high contact pressures well above the flow stress of the billet material at the forging temperature. Also, the amount of surface generation and the sliding velocities at which the billet material moves along the die surface are very large. In addition, in hot forging operations, the die surface is subjected to sudden changes in temperature. These changes are mainly due to contact with the hot workpiece and the cooling and lubrication practices. The high pressures and sliding velocities as well as the sudden changes in temperature may induce die failure due to several mechanisms. When a failure occurs, the production must stop to replace or repair the die. Consequently, the production cost will be significantly reduced if the die life can be increased.

22.2 Classification of Die Failures

In forging, die failures may be classified by one of the following three failure modes:

- The most common cause of die failure is wear. Die wear occurs as the result of the die and the workpiece sliding relative to one another while in contact. Typically, die wear is thought of as removal of material from the die surface, but it also may include buildup of material on the die surface or damage to the die surface. Die wear results in the gradual loss of part tolerances. Eventually, the part tolerances will not meet the customer's specifications, and a new die or die insert will need to be manufactured.
- The second most common cause of die failure is fatigue fracture. Fatigue occurs as the result of the continual stress cycles that the dies are subjected to. The stress cycles are attributed to both mechanical and thermal loading and unloading of the dies. Fatigue is accelerated in the vicinity of stress concentrations, such as small radii.
- The third cause of die failure is plastic deformation. Plastic deformation results from forming pressures that exceed the yield strength of the die material.

Figure 22.1 illustrates a typical precision-forging die and the locations where these various failures may occur. In hot forging, die wear is responsible for nearly 70% of die failures [Schey, 1983]. In cold forging, die failure is primarily the result of fatigue fracture. Thus, the remainder of this chapter focuses solely on wear and fatigue fracture, which are the primary failure modes in hot and cold forging, respectively.

22.3 Fracture Mechanisms

In order to maximize die life, it is important to understand the primary mechanisms of fatigue fracture, summarized as follows:

- Mechanically induced fatigue fracture occurs when a crack is induced and propagates through a die due to mechanical loading. Crack initiation occurs if the tool load exceeds the yield strength of the tool material in the area of a stress concentration, and a localized plastic zone forms. This zone generally forms during the first loading cycle and undergoes plastic cycling during subsequent unloading and reloading. The plastic cycling leads to the initiation of microscopic cracks. These microscopic cracks may grow and lead to crack propagation into the cross section of the tooling if they are subjected to tensile stresses during loading [Knoerr et al., 1994].

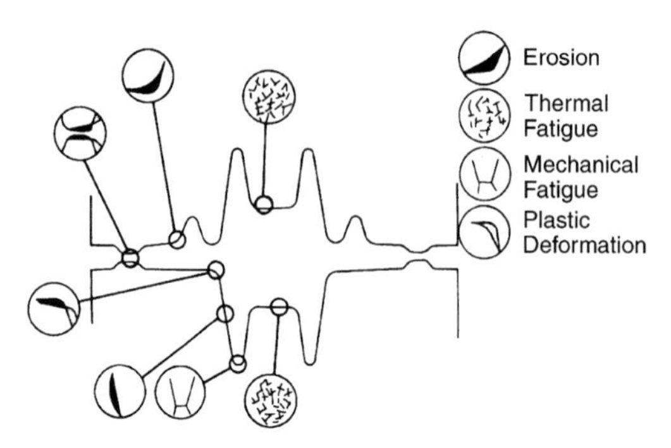

Fig. 22.1 Typical failure modes and locations in precision-forging die [Schey, 1983]

Fig. 22.2 Adhesive wear mechanism [Bay, 2002]

- Thermally induced fatigue fracture occurs when a crack is induced and propagates through a die due to thermal loading. Thermal-induced fatigue fracture is not as common as mechanically induced fatigue fracture, because the bulk die temperature does not fluctuate drastically during typical metal forming operations.

22.4 Wear Mechanisms

In order to maximize die life, it is also important to understand the primary mechanisms of wear. In metal forming, there are four primary mechanisms by which wear may occur. These mechanisms, summarized below, work simultaneously and may be difficult to distinguish in an actual die:

- Adhesive wear occurs when two surfaces slide relative to one another, and material from one surface is sheared and remains adhered to the other surface (Fig. 22.2). When there is metal-to-metal contact between the surface asperities of the workpiece and the die, some small portion of the workpiece surface may be sheared and remain attached to the die surface [Schey, 1983].
- Abrasive wear, sometimes called "plowing," occurs when material is removed from a soft surface due to interaction with a hard asperity on another surface (two-body abrasion) or due to interaction with hard, loose particles trapped between two surfaces (three-body abrasion) (Fig. 22.3). A major concern is the abrasive wear of the die by hard oxides such as scale or intermetallics on the surface of the workpiece.
- Fatigue wear occurs when small cracks form on the surface or subsurface of the die and subsequently break off in the form of small fragments, thus leaving small voids in the die surface (Fig. 22.4). This phenomenon is a result of the repeated mechanical loading and unloading that dies are subjected to. In hot forging, the repeated thermal loading and unloading also contributes to fatigue wear and is often called "heat checking." In general, fatigue wear is affected by surface defects that act as stress-concentration points, surface finish, residual stresses, and the lubricant chemistry. For example, water-based lubricants may accelerate fatigue wear due to hydrogen embrittlement.

- Corrosive/chemical wear occurs when surface films (i.e., oxidation) are removed as a result of the relative sliding between the die and the workpiece in a corrosive environment (Fig. 22.5). In general, corrosive wear is accelerated by the use of lubricants. However, this level of wear is generally accepted in order to prevent the more severe forms of adhesive wear that would result if lubricants were not used.

22.5 Analytical Wear Models

Some of the earliest analytical wear models were developed by Holm (1940s), Archard (1950s), and Rabinowicz (1960s). In general, these models characterized adhesive and abrasive wear as follows:

$$V = \frac{K \cdot P \cdot S}{H} \qquad \text{(Eq 22.1)}$$

where V is the wear volume, K is the experimental wear coefficient, P is the normal pressure, S is the sliding length, and H is the hardness of the softer material. Thus, it was recognized very early that adhesive and abrasive die wear could be minimized by minimizing the wear coefficient, the normal pressure, and the sliding length and by maximizing the die hardness.

Approximate values of the experimental wear coefficient, K, have been determined for many die/workpiece material combinations. For adhesive wear, K_{adh} varies based on the material pair and the lubricant used [Bay, 2002]. Table 22.1 summarizes these values. For abrasive wear, K_{abr} depends on the size, shape, and orientation of the asperity/particle interacting with the die surface and usually varies from 0.02 to 0.2 for two-body abrasive wear and 0.001 to 0.01 for three-body abrasive wear.

22.6 Parameters Influencing Die Failure

There are numerous parameters that influence die failure in cold and hot forging processes. The interaction of these parameters is complex. They may increase or decrease die failure, and the trends are important to understand in order to maximize die life. These parameters are discussed in the following sections and shown in flow chart form in Fig. 22.6.

22.6.1 Incoming bar

Billet Material. The flow stress of the workpiece material influences the normal pressure at the die/workpiece interface. Considering that wear is proportional to interface pressure, a material with a lower flow stress will produce less pressure at the die/workpiece interface and thus less die wear [Dahl et al., 1998].

In addition, adhesive wear is inversely proportional to the workpiece material hardness. Therefore, a material with a high hardness (but less than die hardness) will produce less adhesive die wear [Dahl et al., 1998].

Billet Surface. The scale present on the workpiece surface influences the interface conditions. It acts as an insulator and thus protects the die against increased heating and thermal fatigue, but if it is hard and brittle, it causes abrasive wear. The properties of the scale are influenced by billet heating time and temperature [Dahl et al., 1998, and Schey, 1983].

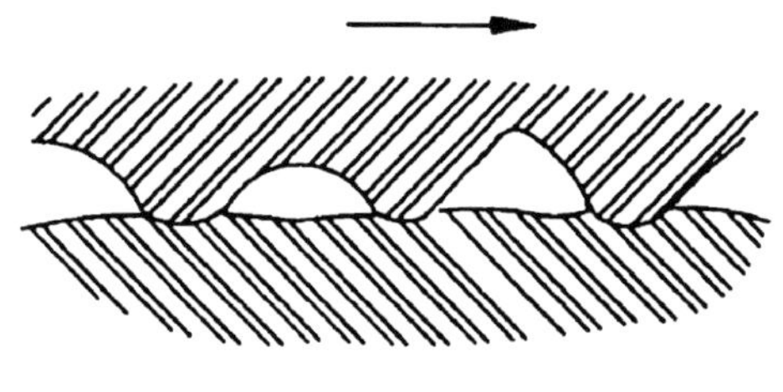

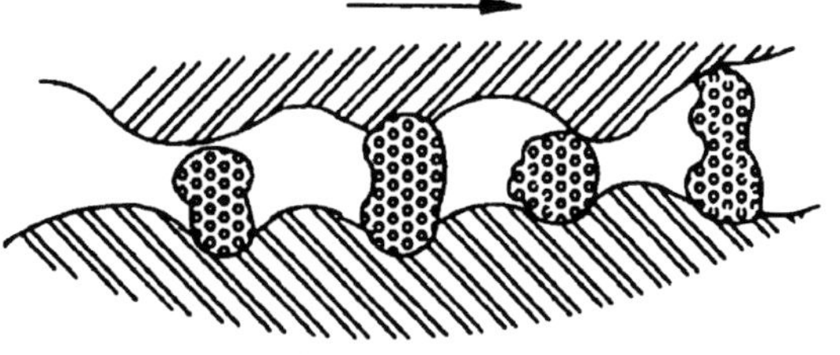

Fig. 22.3 Two-body and three-body abrasive wear mechanisms [Bay, 2002]

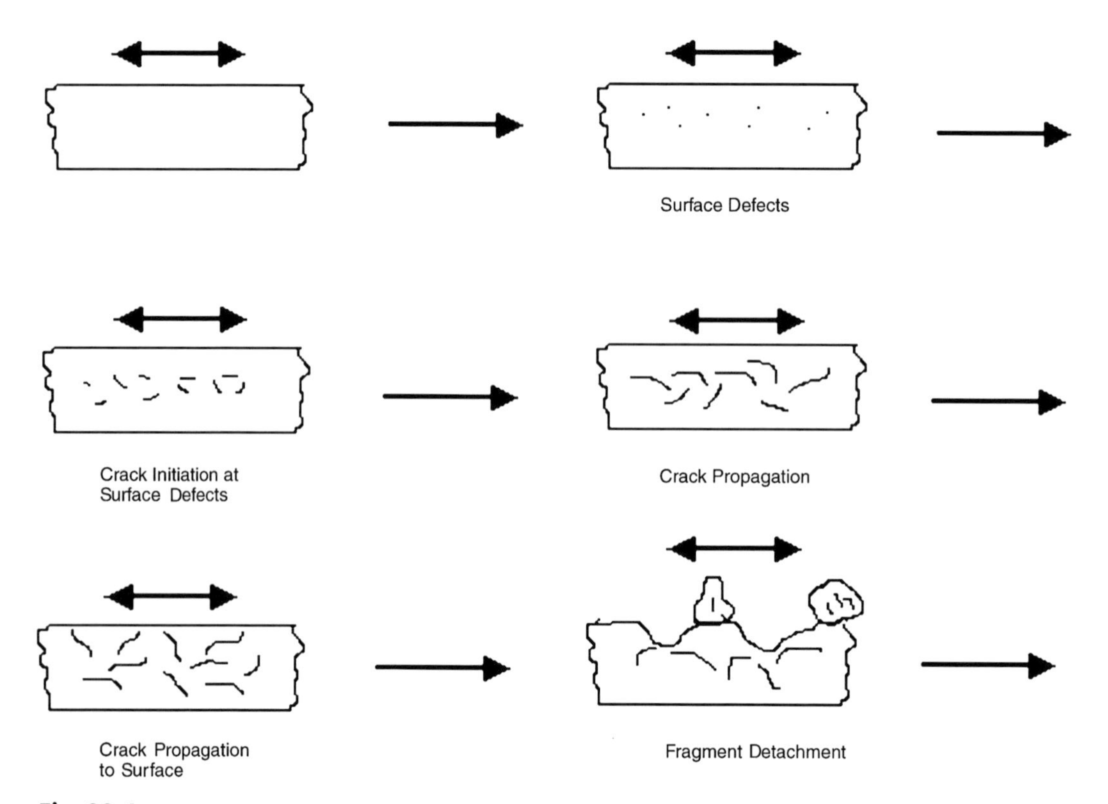

Fig. 22.4 Fatigue wear mechanism [Bay, 2002]

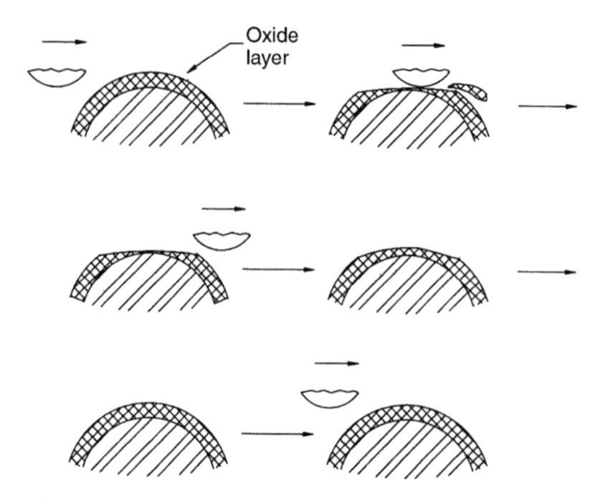

Fig. 22.5 Corrosive wear mechanism [Bay, 2002]

Table 22.1 Summary of wear coefficient, K_{adh}, for adhesive wear [Bay, 2002]

Surface condition	Like metallic pairs	Unlike metallic pairs
Clean	5×10^{-3}	2×10^{-4}
Poor lubrication	2×10^{-4}	2×10^{-4}
Average lubrication	2×10^{-5}	2×10^{-5}
Excellent lubrication	2×10^{-6} to 10^{-7}	2×10^{-6} to 10^{-7}

Billet Geometry. The geometry of the billet influences the amount of sliding that will take place during the forging process. Because abrasive and adhesive wear are proportional to sliding length, the use of preforms where die wear is a concern will increase die life. A billet or preform that requires a small amount of deformation can be forged with a lower load (pressure) and a shorter contact time. In large-volume production, most of the forging deformation is completed in preform or blocker dies. Thus, there is little metal flow in the finisher die, which is used for coining in order to reduce the die wear and maintain part tolerances [Dahl et al., 1998, and Schey, 1983].

Billet Weight and Weight Tolerances. Die wear increases with increasing billet weight. This is likely the effect of increased interface pressure and contact time. In addition, if the billet weight exceeds the weight tolerances (i.e., the volume of the billet is larger than the volume of the die cavity), the die will fill prematurely, causing increased interface pressures and thus increased wear and decreased fatigue life [Dahl et al., 1998].

22.6.2 Billet Separation

Separation Method. The separation method influences the obtainable weight tolerances. The effect of exceeding the weight (volume) tolerance was discussed previously. In general, cropping and sawing are the primary methods by which billets are separated. Cropping is known to produce a geometrically inferior billet to sawing. This method results in more surface and edge defects (i.e., cracks and burrs) than sawing [Dahl et al., 1998].

Sorting of billets either by weight or size can help to extend the useful life of the dies. Billets can be sorted into two or more groups based on weight or size. The smaller billets should be forged first. As the dies begin to wear and the cavities become larger, larger billets may be introduced. The largest billets are forged at the end of the die life when the cavity is the largest. This technique helps to reduce the risk of excessive interface pressures generated as a result of premature die fill [Dahl et al., 1998].

Edge Quality. Angularity and other geometrical imperfections that result from billet separation can cause uneven loading of the dies. If the imperfections are consistent in the billet separation process, the uneven loading becomes a repeated problem occurring at every stroke of the press. This may cause high local wear [Dahl et al., 1998].

22.6.3 Billet Heating

The billet heating process must be closely controlled in order to keep the billets in the optimum forging temperature range and to reduce scale formation. As discussed previously, hard, abrasive scale can be a major source of die wear. Scale may accumulate on the billet surface as a result of burning during the heating process and excessive transportation times between heating and forging. Coatings may be applied to the billets before heating in order to reduce scale formation [Dahl et al., 1998].

22.6.4 Forging Equipment

The press or hammer type influences the length of the contact time. Increased contact times result in increased wear. The reason for this is that the longer the die is in contact with the hot billet, the more the temperature of the dies increases and the temperature of the billet decreases. The increased die temperatures cause decreased die hardness due to thermal softening. Therefore, abrasive wear increases. In addition, the decreased billet temperatures cause increased flow stress, which leads to increased interface pressures. This also leads to increased wear. If the die experiences large temperature oscillations due to long contact times and sub-

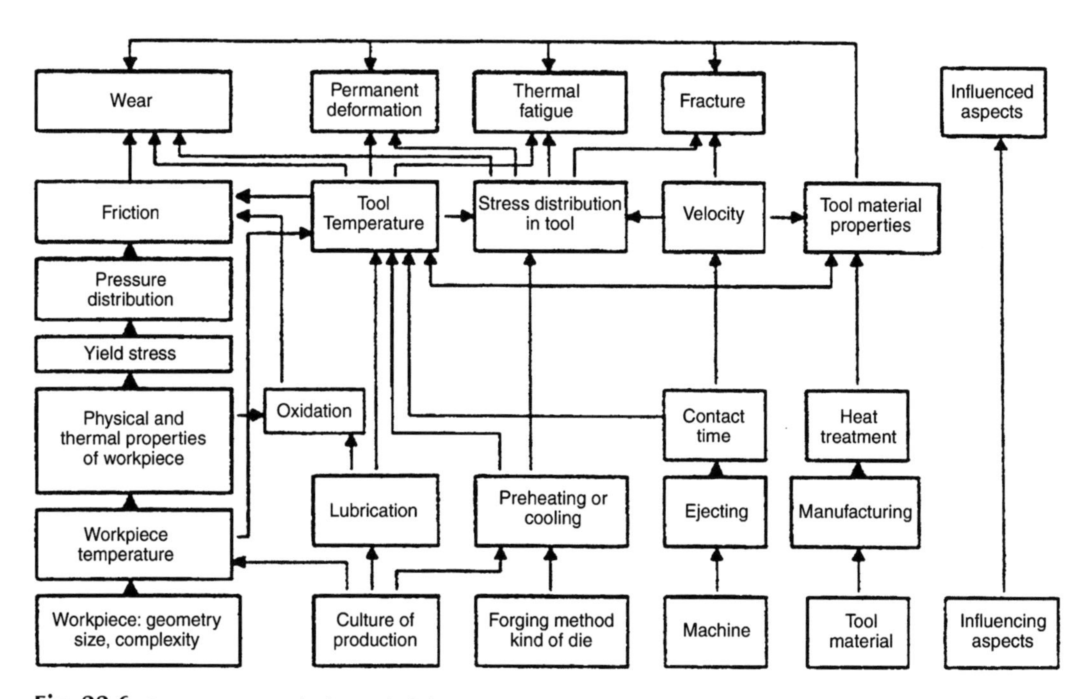

Fig. 22.6 Process parameters leading to die failure [Artinger, 1992]

sequent lubrication/cooling, thermal fatigue wear may also be a concern [Dahl et al., 1998, and Schey, 1983].

In general, hydraulic presses transfer energy from the ram to the workpiece and die slower than other press types. Thus, the contact time is longer. Hammers typically transfer energy very quickly. Thus, the contact time is shorter; however, fatigue fracture is a major concern for dies with low toughness. In addition, the stiffness of the press influences the contact time. If a press has low stiffness, there is more elastic deflection during loading. As a result, the contact time is longer [Kesavapandian et al., 2001].

The press (ram) speed influences the relative sliding velocity between the billet and the dies.

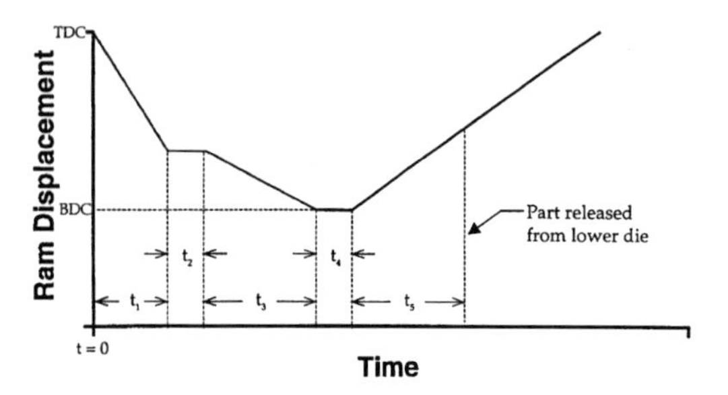

Fig. 22.7 Contact time components in forging processes [Dahl et al., 1998]

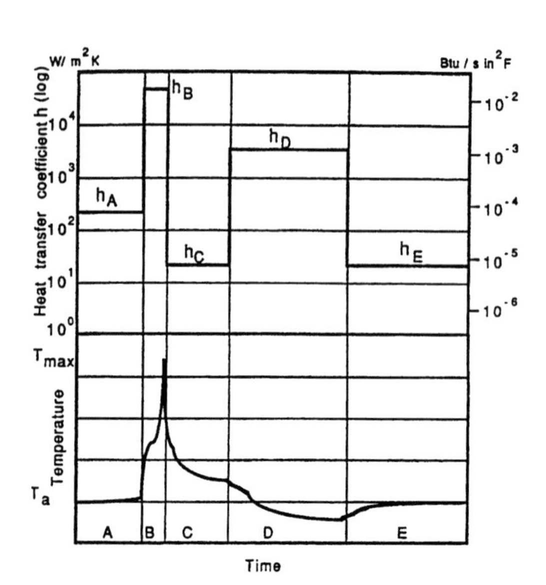

Fig. 22.8 Heat-transfer coefficient and temperature changes in a typical hot forging operation (A, heated billet resting on lower die; B, contact time under pressure; C, forging removed from lower die; D, lubrication of die; E, dwell time with no billet on lower die before next cycle begins) [Knoerr et al., 1989]

Increased sliding velocity results in increased strain rates, which cause increased flow stress and thus increased wear and decreased fatigue life. However, if the press speed it too slow, long contact times may become important [Dahl et al., 1998].

The effect of the press (ram) speed and the contact time is extremely important for die life. The longer the contact time between the billet and the dies, especially under pressure, the shorter is the die life. It is useful to divide the total contact time, t_T, in precision forging into its various components, as follows (using forging in a vertical press as an example) (Fig. 22.7) [Dahl et al., 1998]:

$$t_T = t_1 + t_2 + t_3 + t_4 + t_5 \quad \text{(Eq 22.2)}$$

where:

- t_1 = rest time. (The heated billet or preform is placed on the lower die. The pressure at the die/workpiece interface is due to the workpiece weight only.)
- t_2 = initial dwell time under pressure. (The upper die touches the workpiece. The ram stops for a very short time while the pressure builds up. This is the case in some hydraulic presses, or when a mechanical press ram has some excessive clearances in the eccentric bearings, or the lift-up cylinders are not functioning properly.)
- t_3 = contact time under pressure. (Deformation occurs during this time. This time is influenced by the press stiffness. The effect of elastic deflection, i.e., press stiffness, is especially important in mechanical and screw presses.)
- t_4 = final dwell time under pressure. (The forging stroke is completed. The ram is at bottom dead center (BDC). The ram must be lifted upward. This dwell time is due to elastic deflection of the press in mechanical and

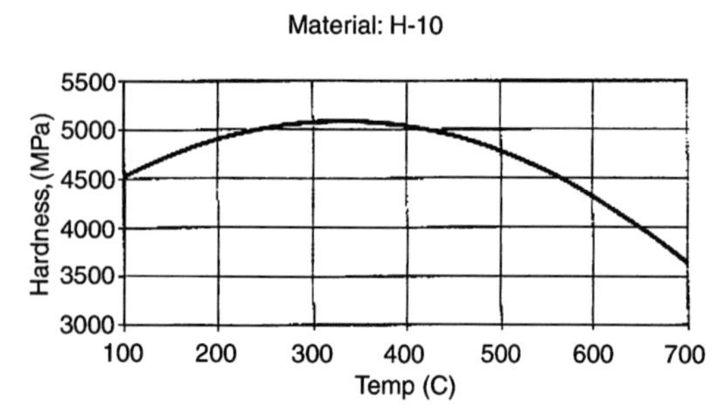

Fig. 22.9 Temperature-hardness curve [Dahl et al., 1999]

screw presses and the time necessary to reverse the hydraulic pressure to lift up the ram in some hydraulic presses.)

- t_5 = final dwell time without pressure. (The upper ram has lifted, but the forging is still in the die before it is removed manually or lifted out by a knockout mechanism.)

As shown in Fig. 22.8, the heat-transfer coefficient at the die/workpiece interface is different for the different contact time components.

22.6.5 Forging Dies

Die Material and Heat Treatment. The strength (i.e., flow stress), toughness, resistance to thermal softening, and hot hardness of the hot forging die material influence its wear resistance. Recall that abrasive and adhesive wear are inversely proportional to the strength/hot hardness of the die material. In addition, good toughness is also important for resistance to fatigue wear/fatigue fracture. Because toughness is a

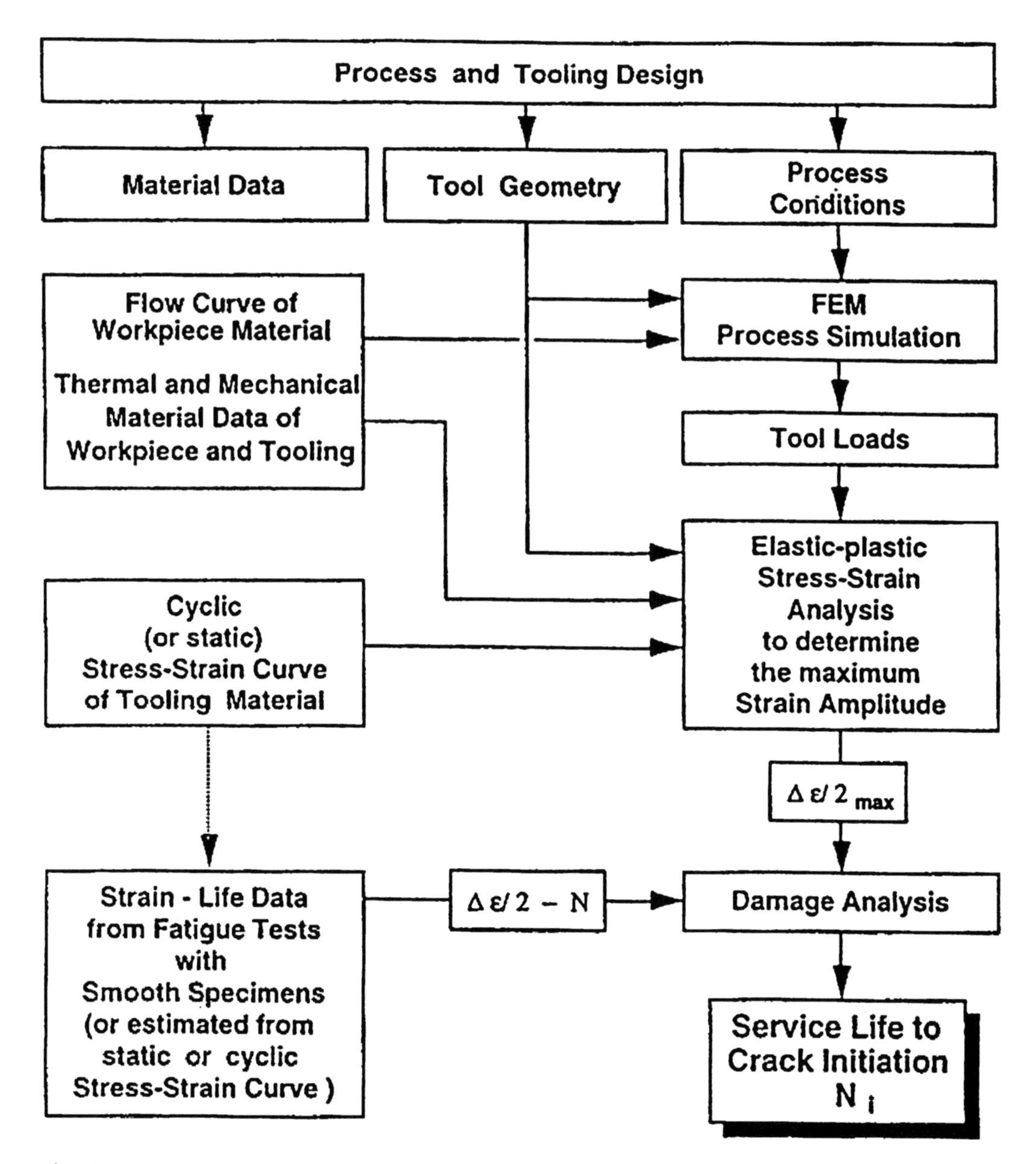

Fig. 22.10 Fatigue analysis method [Knoerr et al., 1994]

function of strength and ductility, it usually increases with increasing temperature [Altan et al., 1983, and Dahl et al., 1998]. A typical temperature-hardness curve is shown in Fig. 22.9. Additional information on die materials is given in Chapter 21, "Die Materials and Die Manufacturing," in this book.

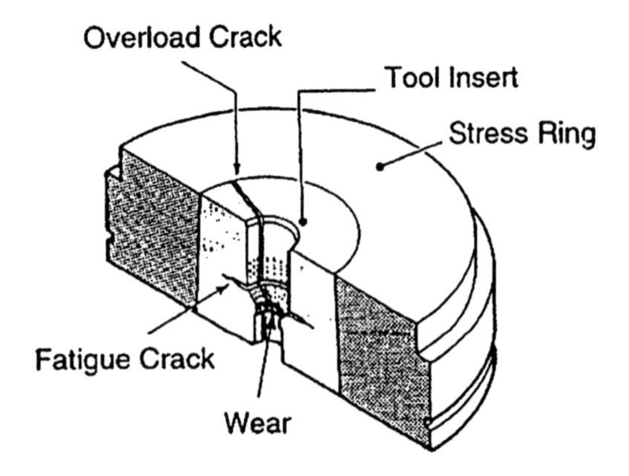

Fig. 22.11 Fatigue fracture in forward extrusion die [Lange et al., 1992a]

Alloying elements such as chromium, tungsten, vanadium, and molybdenum are typically employed in hot forging dies in order to improve the wear resistance of the die. In particular, molybdenum is responsible for resistance to thermal softening at hot forging temperatures. Vanadium improves the resistance to abrasion and thermal fatigue. Tungsten increases toughness (resistance to mechanical fatigue) and resistance to thermal softening [Dahl et al., 1998, and Tulsyan et al., 1993].

In addition, the microstructure resulting from heat treatment influences die wear. It has been shown that microstructure type, grain size, uniformity, and the number of microcracks affect die wear [Shivpuri et al., 1988].

The thermal properties of the die material also influence its wear resistance. High thermal conductivities eliminate the large thermal gradients that lead to thermal fatigue wear by carrying the

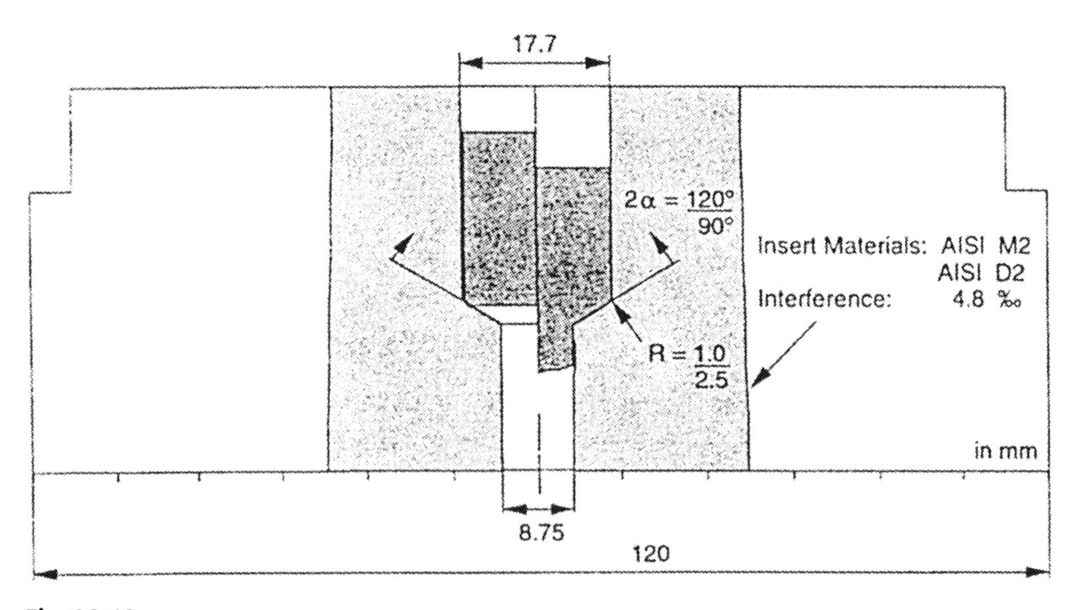

Fig. 22.12 Forward extrusion die cross section used by Hettig for fatigue failure investigation [Hettig, 1990]

Table 22.2 Test conditions and results of Hettig's fatigue failure investigation along with fatigue analysis results [Knoerr et al., 1994]

Case	Insert material	Hardness, HRC	Die opening angle (2α), degrees	Transition radius (R) in.	Transition radius (R) mm	Experimental die life, parts	Predicted die life, parts
I	AISI M2	61	120	0.04	1	50–400	280
II	AISI D2	60	120	0.04	1	65–200	210
III	AISI D2	60	90	0.04	1	900–1000	950
IV	AISI D2	60	90	0.10	2.5	10,000–11,000	10,500

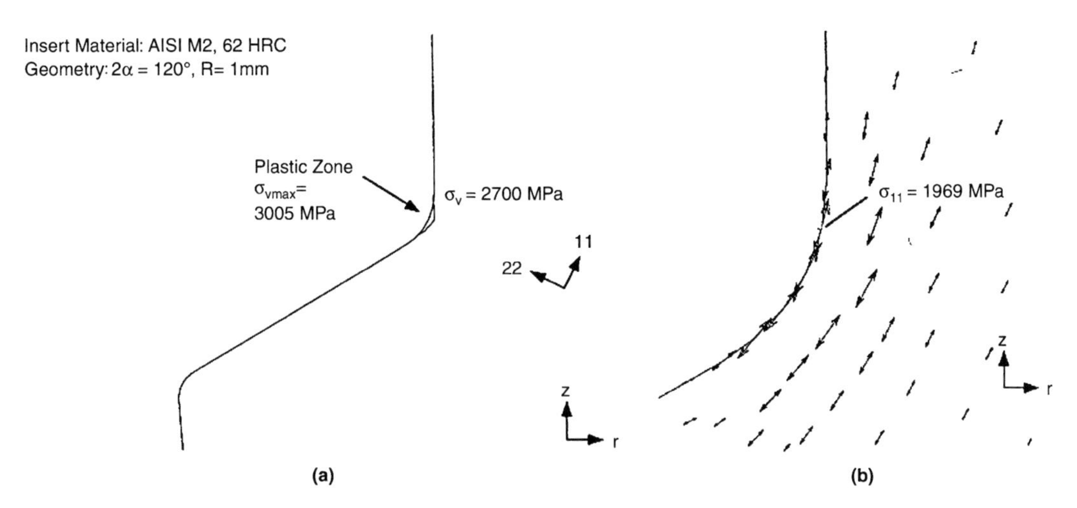

Fig. 22.13 (a) Plastic zone at the transition radius. (b) Tensile maximum principal stress at transition radius [Knoerr et al., 1994]

heat away from the die surface more quickly. Low thermal expansion rates reduce stresses induced by dimensional changes in the die at high temperatures [Dahl et al., 1998].

Surface Treatments. Because dies simultaneously require high hardness to prevent wear and high toughness to prevent fatigue fracture, many dies incorporate a surface treatment, such as nitriding or boriding in hot forging or a coating such as TiN or TiAlN in cold forging, that improves the hardness of the surface while leaving the bulk of the die relatively soft [Dahl et al., 1998].

Die Design. In hot forging, die design parameters, such as flash geometry, fillet radii, draft angles, and die face contact area, influence die wear and fatigue life. It has been found that die wear decreases and fatigue life increases with an increase in flash thickness, because the contact stresses between the die and the flash decrease. In addition, die wear increases and fatigue life decreases with an increase in flash-metal escape, because higher contact stresses are produced by the higher loads required to deliver higher flash-metal escape rates [Aston et al., 1969].

As fillet radii increase, die wear decreases and fatigue life increases, because small radii introduce stress concentrations. Increasing draft angle decreases die life, because higher draft angles require higher loads to fill the die cavity [Knoerr et al., 1989].

Die Manufacturing. The method used to manufacture the die influences the surface characteristics and thus the wear of the die. For example, electrical discharge machining typically produces a very hard and brittle surface that has been shown to be more wear resistant than those produced with other manufacturing methods [Knoerr et al., 1989].

Surface Finish. Generally, the rougher the surface, the more wear is produced. This results because the number of asperities in contact is less for a rough surface. Thus, greater loads per asperity are generated on rough surfaces. However, in cases where the surfaces are very smooth and, more importantly, very clean, adhesive

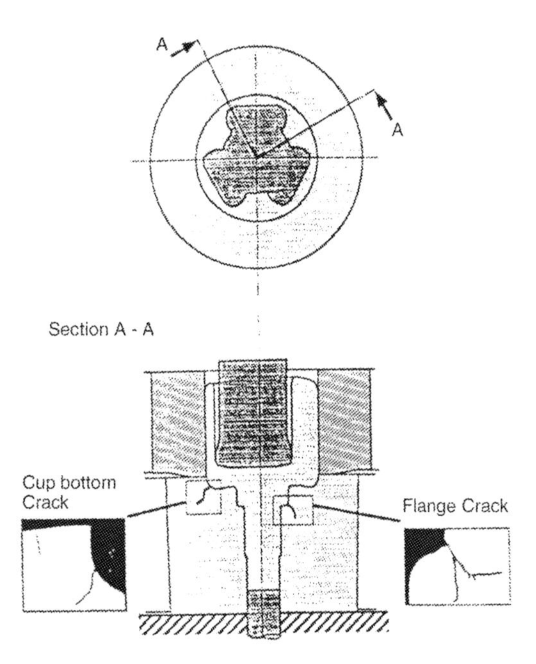

Fig. 22.14 Fatigue fracture in backward extrusion of constant velocity joints [Nagoa et al., 1994]

wear is likely to occur unless coatings are used [Tulsyan et al., 1993].

22.6.6 Lubrication

Die Lubricant Type. The lubricant type influences the interface pressure and the heat transfer between the die and the billet. In general, decreasing lubricity results in increased interface pressure and thus increased die wear and decreased fatigue life. Also, the lubricant acts as an insulator in order to protect the die against extreme temperature changes and thermal fatigue wear [Schey, 1983, and Dahl et al., 1998].

Mode of Application. In hot forging, the lubricant application parameters, i.e., spray time, spray angle, spray distance, and flow rate, influence how the lubricant covers the die. Inadequate lubrication results in increased interface pressure and reduced insulation against extreme die-temperature changes. However, excessive lubricant can result in the buildup of solid lubricant particles in die cavities. This buildup may cause premature die fill and increased interface pressure [Dahl et al., 1998].

22.6.7 Process Conditions

Die Temperature. The die temperature influences the hardness of the die. In general, the die hardness is inversely proportional to the die temperature. Thus, because abrasive and adhesive

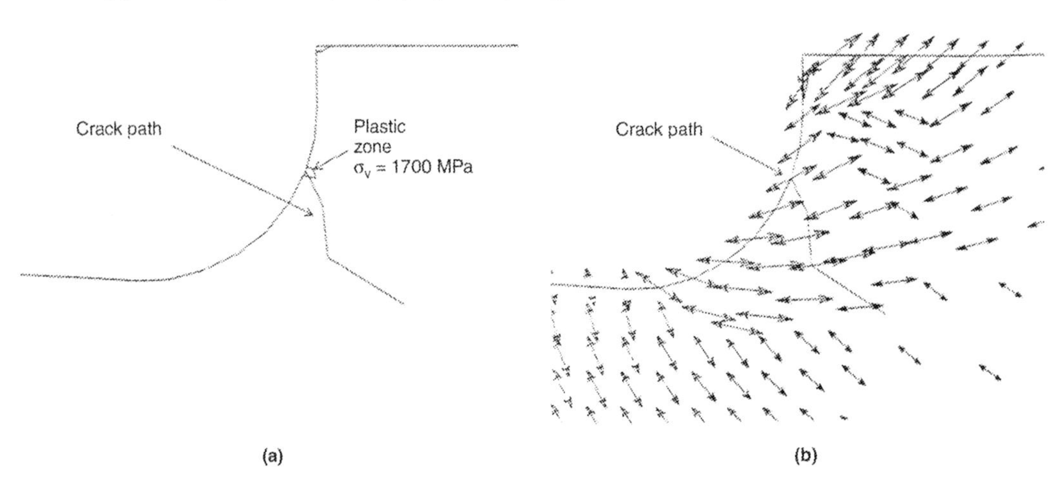

Fig. 22.15 (a) Plastic zone at the cup bottom. (b) Tensile maximum principal stress at cup bottom [Nagao et al., 1994]

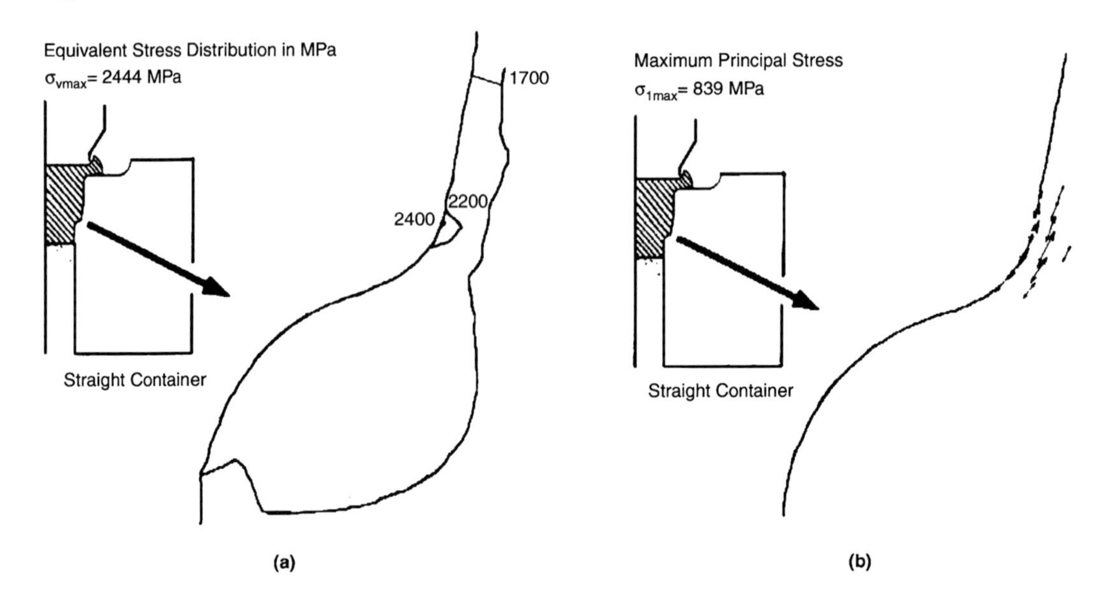

Fig. 22.16 (a) Plastic zone at the flange area. (b) Tensile maximum principal stress at flange area [Nagao et al., 1994]

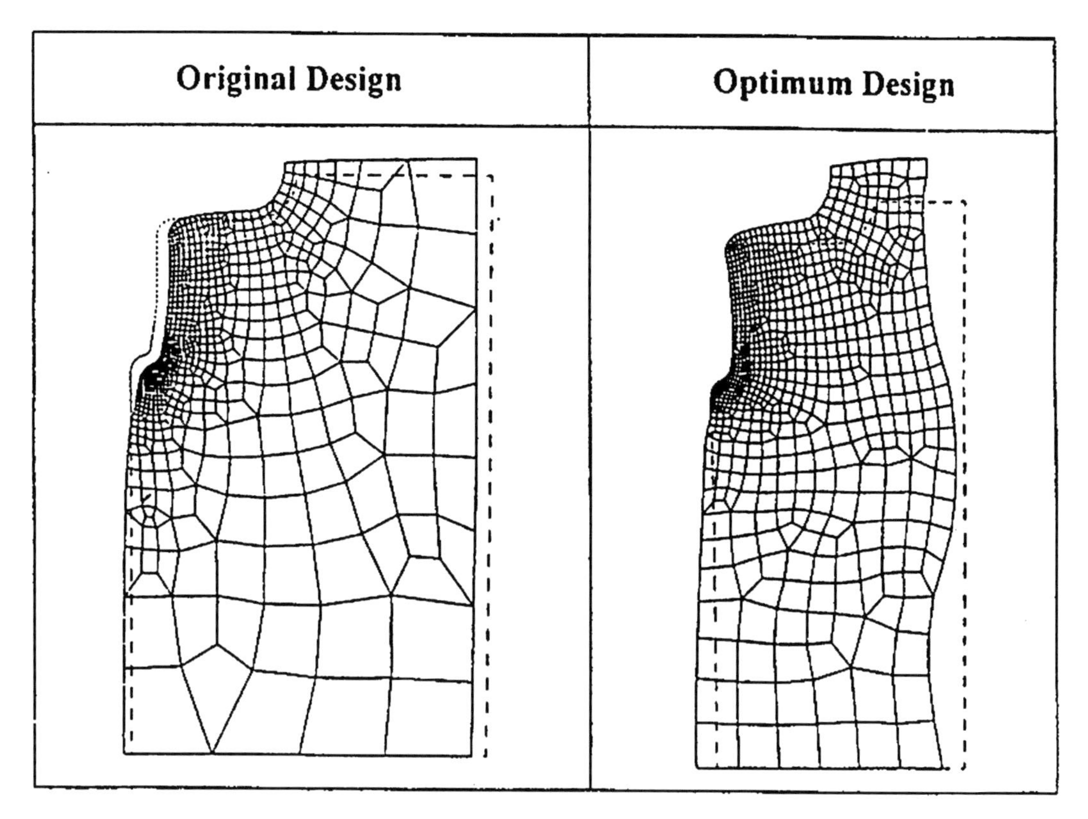

Fig. 22.17 Deformation of the inserts under load for the straight container (original design) and the profiled container (optimum design) [Nagao et al., 1994]

die wear are inversely proportional to die hardness, die wear is proportional to die temperature, which is significantly affected by contact times [Dahl et al., 1998, and Schey, 1983].

Billet Temperature. The billet temperature influences the flow stress of the material. As discussed previously, increased flow stress results in increased interface pressure and thus increased wear and decreased fatigue life [Dahl et al., 1998 and Schey, 1983]. In addition, the billet temperature influences the amount of scale on the workpiece surface, i.e., the amount of scale typically increases with increasing temperature [Tulsyan et al., 1993].

Sliding Velocity. Increasing the sliding velocity between the die and the workpiece increases die wear. In general, the relative sliding between the die and workpiece creates heat. The higher the sliding velocity, the more heat is produced. This heating lowers the hardness of the die [Tulsyan et al., 1993].

Number of Forging Operations. While the exact number of forging operations is not important, what is important is that there are enough operations so that the severity of the deformation at one station is not excessive. Excessive deformation in one die cavity causes large sliding between the billet and the die, which results in increased die wear. Therefore, appropriate and economic preforming operations should

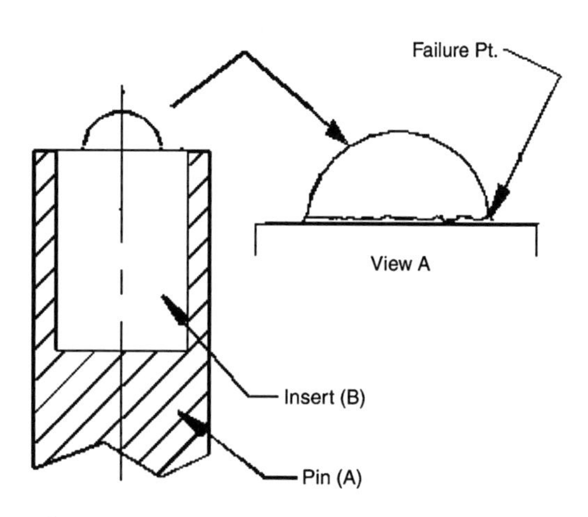

Fig. 22.18 Knockout pin design and insert failure point [Hannan et al., 2001]

be used, especially just prior to the final forging operation where wear rates must be kept low to maintain desired tolerances [Dahl et al., 1998].

Transfer Time. The transfer time, i.e., amount of time required to transfer the billet from the furnace to the press, influences the temperature of the billet. Cooling of the billet will raise the flow stress of the billet. As discussed previously, increased flow stress results in increased interface pressure and thus increased wear and decreased fatigue life [Dahl et al., 1998].

Cycle Time/Production Rate. If the heated billet is allowed to rest on the die for a long period of time before deformation occurs, the billet will cool and the die will heat. Thus, wear will be increased, because the flow stress of the billet will be increased and the hardness of the dies will be decreased [Dahl et al., 1998]. Large production rates in hot forging will increase the overall temperature of the dies, i.e., in high-speed horizontal hot forging machines. In such cases, large amounts of coolants and lubricants

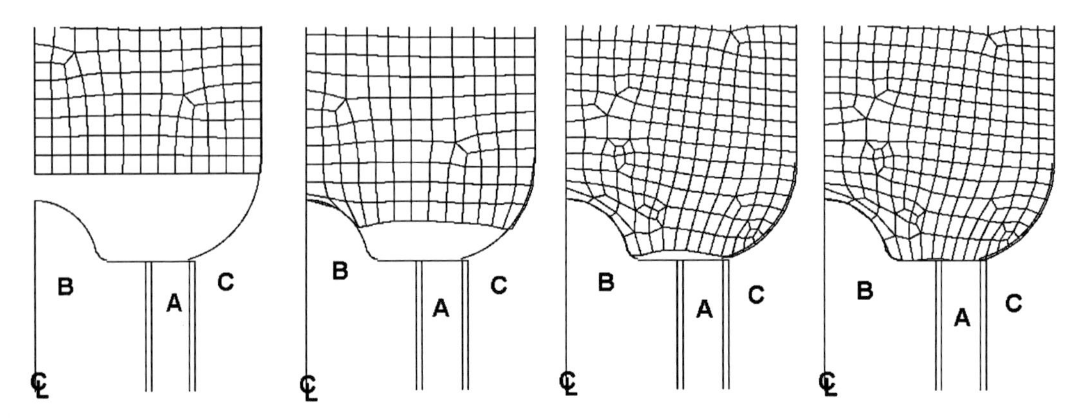

Fig. 22.19 FEM model of knockout pin [Hannan et al., 2001]

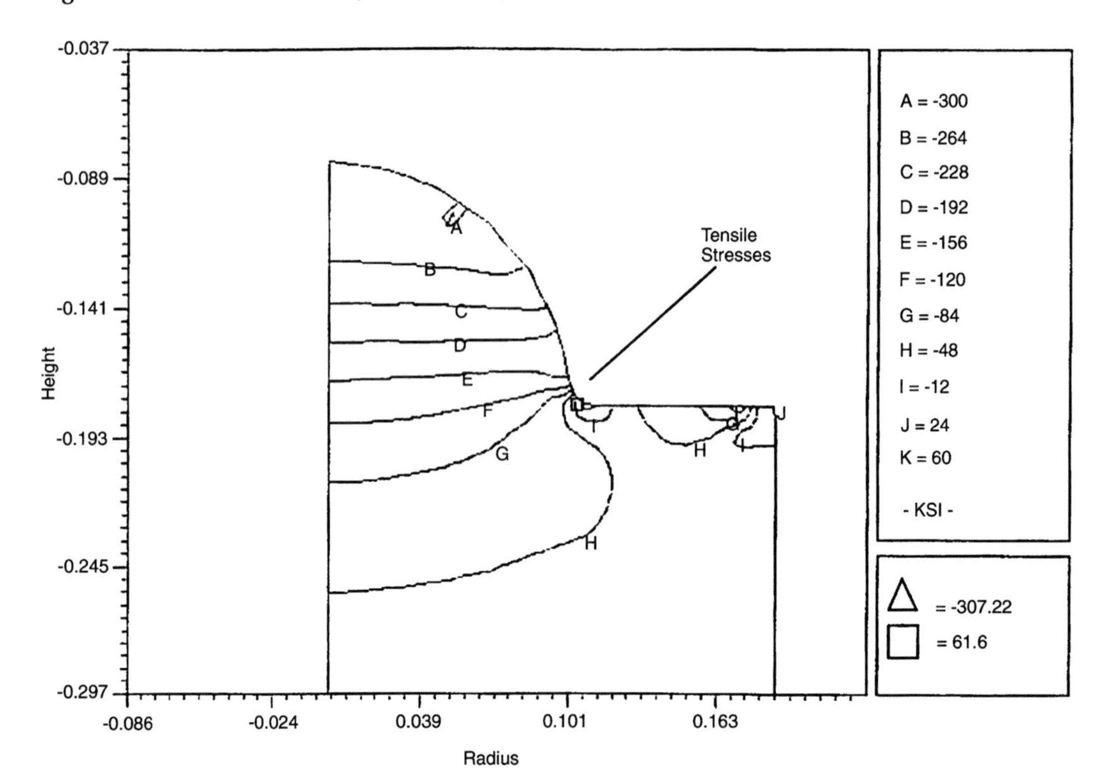

Fig. 22.20 Tensile maximum principal stress in failure area [Hannan et al., 2001]

are needed in order to reduce die temperatures. Alternatively, it is also possible to leave one forging station intermittently empty in order to provide time for the dies to cool.

In order to produce an economically sound part with prolonged die life by either cold or hot forging, a multitude of variables, as discussed above, must be carefully considered. In particular, the effect and importance of each variable based on the nature of the given component must be assessed.

22.7 Prediction of Die Fatigue Fracture and Enhancement of Die Life in Cold Forging Using Finite-Element Modeling (FEM)

A reliable method to analyze, predict, prevent, and/or control die fatigue fracture has long been one of the most important issues in cold forging. Therefore, a fatigue analysis method that can be utilized to estimate tool life has been developed and can be summarized as follows (Fig. 22.10) [Knoerr et al., 1994, and Matsuda, 2002]:

- The forging process is simulated using FEM in order to estimate the tool stresses.
- The tool stress values are used to complete an elastic-plastic stress-strain analysis of the tooling using FEM.
- The stress-strain analysis is used to perform a damage analysis and estimate the number of cycles until fatigue fracture.

If the predicted tool life is insufficient, changes in the process and tooling design must be made in order to reduce the loading conditions. A significant increase in tool life can be achieved by reducing the stresses in the highest loaded zone below the yield strength of the tool material. This may be achieved by the following means [Knoerr et al., 1994]:

- Change material flow in the die to reduce the contact stresses on the tool.
- Redesign the process to avoid drastic changes in the direction of the material flow, which usually leads to peaks in the contact stress.
- Increase the transition radii to reduce the notch effect.
- Split the tooling at the highest loaded zone.
- Increase the interference of the stress ring.
- Apply advanced stress ring techniques, such as strip-wound containers or profiled stress rings.

Several case studies where this fatigue analysis method was employed are presented as follows.

22.7.1 *Forward Extrusion—A Case Study*

In forward extrusion, fatigue cracks initiate at the transition radius to the extrusion shoulder and propagate in the radial direction (Fig. 22.11).

In order to verify the fatigue analysis method for evaluating tool life, an experimental investigation of fatigue fracture of forward extrusion dies, performed at the Institute for Metal Forming (IFUM) at the University of Stuttgart, was used [Hettig, 1990]. Figure 22.12 shows the cross section of the forward extrusion die, and Table 22.2 summarizes the test conditions and the results obtained. In addition, Table 22.2 summarizes the results obtained from the fatigue analysis for the same test conditions. As the table shows, the predicted die lives were in the range of those found experimentally. Thus, the fatigue analysis method was verified.

Table 22.2 illustrates that the die opening angle and the transition radius have a large effect on the die life. With FEM-based fatigue analysis, this effect is noted quickly and economically. Also, the FEM-based fatigue analysis easily determines the location where fatigue fracture will occur. Figure 22.13 shows the local plastic zone at the transition radius found during the FEM-based stress-strain analysis of the die. Figure 22.13 also shows that the maximum principal stresses in this same region are tensile,

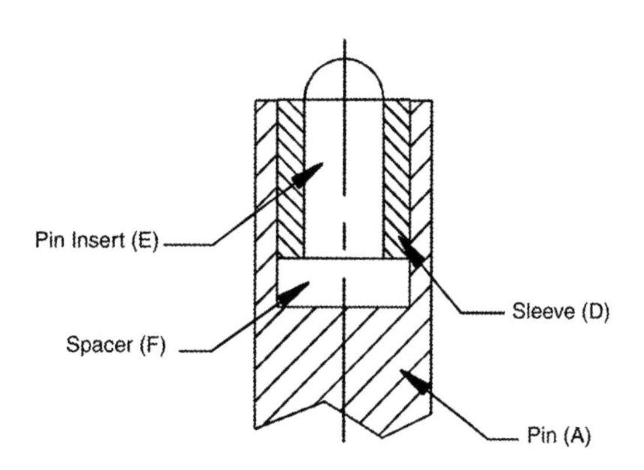

Fig. 22.21 Split knockout pin insert design [Hannan et al., 2001]

which facilitates the propagation of cracks formed in the plastic zone.

22.7.2 Backward Extrusion of Constant Velocity Joints—A Case Study

In cold forging of outer races for constant velocity joints, tool stresses during the backward extrusion operation lead to very high loading conditions in the tool inserts. Therefore, fatigue fracture is the common failure mode for the tooling. Figure 22.14 illustrates the backward extrusion process and the locations where fatigue fracture is experienced [Nagao et al., 1994].

Figure 22.15 shows the results from the fatigue analysis at the cup bottom, where fatigue

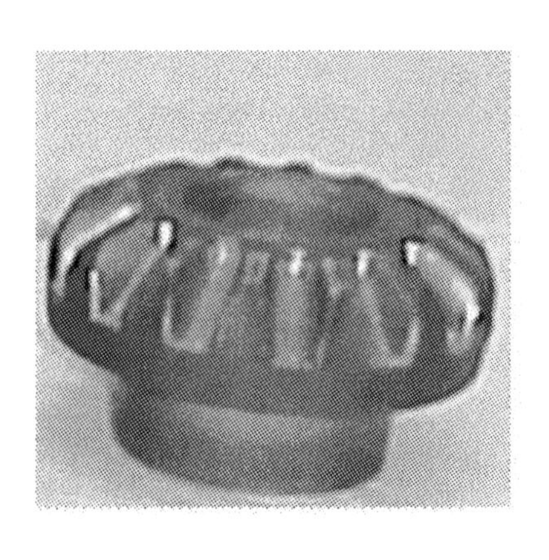

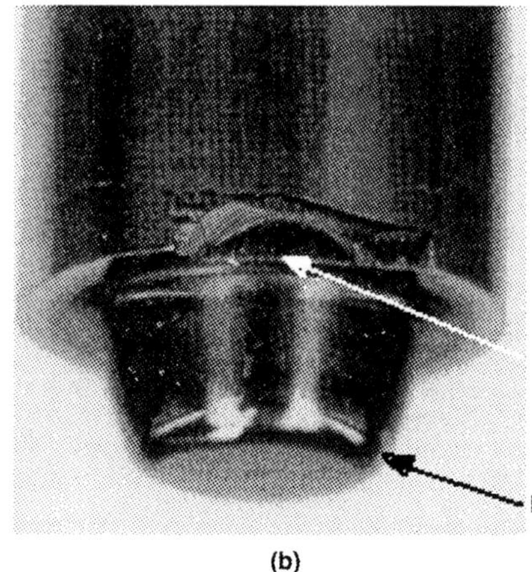

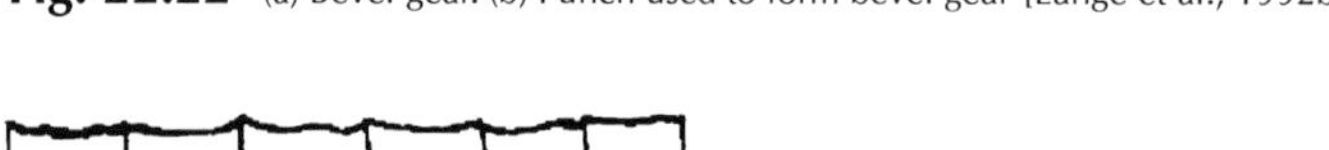

(a) (b)

Fig. 22.22 (a) Bevel gear. (b) Punch used to form bevel gear [Lange et al., 1992b]

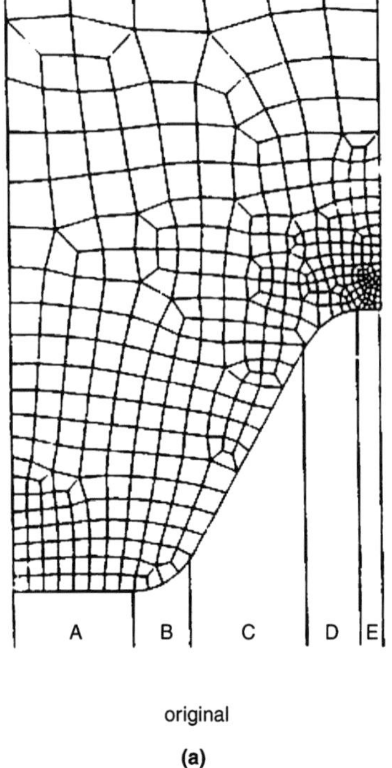

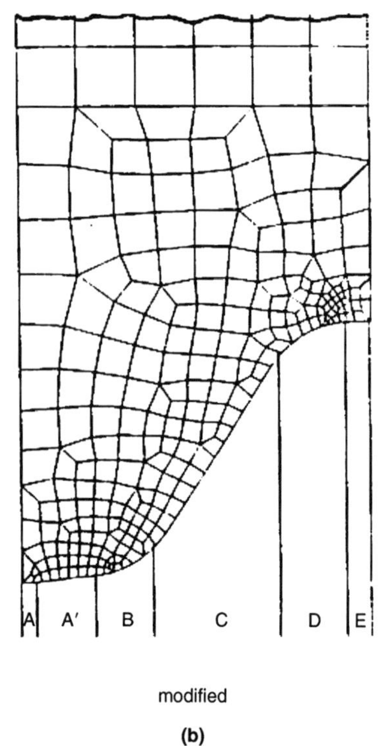

(a) (b)

Fig. 22.23 (a) Finite-element model of original punch geometry. (b) Finite-element model of modified punch geometry (A, face; B, lower punch corner; C, cone angle; D, fillet radius; E, edge) [Lange et al., 1992b]

fracture was experienced under production conditions. Once again, there is a localized plastic zone and tensile maximum principal stresses in this region. Both of these conditions contribute to the onset of fatigue fracture. Similar results were obtained in the flange region [Nagao et al., 1994].

In order to further verify the fatigue analysis method, tooling capable of developing a stress state similar to the actual tooling (Fig. 22.14) was manufactured and tested at Honda Engineering Co.

With this tooling, cracks were experienced in the flange area after 7000 parts. Once again, fatigue analysis of the tooling showed a large plastic zone and tensile maximum principal stresses in the flange area (Fig. 22.16). In order to improve tool life, a new tooling geometry with a profiled container was designed (Fig. 22.17). The results from the fatigue analysis on this tooling showed that there was no plastic zone in the flange area, and the maximum principal stresses remained compressive. With this new tooling geometry, 21,000 parts were forged without experiencing any cracks.

22.7.3 Knockout Pin Insert—A Case Study

This case study analyzed a knockout pin insert that was used to indent a hole on the bottom of a workpiece in a cold heading operation. The

Table 22.3 Upset forging conditions used in die wear tests [Bobke, 1991, and Luig, 1993]

Press
Type: mechanical
Capacity: 3.09 MN (347 tonf)
Stroke: 7 in. (180 mm)
Stroke rate: 2/s
Cycle time(a): 13 s
Billet
Material: 1045 steel
Temperature: 2010 °F (1100 °C)
Diameter: 0.8 in. (20 mm)
Height: 1.2 in. (30 mm)
Dies
Material: H-10
Coating: various
Temperature: 430/570 °F (220/300 °C)
Lubrication: various

(a) The cycle time included (Fig. 22.7) placing the billet on the lower die, the billet resting on the lower die, the forging stroke, the forged billet resting on the lower die, removal of the forged billet, and lubrication (when applied).

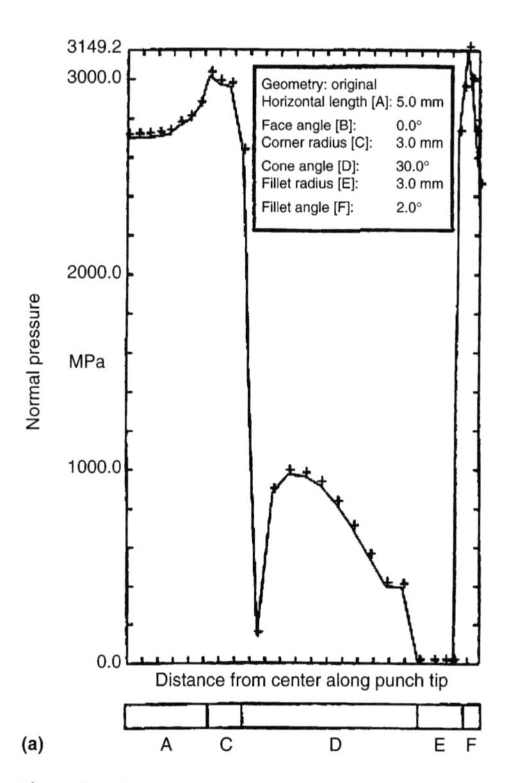

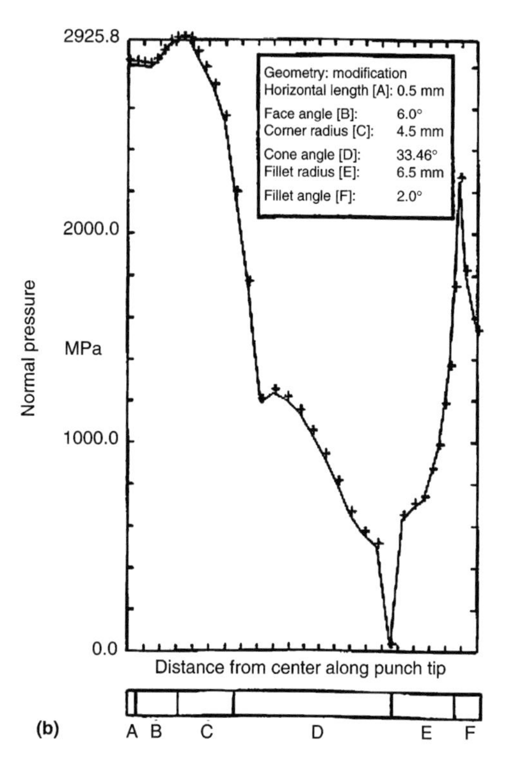

Fig. 22.24 (a) Normal stress distribution for original punch geometry. (b) Normal stress distribution for modified punch geometry (A, face; B, lower punch corner; C, cone angle; D, fillet radius; E, edge) [Lange et al., 1992b]

insert and original pin design are shown in Fig. 22.18 along with the point of failure. The original insert had an average tool life of 53,000 parts [Hannan et al., 2001].

Figure 22.19 shows the FEM model of the knockout pin operation used for the fatigue analysis. Figure 22.20 shows that there is a large tensile maximum principal stress in the region of the failure point [Hannan et al., 2001].

One design change analyzed in this case study was to split the pin insert where it was failing. The design is shown in Fig. 22.21. The radius where failure occurs is eliminated using a straight pin (E) and a sleeve (D). This tool was manufactured for a production run. The resulting tool life increased from 53,000 parts to 200,000 parts. This is about a 300% increase in tool life [Hannan et al., 2001].

22.7.4 *Bevel Gear Punch—A Case Study*

This case study analyzed the fatigue fracture of a punch (Fig. 22.22b) used in the cold forging

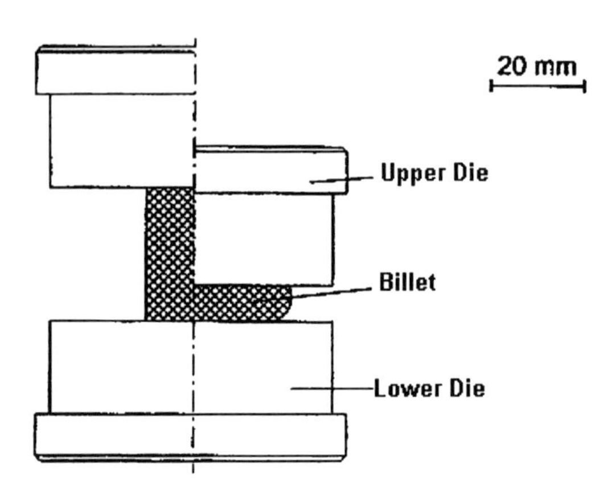

Fig. 22.25 Upset forging process [Luig, 1993]

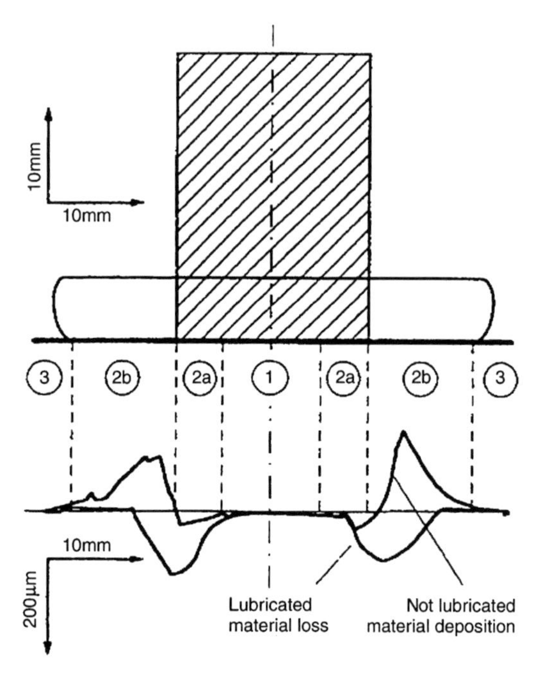

Fig. 22.26 Typical wear profile for upset forging [Doege et al., 1996]

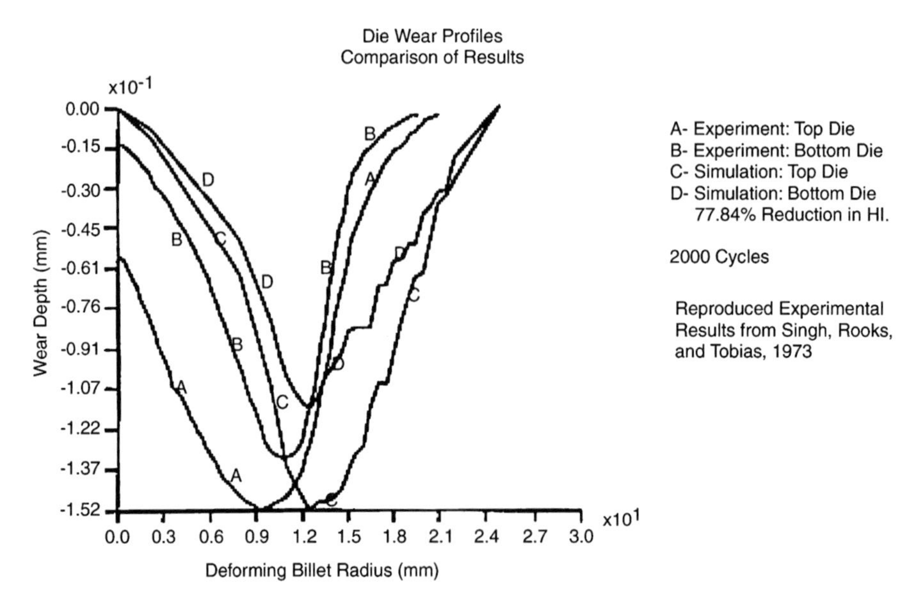

Fig. 22.27 Comparison between predicted and experimental die wear [Liou et al., 1988]

of a straight bevel gear (Fig. 22.22a). The finite-element model is shown in Fig. 22.23(a). Fatigue analysis of the punch showed that very high stresses were generated in the lower punch corner and the upper fillet radius (Fig. 22.24a). By modifying the punch geometry (Fig. 22.23b), it was possible to reduce the peak stresses, especially at the upper fillet radius where the fatigue fracture occurred (Fig. 22.24b). As a result of these geometric changes, the punch life was increased by a factor of 6 to 8 [Lange et al., 1992b].

22.8 Prediction of Die Wear and Enhancement of Die Life in Hot Forging Using FEM

Several studies have been conducted in an attempt to estimate die life in hot forging [Liou et al., 1988]. The ability to predict die wear allows for the optimization of process variables such that die life is improved [Tulsyan et al., 1993].

Many researchers have applied FEM to estimate die wear in hot forging. In general, all of

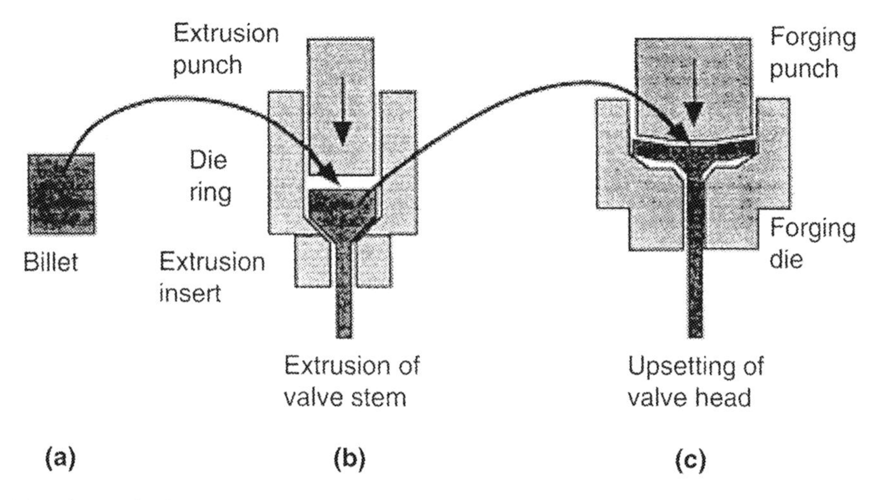

Fig. 22.28 Exhaust valve forming process. (a) Initial billet. (b) Extrusion of valve stem. (c) Upsetting of valve head [Tulsyan et al., 1993]

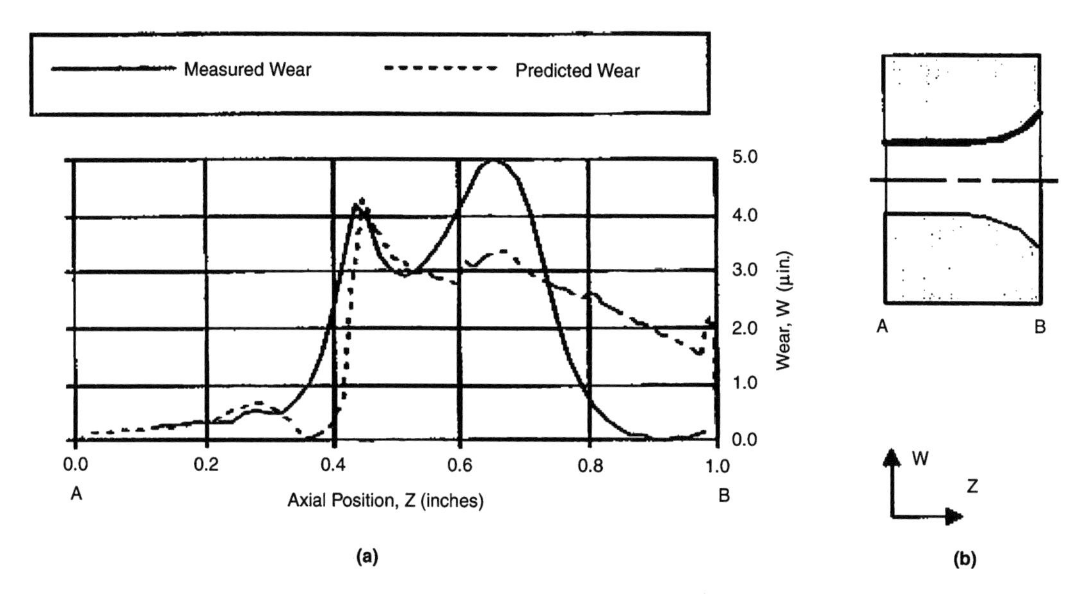

Fig. 22.29 (a) Comparison between predicted and experimental die wear. (b) Axial position along extrusion die throat (picture is rotated 90°) [Painter et al., 1995]

the results have been similar. The general profiles of the predictions have agreed with experiment; however, predicting the exact magnitude of die wear is difficult [Dahl et al., 1999, and Lee and Jou, 2003].

The prediction of die wear (fatigue fracture as well) with FEM has the following specific advantages [Painter et al., 1995]:

- Die changes can be scheduled, based on estimated die lives, in order to reduce unexpected machine downtime.
- Forging parameters such as press speed, die materials, and workpiece and die temperatures can be optimized to increase die life in an economical manner, as opposed to expensive experimental studies.
- The effects of die geometry changes on die wear can be rapidly investigated, again avoiding the high cost of experimental studies.

22.8.1 *High-Speed Hot Upset Forging—A Case Study*

In hot upset forging, the temperature distributions at the top and bottom surfaces are not symmetric. Due to the free resting of the workpiece on the bottom die prior to deformation of the billet, heat is transferred from the billet to the lower die. The different temperatures on the billet surfaces result in an asymmetric flow pattern of the deforming workpiece. This phenomenon, along with the asymmetric temperature distributions on the die surfaces, strongly affects die wear behavior on the two surfaces [Liou et al., 1988].

Experimental Wear Measurement. Measurements of die wear in a simple upsetting operation were made at the IFUM at the University of Hannover in Germany, using the conditions given in Table 22.3 [Bobke, 1991, and Luig, 1993].

Figure 22.25 illustrates the upset forging process. Typical wear profiles after upsetting are shown in Fig. 22.26.

Analytical Wear Prediction. In order to predict die wear using FEM analysis, adhesive wear was considered to be the dominant wear mechanism. Therefore, a differential form of Holm's wear equation was adopted:

$$dV = K_{adh} \cdot \frac{dW \cdot dL}{H} = K_{adh} \cdot \frac{p \cdot dA \cdot U \cdot dt}{H} = dZ \cdot dA \quad \text{(Eq 22.2)}$$

Table 22.4 Summary of K_{abr} and K_{adh} values in extrusion die wear [Painter et al., 1995]

Die set	Die material	K_{abr}	K_{adh}
A	H-11 tool steel	3.2×10^{-6}	2.6×10^{-3}
B	Silicone nitride ceramic	6.0×10^{-7}	7.0×10^{-3}
C	H-11 tool steel	1.5×10^{-6}	2.0×10^{-3}

Table 22.5 Upset forging conditions used in die wear tests [Bobke, 1991, and Luig, 1993]

Press

Type: mechanical
Capacity: 3.09 MN (347 tonf)
Stroke: 7.1 in. (180 mm)
Stroke rate: 2/s
Cycle time(a): 13 s

Billet

Material: 1045 steel
Temperature: 595 °F (1100 °C)
Diameter: 1.2 in. (30 mm)
Heght: 1.6 in. (40 mm)

Dies

Material: H-10, H-12, H-13
Hardness: 0.17/0.22/0.25 psi (1200/1500/1700 Pa)
Temperature: 285/430/570 °F (140/220/300 °C)
Lubrication: various

(a) The cycle time included (Fig. 22.7) placing the billet on the lower die, the billet resting on the lower die, the forging stroke, the forged billet resting on the lower die, removal of the forged billet, and lubrication (when applied).

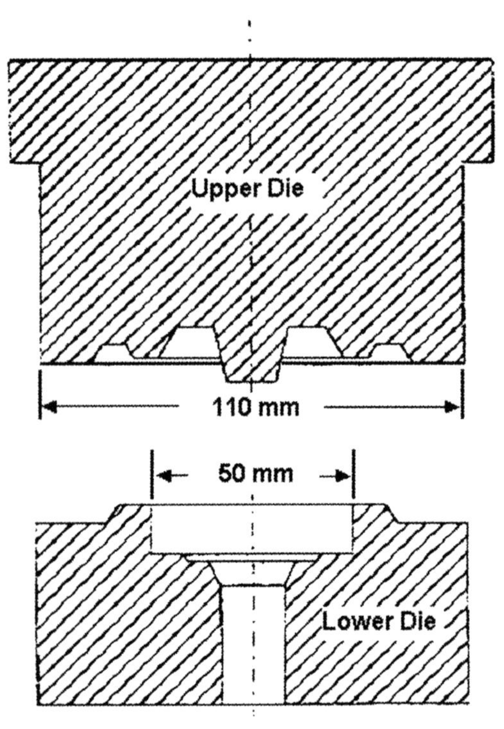

Fig. 22.30 Hot precision forging die [Bobke, 1991]

where dV is the wear volume and is equal to dZ · dA (wear depth · contact area), dW is the normal load to compress the billet and is equal to p · dA (interface pressure · contact area), dL is the sliding distance and is equal to U · dt (sliding velocity · time), H is the die hardness, and K_{adh} is the dimensionless adhesive wear coefficient. It should be noted that the interface pressure and the sliding velocity are local values. Thus, the wear depth can be predicted as:

$$Z_i = \sum_{j=1}^{M} K_{adh} \cdot \frac{p_{ij} \cdot U_{ij} \cdot \Delta t}{H_{ij}} \ldots i = 0, N \qquad \text{(Eq 22.3)}$$

where Z_i is the accumulated wear depth at the ith node, Δt is the time interval of the calculation step, M is the total number of calculation steps, N is the total number of nodes on the die surface, i represents the ith node, and j represents the jth time step [Liou et al., 1988].

In this case, the FEM simulation of the hot upset forging process was divided into ten steps. On completion of each step, the parameters needed for the die wear model (Eq 22.3) were extracted from the FEM simulation. In addition, based on the approximations given in Table 22.1, the wear coefficient was assumed to be $K_{adh} = 7.0 \times 10^{-5}$. With this data, Eq 22.3 was used to calculate the predicted die wear at each contact node on the upper and lower die for the ten simulation steps of one deformation cycle [Liou et al., 1988].

In order to compare the die wear predictions to experimental results, the predicted results were scaled up to 2000 forging cycles. Figure 22.27 shows the comparison between the predicted and experimental die wear.

It is immediately noted that the maximum die wear depth was greater on the top die than on the bottom die. As discussed previously, this may be attributed to asymmetric temperature,

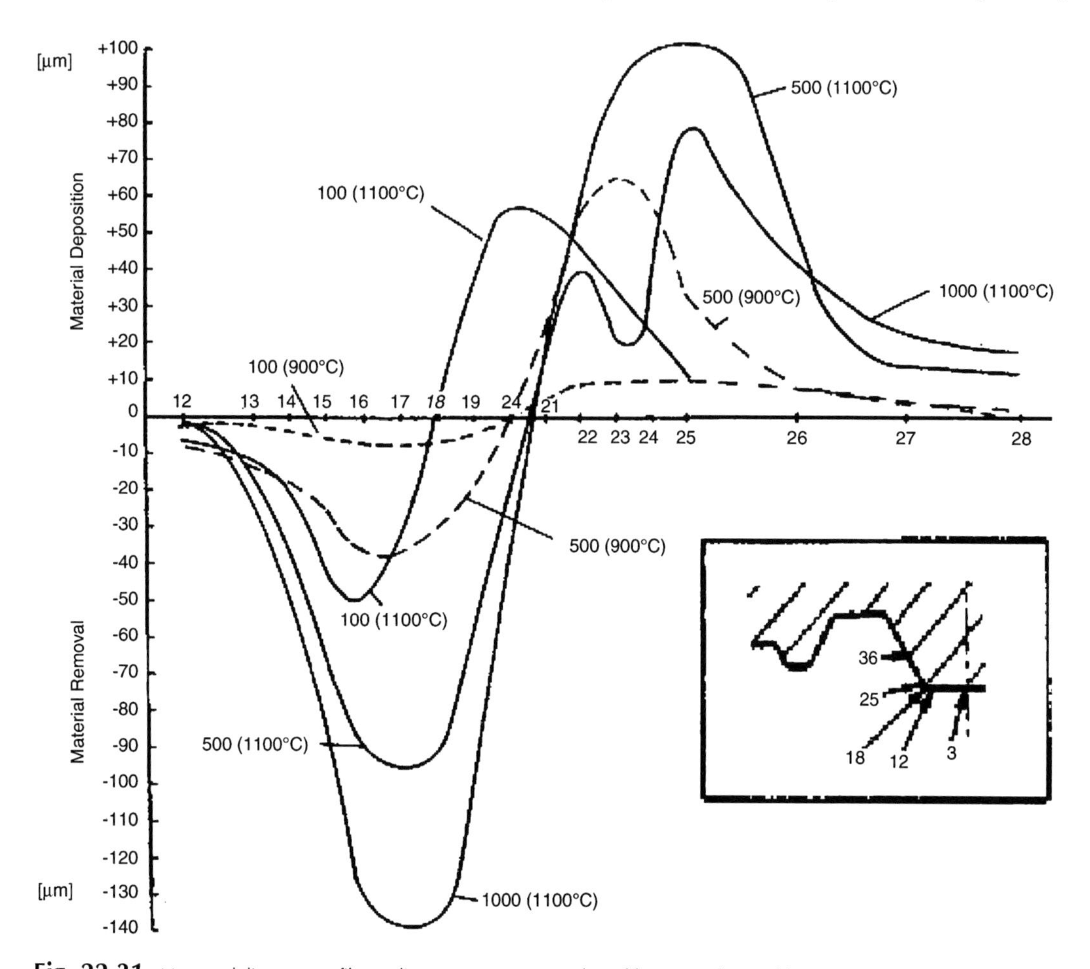

Fig. 22.31 Measured die wear profile vs. die temperature vs. number of forging cycles. [Bobke, 1991]

sliding velocity, and interface pressure distribution on the top and bottom die due to the die-chilling effect of the free resting billet on the lower die prior to deformation. The figure also shows that the die wear profile predicted by the model closely resembled experimental results.

The difference between the maximum wear depth on the bottom die predicted by the model and the experiment was explained by considering the effect of consecutive forging cycles. In this analysis, FEM simulation of the heat transfer for the postfree resting and empty periods of the first upsetting cycle was performed. The temperature fields in the top and bottom die were then used as the initial temperatures for the second upsetting simulation and so on. Simulation of three consecutive forging cycles predicted a temperature increase in the bottom die but not in the top die. Because the temperature of the bottom die increases, its hardness decreases, and thus, wear increases. This explains why scaling the wear predicted on the bottom die for the first forging cycle resulted in a maximum wear depth less than that indicated by experiment.

22.8.2 Extrusion and Forging of Exhaust Valves—A Case Study

As Fig. 22.28 shows, exhaust valves are typically formed in two stages. In the first stage, the valve stem is formed by extruding the bottom of the billet, and in the second stage, the valve head is formed by upsetting the top of the billet. The process is generally completed in a 500 to 700 ton mechanical press at 45 to 60 strokes per minute, with the billet heated to 2000 to 2100 °F (1100 to 1150 °C) [Tulsyan et al., 1993].

Experimental Wear Measurement. Measurements of die wear in the extrusion operation of the exhaust valve forming process were performed. The experiments were performed with three different extrusion dies. Both the die geometry and the die material varied from die to die because it was expected that different wear mechanisms would dominate in each die. Table 22.4 summarizes the die materials used for the experiments. The typical wear profile after extrusion is shown in Fig. 22.29. Because there is relatively little movement between the workpiece and the die during the second stage, the die wear measurement of the extrusion die used in the first stage was the primary focus of this study [Painter et al., 1995].

Analytical Wear Prediction. In high-speed hot extrusion and forging of exhaust valves, abrasive wear is assumed to be the dominant wear mechanism. Because of the success of the Holm's-based adhesive wear equation for high-speed hot upsetting discussed previously, a similar model was established for abrasive wear:

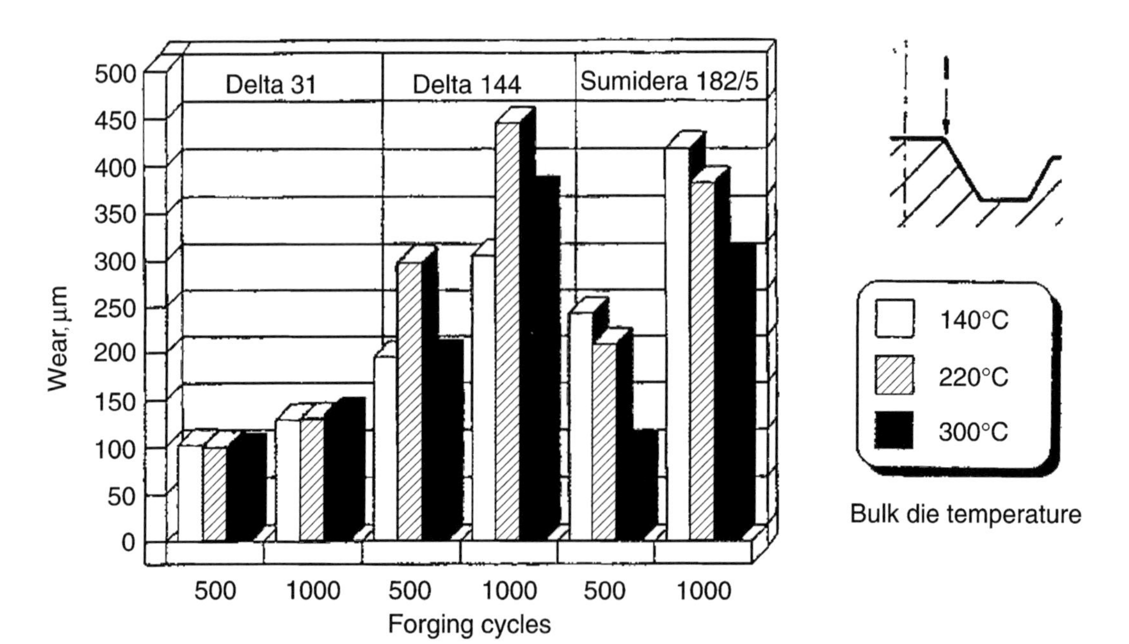

Fig. 22.32 Measured die wear at location I vs. die temperature vs. lubricant vs. number of forging cycles [Bobke, 1991]

$$dV = K_{abr} \cdot \frac{dW \cdot dL}{H^m} = K_{abr}$$

$$\cdot \frac{p \cdot dA \cdot U \cdot dt}{H^m} = dZ \cdot dA \qquad \text{(Eq 22.4)}$$

where all the variables are the same, except K_{abr} is the dimensionless abrasive wear coefficient, and m is an exponent equal to 2 for steels. Thus, the wear depth can be predicted as:

$$Z_i = \sum_{j=1}^{M} K_{abr} \cdot \frac{p_{ij} \cdot U_{ij} \cdot \Delta t}{H_{ij}^m} \ldots i = 0, N \qquad \text{(Eq 22.5)}$$

where all the variables are as defined previously [Tulsyan et al., 1993].

In this case, the FEM simulation was divided into 18 steps. On completion of each step, the parameters needed for both the adhesive wear model (Eq 22.3) and the abrasive wear model (Eq 22.5) were extracted from the FEM simulation, and both the adhesive and the abrasive wear depths were predicted, as discussed previously. Table 22.4 outlines the K_{abr} and K_{adh} values [Painter et al., 1995].

A comparison of the results from the experimental die wear measurements and the FEM die wear predictions showed that abrasive wear was the dominant wear mechanism in extrusion with ceramic dies, while adhesion was the dominant wear mechanism in extrusion with tool steel dies. In order to make this comparison, the experimental results were scaled down to one forging cycle. Figure 22.29 shows the comparison of the results for the ceramic extrusion die. The predicted die wear profile matched the die wear experiment; however, the die wear magnitudes did not exactly match because wear mechanisms other than abrasion were also acting [Painter et al., 1995].

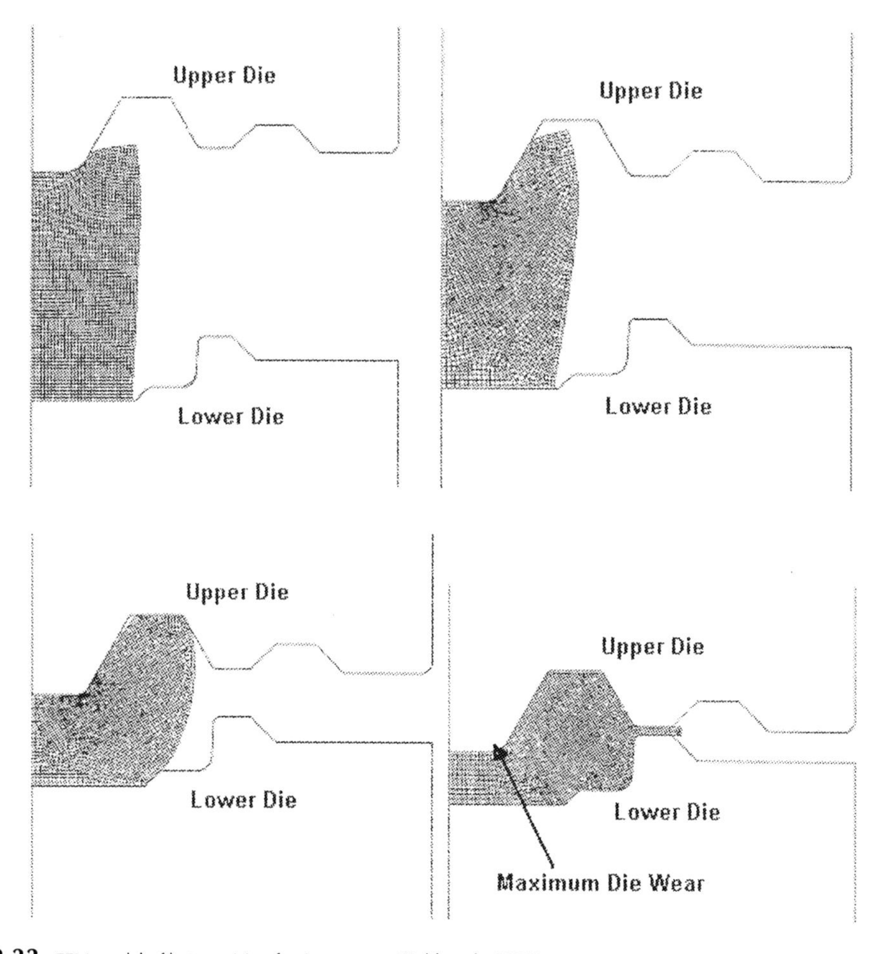

Fig. 22.33 FEM model of hot precision forging process [Dahl et al., 1999]

22.8.3 *Hot Precision Forging—A Case Study*

Hot precision forging of complex shapes is the focus of many manufacturers. The tolerances on these parts are especially critical. Therefore, die wear is exceedingly detrimental to these processes. Thus, prediction of die wear and enhancement of die life is very important to these manufacturers.

Experimental Wear Measurement. Measurement of die wear for the hot precision forging process was carried out at the IFUM at the University of Hannover in Germany, under the conditions given in Table 22.5 [Bobke, 1991, and Luig, 1993].

Figure 22.30 shows the precision forging die used in the experiments. Typical wear profiles after forging are shown in Fig. 22.31. The experimental wear depths for three different lubricants are summarized in Fig. 22.32.

Analytical Wear Prediction. It was assumed that abrasive wear was the dominant wear mechanism in hot precision forging. Therefore, the wear model in Eq 22.5 was used [Dahl et al., 1999].

Figure 22.33 shows the FEM model of the hot precision forging process. On completion of each step, the parameters needed for the die wear model (Eq 22.5) were extracted from the FEM simulation, and the wear depth was predicted, as discussed previously. Based on typical experimental values for two-body abrasion, the wear coefficient was assumed to be $K_{abr} = 0.02$ [Dahl et al., 1999].

The maximum predicted die wear and the maximum measured die wear both occurred at the corner of the upper die (Fig. 22.31, measurement location 18). In general, the model predicted that die wear increases with decreasing corner radius, decreases with increasing friction factor, and increases with decreasing die hardness (i.e., different materials have different temperature-hardness curves). This illustrates the advantage of being able to predict die wear. It is not intuitive that die wear would decrease with increasing friction factor. In the case of the valve extrusion, it was shown that die wear increases with increasing friction factor because of the increased pressure developed on the die. In this case, die wear decreases because increased friction acts to decrease the sliding length [Dahl et al., 1999].

REFERENCES

[Altan et al., 1983]: Altan, T., Oh, S., Gegel, H., *Metal Forming Fundamentals and Applications,* American Society for Metals, 1983.

[Artinger, 1992]: Artinger, I., "Material Science and Tool-Life," Budapest, 1992 (in Hungarian).

[Aston et al., 1969]: Aston, J., and Barry, E., "A Further Consideration of Factors Affecting the Life of Drop Forging Dies," *J. Iron Steel Inst.,* Vol 210 (No. 7), July 1969, p 520–526.

[Bay, 2002]: Bay, N., class notes, Technical University of Denmark, 2002.

[Bobke, 1991]: Bobke, T., "Phenomenon of Edge Layer at Sealing Process of Drop Forge Tools," No. 237, IFUM, University of Hannover, 1991.

[Dahl et al., 1998]: Dahl, C., Vazquez, V., Altan, T., "Effect of Process Parameters on Die Life and Die Failure in Precision Forging," Report No. PF/ERC/NSM-98-R-15, Engineering Research Center for Net Shape Manufacturing, April 1998.

[Dahl et al., 1999]: Dahl, C., Vazquez, V., Altan, T., "Analysis and Prediction of Die Wear in Precision Forging Operations," Report No. PF/ERC/NSM-99-R-21, Engineering Research Center for Net Shape Manufacturing, May 1999.

[Doege et al., 1996]: Doege, E., Seidel, R., and Romanowski, C., "Increasing Tool Life Quantity in Die Forging: Chances and Limits of Tribological Measures," NAMRC, 1996.

[Hannan et al., 2001]: Hannan, D., Ngaile, G., Altan, T., "Elimination of Defects and Improvement of Tool Life in Cold Forging—Case Studies," New Developments in Forging Conference (Stuttgart, Germany), Institute for Metal Forming Technology (IFU), May 2001.

[Hettig, 1990]: Hettig, A., "Influencing Variables of Tool Fracture at Impact Extrusion," No. 106, Reports from the Institute of Metal Forming, University of Stuttgart, Springer, 1990.

[Kesavapandian et al., 2001]: Kesavapandian, G., Ngaile, G., Altan, T., "Die Wear in Precision Hot Forging—Effect of Process Parameters and Predictive Models," Report No. PF/ERC/NSM-01-R-42, Engineering Research Center for Net Shape Manufacturing, Aug 2001.

[Knoerr et al., 1989]: Knoerr, M., and Shivpuri, R., "Failure in Forging Dies," Report No. ERC/NSM-B-89-15, Engineering Research Center for Net Shape Manufacturing, March 1989.

[Knoerr et al., 1994]: Knoerr, M., Lange, K., Altan, T., "Fatigue Failure of Cold Forging Tooling: Causes and Possible Solutions Through Fatigue Analysis," *J. of Materials Processing Technology,* Vol 46, 1994, p 57–71.

[Lange et al., 1992a]: Lange, K., Cser, L., Geiger, M., Kals, J.A.G., Hansel, M., "Tool Life and Tool Quality in Bulk Metal Forming," *Annals of the CIRP,* 1992, p 667–675.

[Lange et al., 1992b]: Lange, K., Hettig, A., Knoerr, M., "Increasing Tool Life in Cold Forging Through Advanced Design and Tool Manufacturing Techniques," *J. of Materials Processing Technology,* Vol 35, 1992, p 495–513.

[Lee and Jou, 2003]: Lee, R.S., and Jou, J.L., "Application of Numerical Simulation for Wear Analysis of Warm Forging Die," *Journal of Material Processing Technology,* Vol. 140, 2003, p 43–48.

[Liou et al., 1988]: Liou, M., and Hsiao, H., "Prediction of Die Wear in High Speed Hot Upset Forging," Report No. ERC/NSM-88-33, Engineering Research Center for Net Shape Manufacturing, Oct 1988.

[Liug, 1993]: Luig, H., "Influence of Wear Protective Coating and Scaling of Raw Parts on Forging Wear," No. 315, IFUM, University of Hannover, 1993.

[Matsuda, 2002]: Matsuda, T., "Improvement of Tool Life in Cold Forging," *Advanced Technology of Plasticity,* Vol 1, 2002, p 73–78.

[Nagao et al., 1994]: Nagao, Y., Knoerr, M., Altan, T., "Improvement of Tool Life in Cold Forging of Complex Automotive Parts," *J. of Materials Processing Technology,* Vol 46, 1994, p 73–85.

[Painter et al., 1995]: Painter, B., Shivpuri, R., Altan, T., "Computer-Aided Techniques for the Prediction and Measurement of Die Wear during Hot Forging of Automotive Exhaust Valves," Report No. ERC/NSM-B-95-06, Engineering Research Center for Net Shape Manufacturing, Feb 1995.

[Schey, 1983]: Schey, J., *Tribology in Metalworking: Friction, Lubrication, and Wear,* American Society of Metals, 1983.

[Shivpuri et al., 1988]: Shivpuri, R., and Semiatin, S., "Wear of Dies and Molds in Net Shape Manufacturing," Report No. ERC/NSM-88-05, Engineering Research Center for Net Shape Manufacturing, June 1988.

[Tulsyan et al., 1993]: Tulsyan, R., Shivpuri, R., Altan, T., "Investigation of Die Wear in Extrusion and Forging of Exhaust Valves," Report No. ERC/NSM-B-93-28, Engineering Research Center for Net Shape Manufacturing, Aug 1993.

CHAPTER 23

Near-Net Shape Forging and New Developments

Manas Shirgaokar
Gracious Ngaile

23.1 Introduction

Net and near-net shape forging companies generally produce for the automotive industry. These components are characterized by close dimensional tolerances, minimal draft, and the absence of flash during forging, with minimal or no postforging machining. This is a major advantage when machining costs are taken into consideration. Net shape forging can be defined as the process of forging components to final dimensions with no postforging machining necessary. Near-net shape forged components, on the other hand, are forged as close as possible to the final dimensions of the desired part, with little machining or only grinding after forging and heat treatment.

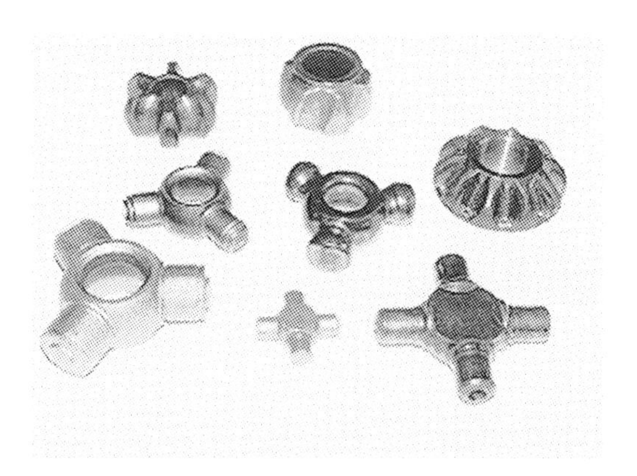

Fig. 23.1 Examples of net shape forged parts [Meidert et al., 2000]

The possibility of eliminating the high costs of machining, combined with the inherent advantages of cold forging regarding material strength and finish, make near-net forging very attractive to the automotive industry.

Examples of typical near-net shape parts are given in Fig. 23.1. These parts are characterized by complex geometries and very close tolerances, in the range of 0.0004 to 0.002 in. (0.01 to 0.05 mm), at complicated and functionally important surfaces. The steering spider seen in the forefront in Fig. 23.1 is a typical net shape part. The other forgings require minor machining or grinding. However, functional areas, such as the gear tooth geometry of the bevel gear, are ready for assembly [Meidert et al., 2000].

Some of the parameters affecting the quality of near-net shape forged components are illustrated in Fig. 23.2.

23.2 Tolerances in Precision Forging

To reduce cost, it is often desirable to replace conventional forging that requires several subsequent machining and finishing operations with precision forging, i.e., cold and warm forging, which yields parts with good dimensional tolerances.

The current minimum dimensional error in practical cold forging is ± 20 to 50 μm, while the error in machining has been reduced over the years to less than ± 1 μm. For cold forging processes to compete with machining, the dimensional accuracy would have to be increased to within ±10 μm. Some of the causes of dimensional variation in precision forging are [Osakada, 1999]:

- *Die manufacturing:* The dimensional accuracy of the dies directly influences that of the parts being produced with them. Hence, die manufacturing is a very crucial part of the manufacturing sequence of precision parts. The use of electrical discharge machining (EDM) and wire EDM machines in manufacturing cold forging dies has considerably improved the accuracy of forging dies.
- *Elastic deflection of the press and tools:* When the forming load is applied, the press and the tools undergo elastic deflection, thus affecting the final tolerances on the part being forged.
- *Variation of process conditions:* In practical situations, process variables such as forming pressure, lubrication conditions, billet dimensions, and material properties do not remain constant. These fluctuations affect the dimensional accuracy of the forming process. The billet dimensions and material properties (flow stress) can be controlled by involving the suppliers in the design process to ensure procurement of billets with consistent dimensions and flow stress. The variation in lubrication conditions can be reduced by using lubricants with consistent friction coefficients.

23.2.1 Precision Die Manufacturing

Electrical Discharge Machining. In precision forging, cemented carbide is the die material of choice, rather than high-speed steel, because of its high stiffness, low thermal effect, and high wear resistance. A flow chart for manufacturing precision dies is shown in Fig. 23.3

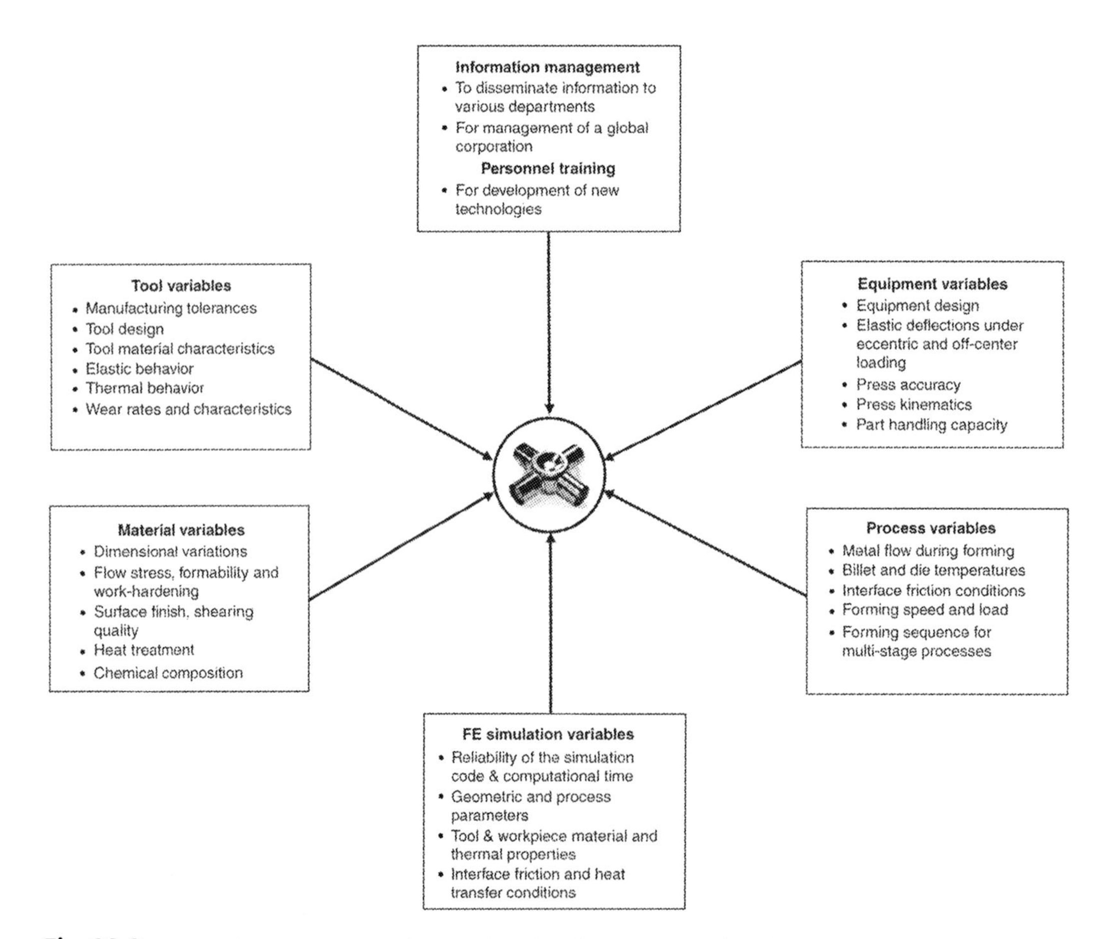

Fig. 23.2 Factors influencing the quality of near-net shape forged components. FE, finite element [Meidert et al., 2000]

[Yoshimura et al., 1997]. The die insert impression is first made by rough grinding, followed by electrical discharge die sinking. The compression ring is produced by machining and milling, followed by hardening or quenching to increase its stiffness. The die insert is then shrink-fitted into the compression ring, and the assembled ring is machined for finishing. The die insert impression is made by die sinking using a master electrode, followed by lapping to obtain a mirrored finish.

Surface coating is done to increase surface hardness on the working portion of the tools, so as to reduce wear and damage in that area. Crack initiation is delayed due to the hardness of the coating, which prevents any lubricant or workpiece material from getting wedged in these cracks. Two commonly used methods of surface coating are [Yoshimura et al., 1997]:

- Chemical vapor deposition: TiC, TiCN, TiN, Al_2O_3
- Physical vapor deposition: TiN

23.2.2 Die Deflection in Net Shape Forging

The stages of process planning, tool design, and manufacture in net shape forming require not only the use of a highly qualified engineer and sophisticated computer-aided design/computer-aided manufacturing (CAD/CAM) packages but also strict control on the tool and press settings. Process design in precision forging usually requires extensive trials before the optimum settings can be identified to yield the desired part tolerances. Thus, it is possible to compensate for the errors in the dimensions by evaluating the various factors affecting the part tolerances. Some of the phenomena affecting die deflection/deformation are listed below in the order that reflects the increasing dependence on production run-time [Kocanda et al., 1996]:

- Elastic deflection is a reversible change in dimension brought about by the applied load. It can be controlled by appropriate calculations and design.

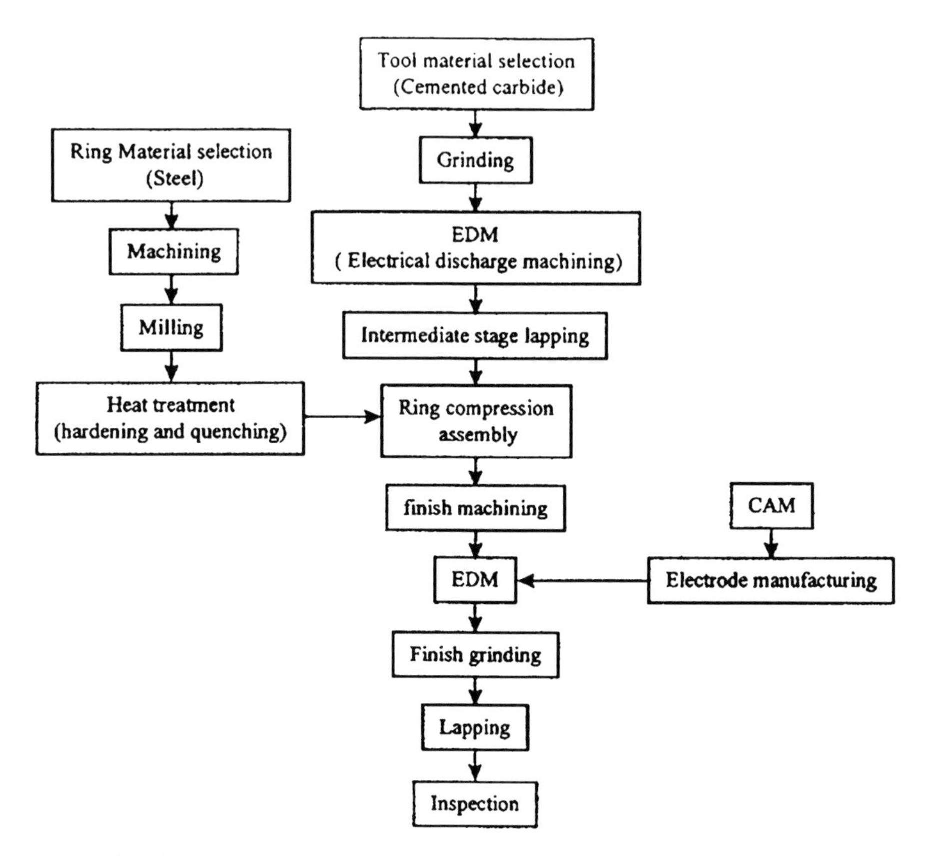

Fig. 23.3 Flow chart of die manufacturing by electrical discharge machining (EDM). CAM, computer-aided manufacturing [Yoshimura et al., 1997]

- Plastic deformation is a permanent deformation as a result of an unintended increase in forming load or stress concentration. This could be caused by improper tool alignment, inadequate lubrication, change in initial properties of the workpiece, etc.
- Cyclic softening or hardening causes significant changes in the stress-strain response of the tool steel. This phenomenon should not be ignored when the die geometry has stress concentration features.
- Thermal expansion of the tooling is a result of the heat generation during deformation.
- Relaxation and creep phenomena considerably decrease the prestresses in the die insert.
- Change in the modulus of elasticity may be caused by change in temperature of the tool steel. Generally, the modulus of elasticity decreases with increasing temperature.
- Cyclic plastic deformation, which occurs near areas of stress concentration, may result in initiation of microcracks.
- Softening of tool steel by unwanted tempering. Tempering is a function of both temperature and time and can occur during cold deformation processing at temperatures lower than the tempering temperature chosen for heat treatment of the tool.

23.2.3 Press Deflection in Net Shape Forging

Major factors that influence the part tolerances are the tool and machine deflections and the various process parameters, such as material properties, lubrication, temperatures, etc. While the process parameters and tool design have been the subject of extensive research, the influence of press deflection on the accuracy of the formed part and methodologies to compensate for these errors have yet to be standardized. Finite-element (FE) simulation has been employed to calculate the stress and deflection behavior of press components such as guide layout, press frame, and the construction of the crosshead [Doege et al., 1990]. However, the nonlinear characteristics of the load-deflection behavior between adjacent components, the clearances in the guiding system, and asymmetry in press assembly make it difficult to analyze the elastic behavior of the press. The horizontal offset and tilting, which occur during multistage forging or forging of complex parts, have a detrimental effect on the part tolerances. In order to quantify the elastic characteristics of the press, it is necessary to measure the load-deflection relationship of the ram under center and off-center loading conditions together with superimposed horizontal forces.

In one particular study [Balendra et al., 1996], a three-dimensional (3-D) FE simulation was conducted to determine the elastic characteristics of a 400 metric ton screw press. The simulation considered the modeling of the press components, appropriate definition of the joints, and the prestressing requirements for assembling the frame. The forming loads were obtained by conducting a two-dimensional closed-die, plane-strain, aerofoil forging simulation. This simulation enabled the determination of the horizontal offset and tilting of the forging dies for various orientations of the die parting line, and methods to minimize the component form errors due to elastic behavior of the press and location and the elastic behavior of the forging dies.

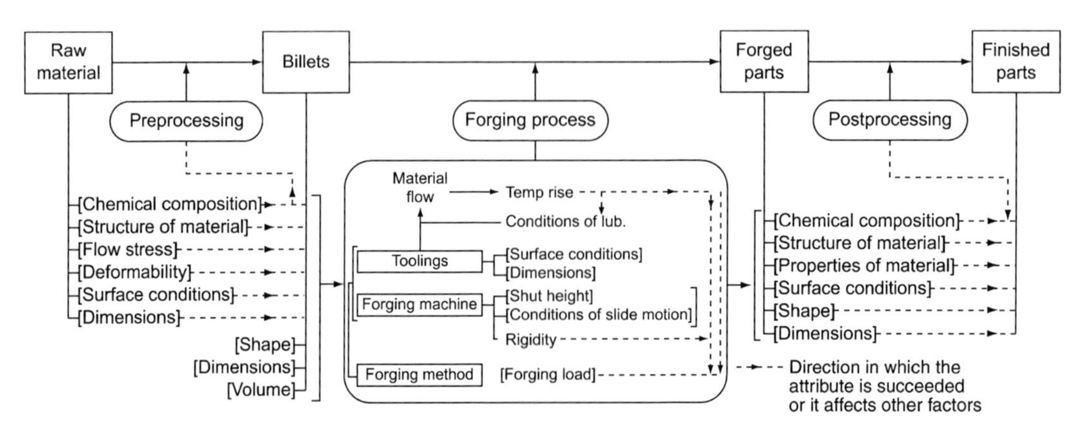

Fig. 23.4 Various stages and influencing factors of the forging process [Ishinaga, 1996]

23.2.4 Process Variables in Forging

Figure 23.4 shows the various stages in forging, with the parameters affecting part attributes and tolerances [Ishinaga, 1996]. The variable factors (enclosed in square brackets) affect each other and cause variations in the dimensions of the forged part. The entire forging process is divided into three stages:

- The preprocessing stage addresses the variables related to the billet. One way to control the part quality is to ensure minimum deviation in each of the billet factors, such as dimensions, flow stress, formability, etc.
- The forging process, which encompasses all the variables related to the forming process (temperature, lubrication, etc.), tooling, and equipment. Variations in the forging pressure would affect the thickness of the forging due to elastic deflection of the press and tools. Off-center loading not only affects the accuracy of the process but is also detrimental to tool life.
- The postprocessing stage deals with machining, surface finishing, or heat treatment of the forged parts and affects the material properties, surface conditions, or chemical composition of the part.

The growing demands for precision cold-forged products in the automotive industry led to the development of economical means to manufacture these parts. Apart from improvements in forging processes, there has been considerable progress in tooling and equipment design, materials engineering, and lubrication techniques in order to produce precise components. Some of these developments and future trends are discussed in the subsequent sections.

23.3 Advances in Tool Design

In order to produce precision-forged parts, factors related to the tooling, namely, elastic deflection, eccentricity, thermal expansion, etc., have to be considered during the process design stage. This is especially true in the case of cold forging, where precision die manufacturing is necessary to ensure the production of parts to close tolerances. The following factors have to be considered in die manufacturing [Yoshimura et al., 1997]:

- Proper equipment and process to ensure high accuracy
- Proper design of die structure to achieve high stiffness
- Proper heat and surface treatments to eliminate the heterogeneity of the die material
- Proper control of heating conditions to maintain constant temperature at the die

Tool life is also a major factor to be considered while quoting a part. Since the cost of manufacturing a die is considerable, simultaneous process design involving process planning and die design are essential to ensure that conditions such as loading pressure, stress concentration, etc. remain within the allowable limits of die performance. Thus, tool wear, tool fracture, and process failure have to be considered for cost reduction.

23.3.1 Enclosed- or Trapped-Die Forging

Enclosed-die forging uses multiple-action tooling as punches press the material in a pre-enclosed die to fill in the die space (Fig. 23.5). By controlling the motion of rams, metal flow can be controlled to obtain the optimum deformation. The ram motions for upper and lower punches can be set as synchronous, asynchronous, or with back pressure to reduce forming load or to improve the filling of material. Some of the advantages of enclosed-die forging are:

- The ability to form complex shapes in one process
- Elimination of flash, thus resulting in material savings. Since parts are formed to close dimensional tolerances, subsequent machining operations are either reduced or, in some cases, eliminated.
- Low forming load is available where the area penetrated by the punch is relatively small.

A Japanese tool supplier has recently developed a special die set for forging radially extruded parts and a family of bevel gears (Fig. 23.6 and 23.7) [Yamanaka et al., 2002]. The advantages obtained from this die set were high productivity, easy installation in nearly all presses, the requirement of only compressed air, and low initial cost.

A similar concept in die design, developed earlier, incorporates a pantograph, as shown in the schematic in Fig. 23.8 [Yoshimura et al., 1997 and 2000]. Hydraulic pressure generated by an external hydraulic unit and an accumulator is used for closing the upper and lower die to form the die cavity. The die-closing pressure is adjustable. However, if this pressure is too low,

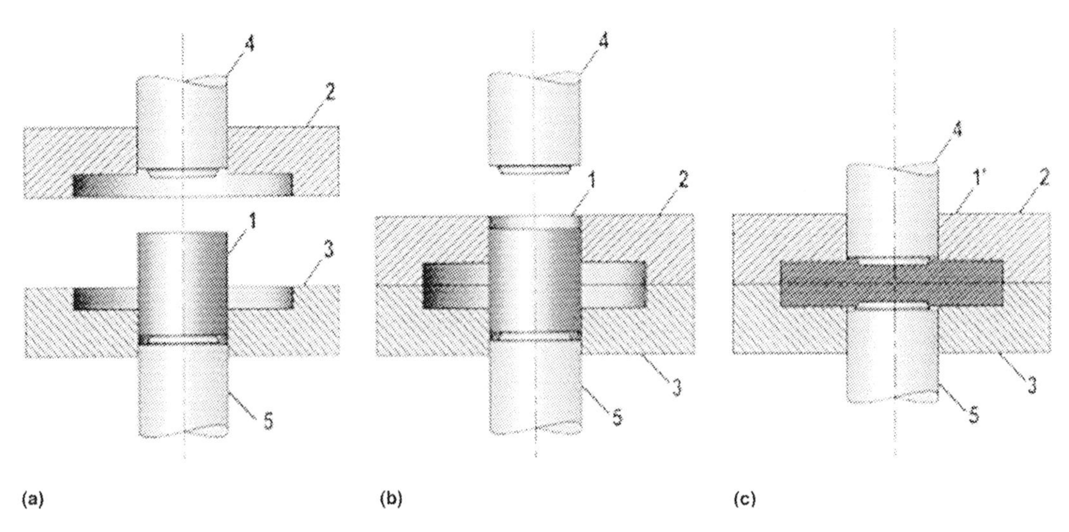

Fig. 23.5 Enclosed-die forging. 1, billet; 2, upper die; 3, lower die; 4, upper punch; 5, lower punch; 1′, finished part [Oudot et al., 2001]

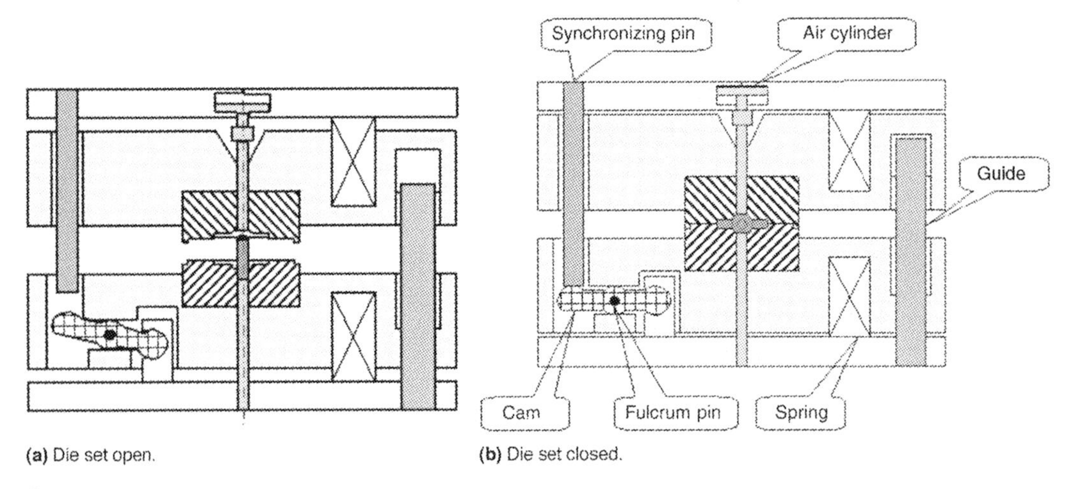

Fig. 23.6 Enclosed forging die set developed by Yamanaka Engineering [Yamanaka et al., 2002]

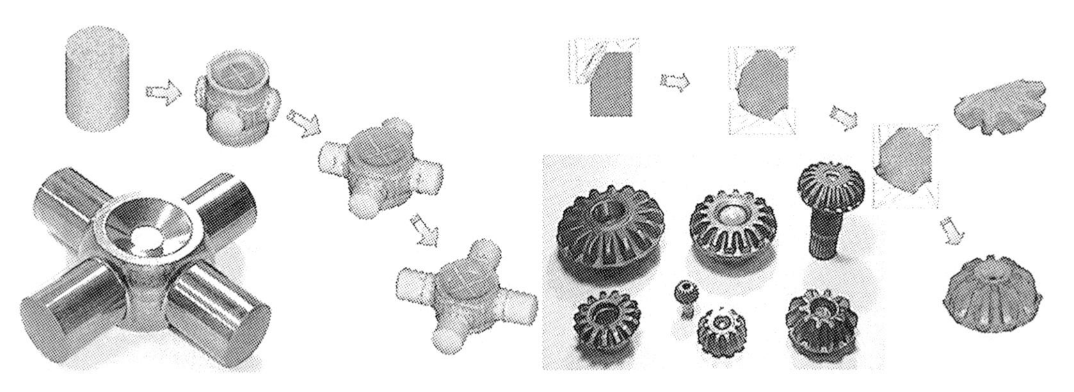

Fig. 23.7 Some example parts forged in an enclosed forging die set [Yamanaka et al., 2002]

it could result in flash formation in the die gap and cause chipping in the die. The die cavity, composed of the upper and lower dies, moves at half the punch speed by the pantograph mechanism in order to obtain uniform deformation of the material in the axial direction [Yoshimura et al., 2000]. Similar die designs have been developed at Institut für Unformtechnik (Institute for Metal Forming Technology) at the University of Stuttgart as well as by European die and press makers [Siegert et al., 2003].

23.3.2 Innovations in Compressive Prestressing of Die Inserts

An important factor in enhancing die life is the design of compressive prestressing container systems, which house the die inserts. The conventional prestressing container normally consists of single or double stress rings. Depending on the complexity of the part and the required tolerances, the compressive prestress generated may be too low. STRECON Technology has developed strip-wound radially prestressed containers with strength that is 2 to 3 times that of conventional stress rings [Groenbaek et al., 1997]. The high strength makes it possible to provide an optimum prestress for the die, leading to two- to tenfold improvement in the die life (Fig. 23.9a). Sometimes, in cold forging dies, fracture may occur because the conventional radial prestressing does not have any appreciable effect on the stress condition in the axial direction. To counteract this effect, STRECON has developed strip-wound containers with integrated axial prestressing (Fig. 23.9b) [Groenbaek et al., 1997].

The strip-wound containers are manufactured by winding a thin strip of high-strength steel around a core of tool steel or tungsten carbide. During the winding process, the steel strip is preloaded with a controlled winding tension. The core material has a structure and a hardness that can withstand high prestress and cyclic working load. The strip steel is developed especially for optimum combination of the physical and mechanical properties. Optimum stress distribution is obtained by varying the winding tensions from layer to layer. The prestressed condition in the coiled strip is equal to that of a conventional construction with 'several hundred' stress rings. Consequently, the strip-wound containers can be loaded with a higher internal pressure than a conventional multiple stress ring set before the material will deform plastically. Thus, it is possible to obtain a higher interference when fitting a die into a strip-wound container than into a conventional multiple stress ring set.

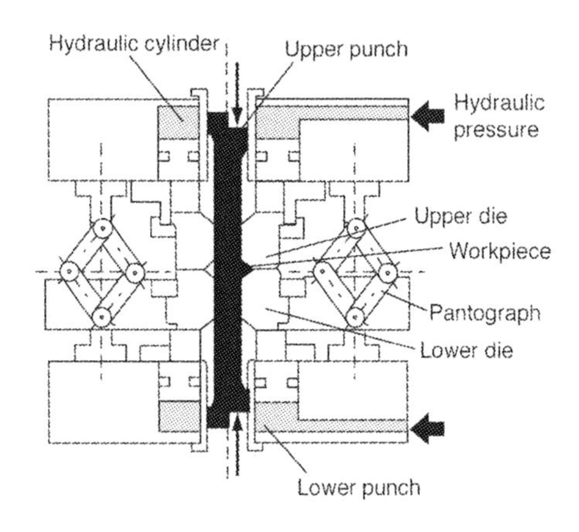

Fig. 23.8 Die set for enclosed-die forging developed by Nichidai Corp. [Yoshimura et al., 1997]

23.3.3 Reduction of Forging Pressure by Divided Flow Method

To reduce forging load and tool stresses, Kondo et al. have developed the divided flow method and applied this concept in forging a variety of parts, such as the gear parts shown in Fig. 23.10. There are two principles of flow relief shown in Fig. 23.10(a) and (b), namely, flow relief hole and flow relief axis. In the former, a blank with a relief hole is compressed by flat tools, resulting in a centripetal flow as an outcome of the hole shrinkage, thus creating divided flow. In the relief axis principle, a reduction in working pressure occurs due to two reasons: extrusion of the axis serves as flow relief and suppresses the increase in fractional reduction in area, and the formation of a friction hill is prevented due to the divided flow. Research shows that the relief hole principle is more suitable for working pressure reduction, since the resistance to flow increases gradually during forming.

Manufacture of a helical gear utilizing flow relief axis with back-up pressure is shown in Fig. 23.11 [Kondo, 1999 and 2002]. The forming process was completed in one step, because the inner plate moved down with the die plate until the preset backup load. The final part with the boss is also shown in Fig. 23.11.

23.4 Advances in Forging Machines

Net shape forging of complex parts, such as helical gears, helical-tooth pinions, etc., requires new concepts in press and tooling design, with consideration of a multitude of interacting variables such as those shown in Fig. 23.12. In order to increase the accuracy of the product and to extend the service life of the tools, press builders have developed multislide and multiaction hydraulic presses [Nakano, 1997, and Ishinaga, 1996].

23.4.1 Multislide Forging Press

Aida Engineering Co. has developed a multislide forging press for the purpose of minimizing the effect of off-center loading during multiprocess transfer forming. The features of this press are [Nakano, 1997]:

- Multislide construction: Each slide has its own independent slide.
- Time difference operation: Each slide strokes with a phase difference.

In model MF-7500, three independent slides operate with a phase difference of 30° and a capacity per slide of 280 tonf (2500 kN). The total capacity is 845 tonf (7500 kN) with different working timing. A schematic of the press with the slide and die area is shown in Fig. 23.13(a),

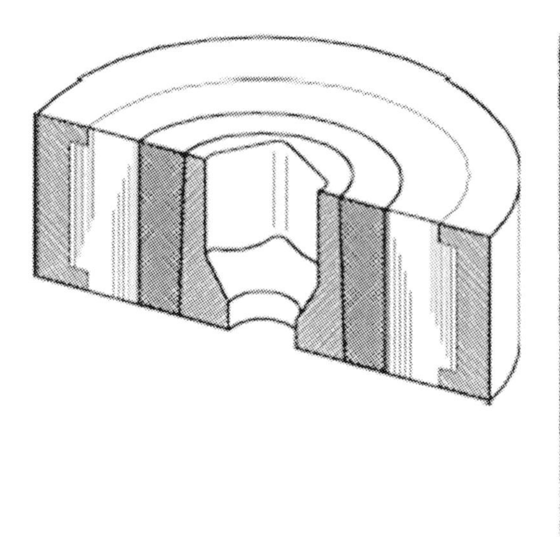

(a) Standard stripwound container.

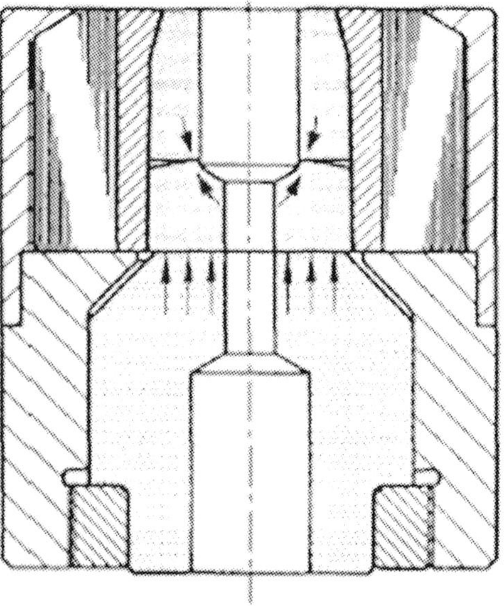

(b) Container with integrated axial prestress.

Fig. 23.9 Prestressed strip-wound containers developed by STRECON Technology [Groenbaek et al., 1997 and 2000]

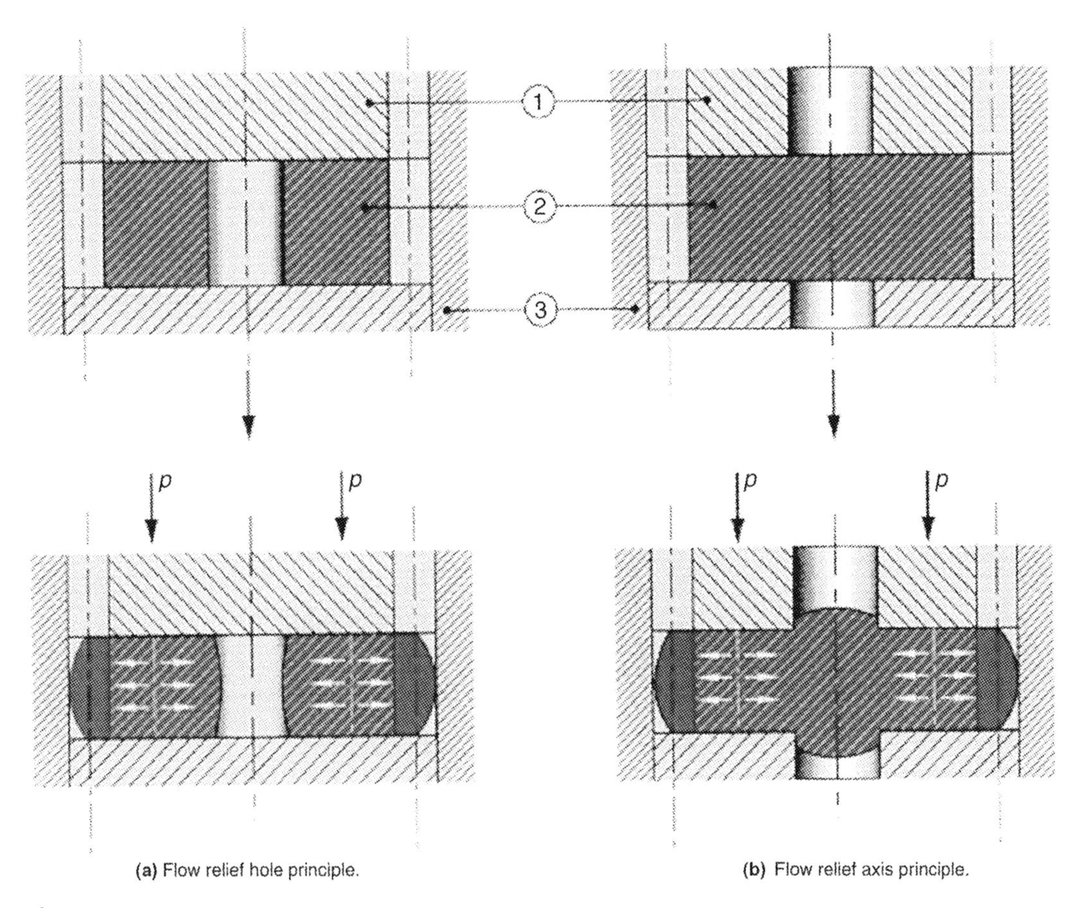

Fig. 23.10 Gear forging process utilizing divided flow [Kondo, 1999]

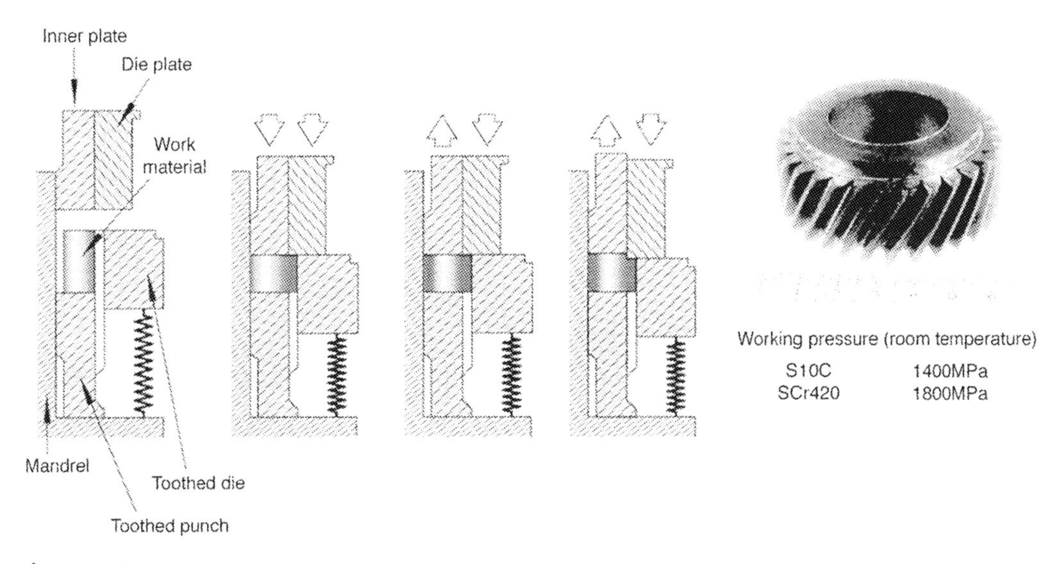

Fig. 23.11 One-step divided flow method with back-up pressure [Kondo, 1999 and 2002]

whereas Fig. 23.13(b) shows the slide motion. As a result of these features it was possible to:

- Reduce the press capacity and reduce facility cost due to reduction of total load and torque of the press
- Reduce slide tilting and thus improve the accuracy of the formed parts, since the slides perform independently without any mutual interference
- Reduce the vibration and noise during press operation

Figure 23.14 shows a helical gear cup formed using the MF-7500 press system [Nakano, 1997]. These cups require a higher forming load in the first and second stage and hence are ideal candidates for production in using this press. The cup, which traditionally required five forming stages, is formed in just three steps. Additionally, two annealing processes and phosphate coating were eliminated.

23.4.2 Multiaction Forging Press

Multiaction forging is an effective means of net shape forging parts with complex features. In multiaction forming, there is more than one pressure source to operate the dies and the slide. Also, the dies make several relative motions during one stroke. A multiaction press for forming helical gears is shown in Fig. 23.15(a). This hydraulically operated press has five cylinders: one for driving the slide, two cylinders in the slide, and two in the bed. Figure 23.15(b) shows the construction of a die for forming the helical gear seen in Fig. 23.15(c) [Nakano, 1997].

23.4.3 Servomotor Press

Figure 23.16 shows a servomotor press that combines a newly developed large-sized, high-torque servomotor drive with a crank mechanism. The press uses computer numerical controls for high functionality. It is thus possible to program the forming motion and speed in order to set the optimum parameters required to form the part (Fig. 23.17). This can be used advantageously to increase die life and produce hard-to-form materials. The press achieves reduced power usage by using a capacitor to store energy (Fig. 23.16).

23.5 Innovative Forging Processes

Besides the conventional precision-forging processes, which are commonly used in the industry, new processes such as microforming and orbital forging are being further developed for practical and economical applications.

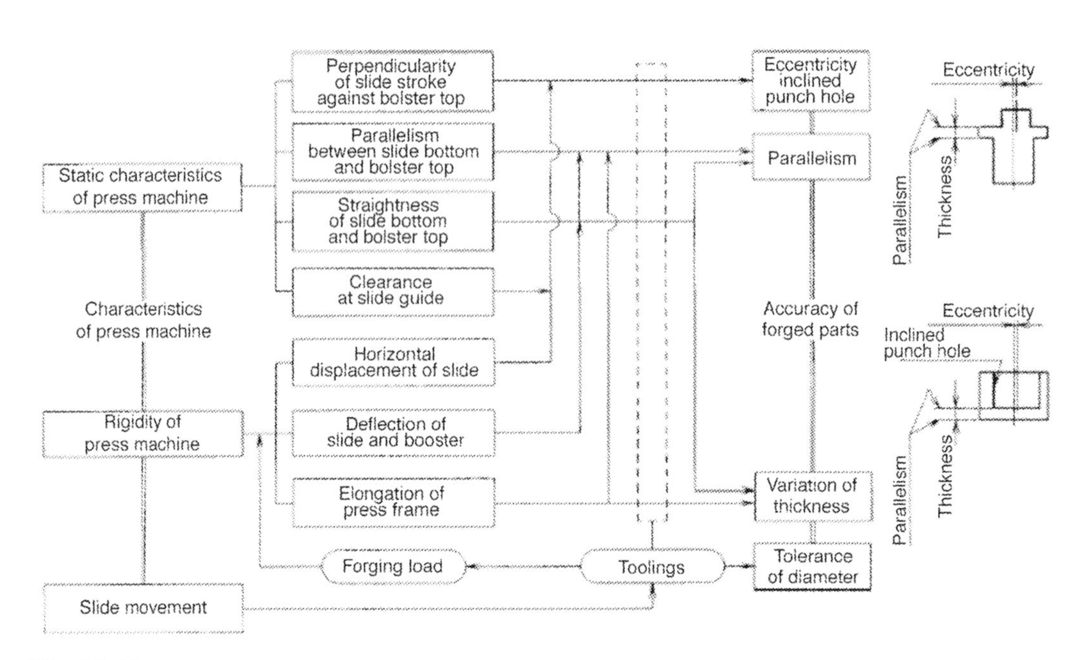

Fig. 23.12 Characteristics of press and accuracy of formed products [Nakano, 1997, and Ishinaga, 1996]

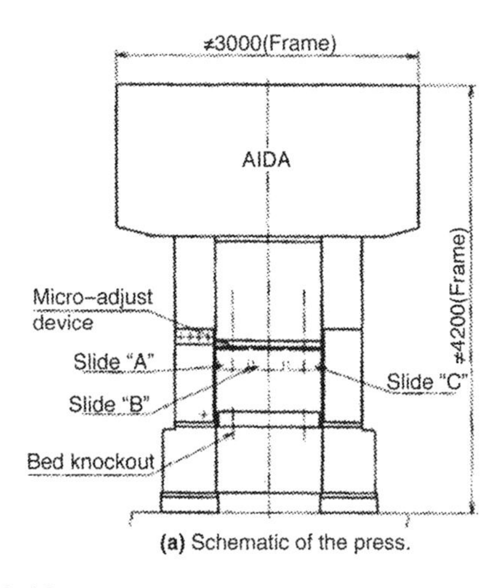

(a) Schematic of the press.

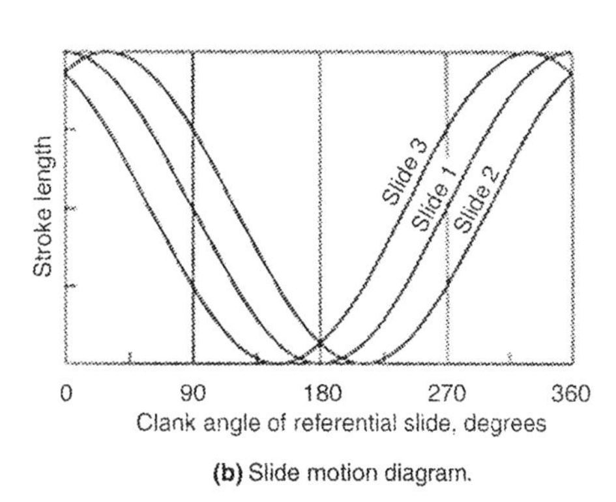

(b) Slide motion diagram.

Fig. 23.13 Multislide forging press MF-7500 [Nakano, 1997]

23.5.1 Microforming

The trend in miniaturization allows the production of cold-forged parts with dimensions less than 0.04 in. (1 mm) range for electronics and biomedical applications. These parts are currently produced by 3-D etching and other metal-removal processes. Microforming is a potential process for mass production of net shape/near-net shape microcomponents. However, for microforming to be cost-effective and competitive, a comprehensive knowledge pertaining to the following factors is needed: scale effects/microplasticity, effect of microstructure on the process, relative stiffness of the tooling, and process control and capability.

With the aid of finite-element modeling (FEM), the interrelationships between these variables can be studied so as to provide guidelines for developing microforming processes. In cooperation with industry, the Engineering Research Center for Net Shape Manufacturing (ERC/NSM) at Ohio State University has developed a microforming process for production of surgical blades. Figure 23.18 shows an example of 3-D FE simulations for a surgical blade with initial blank thickness of 0.004 in. (0.1 mm) and final edge thickness of 0.0004 in. (0.01 mm) [Palaniswamy et al., 2001].

In microforming, due to a high surface-to-volume ratio of parts, friction becomes even more important than in conventional forging. Effects of miniaturization on friction have been investigated by using the double-cup backward extrusion test, with oil as a lubricant (Fig. 23.19) [Tiesler et al., 1999]. Comparisons between experiment and simulation show that friction effects increase with a decrease in the size of the specimen. The friction energy observed for forming a 0.04 in. (1 mm) diameter billet was, in proportion to total forming energy, 4 times larger than that needed for forging a 0.16 in. (4 mm) diameter billet.

23.5.2 Orbital or Rotary Forging

Orbital or rotary forging is a very unique process with a complicated die movement that can be used to reduce axial load requirements for axisymmetric or near-axisymmetric forging operations. This process is not new and has been used for small- or medium-batch production of various round parts. The principle of orbital forming can be seen in Fig. 23.20 and is also discussed in Chapter 12, "Special Machines for Forging," in this book. The main advantage of this process is the reduction of the forging load, since deformation takes place incrementally. Furthermore, a forging process that may require

Fig. 23.14 Helical gear cup formed using the MF-7500 [Nakano, 1997]

multiple-station operations can be done on one machine, although the cycle time is longer than in upset forging.

Recently, orbital forming has been used by various bearing and axle manufacturers for the assembly of spindle drives (NSK Bearings, Japan). At the ERC/NSM at Ohio State University, orbital forging simulations using DEFORM-3D (Scientific Forming Technologies Corp.) were conducted to study and develop a robust assembly process of an automotive spindle and an outer ring. Figure 23.21 shows the simulation progression, which considers elastic and plastic deformation, residual stresses, and quality of assembly. This application illustrates the current capabilities of FEM in simulating complex and incremental cold forging operations to optimize process conditions and product design.

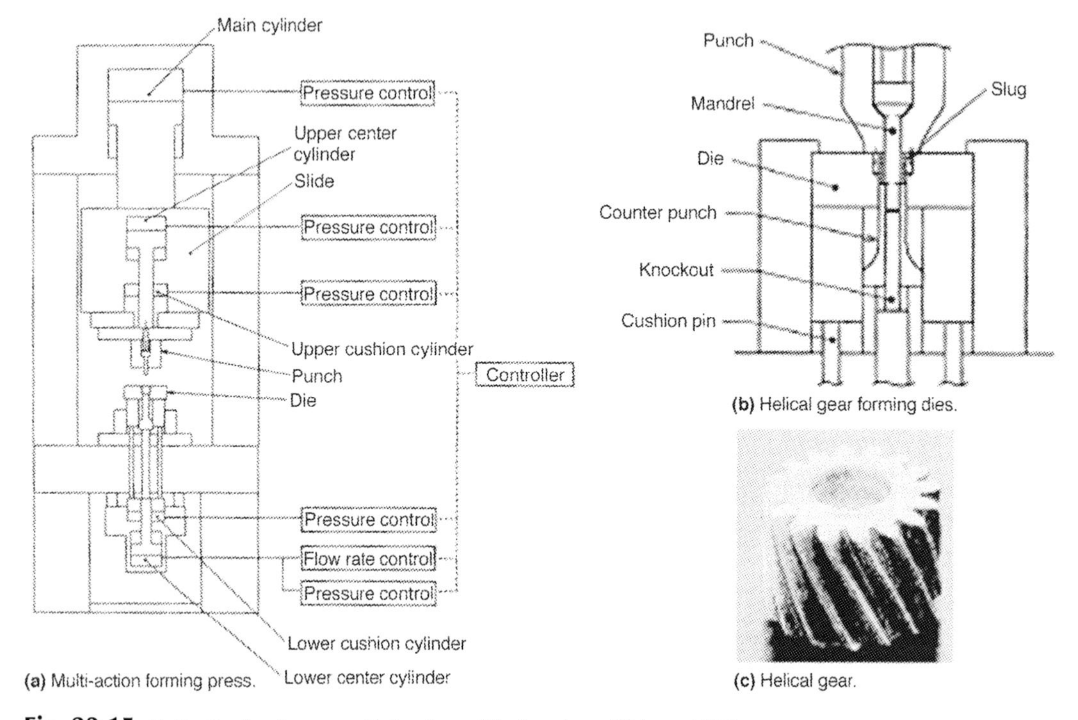

Fig. 23.15 Multiaction forming press with the dies and the forged gear [Nakano, 1997]

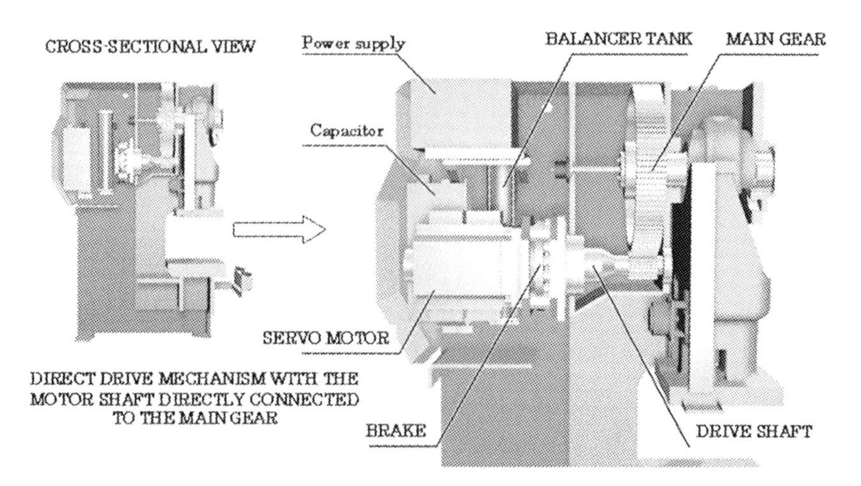

Fig. 23.16 Sectional view of the drive mechanism of the Aida Digital Servo Former. Courtesy of Aida Engineering Co. [Aida Engineering]

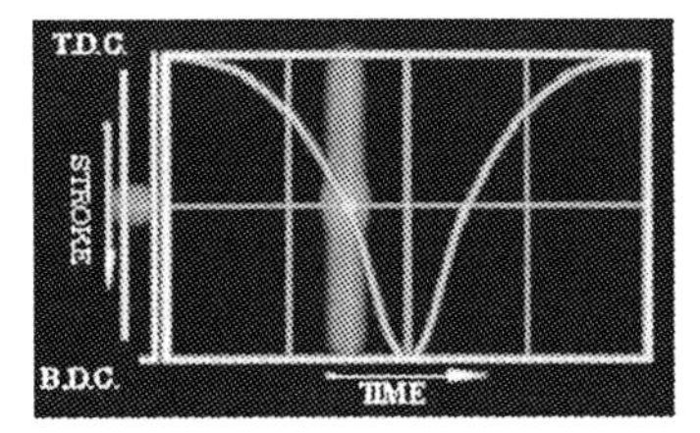

(a) Slow contact, quick return.

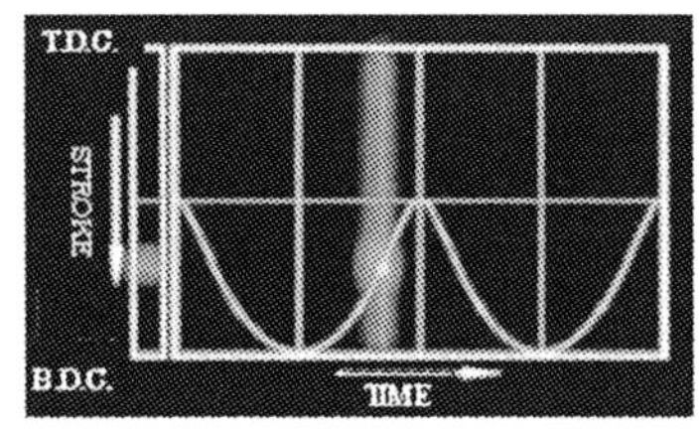

(b) Continuous motion.

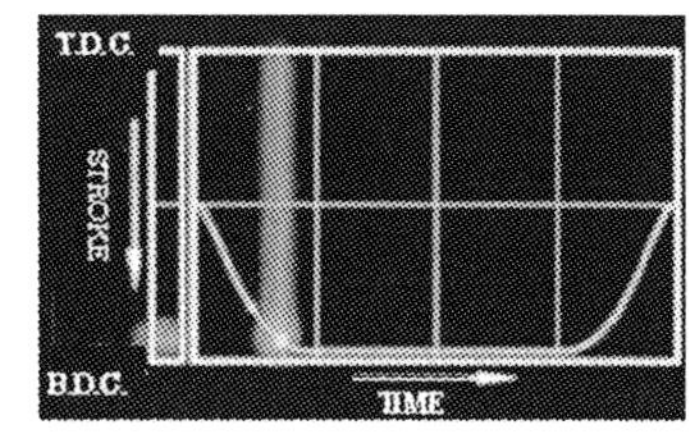

(c) Blanking motion.

Fig. 23.17 Some example settings of forming motions and speeds. Courtesy of Aida Engineering Co. [Aida Engineering]

Machines for hot forging large wheels or rings, using the incremental forming principle of orbital forging, have been developed by several machine tool builders. In these designs, the punch or tool does not move orbitally but rotates around a fixed inclined axis, as shown in Fig. 23.22 [Husmann, 1999]. The part, located in a die, is rotated and pushed vertically upward while the inclined punch rotates. Thus, the machine is similar to a ring rolling mill that has a lower die cavity that is hydraulically pushed upward (Fig. 23.23).

Such machines are used to forge railroad car wheels or forgings for large gears and near-net shape formed spiral bevel ring gears for truck differentials (Fig. 23.24).

23.6 Future of Forging Technology in the Global Marketplace

Automotive companies, which account for a very large portion of forged parts, are rapidly expanding global operations in an effort to produce cost-effective vehicles. This has vast implications for local automotive suppliers. A global economy allows, in many instances, any supplier that can satisfy part design requirements to bid and obtain a contract, irrespective of the geographic boundaries. This implies that the current automotive suppliers will have to change their production strategies to remain competitive in the global marketplace.

Due to globalization, forging technologies are diffusing across firms and across boundaries at increasing rates, through channels such as overseas patenting, licensing, exchange of technology-intensive goods, international investments, and alliances. As a result of inexpensive labor, production costs for simple parts are much less expensive in developing nations. Therefore, conventional cold forging operations will gradually shift to developing nations for production of low-tech components.

In order to succeed in the global marketplace, forging suppliers from developed nations must focus on production of high-value-added forgings, finished parts, and subassemblies with the aid of new developments such as advances in the use of computer modeling in forging process development, advances in press design, use of innovative tool design for complex forging operations, appropriate training in advanced cold forging technologies, and information management in forge shops. While the process tool and equipment design have been the subject of extensive research and development, the advances in information management and personnel training cannot be ignored. These two aspects are just as crucial to ensuring a defect-free part with the desired functionality as the technical developments.

23.6.1 Information Management in the Forge Shop

Management of complex engineering information in the forging environment plays a big role in ensuring that the desired production outputs are met. A forging company should have the ability to organize engineering knowledge and properly disseminate useful information to the respective departments for implementation. It is necessary to store an abundance of data, to be retrieved when and where desired. A number of forging companies have gone global, with

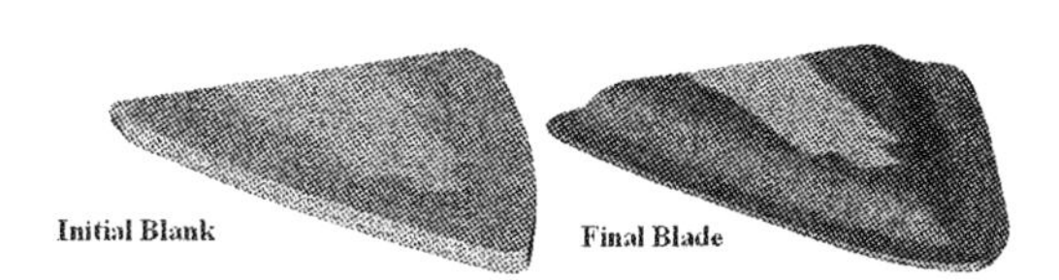

Fig. 23.18 Microforming of surgical blades. Blank thickness = 0.004 in. (0.1 mm); final blade thickness = 0.0004 in. (0.01 mm) [Palaniswamy et al., 2001]

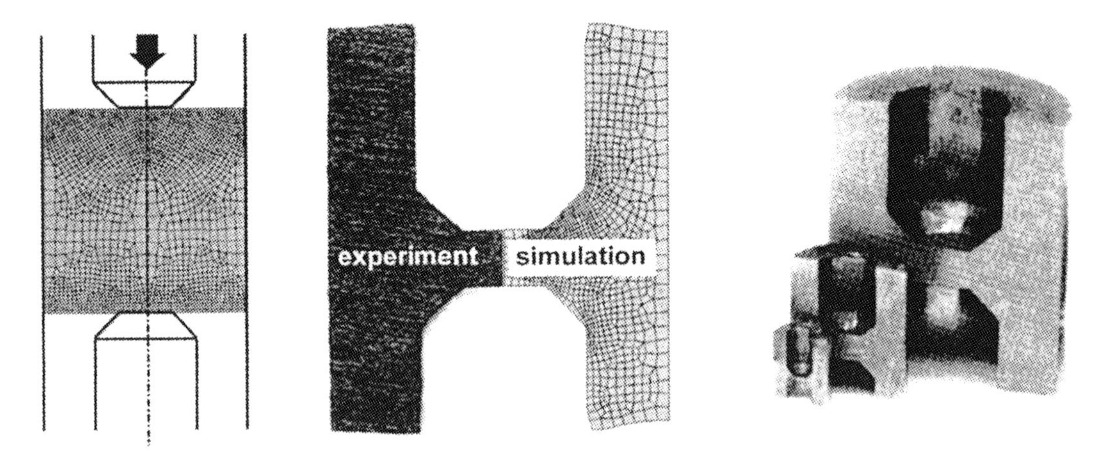

Fig. 23.19 Double-cup extrusion test for determination of friction conditions in microforming [Tiesler et al., 1999]

branches in various locations. Thus, information flow becomes complex and difficult to manage efficiently.

To date, companies such as Plexus Systems have developed computerized systems that can assist the management of information flow at various departmental levels, e.g., inventory tracking, shipping, receiving, engineering, purchasing, etc. [Beatty, 1997]. At the engineering departmental level, information management systems can be helpful for storing and retrieving data pertaining to tool designs, changing tool materials, product specification, process instructions, just-in-time jobs/rush jobs, process control plan, tool life tracking, machine specifications and drawings, press stress analysis, engineering drawing management, and dimensional control plans.

More sophisticated computerized information management systems will continue to emerge, and the forging companies will be compelled to adopt these systems. Failure to do so might hinder their chances of competing on the global marketplace. The steady growth and development of the worldwide web has prompted many forging firms to reassess and redesign the way they share critical business information [Beatty et al., 2001]. The web can provide cold forging companies with both operational and administrative benefits that can improve the firm's overall competitive position.

23.6.2 Training of Personnel

Due to global competition, the training of metal forming engineers, who are expected to plan and supervise the design and production of parts and dies, becomes increasingly important. Continuous improvements in forging technology and application of recently generated research and development results require that engineers be continuously updated in new methods, machines, and process technology. Professional education in forging and metal forming is necessary not only to upgrade the knowledge level of practicing forging engineers but also that of industrial, mechanical, and metallurgical engineers who may be assuming new responsi-

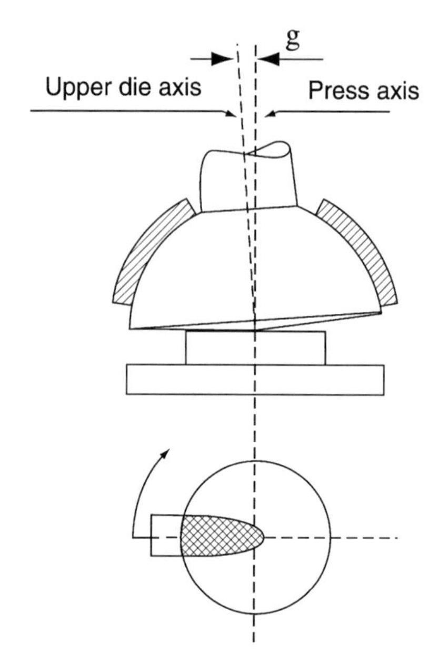

Fig. 23.20 Schematic representation of the orbital forging process [Oudot et al., 2001]

bilities in the forging industry [Vazquez et al., 1999 and 2000].

In order to prepare engineers who will work in high-tech forging companies, the current curriculum given in most universities would have to be restructured. It should be noted that most of the material related to advanced forging technology taught at the university level is very basic and may not be of significant value to the high-tech forging industry. Universities may be required to revise their design courses or develop new ones that address real design issues pertaining to forging. More importantly, for the course to be successful and cost-effective, there is a need to link universities to the forging industry in areas such as process sequence development, die/tool designs, press automation and design, and computational tools such as computer-aided engineering (CAE) software, CAD, and CAM systems.

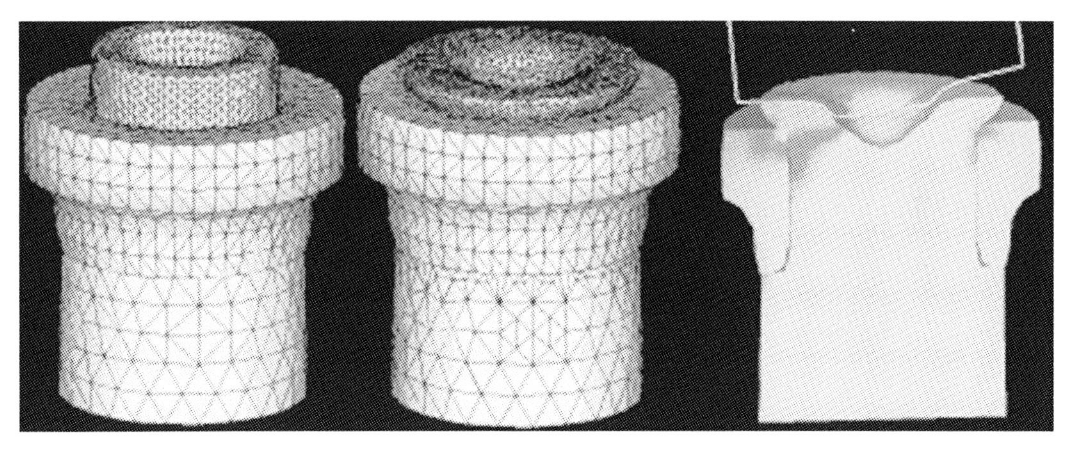

Fig. 23.21 Finite-element simulation of orbital forming (finite-element model and stress distribution) [Altan et al., 2003]

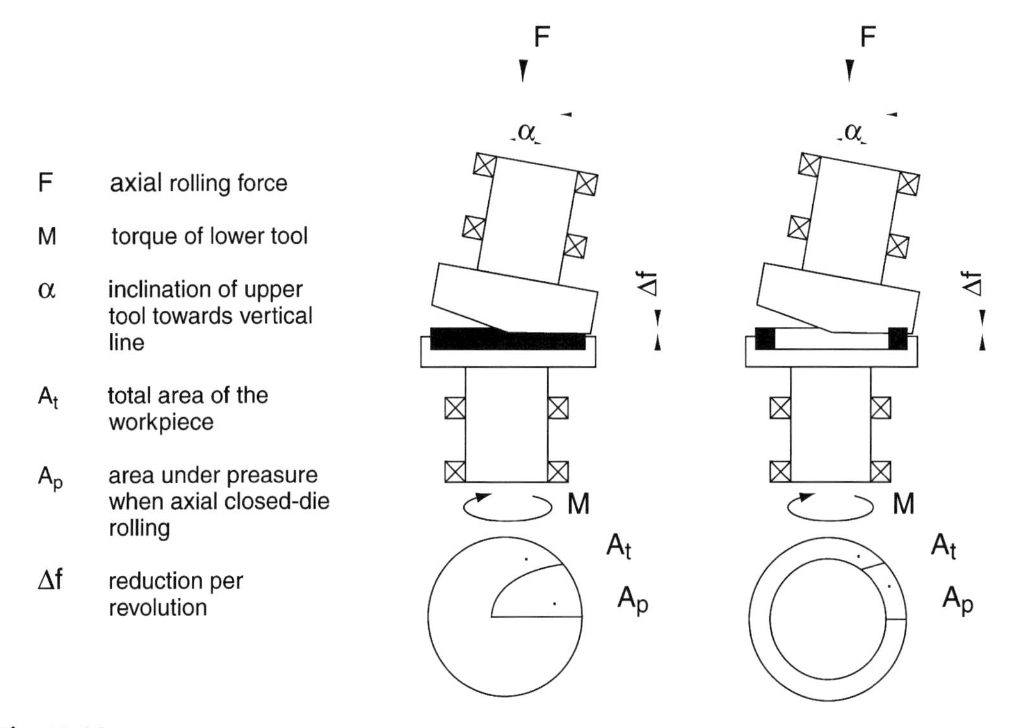

Fig. 23.22 Fundamental principle of axial closed-die rolling [Husmann, 1999]

The following strategies for technology development are being increasingly adopted by competitive world-class companies to respond to increased market globalization:

- Use of CAE for virtual prototyping of complicated high-value parts (microforming, helical gear extrusion, constant velocity joing (CVJ) inner races) and prediction and elimination of forming defects such as laps and chevron cracks
- Design and implementation of advanced forging presses and tooling for production of complex components
- The adoption and development of effective computerized information management systems for cold forging operations

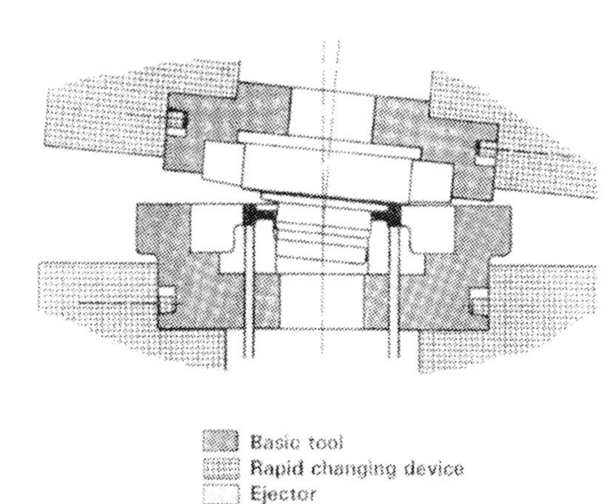

Fig. 23.23 Axial closed-die rolling machine, type AGW, tool section [Husmann, 1999]

Fig. 23.24 Bevel gear with teeth made on an axial closed-die rolling machine [Husmann, 1999]

- Life-long learning and training of engineers to equip them with relevant knowledge that can be readily applied to the dynamic technological environment of the 21st century

23.6.3 Concluding Remarks

The forging industry will continue to focus on the development of tooling and processes for net shape manufacturing of complex forgings, with an emphasis on reduction of press/tool elastic deflection, reduction of variation of process conditions, achieving tight tolerances, and use of special tooling/presses with specific functions that enhance part quality. To facilitate product and process development in minimal time and expense, it is imperative that state-of-the-art CAE capabilities be fully exploited. Furthermore, leading-edge forging companies continue to develop strategies and capabilities for producing "ready-to-assemble" parts and subassemblies for their customers. Thus, it is expected that high-tech cold forging suppliers in developed nations can remain competitive by producing high-value-added parts and subassemblies that require advanced technological expertise.

REFERENCES

[Aida Engineering]: "The Aida Digital Servo Former: NC-1 and NS-1, Hy-Flex D Series," Aida Engineering Ltd.

[Altan et al., 2003]: Altan, T., Ngaile, G., Shirgaokar, M., "Cold Forging Technology in Global Competition," International Conference on New Developments in Forging Technology at IFU (Fellbach/Stuttgart, Germany), June 2–4, 2003.

[Balendra et al., 1996]: Balendra, R., and Ou, H., "Influence of Forming Press Elasticity on the Accuracy of Formed Components," *Advanced Technology of Plasticity,* Vol. 1, Proceedings of the Fifth ICTP (Columbus, OH), Oct. 7–10, 1996.

[Beatty, 1997]: Beatty, R., "Advanced Information Management in the Forge Shop," *Journal of Materials Processing Technology,* Vol. 71, 1997, p 25–29.

[Beatty et al., 2001]: "Factors Influencing Corporate Web Site Adoption: A Time-Based Assessment," *Information and Management,* Vol. 38, 2001, p 337–354.

[Doege et al., 1990]: Doege, E., and Silberbach, G., "Influence of Various Machine Tool Components on Workpiece Quality," *Annals of CIRP,* Vol. 39 (No. 1), 1990.

[Groenbaek et al., 1997]: Groenbaek, J., and Nielsen, E.B., "Stripwound Containers for Combined Radial and Axial Prestressing," *Journal of Materials Processing Technology,* Vol. 71, 1997, p 30–35.

[Groenbaek et al., 2000]: Groenbaek, J., and Birker, T., "Innovations in Cold Forging Die Design," *Journal of Materials Processing Technology,* Vol. 98, 2000, p 155–161.

[Husmann, 1999]: Husmann, J., "High Speed Ring Rolling and Closed-Die Rolling," presented at the FIA Conference (Reno, NV), May 1999.

[Ishinaga, 1996]: Ishinaga, N., "An Advanced Press Design for Cold Forging," Third International Cold and Warm Forging Technology Conference, SME and ERC/NSM (Columbus, OH), Oct. 7–9, 1996.

[Kocanda et al., 1996]: Kocanda, A., Cacko, R., Czyzewski, P., "Some Aspects of Die Deformation in Net-Shape Cold Forging," *Advanced Technology of Plasticity,* Vol. 1, Proceedings of the Fifth ICTP (Columbus, OH), Oct. 7–10, 1996.

[Kondo, 1999]: Kondo, K., "Improvement of Product Accuracy in Cold Die Forging," *Advanced Technology of Plasticity,* Vol. 1, Proceedings of the Sixth ICTP, Sept. 19–24, 1999.

[Kondo, 2002]: Kondo, K., "Some Reminiscences of the Development of Precision Forging Processes," *Advanced Technology of Plasticity,* Vol. 1, Proceedings of the Seventh ICTP (Yokohama, Japan), Oct. 28–31, 2002.

[Meidert et al., 2000]: Meidert, M., and Hansel, M., "Net-Shape Cold Forging to Close Tolerances under QS 9000 Aspects," *Journal of Materials Processing Technology,* Vol. 98, 2000, p 150–154.

[Nakano, 1997]: Nakano, T., "Presses for Cold Forging," Aida Engineering Ltd., JSTP International Seminar of Precision Forging (Osaka, Japan), The Japanese Society for Technology of Plasticity, March 31 to April 1, 1997.

[Osakada, 1999]: Osakada, K., "New Methods for Precision Forging," *Advanced Technology of Plasticity,* Vol. 2, Proceedings of the Sixth ICTP (Nuremberg, Germany), Sept. 19–24, 1999.

[Oudot et al., 2001]: Oudot, H., and Faure, H., "Improvement of Precision in Cold Forged Parts," *La Forge,* No. 4, April 2001, p 27 (in French).

[Palaniswamy et al., 2001]: Palaniswamy, H., Ngaile, G., and Altan, T., "Coining of Surgical Slit Knife," F/ERC/NSM-01-R-26, Engineering Research Center for Net Shape Manufacturing, 2001.

[Siegert et al., 2003]: Siegert, K., and Baur, J., "Cold Lateral Extrusion of Automotive Components," Proceedings of the International Cold Forging Group Conference (Columbus, OH), ERC/NSM, Sept. 2–4, 2003.

[Tiesler et al., 1999]: Tiesler, N., Engel, U., Geiger, M., "Forming of Microparts—Effects of Miniaturization on Friction," *Advanced Technology of Plasticity,* Vol. 2, Proceedings of the Sixth ICTP (Nuremberg, Germany), Sept. 19–24, 1999.

[Vazquez et al., 1999]: Vazquez, V., and Altan, T., "State of the Art Technology for Training Engineers for the Forging Industry," presented at the 1999 FIA/FIERF Conference, 1999.

[Vazquez et al., 2000]: Vazquez, V., and Altan, T., "State of the Art Technology for Training Engineers for the Forging Industry," *Proceedings of Umformtechnik 2000 plus,* M. Geiger, Ed., Meisenbach Verlag, 1999, p 53.

[Yamanaka et al., 2002]: Yamanaka, M., and Sunami, F., "Tool Design for Precision Forging," Cold and Warm Precision Forging Workshop (Canton, MI), Schuler, Inc., Yamanaka Engineering, Scientific Forming Technologies Corp., ERC/NSM, Nov 14, 2002.

[Yoshimura et al., 1997]: Yoshimura, H., and Wang, C.C., "Manufacturing of Dies for Precision Forging," Nichidai Corporation, JSTP International Seminar of Precision Forging (Osaka, Japan), March 31 to April 1, 1997.

[Yoshimura et al., 2000]: Yoshimura, H., Wang, X., Kubota, K., Hamaya, S., "Process Design and Die Manufacturing for Precision Forging," Nichidai Corporation, Second JSTP International Seminar of Precision Forging (Osaka, Japan), May 15–16, 2000.

SELECTED REFERENCE

- **[Yoshimura et al., 1983]:** Yoshimura, H., and Shimasaki, S., "A Study of Enclosed Die Forging," *Journal of the JSTP,* 1983, p 24–30.

Index

A

B

C